AF371257

Translational Research for Zoonotic Parasites: New Findings toward Improved Diagnostics, Therapy and Prevention

Translational Research for Zoonotic Parasites: New Findings toward Improved Diagnostics, Therapy and Prevention

Editor

Vito Colella

MDPI • Basel • Beijing • Wuhan • Barcelona • Belgrade • Manchester • Tokyo • Cluj • Tianjin

Editor
Vito Colella
Veterinary Biosciences
The University of Melbourne
Melbourne
Australia

Editorial Office
MDPI
St. Alban-Anlage 66
4052 Basel, Switzerland

This is a reprint of articles from the Special Issue published online in the open access journal *Pathogens* (ISSN 2076-0817) (available at: www.mdpi.com/journal/pathogens/special_issues/Translational_Research_Zoonotic_Parasites).

For citation purposes, cite each article independently as indicated on the article page online and as indicated below:

LastName, A.A.; LastName, B.B.; LastName, C.C. Article Title. *Journal Name* **Year**, *Volume Number*, Page Range.

ISBN 978-3-0365-2625-6 (Hbk)
ISBN 978-3-0365-2624-9 (PDF)

Contents

About the Editor . . . . **vii**

Vito Colella, Rebecca J. Traub and Robin B. Gasser
Translational Research of Zoonotic Parasites: Toward Improved Tools for Diagnosis, Treatment
and Control
Reprinted from: *Pathogens* **2021**, *10*, 1416, doi:10.3390/pathogens10111416 . . . . **1**

**Lucas G. Huggins, Anson V. Koehler, Bettina Schunack, Tawin Inpankaew and Rebecca J.
Traub**
A Host-Specific Blocking Primer Combined with Optimal DNA Extraction Improves the
Detection Capability of a Metabarcoding Protocol for Canine Vector-Borne Bacteria
Reprinted from: *Pathogens* **2020**, *9*, 258, doi:10.3390/pathogens9040258 . . . . **5**

Shona Chandra and Jan Šlapeta
Biotic Factors Influence Microbiota of Nymph Ticks from Vegetation in Sydney, Australia
Reprinted from: *Pathogens* **2020**, *9*, 566, doi:10.3390/pathogens9070566 . . . . **25**

Handi Dahmana and Oleg Mediannikov
Mosquito-Borne Diseases Emergence/Resurgence and How to Effectively Control It
Biologically
Reprinted from: *Pathogens* **2020**, *9*, 310, doi:10.3390/pathogens9040310 . . . . **49**

**Jakob Thannesberger, Nicolas Rascovan, Anna Eisenmann, Ingeborg Klymiuk, Carina Zittra,
Hans-Peter Fuehrer, Thea Scantlebury-Manning, Marquita Gittens-St.Hilaire, Shane Austin,
Robert Clive Landis and Christoph Steininger**
Highly Sensitive Virome Characterization of *Aedes aegypti* and *Culex pipiens* Complex from
Central Europe and the Caribbean Reveals Potential for Interspecies Viral Transmission
Reprinted from: *Pathogens* **2020**, *9*, 686, doi:10.3390/pathogens9090686 . . . . **75**

**Carina Zittra, Simon Vitecek, Joana Teixeira, Dieter Weber, Bernadette Schindelegger,
Francis Schaffner and Alexander M. Weigand**
Mosquitoes (Diptera: Culicidae) in the Dark—Highlighting the Importance of Genetically
Identifying Mosquito Populations in Subterranean Environments of Central Europe
Reprinted from: *Pathogens* **2021**, *10*, 1090, doi:10.3390/pathogens10091090 . . . . **91**

**Carina Zittra, Günther Wöss, Lara Van der Vloet, Karin Bakran-Lebl, Bita Shahi Barogh,
Peter Sehnal and Hans-Peter Fuehrer**
Barcoding of the Genus *Culicoides* (Diptera: Ceratopogonidae) in Austria—An Update of the
Species Inventory Including the First Records of Three Species in Austria
Reprinted from: *Pathogens* **2020**, *9*, 406, doi:10.3390/pathogens9050406 . . . . **99**

**Edwin Kniha, Vít Dvořák, Petr Halada, Markus Milchram, Adelheid G. Obwaller, Katrin
Kuhls, Susanne Schlegel, Martina Köhsler, Wolfgang Poeppl, Karin Bakran-Lebl, Hans-Peter
Fuehrer, Věra Volfová, Gerhard Mooseder, Vladimir Ivovic, Petr Volf and Julia Walochnik**
Integrative Approach to *Phlebotomus mascittii* Grassi, 1908: First Record in Vienna with New
Morphological and Molecular Insights
Reprinted from: *Pathogens* **2020**, *9*, 1032, doi:10.3390/pathogens9121032 . . . . **111**

Ahyun Hong, Ricardo Andrade Zampieri, Jeffrey Jon Shaw, Lucile Maria Floeter-Winter and Maria Fernanda Laranjeira-Silva
One Health Approach to Leishmaniases: Understanding the Disease Dynamics through Diagnostic Tools
Reprinted from: *Pathogens* **2020**, *9*, 809, doi:10.3390/pathogens9100809 . . . . **129**

Bonnie E. Gulas-Wroblewski, Rebecca B. Kairis, Rodion Gorchakov, Anna Wheless and Kristy O. Murray
Optimization of DNA Extraction from Field-Collected Mammalian Whole Blood on Filter Paper for *Trypanosoma cruzi* (Chagas Disease) Detection
Reprinted from: *Pathogens* **2021**, *10*, 1040, doi:10.3390/pathogens10081040 . . . . **153**

Ahmed M. Zheoat, Samya Alenezi, Ehab Kotb Elmahallawy, Marzuq A. Ungogo, Ali H. Alghamdi, David G. Watson, John O. Igoli, Alexander I. Gray, Harry P. de Koning and Valerie A. Ferro
Antitrypanosomal and Antileishmanial Activity of Chalcones and Flavanones from *Polygonum salicifolium*
Reprinted from: *Pathogens* **2021**, *10*, 175, doi:10.3390/pathogens10020175 . . . . **161**

An Hotterbeekx, Jolien Perneel, Michel Mandro, Germain Abhafule, Joseph Nelson Siewe Fodjo, Alfred Dusabimana, Steven Abrams, Samir Kumar-Singh and Robert Colebunders
Comparison of Diagnostic Tests for *Onchocerca volvulus* in the Democratic Republic of Congo
Reprinted from: *Pathogens* **2020**, *9*, 435, doi:10.3390/pathogens9060435 . . . . **171**

Rossella Panarese, Roberta Iatta, Jairo Alfonso Mendoza-Roldan, Donald Szlosek, Jennifer Braff, Joe Liu, Frédéric Beugnet, Filipe Dantas-Torres, Melissa J. Beall and Domenico Otranto
Comparison of Diagnostic Tools for the Detection of *Dirofilaria immitis* Infection in Dogs
Reprinted from: *Pathogens* **2020**, *9*, 499, doi:10.3390/pathogens9060499 . . . . **181**

K. L. D. Tharaka D. Liyanage, Anke Wiethoelter, Jasmin Hufschmid and Abdul Jabbar
Descriptive Comparison of ELISAs for the Detection of *Toxoplasma gondii* Antibodies in Animals: A Systematic Review
Reprinted from: *Pathogens* **2021**, *10*, 605, doi:10.3390/pathogens10050605 . . . . **189**

Barbara J. Bucher, Gillian Muchaamba, Tim Kamber, Philipp A. Kronenberg, Kubanychbek K. Abdykerimov, Myktybek Isaev, Peter Deplazes and Cristian A. Alvarez Rojas
LAMP Assay for the Detection of *Echinococcus multilocularis* Eggs Isolated from Canine Faeces by a Cost-Effective NaOH-Based DNA Extraction Method
Reprinted from: *Pathogens* **2021**, *10*, 847, doi:10.3390/pathogens10070847 . . . . **213**

Julieta Rousseau, Mónia Nakamura, Helena Rio-Maior, Francisco Álvares, Rémi Choquet, Luís Madeira de Carvalho, Raquel Godinho and Nuno Santos
Non-Invasive Molecular Survey of Sarcoptic Mange in Wildlife: Diagnostic Performance in Wolf Faecal Samples Evaluated by Multi-Event Capture–Recapture Models
Reprinted from: *Pathogens* **2021**, *10*, 243, doi:10.3390/pathogens10020243 . . . . **229**

Meruyert Beknazarova, Harriet Whiley, Rebecca Traub and Kirstin Ross
Opportunistic Mapping of *Strongyloides stercoralis* and Hookworm in Dogs in Remote Australian Communities
Reprinted from: *Pathogens* **2020**, *9*, 398, doi:10.3390/pathogens9050398 . . . . **245**

Ali Taghipour, Ali Rostami, Sahar Esfandyari, Saeed Aghapour, Alessandra Nicoletti and Robin B. Gasser
"Begging the Question"—Does *Toxocara* Infection/Exposure Associate with Multiple Sclerosis-Risk?
Reprinted from: *Pathogens* **2020**, *9*, 938, doi:10.3390/pathogens9110938 . . . . **259**

Mudassar N. Mughal, Qing Ye, Lu Zhao, Christoph G. Grevelding, Ying Li, Wenda Di, Xin He, Xuesong Li, Robin B. Gasser and Min Hu
First Evidence of Function for *Schistosoma japonicumriok-1* and RIOK-1
Reprinted from: *Pathogens* **2021**, *10*, 862, doi:10.3390/pathogens10070862 . . . . **271**

About the Editor

Vito Colella

Dr. Vito Colella is a Lecturer in Medical and Veterinary Parasitology at The University of Melbourne, Australia. His current research focuses on the development of intervention strategies to mitigate the impact of zoonotic parasites on human populations in regions of the Asia Pacific.

Vito obtained a PhD from the University of Bari (Italy) with a thesis on neglected zoonotic parasitic infections of dogs and cats. During the last years, he built a strong network with several public and private institutions through study design, the writing of reports and grant proposals for studies on zoonotic diseases inflicting hardships on animal and human health in low- and middle-income countries. Vito was the recipient of several prestigious awards, including the Soulsby Fellowship in 2020 and the Odile Bain Memorial Prize in 2019.

Editorial

Translational Research of Zoonotic Parasites: Toward Improved Tools for Diagnosis, Treatment and Control

Vito Colella *, Rebecca J. Traub and Robin B. Gasser

Faculty of Veterinary and Agricultural Sciences, The University of Melbourne, Parkville, VIC 3010, Australia; rebecca.traub@unimelb.edu.au (R.J.T.); robinbg@unimelb.edu.au (R.B.G.)
* Correspondence: vito.colella@unimelb.edu.au

Citation: Colella, V.; Traub, R.J.; Gasser, R.B. Translational Research of Zoonotic Parasites: Toward Improved Tools for Diagnosis, Treatment and Control. *Pathogens* **2021**, *10*, 1416. https://doi.org/10.3390/ pathogens10111416

Received: 28 October 2021
Accepted: 28 October 2021
Published: 1 November 2021

Publisher's Note: MDPI stays neutral with regard to jurisdictional claims in published maps and institutional affiliations.

A range of factors, including social, demographic and economic transformation and human-induced environmental changes, are influencing the emergence or re-emergence of zoonoses, posing new challenges in how we detect, treat and prevent such diseases. Fortunately, advanced "next-generation" sequencing (NGS), 'omic' and immunological techniques provide improved tools to explore zoonotic pathogens; for example, the advent of portable genome sequencing technology for the detection and sequencing of the Ebola virus during recent outbreaks in West Africa facilitated and accelerated the implementation of Ebola surveillance to monitor and prevent further outbreaks. The importance of such developments is further exemplified by the rapid and widespread deployment of genomic and informatic tools in many countries for diagnosis and monitoring during the SARS-CoV-2 pandemic. Translational research of zoonotic pathogens is best achieved using a multidisciplinary approach through collaborations, with a clear focus on alleviating the impact of these pathogens and of the diseases that they cause, particularly in underprivileged communities.

This Special Issue brings together a collection of articles on aspects of zoonotic protistan and metazoan parasite translational research, providing insights into areas of epidemiology, pathogen and vector discovery, and therapy, with significant implications for the prevention and control of zoonotic diseases.

Vector-borne pathogens (VBPs) are responsible for life-threatening conditions in humans and animals. NGS-based methods have been used to genetically characterise a diverse range of VBPs, being more sensitive than conventional techniques. Recently, Huggins et al. [1] established an NGS-based approach to identify a wide range of VBP species, enhancing performance and reducing non-specific amplification from host DNA by employing 'blocking primers'. Additionally, using NGS, Chandra et al. [2] showed that geographical origin is a key factor in shaping the microbiota of ticks, influencing the bacteria transferred to their vertebrate hosts and, potentially, disease.

Mosquitoes are major vectors that transmit a diverse array of pathogens, including arboviruses and *Plasmodium*. Dahmana et al. [3] showed how understanding the biology, distribution and vector competence of mosquitoes can assist disease surveillance and guide the development of novel strategies for the control of mosquito-borne diseases, particularly those involving insecticide-resistant vectors. Thannesberger et al. [4] employed an NGS-based approach for virome characterisation in populations of medically important mosquitoes (*Aedes aegypti* and members of the *Culex pipiens* complex) and discovered a rich viral community in a diverse range of host species. Their study recorded novel virus sequences assigned, mainly, to circular Rep-encoding single-stranded (CRESS) DNA viruses. These viruses have been demonstrated to use mosquitoes as "virus mixing vessels", where the genetic recombination or re-assortment of diverse viruses can ensue, potentially resulting in novel pathogens that are able to jump species barriers [4]. Zittra et al. [5,6] utilised molecular methods to identify members of particular mosquito species with phenotypic differences in behaviour, physiology, hosts and habitat preferences; for instance, metabarcoding was applied to genetically characterise *Culicoides* and subterranean mosquito communities that might serve as reservoirs for significant human pathogens. On the other hand, Kniha et al. [7] combined the use of morphological methods, DNA sequencing and mass spectrometry to identify sandfly species, enabling the characterisation of *Phlebotomus mascittii* (a putative vector of *Leishmania infantum*) in metropolitan environments.

Leishmaniases and trypanosomiases are groups of vector-borne diseases caused by protistan parasites affecting millions of people and animals around the world. Hong et al. [8] prioritised the development of an accurate and cost-effective diagnostic test, as well as the use of animals as sentinels, as effective tools for the epidemiological surveillance of leishmaniases. Gulas-Wroblewski et al. [9] developed cost-effective DNA storage and DNA extraction methods for the molecular detection of *Trypanosoma cruzi* in field-based investigations. Furthermore, Zheoat et al. [10] discovered that 2′,4′-dimethoxy-6′-hydroxychalcone from a plant extract of *Polygonum salicifolium* had anti-leishmanial and anti-trypanosomal effects in the absence of toxicity against a human cell line, suggesting its potential use against *Leishmania mexicana*, *Trypanosoma brucei* and *T. congolense*.

Diagnostics is fundamental when monitoring neglected tropical diseases (NTDs), such as onchocerciasis, before, during and after mass drug administration. Hotterbeekx et al. [11] evaluated the diagnostic specificity and sensitivity of detecting microfilariae in skin snips, or specific IgG4 antibodies (anti-OV16) in ELISA or in a rapid diagnostic test (RDT) in the blood of persons with epilepsy living in an onchocerciasis-endemic region in the Democratic Republic of Congo. Of the three methods, ELISA and RDT had the highest respective sensitivities and specificities, resulting in an improved diagnosis of onchocerciasis. Similarly, Panarese et al. [12] reported an increased proportion of dogs diagnosed with *Dirofilaria immitis* (heartworm) when immune complex dissociation was applied prior to ELISA, although the specificity of this approach requires a critical assessment, as indicated by the authors [12].

Evaluating and selecting suitable tests are also critical for the diagnosis of pathogens, such as species of *Toxoplasma* and *Echinococcus*, occurring in different tissues of numerous host species and the environment. A systematic review by Liyanage et al. [13] revealed that the diagnostic performance of ELISA tests for *T. gondii* varied considerably, depending on the type of sample (serum, 'meat juice' and milk), antigen (native, recombinant or chimeric) and antibody-binding reagents used, with the combined use of recombinant and chimeric antigens resulting in better performance compared with native or single recombinant antigens. Bucher et al. [14] proposed two DNA extraction methods for the cost-effective detection of *Echinococcus multilocularis* using a loop-mediated isothermal amplification test, able to be applied in large-scale field studies in low-income countries. Interestingly, Rousseau et al. [15] developed a diagnostic method for the detection of DNA from the ectoparasitic mite *Sarcoptes scabiei* in the faeces of *Canis lupus signatus* (Iberian wolf). This non-invasive method was developed for the surveillance of scabies, a disease that can have a significant adverse impact on wildlife.

Soil-transmitted helminths (STHs) are a group of parasitic worms causing a major disease burden in people living in disadvantaged communities. Of these worms, *Strongyloides stercoralis* and some species of hookworms infect canids, representing reservoirs for transmission to humans. Beknazarova et al. [16] estimated the prevalence of these STHs in dogs in remote Australian communities, with the aim of guiding future intervention and prevention strategies to control these parasites in both humans and canines.

The wealth of literature published over recent decades requires critical systematic appraisals of information on NTDs. Taghipour et al. [17] reviewed available medical literature to test a previous hypothesis of a possible association between *Toxocara* infection/exposure and multiple sclerosis (MS). Despite finding an association between anti-*Toxocara* IgG serum antibodies and MS in the published literature, the authors concluded that well-designed and -controlled studies would be essential for a rigorous assessment of this association [17]. On a distinct topic, focused on the evaluation of the function of a particular molecule, Mughal et al. [18] characterised the right open reading frame protein kinase (*Sj-riok-1*) gene and its gene product (*Sj*-RIOK-1) in *Schistosoma japonicum* (a human blood fluke). The chemical knockdown of *Sj*-RIOK-1 caused a reduction in worm viability and the accumulation of mature oocytes in the seminal receptacle, and of spermatozoa in the sperm vesicle, suggesting an involvement in reproductive and/or developmental processes in *S. japonicum* [18].

In conclusion, we believe that this Special Issue illustrates how multidisciplinary research can deliver improved tools for exploring zoonotic parasites and encourages translational research—from bench to bedside or the field—with a focus on alleviating the impact of zoonotic parasitic diseases, particularly in disadvantaged communities around the world.

Conflicts of Interest: The authors declare no conflict of interest.

References

1. Huggins, L.G.; Koehler, A.V.; Schunack, B.; Inpankaew, T.; Traub, R.J. A Host-Specific Blocking Primer Combined with Optimal DNA Extraction Improves the Detection Capability of a Metabarcoding Protocol for Canine Vector-Borne Bacteria. *Pathogens* **2020**, *9*, 258. [CrossRef] [PubMed]
2. Chandra, S.; Šlapeta, J. Biotic Factors Influence Microbiota of Nymph Ticks from Vegetation in Sydney, Australia. *Pathogens* **2020**, *9*, 566. [CrossRef] [PubMed]
3. Dahmana, H.; Mediannikov, O. Mosquito-Borne Diseases Emergence/Resurgence and How to Effectively Control It Biologically. *Pathogens* **2020**, *9*, 310. [CrossRef] [PubMed]
4. Thannesberger, J.; Rascovan, N.; Eisenmann, A.; Klymiuk, I.; Zittra, C.; Fuehrer, H.-P.; Scantlebury-Manning, T.; Gittens-St.Hilaire, M.; Austin, S.; Landis, R.C.; et al. Highly Sensitive Virome Characterization of *Aedes aegypti* and *Culex pipiens* Complex from Central Europe and the Caribbean Reveals Potential for Interspecies Viral Transmission. *Pathogens* **2020**, *9*, 686. [CrossRef] [PubMed]
5. Zittra, C.; Vitecek, S.; Teixeira, J.; Weber, D.; Schindelegger, B.; Schaffner, F.; Weigand, A.M. Mosquitoes (Diptera: Culicidae) in the Dark—Highlighting the Importance of Genetically Identifying Mosquito Populations in Subterranean Environments of Central Europe. *Pathogens* **2021**, *10*, 1090. [CrossRef] [PubMed]
6. Zittra, C.; Wöss, G.; Van Der Vloet, L.; Bakran-Lebl, K.; Barogh, B.S.; Sehnal, P.; Fuehrer, H.-P. Barcoding of the Genus Culicoides (Diptera: Ceratopogonidae) in Austria—An Update of the Species Inventory Including the First Records of Three Species in Austria. *Pathogens* **2020**, *9*, 406. [CrossRef] [PubMed]
7. Kniha, E.; Dvořák, V.; Halada, P.; Milchram, M.; Obwaller, A.G.; Kuhls, K.; Schlegel, S.; Köhsler, M.; Poeppl, W.; Bakran-Lebl, K.; et al. Integrative Approach to *Phlebotomus mascittii* Grassi, 1908: First Record in Vienna with New Morphological and Molecular Insights. *Pathogens* **2020**, *9*, 1032. [CrossRef] [PubMed]
8. Hong, A.; Zampieri, R.A.; Shaw, J.J.; Floeter-Winter, L.M.; Laranjeira-Silva, M.F. One Health Approach to Leishmaniases: Understanding the Disease Dynamics through Diagnostic Tools. *Pathogens* **2020**, *9*, 809. [CrossRef] [PubMed]
9. Gulas-Wroblewski, B.E.; Kairis, R.B.; Gorchakov, R.; Wheless, A.; Murray, K.O. Optimization of DNA Extraction from Field-Collected Mammalian Whole Blood on Filter Paper for *Trypanosoma cruzi* (Chagas Disease) Detection. *Pathogens* **2021**, *10*, 1040. [CrossRef]
10. Zheoat, A.; Alenezi, S.; Elmahallawy, E.; Ungogo, M.; Alghamdi, A.; Watson, D.; Igoli, J.; Gray, A.; de Koning, H.; Ferro, V. Antitrypanosomal and Antileishmanial Activity of Chalcones and Flavanones from *Polygonum salicifolium*. *Pathogens* **2021**, *10*, 175. [CrossRef] [PubMed]
11. Hotterbeekx, A.; Perneel, J.; Mandro, M.; Abhafule, G.; Fodjo, J.N.S.; Dusabimana, A.; Abrams, S.; Kumar-Singh, S.; Colebunders, R. Comparison of Diagnostic Tests for Onchocerca volvulus in the Democratic Republic of Congo. *Pathogens* **2020**, *9*, 435. [CrossRef]
12. Panarese, R.; Iatta, R.; Mendoza-Roldan, J.A.; Szlosek, D.; Braff, J.; Liu, J.; Beugnet, F.; Dantas-Torres, F.; Beall, M.J.; Otranto, D. Comparison of Diagnostic Tools for the Detection of Dirofilaria immitis Infection in Dogs. *Pathogens* **2020**, *9*, 499. [CrossRef] [PubMed]
13. Liyanage, K.; Wiethoelter, A.; Hufschmid, J.; Jabbar, A. Descriptive Comparison of ELISAs for the Detection of *Toxoplasma gondii* Antibodies in Animals: A Systematic Review. *Pathogens* **2021**, *10*, 605. [CrossRef] [PubMed]
14. Bucher, B.; Muchaamba, G.; Kamber, T.; Kronenberg, P.; Abdykerimov, K.; Isaev, M.; Deplazes, P.; Rojas, C.A. LAMP Assay for the Detection of *Echinococcus multilocularis* Eggs Isolated from Canine Faeces by a Cost-Effective NaOH-Based DNA Extraction Method. *Pathogens* **2021**, *10*, 847. [CrossRef] [PubMed]
15. Rousseau, J.; Nakamura, M.; Rio-Maior, H.; Álvares, F.; Choquet, R.; de Carvalho, L.M.; Godinho, R.; Santos, N. Non-Invasive Molecular Survey of Sarcoptic Mange in Wildlife: Diagnostic Performance in Wolf Faecal Samples Evaluated by Multi-Event Capture–Recapture Models. *Pathogens* **2021**, *10*, 243. [CrossRef] [PubMed]

16. Cooper-Beknazarova, M.; Whiley, H.; Traub, R.; Ross, K. Opportunistic Mapping of Strongyloides stercoralis and Hookworm in Dogs in Remote Australian Communities. *Pathogens* **2020**, *9*, 398. [CrossRef] [PubMed]
17. Taghipour, A.; Rostami, A.; Esfandyari, S.; Aghapour, S.; Nicoletti, A.; Gasser, R.B. "Begging the Question"—Does *Toxocara* Infection/Exposure Associate with Multiple Sclerosis-Risk? *Pathogens* **2020**, *9*, 938. [CrossRef] [PubMed]
18. Mughal, M.; Ye, Q.; Zhao, L.; Grevelding, C.; Li, Y.; Di, W.; He, X.; Li, X.; Gasser, R.; Hu, M. First Evidence of Function for *Schistosoma japonicumriok-1* and RIOK-1. *Pathogens* **2021**, *10*, 862. [CrossRef] [PubMed]

Article

A Host-Specific Blocking Primer Combined with Optimal DNA Extraction Improves the Detection Capability of a Metabarcoding Protocol for Canine Vector-Borne Bacteria

Lucas G. Huggins [1,*], Anson V. Koehler [1], Bettina Schunack [2], Tawin Inpankaew [3] and Rebecca J. Traub [1]

1   Faculty of Veterinary and Agricultural Sciences, University of Melbourne, Parkville, Victoria 3050, Australia; anson.koehler@unimelb.edu.au (A.V.K.); rebecca.traub@unimelb.edu.au (R.J.T.)
2   Bayer Animal Health GmbH, 51373 Leverkusen, Germany; bettina.schunack@bayer.com
3   Faculty of Veterinary Medicine, Kasetsart University, Bangkok 10900, Thailand; tawin.i@ku.th
*   Correspondence: lghuggins@student.unimelb.edu.au; Tel.: +61-3834-48001

Received: 12 March 2020; Accepted: 31 March 2020; Published: 1 April 2020

Abstract: Bacterial canine vector-borne diseases are responsible for some of the most life-threatening conditions of dogs in the tropics and are typically poorly researched with some presenting a zoonotic risk to cohabiting people. Next-generation sequencing based methodologies have been demonstrated to accurately characterise a diverse range of vector-borne bacteria in dogs, whilst also proving to be more sensitive than conventional PCR techniques. We report two improvements to a previously developed metabarcoding tool that increased the sensitivity and diversity of vector-borne bacteria detected from canine blood. Firstly, we developed and tested a canine-specific blocking primer that prevents cross-reactivity of bacterial primer amplification on abundant canine mitochondrial sequences. Use of our blocking primer increased the number of canine vector-borne infections detected (five more *Ehrlichia canis* and three more *Anaplasma platys* infections) and increased the diversity of bacterial sequences found. Secondly, the DNA extraction kit employed can have a significant effect on the bacterial community characterised. Therefore, we compared four different DNA extraction kits finding the Qiagen DNeasy Blood and Tissue Kit to be superior for detection of blood-borne bacteria, identifying nine more *A. platys*, two more *E. canis*, one more *Mycoplasma haemocanis* infection and more putative bacterial pathogens than the lowest performing kit.

Keywords: canine vector-borne disease; blocking primers; blood DNA extraction; next-generation sequencing; kit contaminant bacteria

1. Introduction

Despite recent developments in veterinary care and medicine, canine vector-borne diseases (CVBD) continue to inflict a large burden with regard to morbidity and mortality on dogs across the globe [1–3]. This is especially true in low-socioeconomic countries, that may have little available resources to invest into disease prevention programs [1–3]. In particular, countries spanning the tropics must affront an expansive range of CVBDs that comprise a leading cause of fatality in dogs [3–6]. Bacterial infections can be some of the deadliest CVBDs in such regions with pathogens such as *Ehrlichia canis*, the causative agent of canine monocytic ehrlichiosis (CME), being the biggest contributor to a raft of life-threatening conditions, including pancytopenia, fever, bleeding tendencies and immunosuppression [7–9]. Other disease-causing species include, *Anaplasma platys* which is a cause of recurrent thrombocytopenia in canines [10], haemotropic mycoplasmas that are associated with haemolytic syndrome [6] and *Bartonella* spp. that can produce severe endocarditis [11]. Rates of canine infection with such pathogenic agents can be very high, especially in tropical countries of Southeast Asia, where, for example 25.5% of Malaysian, 21.8% of Cambodian and 9.9% of Thai dogs have previously been found positive for *E. canis* by PCR [3,6,8]. Lower but nonetheless significant levels of infection by haemotropic mycoplasmas (3.7–12.8%) and *A. platys* (3.7–4.4%) have also been detected in these same countries [6,8].

Exploration and surveillance of CVBD is not only important from a veterinary perspective but also a public health standpoint, as several bacterial CVBD agents are zoonotic. For example, *Rickettsia felis* and *Rickettsia conorii*, when transmitted to humans, are the causative agents of flea-borne spotted fever and Mediterranean/Indian/Israeli tick typhus, respectively, with both having dogs as reservoir hosts [12,13]. In addition, dogs are considered sentinel hosts for the zoonotic pathogens *Borrelia burgdorferi* sensu lato, which causes Lyme disease and *Anaplasma phagocytophilum* responsible for the potentially lethal human granulocytic anaplasmosis [14–17].

Thorough characterisation and monitoring of CVBD is crucial to protect the health of dogs and humans, especially in regions of Asia where there has been little explorative research done into established, emerging and novel CVBDs [2,5]. To address these knowledge gaps, state-of-the-art methodologies could be employed such as next-generation sequencing (NGS) based 16S ribosomal RNA (16S rRNA) metabarcoding that can supersede conventional PCR (cPCR) techniques with regard to their ability to detect and discover rare and/or novel organisms [18,19]. 16S rRNA metabarcoding is better able to elucidate diversity in explorative research by not relying on likely pathogen prevalence in a region nor on *a priori* knowledge of a pathogen's target genetic sequences, whilst also being better able to characterise coinfection [18,20]. Previous work has already developed a novel 16S and 18S rRNA metabarcoding methodology to characterise the range of blood-borne bacterial, apicomplexan and kinetoplastid organisms infecting canine hosts in Thailand, a country of substantial CVBD diversity [4,8,21–23]. The current research takes this further, by tackling some of the inherent challenges of NGS microbiome research whilst also improving methods to be better at unearthing pathogen diversity in the context of the canine blood micro-environment [24].

Prior research employing NGS metabarcoding analysis of bacterial pathogens from canines suffered from issues of bacterial primer cross-reactivity with the canine mitochondrial 12S ribosomal RNA (12S rRNA) gene [22]. On average, 47% of total reads during previous experiments were from mitochondrial DNA (mtDNA) cross-reactivity despite poor complementarity between the bacterial primers and 12S rRNA gene sequences [22]. This cross-reactivity is likely due to multiple causes, including a dominance of canine host DNA in the blood [22] in conjunction with the prokaryotic origin of mitochondria producing sequence similarities [25].

Similar challenges have been tackled before by using blocking primers that selectively bind to and prevent the amplification of DNA sequences that would otherwise dominate amplicons in contexts where there is a stacked ratio of low copy number DNA of interest compared to overabundant DNA that is ideally excluded [26–31]. Considering this, we designed and tested a 3′-spacer C3 blocking primer that could prevent amplification of the canine 12S rRNA gene by our previously tested bacterial-universal primers [22]. We then compared our bacterial NGS metabarcoding pipeline with and without this blocking primer to assess its blocking efficacy and elucidate whether it improved bacterial detection capability.

In addition, for NGS-based research, the method of DNA extraction employed can play a critical role in the diversity and sensitivity of 16S rRNA detected, particularly in the context of low biomass samples, such as blood [32–34]. Commercially available DNA extraction kits differ widely in the physical and chemical systems used from which they produce PCR-ready DNA as well as in their required time and complexity [34,35]. Recent attention has been brought to automated DNA extraction systems that simplify the researcher's workload but that may also accrue a cost with regard to DNA quantity and quality for downstream applications [35,36]. Another variable that must be considered when determining the most appropriate DNA extraction kit for 16S rRNA metabarcoding research is that bacteria frequently contaminate kits and reagents [32,37]. Numerous studies have now explored this phenomenon via the use of rigorous negative controls to highlight the prevalence of common kit contaminant bacteria, also demonstrating that levels of kit contamination and the types of contaminants found can vary between kits and even batches within kits [32,33,38].

Taking this into consideration, we tested four DNA extraction systems. Two were a typical spin-column based whole blood DNA extraction procedure and two were automated systems employing the Maxwell® RSC 48 Instrument (Promega, Madison, WI, USA). Of the two automated procedures, one was for extraction of whole blood and the other for extraction from the blood's Buffy Coat layer, which has been reported to increase detection sensitivity of certain vector-borne bacterial species that are typically difficult to detect [39–41]. Both automated kits use magnetic bead extraction and purification. Kit performance was assessed and compared based on DNA yield, sensitivity of pathogen detection and presence of contaminant DNA.

2. Results

2.1. Blocking Primer Performance

Initial experimentation using cPCR analysis with gel electrophoresis assessed the ability of our blocking primer: *Canis*-mito-blk, to bind to canine 12S mitochondrial sequences when it did not have the 3′-spacer C3 modification, which was observable via production of a 137 bp product (data not shown). When the blocking primer had the 3′-spacer C3 added, this 137 bp product was no longer produced even when the bacterial 16S rRNA primers were present in the PCR, indicating complete blocking of canine 12S rRNA amplification. Bacterial primers were not prevented from amplifying bacterial sequences as an approximately 250 bp product was still produced by these primers when used on canine blood samples positive for the pathogens *E. canis*, *A. platys* and *R. felis* (data not shown).

After bioinformatic processing and filtering of NGS data a total of 4,746,542 reads were accrued, a result within the typical range for amplicon-based Illumina sequencing at a high quality Phred score of over thirty. From the total reads, 58% were from samples amplified without blocking primers and 42% from samples with blocking primers.

Analysis of the mean percentage of canine mitochondrial reads relative to total reads on a per sample basis was 59% (±5.1%) for samples amplified without the blocking primer compared to 19% (±3.1%) with the primer (Figure 1). The difference between these two groups was statistically significant at the $p < 0.01$ level, indicating that our blocking primer significantly reduced the amount of canine mitochondrial reads amplified. Mean blocking efficiency across our canine DNA samples was 25% (±3.0%) according to the formula defined by Tan and Liu (2018) [27], with individual sample blocking efficiencies ranging from as low as 2.2% in some samples with low total sequence counts to as high as 100% blocking efficiency in others.

Pathogenic vector-borne bacterial species were detected from the species *E. canis*, *A. platys*, *M. haemocanis*, 'Candidatus Mycoplasma haematoparvum', *Mycoplasma turicensis*, *Bartonella* spp. as had been previously found by Huggins et al. (2019) [22]. However, only the high prevalence of *E. canis*, *A. platys* and *M. haemocanis*, permitted statistically comparable results between the ability of the metabarcoding methodology to detect vector-borne bacterial DNA with and without the use of *Canis*-mito-blk. When our blocking primer was used more *E. canis* and *A. platys* infections were detected by our bacterial metabarcoding method (Table 1). For *E. canis*, 24 infections were found using the blocking primer compared to 19 when they were not used, a similar increase in detection capability was found for *A. platys* with 20 infections detected using the blocking primer compared to 17 without. For detection of *M. haemocanis*, 19 infections were detected both with and without the blocking primer. The mean percentage of vector-borne bacterial reads with the blocking primer was always higher than when no blocking primers was used for; *E. canis* (15% ± 4.0% vs 5.4% ± 2.0%), *A. platys* (35% ± 8.5% vs 23% ± 7.0%) and *M. haemocanis* (53% ± 8.7% vs 35% ± 7.5%) with all differences being statistically significant (Figure 2).

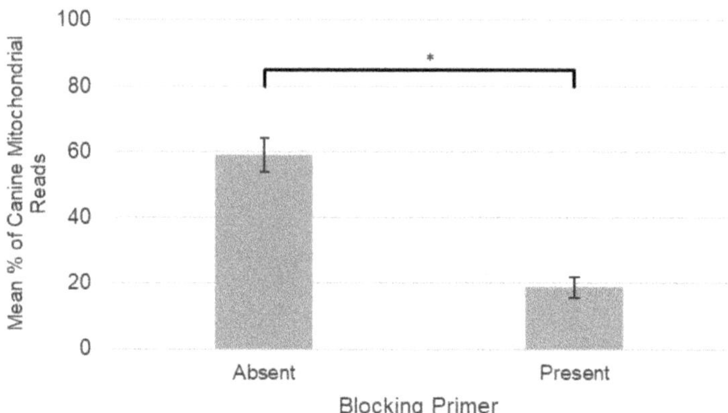

Figure 1. Mean percentage of canine mitochondrial reads across all samples in the absence and presence of the blocking primer. Vertical error bars display standard error, horizontal bar highlights a statistically significant difference at the $p < 0.01$ level (*).

Table 1. Comparison of total number of infections detected when *Canis*-mito-blk was absent or present. Infections were deemed true when the total number of reads of a particular pathogen were over the read cut-off value as defined in the Materials and Methods.

Vector-Borne Bacteria	Total Infections: Blocking Primer Absent	Total Infections: Blocking Primer Present
A. platys	17	20
E. canis	19	24
M. haemocanis	19	19

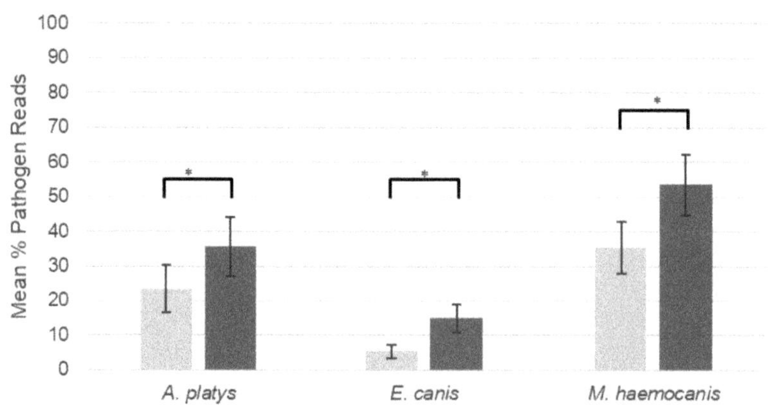

Figure 2. Mean percentage of three different pathogen reads across all pathogen positive samples in the absence of the blocking primer (light grey) and with the blocking primer (dark grey). Vertical error bars display standard error and horizontal bars highlight differences that are significant at the $p < 0.01$ level (*).

For samples with less common vector-borne bacteria, the percentage of pathogen reads compared to total reads was always higher for the samples that utilised the blocking primer. For example, for the two samples that were found to have a *Bartonella* spp. infection and two with a *Mycoplasma turicensis* infection, the average percentage of reads for the former pathogen was 1.5% vs 0.42% (with blocking primer vs without) and 16% vs 2.1% for the latter pathogen. A single sample with '*Candidatus* Mycoplasma haematoparvum' reads had 54% of reads from this pathogen when using *Canis*-mito-blk, compared to 27% without.

Comparing the number of bacterial taxonomic identifications across all samples with the blocking primer, it was found that there were 353 different sequences vs 327 without the blocking primer. Moreover, diversity indices calculated on just the bacterial portion of total sequences for all samples that used blocking primers vs all samples that did not, indicated that using the blocking primer improved characterisation of bacterial diversity. For both indices, a greater species diversity is demonstrated by a larger number. Using Simpson's index, a value of 4.07 was acquired for samples using the blocking primer compared to 3.15 without (Table 2). With the Shannon–Wiener index, which takes into greater consideration rarer species with low read numbers, the index value was 2.3 when the blocking primer was used in contrast to 1.81 for samples without it [42] (Table 2).

Table 2. Comparison of biodiversity indices when blocking primers (BP) were absent or present. The higher the index, the greater the relative diversity.

Bacterial Diversity: Shannon–Wiener (H) Index		Bacterial Diversity: Simpson (D) Index	
BP Absent	**BP Present**	**BP Absent**	**BP Present**
1.81	2.3	3.15	4.07

2.2. DNA Extraction Kit Performance

DNA extraction kit yield was compared which found that the automated extraction methods; Promega Buffy Coat (BC) = 680.2 ng/µL and Promega whole blood (WB) = 483.8 ng/µL obtained much higher average DNA yields than the non-automated kits Qiagen whole blood (QG) = 13.9 ng/µL and Bioline whole blood (BL) = 3.4 ng/µL (Table 3). At the same time, the range of DNA quantified in the extraction negative controls was also higher for the automated methods 0.005–0.024 ng/µL, while for QG, a lower range of 0.001–0.014 ng/µL was found and BL had a DNA yield below the lower limit at which the Qubit^{TM} can detect (Table 3). Such findings demonstrate a potentially higher initial quantity of bacteria or bacterial DNA in the automated extraction kits.

Table 3. Comparison of mean DNA quantity and standard error (S.E.) from the same blood samples extracted with different kits and comparison of contaminant DNA in kit reagents alone. Includes post-bioinformatic analysis results of total raw and filtered reads accrued between different kit types as well as total Amplicon Sequence Variants (ASVs), i.e., diversity of DNA sequences found.

DNA Extraction Kit	Mean DNA Quantity ± S.E. (ng/µL)	DNA Quantity in ExtractionNegative Controls (ng/µL)	Total Raw Reads	Reads Post-Filtering	Total ASVs
Promega Whole Blood (WB)	483.8 (±150.9)	0.005–0.017	5,351,987	3,857,213	435
Promega Buffy Coat (BC)	680.2 (±394.7)	0.012–0.024	5,916,458	4,716,590	428
Bioline Whole Blood (BL)	3.4 (±0.6)	<0.001	5,935,103	4,188,285	1401
Qiagen Whole Blood (QG)	13.9 (±1.1)	0.001–0.014	4,462,788	3,559,370	6683

Following bioinformatic processing, differing numbers of total raw and filtered reads were obtained for the datasets from each extraction method, an inevitable result of samples from each extraction method having to be run on separate NGS flow cells (Table 3). After filtering, the number of reads between datasets varied by a maximum of 1,157,220 reads. The total ASVs found across each kit's dataset also differed greatly with QG kits finding the largest diversity of ASVs at 6683, followed by BL at 1401, WB at 435 and BC at 428 (Table 3). Counterintuitively, total reads post-filtering was not a good predictor of ASV diversity as QG, which had the lowest number of post-filtering reads and also had the highest ASV diversity. Additionally, BC had the highest post-filtering read total and the lowest ASV diversity.

In a similar manner to the data from Thai dogs, the main vector-borne bacterial DNA found was from the species *E. canis*, *A. platys* and *M. haemocanis*, with differences in the numbers of infections elucidated between extraction kit types (Table 4). Across the 43 Cambodian blood samples compared, the QG kit found the highest number of infections for all three main pathogenic bacterial species at 35, followed by WB and BC with 28 and BL finding the least infections at 23 (Table 4). In addition, the QG kits found ASVs taxonomically classified as putative pathogens within the genera *Rickettsia* spp. and *Coxiella* spp., which were not detected by any of the other kit types. These taxonomic classifications could not be resolved to species level through either QIIME2 or a BLASTn search in GenBank due to very high sequence similarities between different species in the same genus.

Table 4. Comparison of total number of infections detected of the three main bacterial pathogens by different extraction kits. Infections were deemed true when the total number of reads of a pathogen were over the read cut-off value as defined in the Materials and Methods.

Vector-Borne Bacteria	Promega Whole Blood (WB)	Promega Buffy Coat (BC)	Bioline Whole Blood (BL)	Qiagen Whole Blood (QG)
A. platys	18	17	13	22
E. canis	7	7	6	8
M. haemocanis	3	4	4	5

The amount of artificial kit-derived contaminant DNA detected by our NGS methodology differed significantly depending on the extraction kit used (Figure 3). Both automated extraction kits had a significantly higher percentage of bacterial contaminant DNA (BC 86% and WB 80% of total reads) compared to 34% of reads for QG and 19% for BL. All differences between kits were significant at the $p < 0.05$ level apart from the mean contaminant reads between the two automated kits; WB and BC.

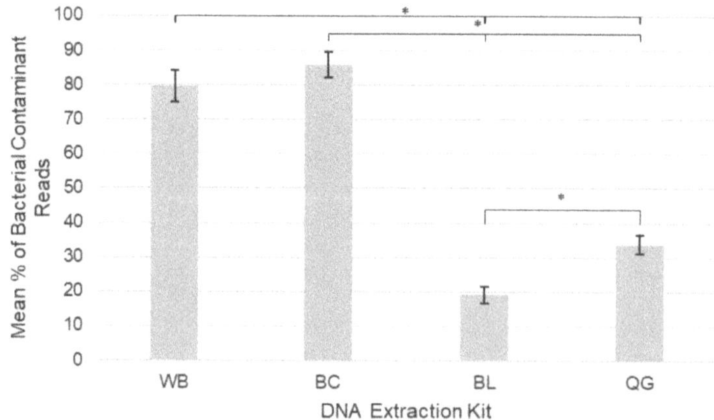

Figure 3. Mean percentages of kit-derived artificial bacterial contaminant reads relative to total reads across four different DNA extraction kits: Promega Whole Blood (WB), Promega Buffy Coat (BC), Bioline Whole Blood (BL), Qiagen Whole Blood (QG). Vertical bars display standard error and asterisks show differences between kit types that are significant at the $p < 0.05$ level (*).

The number of different ASVs assessed as being artificial kit contaminants due to them appearing with a read count of ≥100 in DNA extraction negative controls also varied with WB and BC kits having just four kit-contaminant ASVs, BL (n = 18) and QG (n = 57). Furthermore, the WB and BC kits demonstrated a distinct bacterial contaminant 'fingerprint', (Supplementary Table S1). For example, with the WB kits, results, on average, had a total of 51% of reads classified as *Massilia* spp. and 28% of reads as *Pseudomonas* spp. whilst BC kits on average had a composition of 61% *Massilia* spp. and 24% *Cellulomonas* spp. reads. In comparison, the composition of contaminant bacterial reads was less marked for the non-automated kits with QG having 7.9% of sequences classified as belonging to chloroplast DNA and 4.3% as *Burkholderia* spp., whilst BL kits had 4.8% of total reads on average belonging to *Burkholderia* spp. (Supplementary Table S1).

The amount of cross-reactive amplification between the bacterial primers on canine mitochondrial sequences also differed significantly, depending on the DNA extraction kit used, despite blocking primers being present in all reactions. When samples were extracted using BL kits a statistically significant increase ($p < 0.05$) in the mean percentage of canine mitochondrial reads relative to total reads was observed with an average of 31% of reads being host mitochondrial sequences (Figure 4). This contrasted with much lower percentages of host mitochondrial sequences when samples were extracted with QG (1.5%), WB (1.1%) and BC (0.2%).

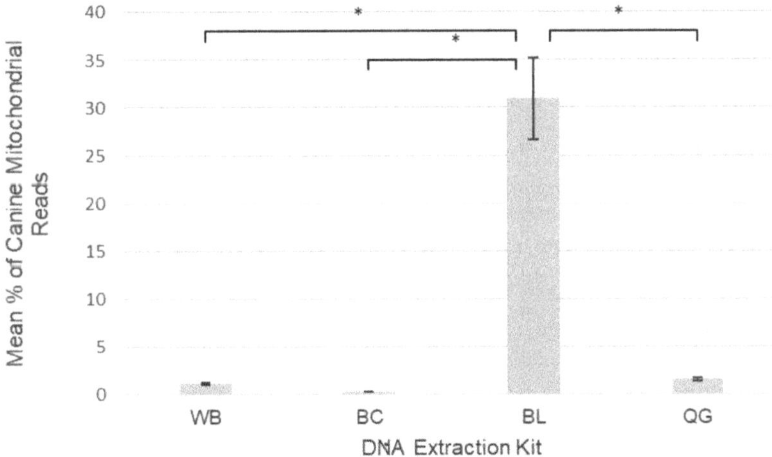

Figure 4. Mean percentages of canine mitochondrial reads relative to total reads across four different DNA extraction kits: Promega Whole Blood (WB), Promega Buffy Coat (BC), Bioline Whole Blood (BL), Qiagen Whole Blood (QG). Vertical bars display standard error and asterisks show differences between kit types that are significant at the $p < 0.05$ level (*).

3. Discussion

3.1. Blocking Primer Performance

The use of our blocking primer (*Canis*-mito-blk) was demonstrated to significantly reduce the amount of bacterial primer cross-reactivity on host canine mitochondrial sequences, significantly increase the relative proportion of vector-borne bacterial DNA sequenced and increase the detection capability for elucidating infections by our NGS metabarcoding method. The increase in detection capability between our metabarcoding method with and without our blocking primer is particularly important as five more *E. canis* infections and three more *A. platys* infections were detected from canines when using the blocking primer in the first-step PCR. Accurate detection and characterisation of such pathogens is key when conducting a range of studies from large-scale surveillance investigations where reductions in sensitivity may impair the ability to provide apparent estimates that match the true prevalence of disease, through to clinical diagnostics on infirm canines where a missed infection may be fatal [7,9,43]. An increase in detected infections when using the blocking primer is similar to several studies which have reported exposition of previously occult DNA of interest using various blocking primers, in the context of both pathogen detection [26] and ancient DNA [30].

The number of detected infections by *M. haemocanis* was the same for samples regardless of whether the blocking primer was used. *M. haemocanis* infections were typically observed to have high numbers of reads, frequently over 10,000 per sample but reaching as high as 149,000 in one sample. This may result in *M. haemocanis* infections being easier to detect than other vector-borne bacterial species and thus minimising the possibility of missed infections even in the absence of blocking primers.

Not only were more samples detected using our blocking primer but the relative proportion of canine vector-borne bacterial reads, relative to the total number of reads, was always higher when using this blocking primer. These results provide greater support for an improved metabarcoding sensitivity when utilising our blocking primer as if more pathogen 16S rDNA is amplified and sequenced relative to total amplified DNA, then the probability that infections with even low levels of circulating bacteria are detectable is also increased. Similar results were obtained by Tan and Liu (2018) [27] who were able to increase the relative abundance of desired protist reads when using metazoan-specific blocking primers in marine environments.

Conversely, the relative proportion of canine mitochondrial sequences that were amplified by our bacterial primers was lower by an average of three times when using our blocking primer, a result that is further supported by the primer's mean blocking efficiency of 25%. Not only do such results demonstrate the blocking primer's ability to prevent amplifying primer cross-reactivity on host 12S rRNA sequences but they also signify an increase in relevant data obtainable when deep sequencing with our blocking primer. A typical MiSeq v3 chemistry run outputs between 3.3 to 15 Gb of data, [44] and if, as found in the current research, as much as 59% of the data output is the result of primer cross-reactivity on host (mtDNA), then a substantial cost is accrued in terms of labour, time and money to the researcher. Furthermore, if a large proportion of sampling effort is due to cross-reactive amplification then there is the potential for a significant masking effect whereby bacterial pathogen DNA may be missed as disproportional quantities of canine mtDNA are sequenced instead.

The concentration of blocking primer used throughout this study was informed by prior experimentation elucidating a roughly 1:1 ratio of target to non-target DNA amplification. When this information was assessed, in conjunction with comparison from studies that had dealt with similar proportions of non-target DNA [26,27,45], a ratio of 3:1 blocking primer to amplifying primers was deemed sufficient. Experimentation using cPCR and evidence from a reduction of canine mitochondrial sequences demonstrates our chosen ratio to be effective, however, even with the addition of a blocking primer at this ratio there was still some continued amplification of canine mtDNA. On average 18% of each sample's total reads was comprised of canine mitochondrial reads, with great variability on a sample by sample basis ranging from 0% to 64%. Taking this into consideration, an even higher ratio of blocking primer to amplifying primers may have reduced this cross-reactivity even more. Nonetheless, there is also a possibility that blocking primers could be generating off-target blocking effects on bacterial sequences, potentially reducing the detection of certain bacterial groups. For example, the 5′ end of our blocking primer has a run of at least seven bp that are capable of binding to bacterial 16S rRNA sequences, indicating there may be some potential for this blocking primer to anneal to bacterial sequences and reduce Wehi_Adp_515F efficiency. Therefore, it was important to keep blocking primer concentration as low as possible, whilst remaining effective at blocking desired targets [27,30].

3.2. DNA Extraction Kit Performance

Differences between results acquired throughout the range of parameters investigated were substantial between DNA extraction kit types, despite the same canine blood samples being used. Considering all kit performance results, the Qiagen DNeasy Blood and Tissue Kit performed the best within our NGS metabarcoding framework as it was able to detect the highest number of vector-borne bacterial infections (Table 4). This kit also elucidated a greater diversity of bacterial sequences, including some from putative pathogens, such as *Rickettsia* spp. and *Coxiella* spp., that were not found using the other extraction kits. Such results are of importance to our study given that the strength of our NGS-methodology is its ability to detect unusual or novel pathogens without *a priori* information regarding likely pathogens or their respective 16S rRNA sequences. Our findings, mirror those of other NGS kit comparison assessments that have found that DNA extraction kit employed can have a significant effect on the abundance and diversity of bacterial groups found [38,46]. For example, similar to our current results, Hart et al. (2015) found that a Qiagen spin-column based extraction method was better able to detect rare phyla when compared to four other extraction methods [46].

The considerably higher number of ASVs detected when using the QG kits was over 15 times the amount found by the automated WB and BC kits and almost five times that found by the BL kits, demonstrating the strength of the QG kits to liberate a higher diversity of bacterial sequences and allow for better blood microbiome characterisation (Table 3). Moreover, of all the kit types tested, the QG kit used the lowest quantity of initial starting material at just 100 µL of whole blood compared to 200 µL for both the BL and BC kits and 500 µL for the WB kit. Such findings provide weight to the argument that quantity of starting material alone does not necessarily improve sensitivity of pathogen detection and that DNA extraction methodology and chemistry may play a more significant role. This is further supported by the data on the number of raw and filtered reads obtained for the different kits, which found that the QG kits had less than that obtained for WB, BC and BL, an inevitable result of the samples from different kits having to be run on separate NGS flow cells to permit an adequate sampling depth (Table 3). Therefore, despite an initially smaller sample volume and a lower quantity of NGS read data, the superiority of the QG kit extraction process appears to have overcome such potential disadvantages and highlight the greatest number of infections and bacterial diversity of the four kits tested.

A range of similar studies have explored the impacts of DNA extraction kits, either manual or automated, for pathogen detection using various downstream molecular methods, including cPCR, real-time PCR, restriction digest analysis and deep sequencing [34–36,46–49]. Overall, there are few consistent DNA extraction types that perform well in all contexts, with optimal extraction methods typically being very host, pathogen and sample-type specific [34,48]. For example, large differences in kit performance may be observed depending on whether the source material contains lots of PCR inhibitors, such as in faeces and blood, or on the biological properties of the pathogen of interest, e.g., viruses that are easy to lyse vs resilient spore-forming bacteria [34,36,38,49]. Moreover, the ability of different extraction kits to remove PCR-inhibitory molecules may play a large impact on obtainable downstream results. Blood contains potent PCR inhibitors such as haemoglobin that affects DNA polymerase activity and hence observed differences in kit performance could be due to how well different extraction processes remove such molecules [50].

In the current research, differences in extraction chemistry may be responsible for differences in bacterial pathogen detection between kits [35]. For example, neither of the automated WB and BC kits used a proteinase K digestion step, whilst both manual QG and BL kits did use this digestion process. Proteinase-based degradation may release DNA sequestered by peptides or from within bacterial cell walls, thus liberating more pathogen DNA, making it more readily available for PCR amplification [36]. Nonetheless, whilst both manual kits used proteinase K digestion only, the QG kit identified more infections than the automatic kits, meaning that this biochemical variable is likely just one of many responsible for differences in pathogen detection capability between extraction kits.

Depending on the extraction method utilised, the average DNA yield obtained also demonstrated marked differences. The automated BC and WB methodologies consistently extracted DNA yields in the range of hundreds of ng/µL with large variability between samples in the amount of DNA extracted (Table 3). In contrast, QG kits extracted an average of 13.9 ng/µL and BL kits extracted the lowest quantities at 3.4 ng/µL, with both spin-column based methods displaying much lower variability in nucleic acid yield. Such substantial differences in yield are due to the physical extraction methodology and chemistry employed by a given kit [48,51]. QG and BL kits are dependent on retention of DNA in a silica column matrix which becomes saturated, capping the maximum amount of DNA obtainable [46,48]. The Promega WB and BC automated methods use magnetic-bead based extraction whereby the large quantity of bead surface area available permits much larger amounts of DNA to be carried over between extraction stages [35,47]. Nonetheless, as previously explored, the higher DNA yields gained by the WB and BC kits did not necessarily translate to a greater number of vector-borne bacterial infections found. One explanation for this may be that the larger quantity of DNA extracted by the WB and BC kits meant total host DNA was in great excess, making efficient blocking by our *Canis*-mito-blk primer ineffective, resulting in bacterial infections being missed.

With the advent of 16S rRNA bacterial metabarcoding, there has been a growing number of studies investigating the existence of endogenous bacterial contamination within DNA extraction kits and laboratory reagents [32,33,37]. Such findings underscore the importance of conducting no sample/reagent only negative controls within metabarcoding studies to correctly identify such contaminant species and thereby discern true species present within samples from those introduced by extraction kits [24,38,52,53]. In NGS studies that use a source material with a high density of bacteria, such as soil or faeces, this contamination typically has little effect as kit contaminant bacterial DNA is masked by that extracted from the source material [32,33]. However, when the source material has a low bacterial biomass, such as in blood, then kit contaminant bacterial DNA may be amplified and sequenced at similar levels to that of the source material, potentially competing and obscuring the microbiome signature of true bacterial inhabitants of the blood compartment [32,33]. An ever-growing list of common contaminant bacteria of kits highlights the tenacity of such organisms to survive and grow in a variety of environments, including ultra-pure water systems and other reagent manufacturing equipment and processes [32,37]. This creates a necessity to explore and identify differences in bacterial contamination between extraction kits which may influence the detection of true sample bacteria and demonstrate whether some kits are better suited to NGS applications.

In this study, bacterial taxa were considered an artificial kit contaminant if they had a read count of 100 or over in either of the two no sample/reagent only negative controls, extracted alongside the blood samples for each kit type. Large differences in the average percentage of all contaminant bacterial reads compared to total reads were observable between kit types with the automated BC (86%) and WB (80%) kits having on average over two times as much contamination as the QG kits and just under four times as much contamination as the BL kits (Figure 3). Such findings are supported by the DNA yield data found within extraction negative controls, from which WB and BC kits had the highest levels of negative control DNA, especially compared to the BL controls which had levels below the detection limit of the Qubit™ 3.0 fluorometer. Kit-derived bacterial contaminant DNA was dominated by just a few key genera in the WB (51% *Massilia* spp. and 28% *Pseudomonas* spp.) and BC kits (61% *Massilia* spp. and 24% *Cellulomonas* spp.) whilst the QG and BL kits demonstrated a larger diversity of kit contaminants, none of which comprised more than 9% of total reads (Supplementary Table S1).

Disparate levels of kit contamination may be due to the physical properties of the extraction protocol used by each kit type. The automated WB and BC extraction methods utilise a series of seven wells all filled with between 700 to 1000 µL of lysis and wash buffers providing a large reagent pool from which contaminant bacteria could be introduced. In contrast, less than 600 µL of each reagent are added at any one time to the spin-column based QG and BL protocols with fewer wash steps from which endogenous bacteria could be added, lysed and their DNA extracted. The preponderance of a few key bacteria in the automated kit types may point to the growth or introduction of these species within the manufacturing process of these kits, whilst a larger diversity of less prevalent bacterial species in the QG and BL kits may be the product of small contamination events from the laboratory environment [32]. The large percentage of contaminant reads from the WB and BC kits may explain, in some part, the reduced number of vector-borne bacterial infections elucidated when compared to those found with the QG kits. Endogenous bacterial DNA may be concealing the detection of pathogen DNA when using these kits, particularly in blood samples with low levels of circulating pathogens [32,33].

An unexpected result identified by our assessment of DNA extraction kit performance was a significant difference in the amount of bacterial primer cross-reactivity on canine mitochondrial sequences between the kit types utilised. The BL kit obtained results that had a significantly higher amount of canine mitochondrial reads compared to the total, despite the use of the *Canis*-mito-blk primer, whilst the other three kit types had as low as 0.2% to 1.5% of total reads generated from cross-reactivity (Figure 4). An explanation may be the much lower DNA yields from the BL kits, that had, on average, just 3.4 ng/μL of DNA, a large proportion of which would be canine DNA. Such low sample yields may not have provided sufficient bacterial DNA template for 16S rRNA amplification, leading to more off-target amplification and sequencing of host mitochondrial sequences. Moreover, such low starting DNA yields may also be the reason for the poorer performance of the BL kits with respect to the number of vector-borne bacterial infections found, compared to the other kit types [51,54].

4. Conclusions

Taken together, the current study provides two complementary methodological amendments to improve the vector-borne bacterial detection capability of a previously developed metabarcoding protocol [22]. Firstly, employment of a 3'-spacer C3 blocking primer (*Canis*-mito-blk) to prevent bacterial amplification primers from cross-reacting on host mtDNA was demonstrated to improve the number of infections detected and the relative quantity of pathogen DNA relative to total DNA. Secondly, use of Qiagen's DNeasy Blood and Tissue Kit appears superior to two automated methods utilising Promega's Maxwell® RSC 48 Instrument and Bioline's manual ISOLATE II Genomic DNA Kit. This superiority was demonstrated via the DNeasy Blood and Tissue Kit's ability to detect more infections as well as its lower quantities of kit contaminant bacterial DNA and generation of less off-target cross-reactivity on host mtDNA. Such methodological optimisation improves the power of our developed technique to unearth potentially latent vector-borne bacterial infections, a common phenomenon of many CVBDs where clinical signs may be followed by cyclical periods of remission when infections are hard to detect [7,55]. Furthermore, both amendments increased the diversity of different bacterial sequences elucidated, augmenting our protocol's possibility of identifying rare or novel bacterial pathogens. To conclude, together these are key improvements to our metabarcoding method allowing it to mine and more fully characterise the range of bacterial pathogens in regions of the world where there is a plethora of different CVBDs alongside a dearth of applicable research [2,5,56].

5. Materials and Methods

5.1. Sampling and DNA Extraction

For blocking primer experiments, blood extracted DNA from 50 temple community dogs in Thailand were used as described in Huggins et al. (2019) [22] under Ethics Permit: OACKU-00758 and extracted using the E.Z.N.A.® Blood DNA Mini Kit (Omega Biotek Inc., Norcross, GA, USA). For the comparison of DNA extraction kits in NGS metabarcoding applications, 50 whole blood samples were collected at four sites across Phnom Penh, Cambodia, from a mixture of pagoda temple communities, and locally owned and semi-domesticated dogs. Blood was only taken after obtaining informed consent from the relevant monk in the case of temple community dogs or the owner from local pet dogs. A qualified veterinarian conducted collection of two 1 mL blood samples per dog via cephalic puncture into anti-coagulation EDTA tubes on ice. These were then transferred to a −20 °C freezer upon return from the field. Approximately, 700 μL of blood from one sample, per dog, was then aliquoted to a separate Eppendorf tube and centrifuged at 12,000 rpm for 5 min in a portable LW-ZipCombo-C microcentrifuge (LW Scientific, Lawrenceville, GA, USA). From this, the blood plasma was disposed of and 250 μL of the Buffy Coat layer and surrounding erythrocytes was aliquoted and stored separately at −20 °C. Work in Cambodia was conducted under Ethics Permit: 1814620.1 from the University of Melbourne.

DNA extraction from the same 50 Cambodian canine whole blood or Buffy Coat samples was then carried out via one of four methods. Extractions were conducted according to manufacturer's instructions in the same laboratory using the maximum possible starting quantity of sample for each kit. This was done to elucidate the pathogen detection capability and level of bacterial kit contaminants without subsequent researcher modifications to kit protocol and to assess kit utility for 16S rRNA metabarcoding as developed by the manufacturer. Extraction methods were as follows:

(1) Extraction of 500 μL of whole blood using the Maxwell® RSC Whole Blood DNA Kit (Promega, Madison, WI, USA) on the automated Maxwell® RSC 48 Instrument DNA extraction robot. This was conducted as per the manufacturer's instructions and eluted in 100 μL of Ambion Nuclease-Free Water (Life Technologies, Carlsbad, CA, USA).

(2) Extraction of 250 μL of Buffy Coat layer using the Maxwell® RSC Buffy Coat DNA Kit (Promega, Madison, WI, USA) on the Maxwell® RSC 48 Instrument, conducted as per the manufacturer's instructions and eluted in 100 μL of Ambion Nuclease-Free Water.

(3) Manual extraction of 200 μL of whole blood, using the ISOLATE II Genomic DNA Kit (Bioline, Memphis, TN, USA), using the manufacturer's protocol with a final elution step in 50 μL.

(4) Manual extraction of 100 μL of whole blood, using the DNeasy Blood and Tissue Kit (Qiagen, Hilden, Germany) using the manufacturer's protocol with a final elution step in 50 μL.

With all DNA extraction methods, two no blood/reagent only, DNA extraction negative controls were run per method to assess for levels of bacterial contamination in kits. Comparison of whole blood vs Buffy Coat based extraction was conducted due to a body of research suggesting that testing of specific blood fractions may improve sensitivity of detection for some blood-borne pathogens [40,41].

5.2. DNA Quantification

All DNA extractions were quantified on a Qubit™ 3.0 fluorometer (Life Technologies) using the Qubit™ dsDNA HS Assay Kit and means determined. Total DNA was also tested for in no blood/reagent only negative controls.

5.3. Design of Blocking Primer

Previous work by Huggins et al. (2019) [22] testing a variety of bacteria-universal 16S rRNA targeting primer pairs at hypervariable regions V1 to V6 identified cross-reactivity on canine host DNA as a recurrent problem. Even the best performing primer pair; Wehi_Adp_515F (5′-GTGYCAGCAGCCGCGGTAA-3′) and Wehi_Adp_806R (5′-GGACTACNVGGGTATCTAAT-3′) still generated as many as 97% of total reads in some samples from primer cross-reactivity against canine 12S rRNA sequences, with an average of 47% across all samples found [22]. This was despite prior modifications to both primers to reduce degeneracy and thereby likelihood of cross-reactivity with canine 12S rRNA sequences [22]. Hence, to further prevent this, a *Canis lupus* 12S rRNA specific blocking primer was designed (*Canis*-mito-blk) with a 3′-spacer C3 to prevent DNA polymerase elongation of such mitochondrial sequences when used alongside the Wehi_Adp_515F and Wehi_Adp_806R, 16S rRNA primers. Blocking primer design was conducted in Geneious v. 11.1.5 (Biomatters Ltd.) using 26 16S rRNA sequences from pathogenically relevant bacterial species and an *Escherichia coli* reference sequence. Moreover, eight sequences of the *C. lupus* mitochondrial 12S rRNA gene (which cross-reacted with bacterial primers in prior experiments) were added, alongside the sequences for the bacterial 16S rRNA primers [22] (Figure 5). The 5′ end of our designed *Canis*-mito-blk primer has an 8 bp overlap with the bacterial forward primer Wehi_Adp_515F, then extending 11 bp downstream of this primer's binding site, ending with a 3′ C3 spacer that prevents DNA polymerisation (Figure 1). The primer's high complementarity to canine 12S rRNA sequences suggests it may block amplification of these sequences. Furthermore, base pair (bp) differences with bacterial 16S rRNA sequences prevent this blocking primer from efficiently binding to this region, allowing amplification of bacterial sequences to continue. The final design of *Canis*-mito-blk was 5′-CCGCGGTCATACGATTAACC/3SpC3/-3′.

Figure 5 shows the location of the three bp differences between the Wehi_Adp_515F primer and the canine 12S rRNA mitochondrial sequences, where cross-reactivity can still occur.

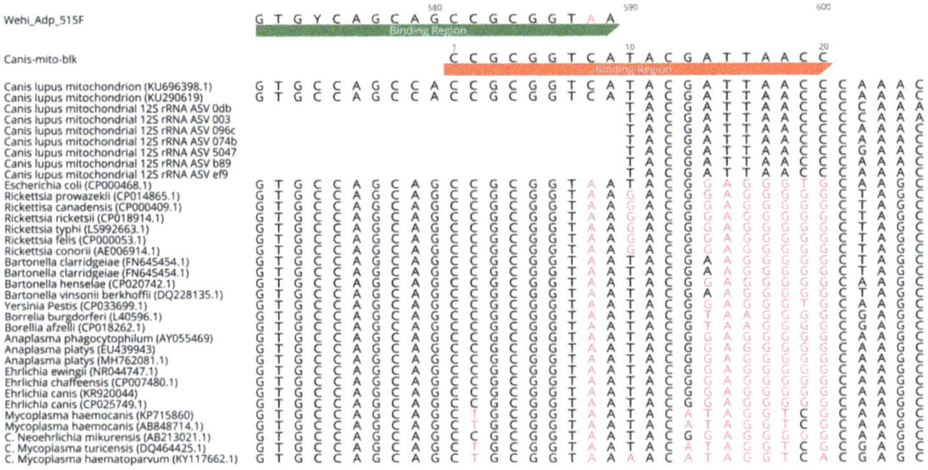

Figure 5. Alignment of pathogenic bacterial 16S rRNA sequences, canine 12S rRNA mitochondrial sequences, the forward bacterial primer Wehi_Adp_515F (green bar) and the *Canis*-mito-blk blocking primer (red bar) in the 5′ to 3′ direction. Nucleotide bases that differ from those at the relevant position in the *Canis*-mito-blk primer are coloured in light red, demonstrating dissimilarity between our designed blocking primer and bacterial 16S rRNA sequences but perfect complementarity with the 12S rRNA mitochondrial counterparts.

5.4. Validation of Canis-mito-blk

To test for the specificity of *Canis*-mito-blk in vitro to canine 12S rRNA sequences, cPCR was conducted with and without the addition of the 3′ C3 spacer, plus a reverse canine 12S rRNA specific primer (5′-AATCCCAGTTTGGGTCTTAGC-3′) producing a 137 bp product. PCRs were 20 μL in total, comprising 10 μL of OneTaq® 2X Master Mix with Standard Buffer (New England Biolabs, Ipswich, MA, USA) 0.4 μM of both forward and reverse 16S bacterial primers, 1.2 μM of both *Canis*-mito-blk and the 12S rRNA reverse primer, 1 μL of template DNA and Ambion Nuclease-Free Water. The ratio of blocking primer to bacterial primer should be informed by the ratio of cross-reactive 12S rRNA reads to 16S rRNA bacterial sequences [45], which is at a ratio of approximately 1:1 as calculated using prior data collected by Huggins et al. (2019) [22]. Therefore, to ensure blocking primer was in excess we used a 1:3 ratio of bacterial primer to blocking primer, as reported by similar studies, utilising blocking primers [26,27,29]. Thermocycling conditions were taken from Huggins et al. (2019) [22] and were 95 °C for 3 min, 35 cycles of 95 °C for 45 s, 56 °C for 60 s and 72 °C for 90 s with a final elongation at 72 °C for 10 min. Specific amplification of 12S rRNA sequences was conducted by visual observation of 137 bp bands on a 1.5% agarose gel using a GelDoc™ XR + System (Bio-Rad, Hercules, CA, USA). PCR combinations using the blocking primer with and without bacterial 16S rRNA primers and the 3′ C3 spacer on DNA from non-infected and infected blood were conducted to ensure that *Canis*-mito-blk could amplify 12S rRNA sequences in the absence of the 3′ C3 spacer and completely block amplification when the 3′ C3 spacer was present.

5.5. Bacterial 16S rRNA Metabarcoding Assessment of Canis-mito-blk Efficacy

Fifty previously characterised blood-extracted DNA samples from Thai dogs [22] were deep-sequenced with and without the use of the *Canis*-mito-blk primer to assess its efficacy at preventing bacterial primer cross-reactivity and improving pathogen diversity characterisation.

All first-step and second-step PCRs for deep-sequencing library preparation were conducted within a PCR hood after UV sterilisation. The first-step PCR consisted of 10 µL of OneTaq® 2X Master Mix, 0.4 µM of both forward and reverse 16S bacterial primers with NGS overhang sequences, 1.2 µL of *Canis*-mito-blk (for the 50 samples using the blocking primer), 1 µL of template DNA with the whole reaction made up to a total volume of 20 µL with Ambion Nuclease-Free Water. Bacteria primer sequences with the addition of NGS overhangs (underlined) were WehiNGS_Adp_F (5′-GTGACCTATGAACTCAGGAGTCGTGYCAGCAGCCGCGGTAA-3′) and WehiNGS_Adp_R (5′-CTGAGACTTGCACATCGCAGCGGACTACNVGGGTATCTAAT-3′). The thermocycling profile was used with just 20 amplification cycles, instead of 35. PCR product was then cleaned using 1X Ampure Beads (Beckman Coulter, Brea, CA, USA) and checked for quality on an Agilent 2200 Tape Station (Agilent Technologies, Santa Clara, CA, USA) before proceeding. Two no-template PCR negative controls and four uniquely identifiable positive controls were also included in this first-step PCR. Due to the need to assess for possible cross-contamination between samples during deep-sequencing library preparation or Illumina indexing errors, artificial positive control constructs were designed and synthesised as recommended by Kim et al. (2017) [24]. Positive controls were a gBlock synthetic DNA construct (Integrated DNA Technologies, Coralville, IA, USA) comprised of a uniquely identifiable 253 bp 16S rRNA sequence from *Aliivibrio fischeri*, a bacterial endosymbiont found within the Bobtail squid, *Euprymna scolopes* and its environment [57,58]. The *A. fischeri* sequence is flanked by the relevant 16S rRNA primer binding sites to allow for positive control amplification and is 253 bp long to match the typical length of fragments amplified by the 16S rRNA primers on naturally occurring bacterial species. This bacterial sequence was chosen as there is very little possibility that *A. fischeri* would be found as an environmental or laboratory contaminant, nor would it be found in canine blood, making it uniquely identifiable. Supplementary file 2 has the sequence of this positive control construct. The second-step PCR for addition of library indices and deep sequencing was then conducted according to Huggins et al. (2019) [22] using an Illumina MiSeq (Illumina, San Diego, CA, USA) with paired-end 600-cycle v3 chemistry at the Walter and Eliza Hall Institute (WEHI) Proteomics Facility, Parkville, Australia. Paired-end sequencing was conducted to assure read quality as only reads that had identical overlapping R1 and R2 regions were passed on to downstream analysis.

5.6. Bacterial 16S rRNA Metabarcoding Comparison of DNA Extraction Kit Performance

The 50 Cambodian dog blood samples and two reagent-only extractions per method were conducted using the previously described protocols on the Maxwell® RSC Whole Blood DNA Kit (WB), the Maxwell® RSC Buffy Coat DNA Kit (BC), the Bioline ISOLATE II Genomic DNA Kit (BL) and the Qiagen DNeasy Blood and Tissue Kit (QG). Samples were then prepared for deep-sequencing in the same manner as those used in the blocking primer comparison, with 112 samples indexed and multiplexed per flow cell. Each batch of 50 DNA extractions per kit included an additional seven controls that were deep sequenced. These were; two reagent-only DNA extraction negative controls, two no-template PCR negative controls, two uniquely identifiable *A. fischeri* positive controls and one ZymoBIOMICS Microbial Community DNA Standard (Zymo Research, Irvine, CA, USA) positive control to assess for PCR amplification bias.

5.7. Bioinformatics

Raw NGS data was demultiplexed using in-house software at WEHI and then imported into the QIIME 2 (v. 2019.1) environment for bioinformatic analysis [59]. Raw data were processed according to Huggins et al. (2019) [22] with the only adaptation being that amplicon sequence variants (ASVs) were generated and taxonomically assigned as opposed to sequence clustering and formation of operational taxonomic units (OTUs), reflecting recent research into microbiome analysis best practice [60,61]. Alpha rarefaction plots were generated, using the functions MAFFT [62] and FastTree 2 [63] to ensure that OTU diversity plateaued and hence a sufficient sequencing depth had been achieved. All NGS data generated exploring the use of our blocking primer is available from the NCBI BioProject database, with BioProject ID: PRJNA528154 and SRA data accession numbers SRR10895013 to SRR10895109. All data that was obtained to compare the performance of DNA extraction kits is available under BioProject ID: PRJNA601241 and SRA data accession numbers SRR10959940 to SRR10960037 (WB and BC), SRR10972728 to SRR10972776 (BL) and SRR10972824 to SRR10972872 (QG).

The read threshold for a true infection by a bacterial pathogen was determined as the mean reads of non-control samples that were identified as having sequences from *A. fischeri* (the positive control construct) within them. The appearance of these positive control sequences in other samples could be due to occasional index misreading or hybridisation errors during Illumina sequencing as well as low-level cross-contamination during library preparation [24]. Read threshold values were never higher than 62 reads, which was the NGS run found to have the highest amount of cross-contamination. Appearance of contaminant sequences across the 96-well plate was scattered and not concentrated near the positive control samples, implying that library preparation was unlikely to be the major cause of such cross-contamination.

Taxa found in no-reagent DNA extraction negative controls and no-template PCR negative controls were deducted from the overall dataset used for comparison of pathogen detection capability. This negative control data was used for comparison of contaminant bacterial DNA arising from extraction kits.

The results from the ZymoBIOMICS Microbial Community DNA Standards were compared to the expected read composition. A high level of similarity was found, demonstrating no PCR amplification bias by our chosen bacterial 16S rRNA primers.

5.8. Data Analysis of Blocking Primer Performance

Data analysis to calculate blocking primer performance, including biodiversity index calculations and blocking efficiency was conducted in Excel 2016 v. 1803 (Microsoft, Redmond, WA, USA). Statistical analyses were carried out using Minitab v. 19.2 (Minitab LLC, State College, PA, USA).

The percentage of relevant vector-borne bacterial reads (*A. platys*, *E. canis* and *Mycoplasma haemocanis*) compared to the total reads for each sample was calculated, to account for differences in sampling effort between samples, observable as differing numbers of total sample reads. Data normality was tested for using the Anderson–Darling test. The mean percentage of bacterial reads compared to total reads, standard error and Wilcoxon paired t-test statistics were calculated to determine if differences in means between paired samples, i.e., the same sample with and without the blocking primer, were statistically significant. Results were interpreted as significant at the $p < 0.01$ level. The same analysis was carried out to elucidate the average proportion of canine mitochondrial cross-reactivity between the same samples with and without the blocking primer. In addition, blocking primer efficiency was calculated as defined by Tan and Liu (2018) [27] in the equation $(X - Y) \div Y$. Here, X is defined as the ratio of mitochondrial cross-reactive reads to the total of all bacterial reads without the blocking primer for a sample and Y is the ratio of mitochondrial reads to total bacterial reads with the blocking primer for each sample. This equation works on the assumptions that *Canis*-mito-blk is specific to canine mitochondrial sequences and, therefore, that non-mitochondrial sequences will not be inhibited. Following this, the mean and standard error of all sample blocking efficiencies was calculated to get an approximation of overall *Canis*-mito-blk blocking efficiency.

Read cut-off values to ascertain if a sample had a particular vector-borne bacterial species present, i.e., was a true infection were defined according to Huggins et al. (2019) [22]. In brief, the mean number of unique *A. fischeri* positive control construct reads that appeared in non-positive control samples was calculated across all samples, with and without the blocking primer. The means of these reads was taken to demonstrate the average quantity of Illumina indexing errors or potential cross-contamination between samples during library preparation. Therefore, read values under these means were not counted as true infections while those above were counted. The total number of infections was tallied across the same 50 samples with and without the blocking primer to assess sensitivity of pathogen detection.

Biodiversity indices were calculated for all bacterial reads across all samples with and without the blocking primer to elucidate whether their addition improved the ability of our methodology to characterise bacterial species diversity and, therefore, the potential to detect rare or novel species. We used two biodiversity indices; Shannon–Wiener (H) and Simpson's (D) index, which both account for species richness (number of bacterial species) and species abundance (approximated via number of reads of each species) [64–68].

5.9. Data Analysis of DNA Extraction Kit Performance

DNA extraction kit performance was assessed using 43 blood samples that were successfully deep-sequenced from all four kit types used. DNA yield, number of vector-borne bacterial infections detected, mean percentages of contaminant bacterial reads and mean percentage of bacterial primer cross-reactivity on canine mitochondrial sequences, relative to total reads were all calculated. Statistical analysis to compare mean percentages of cross-reactivity and mean percentages of bacterial contaminant sequences were calculated using Minitab v. 19.2 using one-way ANOVAs and Tukey pairwise comparisons to assess which means were statistically significant between kit types at the $p < 0.05$ level.

Regarding the assessment of levels of bacterial contamination from kits, i.e., artificial contamination, a bacterial taxon was deemed to be an artificial contaminant if it had a read count of ≥ 100 in either of the no blood/reagent only DNA extraction negative controls, run for that kit type. For every sample, the number of reads for all artificial kit contaminant taxa was totalled and a percentage calculated relative to the total number of reads. Artificial kit contaminants are different from natural contaminants, such as bacteria residing on canine skin, that may be picked up during sampling via venepuncture or the sampling environment, these naturally derived contaminants were not assessed.

Supplementary Materials: The following are available online at http://www.mdpi.com/2076-0817/9/4/258/s1, Table S1: Most abundant bacterial ASVs found within four different DNA extraction kits' reagents. Supplementary file 2: Sequence of our unique positive control gBlock DNA construct.

Author Contributions: Conceptualization, L.G.H., A.V.K. and R.J.T.; Data curation, L.G.H. and T.I.; Formal analysis, L.G.H.; Funding acquisition, B.S. and R.J.T.; Investigation, L.G.H., A.V.K. and R.J.T.; Methodology, L.G.H., A.V.K. and R.J.T.; Project administration, B.S. and R.J.T.; Resources, B.S., T.I. and R.J.T.; Supervision, R.J.T.; Validation, L.G.H. and R.J.T.; Visualization, L.G.H.; Writing–original draft, L.G.H.; Writing–review & editing, L.G.H., A.V.K., T.I. and R.J.T. All authors have read and agreed to the published version of the manuscript.

Funding: This research was funded by an Australian Research Council Linkage grant LP170100187 with Bayer Animal Health GmbH and Bayer Australia as industry partners. Financial support was also provided by the University of Melbourne postgraduate scholarship scheme.

Acknowledgments: The authors are grateful for the support of Ross Hall (Melbourne Veterinary School) who assisted with bioinformatic processing and analysis as well as Stephen Wilcox (Walter and Eliza Hall Institute) who aided generation of NGS data.

Conflicts of Interest: The authors declare no conflict of interest.

References

1. Maggi, R.G.; Krämer, F. A review on the occurrence of companion vector-borne diseases in pet animals in Latin America. *Parasit. Vectors* **2019**, *12*, 145. [CrossRef]

2. Traub, R.J.; Irwin, P.; Dantas-Torres, F.; Tort, G.P.; Labarthe, N.V.; Inpankaew, T.; Gatne, M.; Linh, B.K.; Schwan, V.; Watanabe, M.; et al. Toward the formation of a Companion Animal Parasite Council for the Tropics (CAPCT). *Parasit. Vectors* **2015**, *8*, 271. [CrossRef]

3. Koh, F.X.; Panchadcharam, C.; Tay, S.T. Vector-borne diseases in stray dogs in peninsular Malaysia and molecular detection of *Anaplasma* and *Ehrlichia* spp. from *Rhipicephalus sanguineus* (Acari: Ixodidae) ticks. *J. Med. Entomol.* **2016**, *53*, 183–187. [CrossRef]

4. Rucksaken, R.; Maneeruttanarungroj, C.; Maswanna, T.; Sussadee, M.; Kanbutra, P. Comparison of conventional polymerase chain reaction and routine blood smear for the detection of *Babesia canis*, *Hepatozoon canis*, *Ehrlichia canis*, and *Anaplasma platys* in Buriram Province, Thailand. *Vet. World* **2019**, *12*, 700–705. [CrossRef]

5. Irwin, P.J.; Jefferies, R. Arthropod-transmitted diseases of companion animals in Southeast Asia. *Trends Parasitol.* **2004**, *20*, 27–34. [CrossRef]

6. Inpankaew, T.; Hii, S.F.; Chimnoi, W.; Traub, R.J. Canine vector-borne pathogens in semi-domesticated dogs residing in northern Cambodia. *Parasit. Vectors* **2016**, *9*, 253. [CrossRef]

7. Mylonakis, M.E.; Harrus, S.; Breitschwerdt, E.B. An update on the treatment of canine monocytic ehrlichiosis (*Ehrlichia canis*). *Vet. J.* **2019**, *246*, 45–53. [CrossRef]

8. Kaewmongkol, G.; Lukkana, N.; Yangtara, S.; Kaewmongkol, S.; Thengchaisri, N.; Sirinarumitr, T.; Jittapalapong, S.; Fenwick, S.G. Association of *Ehrlichia canis*, Hemotropic *Mycoplasma* spp. and *Anaplasma platys* and severe anemia in dogs in Thailand. *Vet. Microbiol.* **2017**, *201*, 195–200. [CrossRef]

9. Little, S.E. Ehrlichiosis and anaplasmosis in dogs and cats. *Vet. Clin. North Am. Small Anim. Pract.* **2010**, *40*, 1121–1140. [CrossRef]

10. Carrade, D.D.; Foley, J.E.; Borjesson, D.L.; Sykes, J.E. Canine granulocytic anaplasmosis: A review. *J. Vet. Intern. Med.* **2009**, *23*, 1129–1141. [CrossRef]

11. Chomel, B.B.; Mac Donald, K.A.; Kasten, R.W.; Chang, C.C.; Wey, A.C.; Foley, J.E.; Thomas, W.P.; Kittleson, M.D. Aortic valve endocarditis in a dog due to *Bartonella clarridgeiae*. *J. Clin. Microbiol.* **2001**, *39*, 3548–3554. [CrossRef]

12. Hii, S.F.; Kopp, S.R.; Abdad, M.Y.; Thompson, M.F.; O'Leary, C.A.; Rees, R.L.; Traub, R.J. Molecular evidence supports the role of dogs as potential reservoirs for *Rickettsia felis*. *Vector-Borne Zoonotic Dis.* **2011**, *11*, 1007–1012. [CrossRef]

13. Ng-Nguyen, D.; Hii, S.-F.; Hoang, M.-T.T.; Nguyen, V.-A.T.; Rees, R.; Stenos, J.; Traub, R.J. Domestic dogs (*Canis familiaris*) are natural mammalian reservoirs for flea-borne spotted fever caused by *Rickettsia felis*. *Sci. Rep.* **2020**, *10*, 4151. [CrossRef]

14. Fourie, J.J.; Evans, A.; Labuschagne, M.; Crafford, D.; Madder, M.; Pollmeier, M.; Schunack, B. Transmission of *Anaplasma phagocytophilum* (Foggie, 1949) by *Ixodes ricinus* (Linnaeus, 1758) ticks feeding on dogs and artificial membranes. *Parasit. Vectors* **2019**, *12*, 136. [CrossRef]

15. Krämer, F.; Hüsken, R.; Krüdewagen, E.M.; Deuster, K.; Blagburn, B.; Straubinger, R.K.; Butler, J.; Fingerle, V.; Charles, S.; Settje, T.; et al. Prevention of transmission of *Borrelia burgdorferi* sensu lato and *Anaplasma phagocytophilum* by *Ixodes* spp. ticks to dogs treated with the Seresto® collar (imidacloprid 10% + flumethrin 4.5%). *Parasitol. Res.* **2019**, *119*, 299–315.

16. Krupka, I.; Straubinger, R.K. Lyme Borreliosis in dogs and cats: Background, diagnosis, treatment and prevention of infections with *Borrelia burgdorferi* sensu stricto. *Vet. Clin. North Am. Small Anim. Pract.* **2010**, *40*, 1103–1119. [CrossRef]

17. Bakken, J.S.; Dumler, J.S. Human granulocytic anaplasmosis. *Infect. Dis. Clin. North Am.* **2015**, *29*, 341–355. [CrossRef]

18. Takhampunya, R.; Korkusol, A.; Pongpichit, C.; Yodin, K.; Rungrojn, A.; Chanarat, N.; Promsathaporn, S.; Monkanna, T.; Thaloengsok, S.; Tippayachai, B.; et al. Metagenomic approach to characterizing disease epidemiology in a disease-endemic environment in northern Thailand. *Front. Microbiol.* **2019**, *10*, 319. [CrossRef]

19. Vayssier-Taussat, M.; Moutailler, S.; Michelet, L.; Devillers, E.; Bonnet, S.; Cheval, J.; Hébert, C.; Eloit, M. Next generation sequencing uncovers unexpected bacterial pathogens in ticks in western Europe. *PLoS ONE* **2013**, *8*, e81439. [CrossRef]

20. Greay, T.L.; Gofton, A.W.; Paparini, A.; Ryan, U.M.; Oskam, C.L.; Irwin, P.J. Recent insights into the tick microbiome gained through next-generation sequencing. *Parasit. Vectors* **2018**, *11*, 12. [CrossRef]

21. Huggins, L.G.; Koehler, A.V.; Ng-Nguyen, D.; Wilcox, S.; Schunack, B.; Inpankaew, T.; Traub, R.J. A novel metabarcoding diagnostic tool to explore protozoan haemoparasite diversity in mammals: A proof-of-concept study using canines from the tropics. *Sci. Rep.* **2019**, *9*, 12644. [CrossRef]

22. Huggins, L.G.; Koehler, A.V.; Ng-Nguyen, D.; Wilcox, S.; Schunack, B.; Inpankaew, T.; Traub, R.J. Assessment of a metabarcoding approach for the characterisation of vector-borne bacteria in canines from Bangkok, Thailand. *Parasit. Vectors* **2019**, *12*, 394. [CrossRef] [PubMed]

23. Liu, M.; Ruttayaporn, N.; Saechan, V.; Jirapattharasate, C.; Vudriko, P.; Moumouni, P.F.A.; Cao, S.; Inpankaew, T.; Ybañez, A.P.; Suzuki, H.; et al. Molecular survey of canine vector-borne diseases in stray dogs in Thailand. *Parasitol. Int.* **2016**, *65*, 357–361. [CrossRef] [PubMed]

24. Kim, D.; Hofstaedter, C.E.; Zhao, C.; Mattei, L.; Tanes, C.; Clarke, E.; Lauder, A.; Sherrill-Mix, S.; Chehoud, C.; Kelsen, J.; et al. Optimizing methods and dodging pitfalls in microbiome research. *Microbiome* **2017**, *5*, 52. [CrossRef] [PubMed]

25. Taanman, J.W. The mitochondrial genome: Structure, transcription, translation and replication. *Biochim. Biophys. Acta-Bioenerg.* **1999**, *1410*, 103–123. [CrossRef]

26. Gofton, A.W.; Oskam, C.L.; Lo, N.; Beninati, T.; Wei, H.; McCarl, V.; Murray, D.C.; Paparini, A.; Greay, T.L.; Holmes, A.J.; et al. Inhibition of the endosymbiont "*Candidatus* Midichloria mitochondrii" during 16S rRNA gene profiling reveals potential pathogens in *Ixodes* ticks from Australia. *Parasit. Vectors* **2015**, *8*, 345. [CrossRef] [PubMed]

27. Tan, S.; Liu, H. Unravel the hidden protistan diversity: Application of blocking primers to suppress PCR amplification of metazoan DNA. *Appl. Microbiol. Biotechnol.* **2018**, *102*, 389–401. [CrossRef]

28. Vestheim, H.; Jarman, S.N. Blocking primers to enhance PCR amplification of rare sequences in mixed samples – a case study on prey DNA in Antarctic krill stomachs. *Front. Zool.* **2008**, *5*, 12. [CrossRef]

29. Leray, M.; Agudelo, N.; Mills, S.C.; Meyer, C.P. Effectiveness of annealing blocking primers versus restriction enzymes for characterization of generalist diets: Unexpected prey revealed in the gut contents of two coral reef fish species. *PLoS ONE* **2013**, *8*, e58076. [CrossRef]

30. Boessenkool, S.; Epp, L.S.; Haile, J.; Bellemain, E.; Edwards, M.; Coissac, E.; Willerslev, E.; Brochmann, C. Blocking human contaminant DNA during PCR allows amplification of rare mammal species from sedimentary ancient DNA. *Mol. Ecol.* **2012**, *21*, 1806–1815. [CrossRef]

31. Hino, A.; Maruyama, H.; Kikuchi, T. A novel method to assess the biodiversity of parasites using 18S rDNA Illumina sequencing; parasitome analysis method. *Parasitol. Int.* **2016**, *65*, 572–575. [CrossRef] [PubMed]

32. Eisenhofer, R.; Minich, J.J.; Marotz, C.; Cooper, A.; Knight, R.; Weyrich, L.S. Contamination in low microbial biomass microbiome studies: Issues and recommendations. *Trends Microbiol.* **2019**, *27*, 105–117. [CrossRef] [PubMed]

33. Glassing, A.; Dowd, S.E.; Galandiuk, S.; Davis, B.; Chiodini, R.J. Inherent bacterial DNA contamination of extraction and sequencing reagents may affect interpretation of microbiota in low bacterial biomass samples. *Gut Pathog.* **2016**, *8*, 24. [CrossRef] [PubMed]

34. Podnecky, N.L.; Elrod, M.G.; Newton, B.R.; Dauphin, L.A.; Shi, J.; Chawalchitiporn, S.; Baggett, H.C.; Hoffmaster, A.R.; Gee, J.E. Comparison of DNA extraction kits for detection of *Burkholderia pseudomallei* in spiked human whole blood using real-time PCR. *PLoS ONE* **2013**, *8*, e58032. [CrossRef]

35. Yang, G.; Erdman, D.E.; Kodani, M.; Kools, J.; Bowen, M.D.; Fields, B.S. Comparison of commercial systems for extraction of nucleic acids from DNA/RNA respiratory pathogens. *J. Virol. Methods* **2011**, *171*, 195–199. [CrossRef]

36. Riemann, K.; Adamzik, M.; Frauenrath, S.; Egensperger, R.; Schmid, K.W.; Brockmeyer, N.H.; Siffert, W. Comparison of manual and automated nucleic acid extraction from whole-blood samples. *J. Clin. Lab. Anal.* **2007**, *21*, 244–248. [CrossRef]

37. Laurence, M.; Hatzis, C.; Brash, D.E. Common contaminants in next-generation sequencing that hinder discovery of low-abundance microbes. *PLoS ONE* **2014**, *9*, e97876. [CrossRef]

38. Velásquez-Mejía, E.P.; de la Cuesta-Zuluaga, J.; Escobar, J.S. Impact of DNA extraction, sample dilution, and reagent contamination on 16S rRNA gene sequencing of human feces. *Appl. Microbiol. Biotechnol.* **2018**, *102*, 403–411. [CrossRef]

39. Werren, J.H.; Baldo, L.; Clark, M.E. *Wolbachia*: Master manipulators of invertebrate biology. *Nat. Rev. Microbiol.* **2008**, *6*, 741–751. [CrossRef]

40. Alexandre, N.; Santos, A.S.; Núncio, M.S.; de Sousa, R.; Boinas, F.; Bacellar, F. Detection of *Ehrlichia canis* by polymerase chain reaction in dogs from Portugal. *Vet. J.* **2009**, *181*, 343–344. [CrossRef]

41. Mylonakis, M.E.; Koutinas, A.F.; Billinis, C.; Leontides, L.S.; Kontos, V.; Papadopoulos, O.; Rallis, T.; Fytianou, A. Evaluation of cytology in the diagnosis of acute canine monocytic ehrlichiosis (*Ehrlichia canis*): A comparison between five methods. *Vet. Microbiol.* **2003**, *91*, 197–204. [CrossRef]

42. Morris, E.K.; Caruso, T.; Buscot, F.; Fischer, M.; Hancock, C.; Maier, T.S.; Meiners, T.; Müller, C.; Obermaier, E.; Prati, D.; et al. Choosing and using diversity indices: Insights for ecological applications from the German Biodiversity Exploratories. *Ecol. Evol.* **2014**, *4*, 3514–3524. [CrossRef] [PubMed]

43. Greiner, M.; Gardner, I.A. Epidemiologic issues in the validation of veterinary diagnostic tests. *Prev. Vet. Med.* **2000**, *45*, 3–22. [CrossRef]

44. Ravi, R.K.; Walton, K.; Khosroheidari, M. Miseq: A next generation sequencing platform for genomic analysis. In *Methods in Molecular Biology*; Humana Press: New York, NY, USA, 2018; Volume 1706, pp. 223–232.

45. Vestheim, H.; Deagle, B.E.; Jarman, S.N. Application of blocking oligonucleotides to improve signal-to-noise ratio in a PCR. In *Methods in Molecular Biology*; Humana Press: New York, NY, USA, 2010; Volume 687, pp. 265–274.

46. Hart, M.L.; Meyer, A.; Johnson, P.J.; Ericsson, A.C. Comparative evaluation of DNA extraction methods from feces of multiple host species for downstream next-generation sequencing. *PLoS ONE* **2015**, *10*, e0143334. [CrossRef] [PubMed]

47. Wiesinger-Mayr, H.; Jordana-Lluch, E.; Martró, E.; Schoenthaler, S.; Noehammer, C. Establishment of a semi-automated pathogen DNA isolation from whole blood and comparison with commercially available kits. *J. Microbiol. Methods* **2011**, *85*, 206–213. [CrossRef] [PubMed]

48. Schiebelhut, L.M.; Abboud, S.S.; Gómez Daglio, L.E.; Swift, H.F.; Dawson, M.N. A comparison of DNA extraction methods for high-throughput DNA analyses. *Mol. Ecol. Resour.* **2017**, *17*, 721–729. [CrossRef]

49. Phillips, K.; McCallum, N.; Welch, L. A comparison of methods for forensic DNA extraction: Chelex-100® and the QIAGEN DNA Investigator Kit (manual and automated). *Forensic Sci. Int. Genet.* **2012**, *6*, 282–285. [CrossRef]

50. Sidstedt, M.; Hedman, J.; Romsos, E.L.; Waitara, L.; Wadsö, L.; Steffen, C.R.; Vallone, P.M.; Rådström, P. Inhibition mechanisms of hemoglobin, immunoglobulin G, and whole blood in digital and real-time PCR. *Anal. Bioanal. Chem.* **2018**, *410*, 2569–2583. [CrossRef]

51. Hall, C.M.; Jaramillo, S.; Jimenez, R.; Stone, N.E.; Centner, H.; Busch, J.D.; Bratsch, N.; Roe, C.C.; Gee, J.E.; Hoffmaster, A.R.; et al. *Burkholderia pseudomallei*, the causative agent of melioidosis, is rare but ecologically established and widely dispersed in the environment in Puerto Rico. *PLoS Negl. Trop. Dis.* **2019**, *13*, e0007727. [CrossRef]

52. Zinger, L.; Bonin, A.; Alsos, I.G.; Bálint, M.; Bik, H.; Boyer, F.; Chariton, A.A.; Creer, S.; Coissac, E.; Deagle, B.E.; et al. DNA metabarcoding—Need for robust experimental designs to draw sound ecological conclusions. *Mol. Ecol.* **2019**, *28*, 1857–1862. [CrossRef]

53. Tedersoo, L.; Drenkhan, R.; Anslan, S.; Morales-Rodriguez, C.; Cleary, M. High-throughput identification and diagnostics of pathogens and pests: Overview and practical recommendations. *Mol. Ecol. Resour.* **2019**, *19*, 47–76. [CrossRef] [PubMed]

54. Vijayvargiya, P.; Jeraldo, P.R.; Thoendel, M.J.; Greenwood-Quaintance, K.E.; Esquer Garrigos, Z.; Sohail, M.R.; Chia, N.; Pritt, B.S.; Patel, R. Application of metagenomic shotgun sequencing to detect vector-borne pathogens in clinical blood samples. *PLoS ONE* **2019**, *14*, e0222915. [CrossRef] [PubMed]

55. Beck, A.; Huber, D.; Antolić, M.; Anzulović, Ž.; Reil, I.; Polkinghorne, A.; Baneth, G.; Beck, R. Retrospective study of canine infectious haemolytic anaemia cases reveals the importance of molecular investigation in accurate postmortal diagnostic protocols. *Comp. Immunol. Microbiol. Infect. Dis.* **2019**, *65*, 81–87. [CrossRef] [PubMed]

56. Colella, V.; Nguyen, V.L.; Tan, D.Y.; Lu, N.; Fang, F.; Zhijuan, Y.; Wang, J.; Liu, X.; Chen, X.; Dong, J.; et al. Zoonotic vectorborne pathogens and ectoparasites of dogs and cats in Asia. *Emerg. Infect. Dis.* **2020**, *26*. [CrossRef]

57. Ruby, E.G.; Urbanowski, M.; Campbell, J.; Dunn, A.; Faini, M.; Gunsalus, R.; Lostroh, P.; Lupp, C.; McCann, J.; Millikan, D.; et al. Complete genome sequence of *Vibrio fischeri*: A symbiotic bacterium with pathogenic congeners. *Proc. Natl. Acad. Sci. USA* **2005**, *102*, 3004–3009. [CrossRef]

58. Nyholm, S.V.; McFall-Ngai, M.J. The winnowing: Establishing the squid-*Vibrio* symbiosis. *Nat. Rev. Microbiol.* **2004**, *2*, 632–642. [CrossRef]

59. Bolyen, E.; Rideout, J.R.; Dillon, M.R.; Bokulich, N.A.; Abnet, C.C.; Al-Ghalith, G.A.; Alexander, H.; Alm, E.J.; Arumugam, M.; Asnicar, F.; et al. Reproducible, interactive, scalable and extensible microbiome data science using QIIME 2. *Nat. Biotechnol.* **2019**, *37*, 852–857. [CrossRef]

60. Langille, M.G.I.; Nearing, J.T.; Douglas, G.M.; Comeau, A.M. Denoising the Denoisers: An independent evaluation of microbiome sequence error-correction approaches. *PeerJ* **2018**, *8*, e5364.

61. Knight, R.; Vrbanac, A.; Taylor, B.C.; Aksenov, A.; Callewaert, C.; Debelius, J.; Gonzalez, A.; Kosciolek, T.; McCall, L.I.; McDonald, D.; et al. Best practices for analysing microbiomes. *Nat. Rev. Microbiol.* **2018**, *16*, 410–422. [CrossRef]

62. Katoh, K.; Standley, D.M. MAFFT multiple sequence alignment software version 7: Improvements in performance and usability. *Mol. Biol. Evol.* **2013**, *30*, 772–780. [CrossRef]
63. Price, M.N.; Dehal, P.S.; Arkin, A.P. FastTree 2—Approximately maximum-likelihood trees for large alignments. *PLoS ONE* **2010**, *5*, e9490. [CrossRef] [PubMed]
64. Spellerberg, I.F.; Fedor, P.J. A tribute to Claude-Shannon (1916–2001) and a plea for more rigorous use of species richness, species diversity and the "Shannon-Wiener" Index. *Glob. Ecol. Biogeogr.* **2003**, *12*, 177–179. [CrossRef]
65. Gill, C.A.; Joanes, D.N. Bayesian estimation of Shannon's index of diversity. *Biometrika* **1979**, *66*, 81–85. [CrossRef]
66. Hunter, P.R.; Gaston, M.A. Numerical index of the discriminatory ability of typing systems: An application of Simpson's index of diversity. *J. Clin. Microbiol.* **1988**, *26*, 2465–2466. [CrossRef]
67. Ogedengbe, M.E.; El-Sherry, S.; Ogedengbe, J.D.; Chapman, H.D.; Barta, J.R. Phylogenies based on combined mitochondrial and nuclear sequences conflict with morphologically defined genera in the eimeriid coccidia (Apicomplexa). *Int. J. Parasitol.* **2018**, *48*, 59–69. [CrossRef]
68. Allaby, M. *A Dictionary of Ecology*, 4th ed.; Oxford University Press: Oxford, UK, 2010.

Article

Biotic Factors Influence Microbiota of Nymph Ticks from Vegetation in Sydney, Australia

Shona Chandra and Jan Šlapeta *

Sydney School of Veterinary Science, Faculty of Science, University of Sydney, Sydney, NSW 2006, Australia;
shona.chandra@sydney.edu.au
* Correspondence: jan.slapeta@sydney.edu.au

Received: 19 June 2020; Accepted: 10 July 2020; Published: 13 July 2020

Abstract: Ticks are haematophagous ectoparasites of medical and veterinary significance due to their excellent vector capacity. Modern sequencing techniques enabled the rapid sequencing of bacterial pathogens and symbionts. This study's aims were two-fold; to determine the nymph diversity in Sydney, and to determine whether external biotic factors affect the microbiota. Tick DNA was isolated, and the molecular identity was determined for nymphs at the *cox*1 level. The tick DNA was subjected to high throughput DNA sequencing to determine the bacterial profile and the impact of biotic factors on the microbiota. Four nymph tick species were recovered from Sydney, NSW: *Haemaphysalis bancrofti*, *Ixodes holocyclus*, *Ixodes trichosuri* and *Ixodes tasmani*. Biotic factors, notably tick species and geography, were found to have a significance influence on the microbiota. The microbial analyses revealed that Sydney ticks display a core microbiota. The dominating endosymbionts among all tick species were *Candidatus* Midichloria sp. Ixholo1 and *Candidatus* Midichloria sp. Ixholo2. A novel *Candidatus* Midichloria sp. OTU_2090 was only found in *I. holocyclus* ticks (nymph: 96.3%, adult: 75.6%). *Candidatus* Neoehrlichia australis and *Candidatus* Neoehrlichia arcana was recovered from *I. holocyclus* and one *I. trichosuri* nymph ticks. *Borrelia* spp. was absent from all ticks. This study has shown that nymph and adult ticks carry different bacteria, and a tick bite in Sydney, Australia will result in different bacterial transfer depending on tick life stage, tick species and geography.

Keywords: bacterial profile; *cox*1; *Haemaphysalis bancrofti*; *Ixodes holocyclus*; *Ixodes trichosuri*; *Ixodes tasmani*; V3-V4 *16S* rRNA gene

---

1. Introduction

Ticks (Acari: Ixodida) are important vectors for diseases and the application of high throughput DNA sequencing enables the study of the bacterial symbionts, commensals and pathogenic microorganisms that ticks may carry. The bacterial community within ticks have been well studied using next-generation sequencing (NGS) technologies, enabling the rapid sequencing of their microbiota and microbiome [1–3]. Globally, there have been studies implementing NGS that look at ticks, but few that involved a comparison between questing ticks across at least two life stages [4–7]. Such studies have found that bacterial composition diversity differs with host species, geographic distribution and tick life stage [5,8].

In Australia, the most important tick in the medical and small animal veterinary context is the Australian paralysis tick *Ixodes holocyclus* Neumann, 1899. Found along the eastern coastline, *I. holocyclus* possesses a potent neurotoxin that causes fatal paralysis eventuating in death for many domestic animals, and can have a flaccid paralysis effect in humans, particularly infants [9–12]. In humans, salivary secretions from *I. holocyclus* bites have been implicated with causing a red meat allergy (alpha-1, 3-galactose allergy) and hypersensitivity for allergic reactions [13,14]. Despite the medical and veterinary significance of ticks, there has been no systematic survey and/or study involving

NGS methods to determine the diversity of the bacterial microbial communities within the common ticks which have been flagged in Sydney.

Previous studies using NGS methods have determined that the bacterial diversity between the tick life stages are different and are altered by previous life stage's host-blood meal [4,5,8]. Studies which have compared the bacterial microbiota between adults and nymphs have indicated that nymph ticks have higher bacterial diversity compared to the adult tick life stage [4,5]. However, with few studies published in Australia, there has been a gap in scientific literature regarding the microbial comparison between Australian tick life stages [1,15]. Further, as *Borrelia* spp. have been identified in Sydney, Australia [2], it is crucial to survey Sydney's questing nymph ticks to determine the diversity of pathogens vectored by the ticks in the area.

In Sydney, the most common tick species found in the environment is *I. holocyclus*, however, other tick species including the wallaby tick *Haemaphysalis bancrofti* Nuttall & Warburton, 1915, the bush tick *Haemaphysalis longicornis* Neumann, 1901, the common marsupial tick *Ixodes tasmani* Neumann, 1899, and the possum tick *Ixodes trichosuri* Roberts, 1960 have been anecdotally recorded [16,17]. Further, *H. bancrofti* and *H. longicornis* have been known to seek humans as a possible host species [1]. With more ticks than just *I. holocyclus* present in the environment, adult humans, children and their companion animals are potentially at risk of being exposed to more pathogens than commonly thought.

Most ticks of veterinary, medical and public health significance are three-host ticks, including *I. holocyclus* [18]. In Europe and North America, *Ixodes ricinus* Linnaeus, 1758, *Ixodes scapularis* Say, 1821 (syn. *Ixodes dammini*), and *Ixodes pacificus* Cooley & Kohls, 1943 are three-host ticks which are recognised vectors of the bacterial spirochaete, *Borrelia burgdorferi* sensu lato (s.l.), the causative agent of Lyme borreliosis [19–22]. The spirochaete is not transmitted trans-ovarially, so hatched larval ticks are free from *B. burgdorferi* s.l. [23,24]. Larvae of *Ixodes* spp. will then feed on an infected wildlife host, causing the larvae to become infected and it will moult into a nymph tick where the bacteria will undergo rapid replication [25,26]. The *Borrelia* spirochaete will actively migrate out of an infected nymph during feeding, and then moult into an adult tick with lower burdens of *B. burgdorferi* s.l. [24,26,27]. Both nymph and adult ticks can parasitise humans, but due to the biology of *B. burgdorferi* s.l. and the small size of nymph ticks, unfed nymphs are more efficient vectors and are primarily responsible for causing Lyme borreliosis [19,24–27].

The aims of this study were: to determine the nymph diversity of ticks found around the Sydney region, and to determine whether external biotic factors influence the tick's microbiota. To do so, unfed nymphs were collected by flagging vegetation around the Northern Beaches, and North Shore areas in Sydney. Two DNA isolation methods and two *16S* ribosomal RNA (rRNA) hypervariable regions were initially compared to determine which combination afforded the greatest bacterial diversity. The nymph ticks were subject to conventional PCR and Sanger sequencing for speciation and modern NGS tools to determine their *16S* rRNA bacterial profile. The microbial contents of the tick nymphs were evaluated using the V3-V4 *16S* rRNA gene amplicon profile and compared to adult ticks collected from dogs to determine the differences between external factors, including the tick itself as a host for the bacteria, the life stages, tick species, tick host and geographic location.

2. Results

2.1. Pilot Study Determined gDNA Isolation and V3-V4 16S rRNA Gene Diversity Profiling Assay Optimal Combination for Tick Microbiota Studies

For the pilot study, six adult female *Ixodes holocyclus* (SC0063-1 to SC0063-6) of varying levels of engorgement were collected from a single dog host. The single host was important to reduce bias and variability from the host and the blood meal. As the ticks were bisected longitudinally, each half subject to different DNA isolation method (genomic versus faecal DNA isolation procedure) and *16S* rRNA gene diversity profiling assays (V1-V3 versus V3-V4 hypervariable region), it would allow for the DNA isolation biases and diversity profiling biases to be seen. We discovered there was no significant difference between the OTUs generated at either the DNA isolation method variable, or the hypervariable region selected (Supplementary Figure S1). However, while there were more OTUs generated in the V1-V3 hypervariable region, we noted that there were less unassigned OTUs in the V3-V4 hypervariable region than in the V1-V3 region. At the bacterial genus level, there were more bacteria identified at the V3-V4 hypervariable region than the V1-V3, where many bacteria could only be identified to the family level. From this, it was decided that moving forwards, we would continue with the genomic DNA (gDNA) isolation procedure and the V3-V4 hypervariable region as the optimal approach for our tick microbial study.

Morphologically, SC0063-1 to SC0063-6 were identified as *I. holocyclus* and gDNA analysis at the *cox*1 mtDNA marker confirmed that all were molecularly *I. holocyclus* (99.8% identity, 603/604 nt, AB075955). All six ticks were molecularly identical to each other, and the was a distinct A/G polymorphism at residue 310 across the 604 nt sequence from there reference *I. holocyclus cox*1 sequence (AB075955).

Both V1-V3 and V3-V4 *16S* rRNA hypervariable regions were dominated by OTU_1 belonging to *Candidatus* Midichloria sp. Ixholo1 (Supplementary Figure S1). At the class level, both hypervariable regions were dominated by Alphaproteobacteria followed by Gammaproteobacteria and Betaproteobacteria. At the order level, both hypervariable regions were dominated by Rickettsiales and Enterobacteriales (Supplementary Figure S1). However, the V1-V3 region detected Aeromondales and Campylobacterales, which was not detected by the V3-V4 region (Supplementary Figure S1). At the family level, Candidatus Midichloriaceae and Enterobacteriaceae dominated both hypervariable regions, followed by Comamonadaceae, Aeromonadaceae and Campylobacteraceae at the V1-V3 hypervariable region and Rhizobiaceae, Comamonadaceae and Oxalobacteraceae at the V3-V4 region. (Supplementary Figure S1). More genera were described at the V3-V4 hypervariable region than at the V1-V3 region, where many OTUs were only identifiable up to the family level (Supplementary Figure S1). While the two dominating bacteria were identical across both hypervariable regions, variability was seen in the non-dominant bacteria.

At the OTU level for the V1-V3 hypervariable region, multivariate analysis by non-metric multidimensional scaling (nMDS) methods with Bray-Curtis similarity at 35% revealed clustering for the faecal and genomic DNA kits. The permutation based hypothesis testing approach, analysis of similarities (ANOSIM), revealed the OTUs generated was significantly different (R = 0.741, *p* = 0.002). At the OTU level for the V3-V4 hypervariable region, multivariate analysis by nMDS with Bray-Curtis similarity at 35% revealed clustering for the faecal and genomic DNA kits. ANOSIM revealed the OTUs generated was significantly different (R = 0.976, *p* = 0.002). Similarly, at the genus level for the V1-V3 hypervariable region, multivariate analysis by nMDS with Bray-Curtis similarity at 40% revealed clustering for the faecal and genomic DNA kits. ANOSIM revealed the OTUs generated was significantly different (R = 0.739, *p* = 0.002). At the genus level for the V3-V4 hypervariable region, multivariate analysis by nMDS with Bray-Curtis similarity at 35% revealed clustering for the faecal and genomic DNA kits. ANOSIM revealed the OTUs generated was significantly different (R = 0.978, *p* = 0.002). At the genus level, it was determined that there were 12 OTUs that were overlapped between the two *16S* rRNA hypervariable regions (V1-V3: 12/68, V3-V4: 12/63), indicating that the hypervariable regions displayed some bias towards certain genera of bacteria. Following, the Mann-Whitney Test for significance revealed there was no significant difference (*p* = 0.470) between OTUs generated for *16S* rRNA hypervariable regions, V1-V3 or V3-V4, indicating no significant difference between hypervariable regions used (Supplementary Figure S1).

2.2. Ixodes holocyclus, Ixodes trichosuri, Ixodes tasmani and Haemaphysalis bancrofti Nymph Ticks Recovered from Vegetation in Sydney

Nymph ticks were collected from 39 locations on the Northern Beaches and North Shore of Sydney, New South Wales (NSW) and one location from the South Coast of NSW (Figure 1a). For the nymphs, all ticks (*n* = 175) were morphologically identified to the tick genus level, which revealed *Ixodes* spp. (*n* = 164) and *Haemaphysalis* spp. (*n* = 11) was flagged from vegetation (Figure 1b,c).

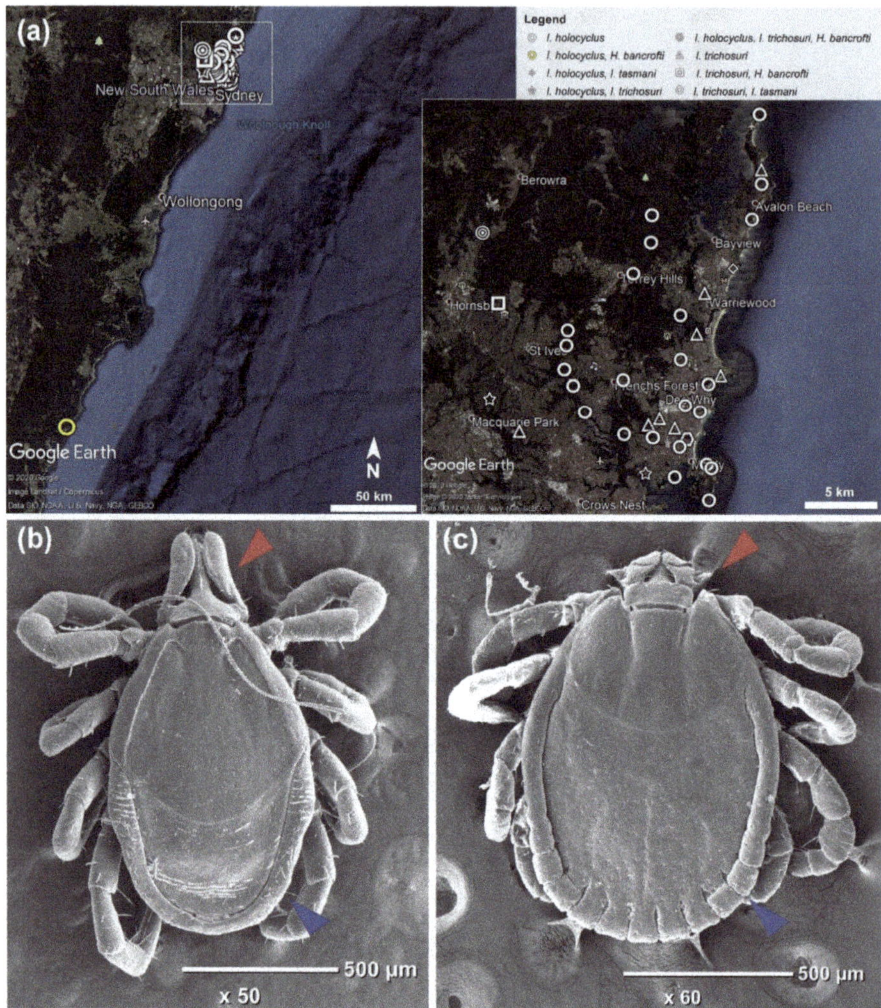

Figure 1. (**a**) Collection points and flagging location for nymph ticks from the eastern coastline in Australia. Zoomed-in segment depicts collection points for nymph ticks from the Northern Beaches and North Shore regions in Sydney, Australia. Map sourced from data imported into Google Earth, 2020. (**b**) Scanning electron microscopy of SC790-8 *Ixodes* spp. nymph, collected from Murramarang National Park, South Durras, NSW. Red arrow pointing to long, elongate palps; blue arrow pointing to lack of festoons, and the narrow body profile at the posterior margin. (**c**) Scanning electron microscopy of SC790-9 *Haemaphysalis bancrofti* nymph, collected from Murramarang National Park, South Durras, NSW. Red arrow pointing to short, wide palps, and palpal article 2 with large lateral projection; blue arrow pointing to prominent festoons, and the wider body profile at the posterior margin.

For more in-depth identification, 108 were selected for molecular identification to the species level at *cox*1 mtDNA region (Supplementary Table S1). Of these, it was determined that in addition to *I. holocyclus* (80/108, 99.1–100% identity, AB075955), there were *Ixodes trichosuri* Roberts, 1960 (20/108, 99–100% identity, KY213778), *Ixodes tasmani* Neumann, 1899 (2/108, 97.4–98.8% identity, MH043269) and *Haemaphysalis bancrofti* Nuttall & Warburton, 1915 (6/108, 97.9–99.8% identity, MH043268) present in the area (Figure 2).

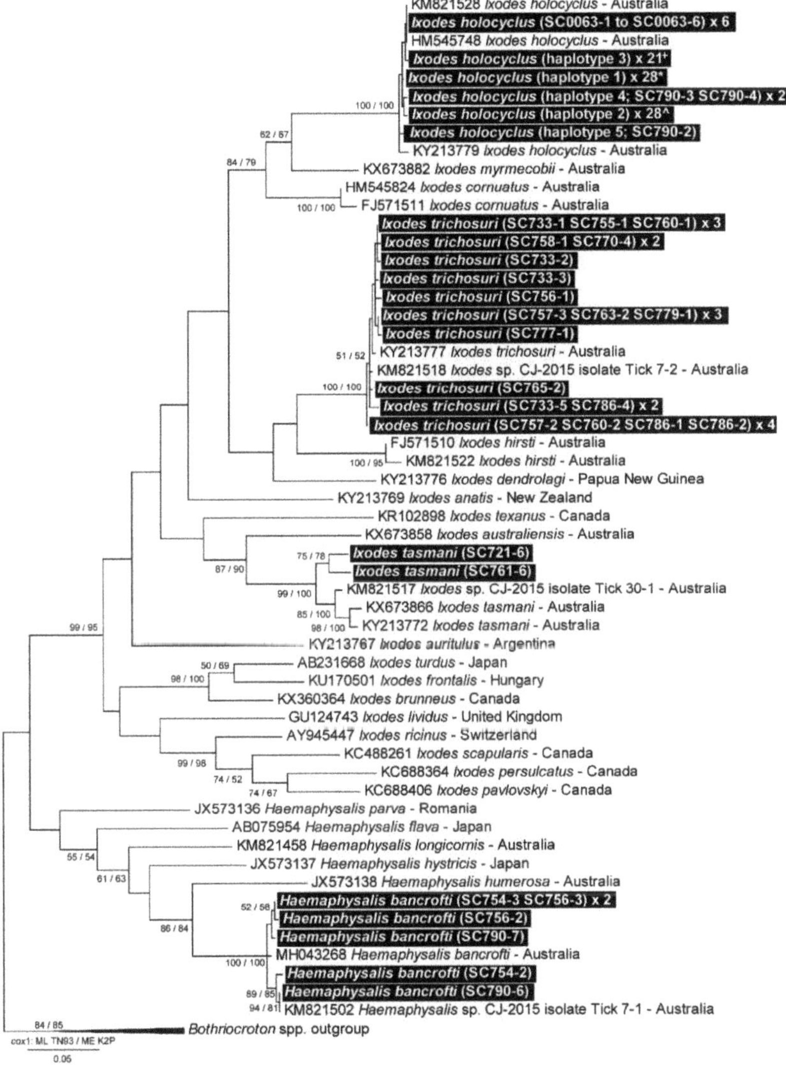

* SC 721-1, 754-1, 759-1, 760-3, 761-4, 762-1, 762-2, 763-3, 764-1, 764-3, 765-1, 766-2, 767-1, 767-2, 768-1, 768-3, 769-1, 769-3, 770-3, 771-1, 772-2, 776-1, 778-2, 781-1, 787-3, 787-4, 787-5, 788-4

^ SC 721-2, 721-3, 721-4, 721-5, 733-4, 753-1, 753-2, 753-3, 753-4, 753-5, 757-1, 758-3, 761-5, 782-3, 764-2, 774-1, 776-2, 778-1, 780-1, 785-1, 785-4, 787-2, 788-1, 788-2, 788-3, 788-5, 790-1, 790-5

⁺ SC 760-4, 761-1, 761-2, 761-3, 763-1, 766-1, 758-2, 769-2, 770-1, 770-2, 771-1, 772-1, 773-1, 773-2, 775-1, 782-1, 783-1, 785-2, 785-3, 786-3, 787-1

Figure 2. Molecular phylogenetic analysis of *Ixodes* spp. and *Haemaphysalis* spp. ticks at the cytochrome c oxidase subunit I (*cox1*) ribosomal DNA locus. The evolutionary history, evolutionary distance and evolutionary rate differences were inferred by using the following methods: maximum likelihood method based on the Tamura-Nei (ML TN93) and minimum evolution with Kimura-2 Parameters (ME K2P) models. Test of phylogeny was implicated through the Bootstrap method (1000 replications). Codon positions included were 1st and 2nd. The bootstrap confidence intervals (%) have been grouped as follows: ML TN93/ME K2P, and values < 50% have been hidden. The evolutionary analyses were conducted in MEGA 7 [28].

2.3. Only Adult Ixodes holocyclus Ticks Recovered from Dogs in Sydney Veterinary Clinics

The adult ticks recovered from veterinary clinics were from 26 distinct localities across the eastern coastline of NSW, ranging from the Northern Beaches to the South Coast (Figure 3). All specimens (n = 92) were weighed prior to DNA isolation (Supplementary Table S2). The ticks were morphologically identified as *I. holocyclus* and displayed the characteristic features including the leg colour pattern, the scutum proportion was wider than long, there was the notable absence of the cornua, short cervical grooves and distinct anterior position of the anal groove compared to the anus.

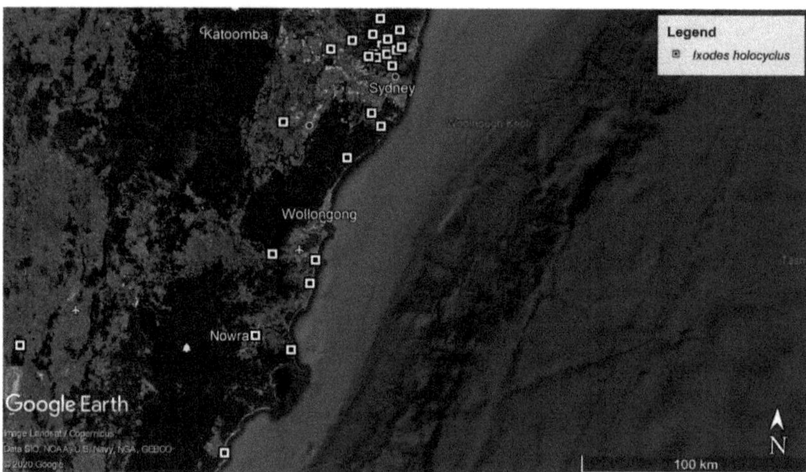

Figure 3. Collection points and flagging location for adult ticks from the eastern coastline in Australia. Map sourced from data imported into Google Earth, 2020.

2.4. Pre-Treatment of Adult and Nymph Tick Bacterial Diversity at the 16S rRNA Hypervariable Region V3-V4

The dominating OTUs in V3-V4 bacterial profile were *Ca.* Midichloria sp. Ixholo1 (OTU_1, 69.4%), and *Ca.* Midichloria sp. Ixholo2 (OTU_1948, 21.5%). The two dominating OTUs (OTU_1, OTU_1948) were excluded, and reads belonging to phylum Proteobacteria (82.5%) dominated the V3-V4 hypervariable region. Reads in phylum Actinobacteria (12.3%), Bacteriodetes (0.6%), and Firmicutes (4.1%) were also present. Of the reads within the phylum Proteobacteria, most were of class Gammaproteobacteria (60.2%), followed by Alphaproteobacteria (33.3%) and Betaproteobacteria (6.4%). Shannon diversity indices (H'; t-test: t = 93.77, p = 0.007) and the Simpsons index (1-λ'; t-test: t = 48.68, p = 0.013) indicated that there was a significant difference between the bacterial composition in nymph and adult ticks (Table 1).

Table 1. Bacterial diversity measures for nymph *Ixodes holocyclus*, *Ixodes trichosuri*, *Ixodes tasmani* and *Haemaphysalis bancrofti*, and adult *I. holocyclus* ticks from Sydney. The following abbreviations are used in the table: S total number of OTUs/species; H' Shannon diversity index using the natural logarithm, logarithm to the base e; 1-λ' Simpson's diversity index; SD standard deviation.

	S	$H' \pm$ **SD**	1-$\lambda' \pm$ **SD**
Nymphs	2936	2.938 ± 0.59	0.919 ± 0.07
Adults	2512	2.876 ± 0.82	0.882 ± 0.11
Nymphs + Adults	5448	2.912 ± 0.69	0.903 ± 0.09

2.5. Presence of Candidatus Midichloria spp. in Adult and Nymph Ticks

In nymph ticks, the nMDS ordination plots for bacterial abundance, shown at the OTU level reveal an apparent clustering among the four tick species (Figure 4a). Visualising the bubble plots for OTU_1 (*Ca.* Midichloria sp. Ixholo1) and OTU_1948 (*Ca.* Midichloria sp. Ixholo2) over the alignment of samples with nMDS distribution revealed that these two OTUs were abundant in most nymph ticks, and not just in nymphs identified as *I. holocyclus* (Figure 4b). Interestingly, a novel *Ca.* Midichloria sp. was found (OTU_2090) which had 97.3% identity to *Ca.* Midichloria sp. Ixholo2 (FM992373). A bubble plot was generated for OTU_2090, which was unique to *I. holocyclus* nymphs (77/80, 96.3%), and was absent from the other nymph species found (Figure 4c). The bubble plot generated for OTU_12 (*Candidatus* Neoehrlichia arcana) and OTU_112 (*Candidatus* Neoehrlichia australis) revealed that these OTUs were only present in *I. holocyclus*, with the exception of one *I. trichosuri* (JS3333; OTU_112) (Figure 4d).

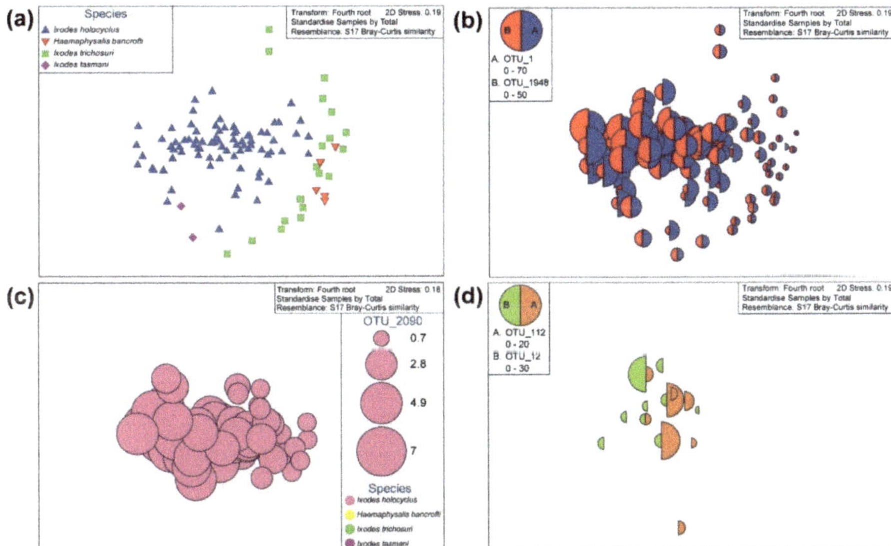

Figure 4. Multivariate statistical analysis of the V3-V4 bacterial profile for *Ixodes holocyclus*, *Ixodes trichosuri*, *Ixodes tasmani* and *Haemaphysalis bancrofti* nymph ticks from Sydney, Australia. Non-metric multidimensional scaling (nMDS) ordination plots for bacterial abundance and bubble plots. Outliers have been removed for the nMDS plot, and from further analyses. The outliers which were removed were: JS3368 and JS3299 (low reads). (**a**) nMDS ordination plots for bacterial abundance, shown at the operational taxonomic unit (OTU) level. (**b**) Plot for OTU_1 (*Ca.* Midichloria sp. Ixholo1) and OTU_1948 (*Ca.* Midichloria sp. Ixholo2) reveal that these endosymbionts are prevalent across all four nymph tick species collected. (**c**) Plot for OTU_2090 (97.3% identity to *Ca.* Midichloria sp. Ixholo2 (FM992373)) reveal that this endosymbiont is unique to *Ixodes holocyclus* nymph ticks. (**d**) Plot for OTU_12 (*Candidatus* Neoehrlichia arcana) and OTU_112 (*Candidatus* Neoehrlichia australis) reveal that these endosymbionts were mostly present in *Ixodes holocyclus*, except for one *Ixodes trichosuri* nymph tick (JS3333).

In adult ticks, the nMDS ordination plot for bacterial abundance, shown at the OTU level (Figure 5a) display a clustering of OTUs generated in adult *I. holocyclus* ticks. Plotting bubble plots for OTU_1 (*Ca.* Midichloria sp. Ixholo1) over the weight factor indicates there is little relationship with weight of the tick and abundance of OTU_1 (Figure 5b). Plotting bubble plots for OTU_1 (*Ca.* Midichloria sp. Ixholo1 and OTU_1948 (*Ca.* Midichloria sp. Ixholo2) over the alignment of samples with nMDS distribution revealed that these two OTUs were abundant in most adult ticks, all of which were morphologically identified as *I. holocyclus* (Figure 5c). Like in the nymph ticks, a bubble plot was generated for OTU_2090 (97.3% identity *Ca.* Midichloria sp. Ixholo2; FM992373), which was found in adult *I. holocyclus* (59/78, 75.6%) ticks (Figure 5d). There was an absence of OTU_12 (*Ca.* Neoehrlichia arcana) and OTU_112 (*Ca.* Neoehrlichia australis) in adult *I. holocyclus* specimens, despite their presence in nymphs.

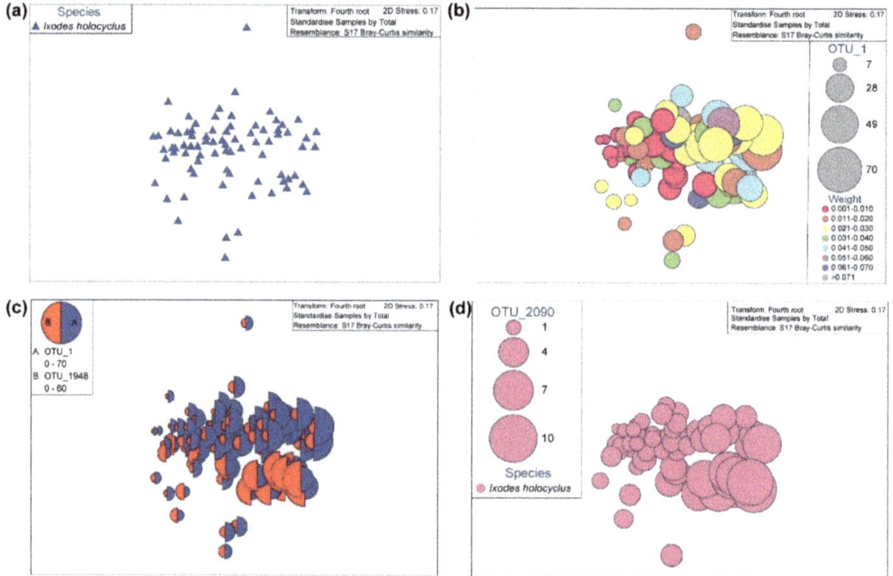

Figure 5. Multivariate statistical analysis of the V3-V4 bacterial profile for adult *Ixodes holocyclus* ticks from Sydney, Australia. Non-metric multidimensional scaling (nMDS) ordination plots for bacterial abundance, bootstrap averages plot and bubble plots. (**a**) nMDS ordination plots for bacterial abundance, shown at the operational taxonomic unit (OTU) level. (**b**) Bubble plot for OTU_1 (Ca. Midichloria sp. Ixholo1) with the weight factor overlayed. (**c**) Bubble plot for OTU_1 (*Ca.* Midichloria sp. Ixholo1) and OTU_1948 (*Ca.* Midichloria sp. Ixholo2) reveal that these endosymbionts are prevalent across the *Ixodes holocyclus* adult tick specimens. (**d**) Bubble plot for OTU_2090 (97.3% identity to Ca. Midichloria sp. Ixholo2 (FM992373)) reveal that this endosymbiont is present in most *Ixodes holocyclus* adult ticks.

To determine the bacterial diversity within the nymph ticks, we compiled the OTUs and sorted the reads into four main taxonomic ranks (class, order, family, genus) and analysed the bacterial diversity at each level. At the class level, nymph ticks in Sydney are dominated by Alphaproteobacteria, while at the order level, they are mostly dominated by Rickettsiales which were visualised on bar graphs (Supplementary Figure S2a,b). When the two dominating OTUs (OTU_1, OTU_1948; *Ca.* Midichloria spp.) were hidden from the analysis, it still revealed high bacterial diversity among nymph ticks at the bacterial family (55 taxa identified) and genus (78 taxa identified) levels, visualised on bar graphs (Supplementary Figure S2c,d).

2.6. Permutation-Based Hypothesis Testing Reveals External Factors Can Influence the Tick's Microbiota

Ticks species were collected from greater Sydney and eastern New South Wales (Figure 1). To determine whether the geographical location of collection has a significant effect on tick samples we evaluated it using the permutation-based hypothesis testing approach (analysis of similarities, ANOSIM) on Bray-Curtis dissimilarity matrix (Figure 6). Location was defined as the suburb of which the ticks were collected from, or where the dog host of the adult tick was living. The factor 'region' was defined as the NSW government's division of state's political and administrative regions and localities (i.e. Northern Beaches, North Shore, South Coast, etc.). The north–south orientation was determined in relation to Sydney CBD and for coastal proximity we assigned each samples with a category (1–5) based on distance from coast (1: 0–4.99 km, 2: 5–9.99 km, 3: 10–14.99 km, 4: 15–19.99 km, 5: ≥ 20 km).

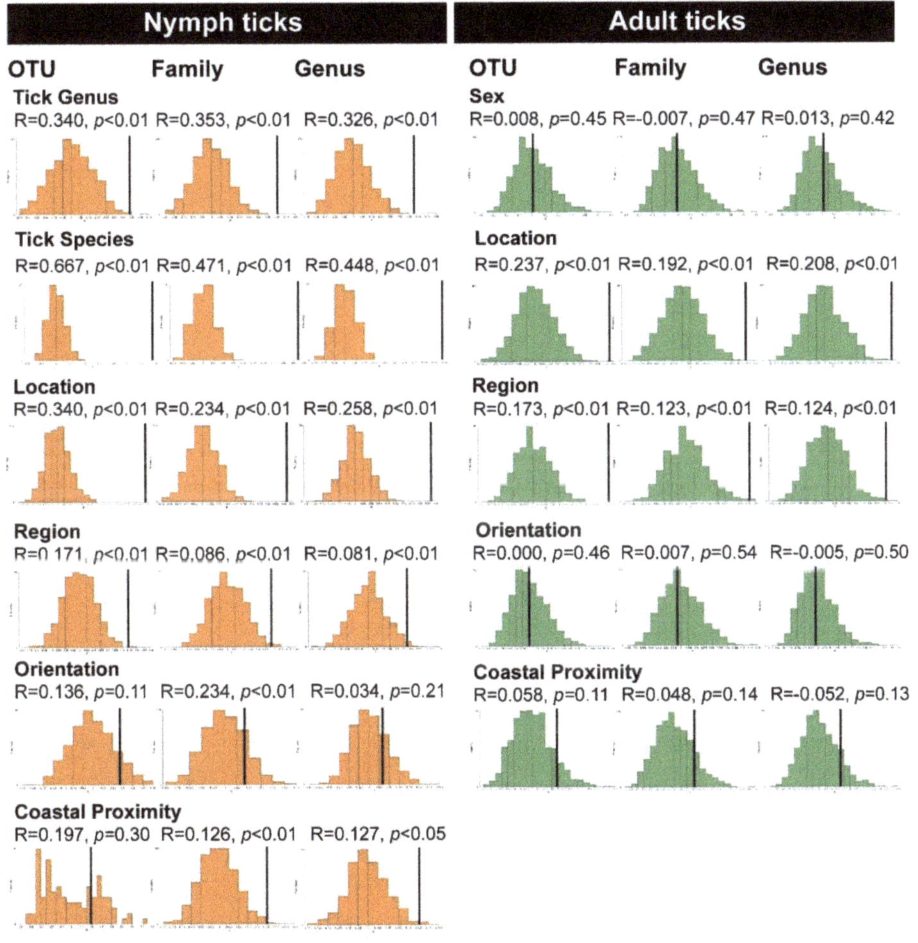

Figure 6. Analysis of similarities (ANOSIM) histograms for nymph *Ixodes holocyclus*, *Ixodes trichosuri*, *Ixodes tasmani* and *Haemaphysalis bancrofti* ticks (orange) and adult *I. holocyclus* ticks (green) at the bacterial OTU, family and genus levels.

Taking into account nymph genus as the external factor and bacterial OTU, genus and family dissimilarity matrix, the ANOSIM histograms revealed significant difference between the unordered tick genera sub-groups (*Ixodes, Haemaphysalis*) (OTU: $R = 0.340$, $p = 0.003$; family: $R = 0.353$, $p = 0.003$; genus: $R = 0.328$, $p = 0.002$;) (Figure 6). Similarly, sub-groups based on nymph species as the external factor were significant using ANOSIM (OTU: $R = 0.667$, $p = 0.001$; family: $R = 0.471$, $p = 0.001$; genus: $R = 0.488$, $p = 0.001$) (Figure 6).

When the location of collection was selected as the external factor and bacterial OTU, genus and family dissimilarity matrix, the ANOSIM histograms revealed significant difference between the unordered location subgroups (Mona Vale, Frenchs Forest, Manly, etc.) (OTU: R = 0.340, p = 0.001; family: R = 0.234, p = 0.001; genus: R = 0.258, p = 0.001) in nymph ticks (Figure 6). Similarly, nymph sub-groups with geographical region (Northern Beaches, North Shore, South Coast) as the external factors were significant using ANOSIM (OTU: R = 0.171, p = 0.006; family: R = 0.086, p = 0.010; genus: R = 0.081, p = 0.013), (Figure 6). However, nymph sub-groups with geographical orientation (north or south of Sydney) as the external factors were not significant by ANOSIM (OTU: R = 0.136, p = 0.113; family: R = 0.041, p = 0.169; genus: R = 0.034, p = 0.210) (Figure 6). While the p-value recorded for the geographical orientation was not traditionally regarded as significant (p > 0.05), the ANOSIM histogram curve still indicates a normal distribution. Sub-groups for nymph ticks with the coastal proximity (1: 0–4.99 km, 2: 5–9.99 km, 3: 10–14.99 km, 4: 15–19.99 km, 5: ≥ 20 km) as the external factors were not significant by ANOSIM at the OTU-level (R = 0.197, p = 0.302), but were significant at the bacterial family and genus levels (family: R = 0.126, p = 0.007; genus: R = 0.127, p = 0.011) (Figure 6).

Taking into account sex as the external factor for adult ticks and bacterial OTU, genus and family dissimilarity matrix, the ANOSIM histograms revealed significant difference between the unordered sub-groups (male/female) (OTU: R = 0.008, p = 0.446; family: R = −0.007, p = 0.474; genus: R = 0.013, p = 0.416) (Figure 6). Sub-groups based on the location of collection for adult ticks as the external factor were significant using ANOSIM (OTU: R = 0.237, p = 0.001; family: R = 0.192, p = 0.002; genus: R = 0.208, p = 0.002) (Figure 6). Similarly, adult tick sub-groups with geographical region (Northern Beaches, North Shore, South Coast, etc.) as the external factors were significant using ANOSIM (OTU: R = 0.173, p = 0.001; family: R = 0.123, p = 0.005; genus: R = 0.124, p = 0.003), (Figure 6). However, among adult *I. holocyclus*, sub-groups with geographical orientation (north or south of Sydney) as the external factors were not significant by ANOSIM (OTU: R = 0.000, p = 0.455; family: R = −0.007, p = 0.544; genus: R = −0.005, p = 0.499) and had no relationship with the microbiota (Figure 6). Sub-groups for adult ticks with the coastal proximity (1: 0–4.99 km, 2: 5–9.99 km, 3: 10–14.99 km, 4: 15–19.99 km, 5: ≥ 20 km) as the external factors were not significant by ANOSIM (OTU: R = 0.058, p = 0.112; family: R = 0.048, p = 0.141; genus: R = 0.052, p = 0.134) (Figure 6).

2.7. Absence of Borrelia spp. in Ixodes holocyclus, Ixodes trichosuri, Ixodes tasmani and Haemaphysalis bancrofti from Sydney

The *Borrelia* spp. nested PCR assay at the *16S* rRNA gene was negative for all nymph and adult ticks examined (n = 200; nymph: n = 100, adult: n = 92). The negative controls did not amplify any *Borrelia* spp. DNA. Positive controls were omitted to prevent the risk of any cross-contamination.

Implementing NGS technologies, the pilot study using two *16S* rRNA diversity profiling assays (V1-V3 and V3-V4 hypervariable regions) did not record the presence of *Borrelia* spp. in the adult *I. holocyclus* ticks. Similarly, nymph and adult ticks were free of *Borrelia* spp. at the *16S* rRNA V3-V4 hypervariable region.

3. Discussion

This study has shown that in Sydney, NSW, people and their companion animals are exposed to *Ixodes holocyclus*, *Ixodes trichosuri*, *Ixodes tasmani* and *Haemaphysalis bancrofti* nymph ticks, which has been confirmed at the morphological and molecular (*cox*1) levels. It is curious that the species diversity of questing nymphs was notably higher than the species diversity of adult ticks from dogs in veterinary clinics in the surrounding areas, where only *I. holocyclus* was recovered. The presence of only *I. holocyclus* on dogs in veterinary clinics is indicative that it is only *I. holocyclus* that has a paralytic effect on hosts and does not support previously purported claims that *I. trichosuri* may contribute to tick paralysis [29]. Interestingly, a study done in Sydney between January 1990 and December 1992 found that *I. holocyclus* was the most common nymph and adult tick, followed by *Haemaphysalis longicornis* and *H. bancrofti*, which were only confirmed morphologically [16]. Nearly three decades later, *H. longicornis* was not recovered from the environment or from dogs in veterinary clinics, while

I. holocyclus remains to be the most common tick in the area. This has been the first study to the authors' knowledge, of this scale, to have flagged for questing nymphs, identified them morphologically and molecularly, and used modern NGS techniques to determine the tick's bacterial diversity from the Sydney region.

In recent years, NGS technologies has been used to detect and study the presence of pathogenic bacteria and those significant in human and animal health around the world, but few have compared the microbial diversity between tick's life cycles, especially in Australia [3,5,6]. Further, considering the interest in the detection of *Borrelia* spp. in Australian ticks, few studies have looked into finding *Borrelia* spirochaetes in unengorged Australian nymphs in the past 25 years [16]. In North America and Europe, typically, the nymph stages of *Ixodes scapularis*, *Ixodes pacificus* and *Ixodes ricinus*, are the main vectors for transmitting and infecting humans with *Borrelia burgdorferi* sensu lato (s.l.) [19,30,31]. The *Borrelia* spirochaete population within the nymph tick is amplified following the larva-to-nymph moult and is higher in unfed nymph ticks compared to unfed adult ticks [25,32]. It has been found that nymph density and *B. burgdorferi* s.l. infections were positively correlated, suggestive of the importance of the role of nymph ticks as a vector [33–35]. Hence, to accurately determine the possibility of a local infection of Lyme disease in Sydney, it was crucial that nymph ticks of the *Ixodes* genus were screen for the causative agent, *B. burgdorferi* sensu lato.

While *Borrelia* spp. has been detected in Australian ticks in Sydney, *B. burgdorferi* s.l. had not been identified in Sydney, Australia [2,16]. In 2017, Panetta et al. [2] found a reptile associated *Borrelia* spp. from the goanna tick, *Bothriocroton undatum* Fabricius, 1775, in Sydney by sequencing both *16S* rRNA hypervariable regions, V1-V3 and V3-V4, and confirmed the presence of *Borrelia* spp. in *B. undatum* using a *Borrelia* specific nested PCR. The reptile associated *Borrelia* detected by Panetta et al. [2] was evolutionarily distinct from the monophyletic clade for *B. burgdorferi* s.l., the causative agent of Lyme disease, and it only clustered with other reptile associated *Borrelia*. Our studies utilised the same NGS technologies and nested PCR assay and did not detect any *Borrelia* spp. in our tick samples. In one other Sydney-based study, the midgut from 10,970 ticks (2379 nymphs, 528 males, 1832 females) was dissected out and examined by dark field microscopy for spirochaetes and was cultured using *B. burgdorferi* culture media (BSKII) over three months [16]. Spirochaete-like objects were found, but it was determined that these were not *Borrelia* spirochaetes. Additionally, 1038 ticks (576 nymphs, 62 males, 396 females) were subject to molecular screening for *B. burgdorferi* s.l. using three *Borrelia* specific primer sets, which targeted different regions—outer surface protein A (*ospA*), flagellin (*fla*) and *Borrelia 16S* rDNA that had failed to detect any *Borrelia* DNA in the samples. Utilising modern NGS technologies, Panetta et al. [2] detected reptile associated *Borrelia*, which could have been missed by Russell et al. [16], as the primers at the time may not have been able to pick up reptile or monotreme associated *Borrelia*. Especially considering the presence of the echidna tick, *Bothriocroton concolor* Neumann, 1899, which a monotreme associated *Borrelia*, *Candidatus* Borrelia tachyglossi, has been detected from previously [36]. It was necessary for our study to utilise NGS technologies and the *Borrelia* specific nested PCR to screen for *Borrelia* spp. in our samples. However, this study did not find any *Borrelia* spp. in the flagged nymphs from Sydney, or the adult *I. holocyclus* on dogs in the surrounding areas. For over 25 years, there has been a lot of debate surrounding the capacity for *I. holocyclus* to vector the causative agent of Lyme disease, *Borrelia burgdorferi* s.l. [37]. While we cannot comment on the vector capacity of *I. holocyclus* with regards to *Borrelia* spp., we confirm other similar findings that there has not been any tangible evidence suggesting that classical Lyme disease can be contracted locally, as *B. burgdorferi* s.l. has not been found in Australia to this date [38,39].

Like most other living organisms, ticks require bacteria for their survival and success [40]. These bacteria can be pathogenic or symbionts, and some pest insects, like aphids, have a nutritional dependence on the symbiont *Buchnera* [41]. In the soft tick, *Ornithodoros moubata*, Murray, 1877, one *Francisella* strain, *Francisella* F-Om, has been reported to be an obligate nutritional mutualist symbiont which enables *O. moubata* to feed on vertebrate hosts, and was necessary for the life cycle [42]. Further, *Coxiella* is a known symbiont, particularly in Ixodid ticks, and studies have detected *Coxiella* in *Rhipicephalus sanguineus* and *Rhipicephalus* (*Boophilus*) *microplus*, Canestrini, 1888, from different countries [43–45]. Further, while the purpose of *Candidatus* Midichloria mitochondrii (CMM) is still unknown in ticks, it is known to have an intra-mitochondrial lifestyle and appears to play a role in the biology of female *I. ricinus* [46]. It has been purported that CMM may have some relationship with ticks as a nutritional mutualist, and it is thought that it is linked to the tick's capacity as being a vector for disease [40].

The main benefit of NGS technologies, is the ability to gain rapid insight into the *16S* rRNA bacterial endosymbionts of ticks [47]. One study using pyrosequencing to determine the bacterial diversity in *Rhipicephalus microplus* Canestrini, 1888 found shared bacterial genera among *R. microplus* collected from various countries, implying there is a core microbiota associated with the tick species [45]. The core microbiota of ticks is the known microbial symbionts within tick species. It is expected to be unique for each tick species but common among ticks of that species, and the core microbiota includes symbionts which have ecological or biological functional roles [40,48,49]. Geography and sex can influence the core microbiota of ticks [50–52]. It is known that while CMM is present in the reproductive tissues of female *I. ricinus*, and it is absent from *I. holocyclus* [46,53]. The ovarian tissues of all female *I. ricinus* are known to be dominated by CMM, which forms part of the core microbiota [46,54–57]. It was found that CMM are lacking in the midgut of *I. ricinus*, and the midgut did not reflect the presence of a core microbiota and was heavily influenced by host-blood meal [54]. In *I. holocyclus* ticks, *Ca.* Midichloria sp. Ixholo1 and *Ca.* Midichloria sp. Ixholo2 are found as endosymbionts [53]. Further, *Ca.* Midichloria sp. Ixholo1 and *Ca.* Midichloria sp. Ixholo2 have been isolated from *I. holocyclus* ovaries [53]. This, and the abundance of *Ca.* Midichloria sp. Ixholo1 and *Ca.* Midichloria sp. Ixholo2 in *I. holocyclus*, supports the finding that they are part of the core microbiota of the ticks from this study. From the species clustering seen in our microbial analyses, it implies that the ticks in Sydney do have a distinct, core microbiota, which appears to be unique to the species observed. Microbial overlap is apparent, with *Ca.* Midichloria sp. Ixholo1 and *Ca.* Midichloria sp. Ixholo2 found indiscriminately across all four tick species in the study, hence, *Ca.* Midichloria spp. are obligate symbionts of the endemic Australian ticks in this study and are not exclusive to *I. holocyclus* [15,53]. Ticks in Sydney during 2017–2018 were dominated by *Ca.* Midichloria sp. Ixholo1 (69.41%) and *Ca.* Midichloria sp. Ixholo2 (21.45%), so it is part of the core microbiota of ticks in the North Shore and Northern Beaches in Sydney, NSW. Additionally, this study reports the presence of a previously undescribed bacterial taxon, OTU_2090. The unique taxon, OTU_2090, which had 97.3% identity to *Ca.* Midichloria sp. Ixholo2 (FM992373) found in *I. holocyclus* nymph and adult ticks only. It was absent from *I. trichosuri*, *I. tasmani* and *H. bancrofti* nymph ticks, despite sharing the same environment. From this, we conclude that OTU_2090 is a novel *Candidatus* Midichloria sp., which is unique to and is part of the core microbiota of *I. holocyclus* ticks. Other common tick endosymbionts including *Coxiella*, *Francisella* and *Rickettsia* were absent from our ticks [40,43,44,47]. Like other ticks within the *Ixodes* genus, *Ca.* Midichloria spp. remains to be the dominant bacterial genus in our *I. holocyclus*, *I. tasmani* and *I. trichosuri* specimens [15,58]. Notably, this study confirms the presence of *Ca.* Neoehrlichia arcana and *Ca.* Neoehrlichia australis in *I. holocyclus* nymphs and reports the presence of *Ca.* Neoehrlichia australis in an *I. trichosuri* for the first time. The presence of the *Ca.* Neoehlichia spp. in nymph ticks only highlights the significance of nymph ticks as vectors for disease. Previous studies [59] determined that *Ca.* Neoehrlichia arcana and *Ca.* Neoehrlichia australis was closely related to the *Candidatus* Neoehrlichia mikurensis, the causative agent of Neoehrlichiosis in humans [60]. As nymph ticks feed, host and tick bacterial exchange occurs during the uptake of the blood meal, which is an explanation for the absence of *Ca.* Neoehrlichia spp. from adult ticks in our study. Previous findings have shown that host species affects the tick microbiota [8,61]. The absence of *Ca.* Neoehrlichia arcana and *Ca.* Neoehrlichia Australia in our adult tick samples highlights the importance of screening nymph ticks as vectors for disease. While engorged adult ticks still may carry zoonotic bacteria, our findings are suggestive that the non-commensal bacteria leave the tick host during the nymph tick's blood meal.

Previous studies in North America and Europe have demonstrated that the location of the collection site [5,52,62,63], sex [4,51,52,64], life stage [5,62] and tick species [52,63] had a strong influence on the diversity of the bacterial taxa within ticks. This study has shown that biotic factors including nymph genera and species, geographical location of collection and geographical region of collection, have an influence on the tick's microbiota. Like other studies, our study found that the tick species was a factor which effected the bacterial taxa [52,63]. While other studies have compared location of collection, they were based on distances hundreds of kilometres apart [5,52,62,63], this study compares nymph ticks at a much finer scale within a distance of approximately 28 km on the North Shore and Northern Beaches regions of Sydney, Australia. Many collection sites were < 1 km apart, and most collection sites were not separated by more than 10 km. Despite the proximity, our study determined that two geographical factors—location and region—had an influence on the tick's microbiota. One other study [51] has compared collection site at a similarly small scale, whereby the ticks were collected from the Ames Plantation in Western Tennessee, USA, which determined that soil type influenced the microbiota, however it was stated that this was "possibly due to contamination". As the ticks recovered in this study are known to parasitise humans and companion animals, the community in the Northern Beaches and North Shore regions in Sydney, Australia are exposed to different bacteria of unknown pathogenicity, which varies depending on the geography and tick species [1,18]. Other factors which affected the bacterial profile of ticks elsewhere included sex, life stage and host species was not reflected in our study [7,52,65]. To eliminate the potential bias for host species to affect the microbiota of the ticks, all adult ticks were collected from dogs in veterinary clinics and all nymph ticks were sourced from the environment. The microbial contents for adults and nymphs were analysed separately, however, Shannon's and Simpson's indices indicated there were no significant difference between the life stages. Unlike other microbial studies comparing between at least two life stages, we did not find this to be a significant biotic factor [5,49,51]. While other studies [4,51,52,64] have determined there was a difference in the microbiota of male and female adult ticks, we could not make this comparison due to the lack of male specimens in our study. Further, our study could not make a direct comparison between unfed adult and nymph ticks due to the absence of flagged adult ticks. Flagging for adult ticks could see an increase in the number of male ticks, which would address potential differences between the adult male and female tick microbiota in Australia. Future studies would benefit from flagging for adult ticks to determine the impact of the host-blood meal on the microbiota of ticks.

The decision to employ a pilot study to determine the best method for DNA isolation, and the best hypervariable region to obtain the *16S* rRNA gene amplicon profile for our ticks enabled us to find the optimal combination of DNA isolation method and target hypervariable region for the downstream applications for nymph and adult ticks in Sydney. Other studies have shown that using different DNA isolation protocols can impact the bacterial communities generated for microbiota analyses [2,66,67]. In the study by Panetta et al. [2], the use of two DNA isolation protocols and *16S* bacterial profiling target regions (V1-V3 and V3-V4) found that the V1-V3 hypervariable region and Method 1 (ISOLATE Faecal DNA Kit, Bioline, Eveleigh, Australia), afforded the greatest capacity to detect *Borrelia* spp. in their samples. However, the V3-V4 hypervariable region is favoured as it is associated with having the highest diversity estimates in the sequenced bacterial communities [63]. Thus, analysing different DNA isolation protocols and hypervariable regions was necessary in our pilot study to determine which was the ideal combination to determine the most complete rigorous bacterial profile for the nymph and adult ticks in Sydney, NSW. As we did not determine a significant difference between our chosen DNA isolation protocols, due to the lower number of unassigned OTUs, the V3-V4 *16S* rRNA hypervariable region with the ISOLATE II Genomic DNA Kit (Bioline, Eveleigh, Australia) was determined to be the ideal combination to determine the tick's bacterial diversity in this study.

4. Materials and Methods

4.1. Tick Specimens

A total of 159 engorged and un-engorged adult ticks were collected from veterinary practices from the Northern Beaches area of Sydney to the South Coast region of New South Wales for this study between the Australian 2016–2017 summer season (November 2016 to February 2017) (Supplementary Table S1). All ticks were forcibly removed by veterinarians or veterinary nurses at the veterinary practices using one of the following methods: Tick Twister®(O'TOM, Lavancia-Épercy, France), tapered tip forceps or removed manually. After removal, ticks were immediately placed in 96–100% ethanol and stored at −20 °C. Ticks were grouped according to host animal and the host species, number of ticks per host, location of host and date of collection were obtained for accuracy.

A total of 148 unfed nymphs were collected from the Northern Beaches and North Shore areas of Sydney and 27 unfed nymphs and 20 unfed larvae were collected from the South Coast region of New South Wales for this study between July and September 2017 (Figure 1, Supplementary Table S1). All larvae and nymphs were collected using a tick flagging/dragging method, whereby a 1 m × 0.7 m white towel was dragged through shrubs or bush and was visually inspected every 5–10 metres. Ticks which were present were immediately removed using forceps and placed into a 1.5 mL microcentrifuge tube with 96–100% ethanol and later stored at −20 °C. Ticks were grouped according to flagging site and the number of ticks per site, GPS co-ordinate of site and date of collection were obtained for accuracy.

4.2. Morphological and Molecular Identity of the Same Ticks Using Two Different Methods

As a pilot study, six female ticks (SC0063-1 to SC0063-6) morphologically identified as *I. holocyclus*, using keys and guides [17,18], were of varying levels of engorgement and selected from a single host ('Bear', German Shepherd Dog, *Canis lupus familiaris*). All ticks were bisected longitudinally using a new, sterilised no. 15 scalpel blade. Each half was subject to different DNA isolation methods, using the ISOLATE Faecal DNA Kit (Bioline, Eveleigh, Australia) or the ISOLATE II Genomic DNA Kit (Bioline, Eveleigh, Australia) (Figure 7). One half was used for faecal DNA isolation, in accordance with the ISOLATE Faecal DNA Kit (Bioline, Eveleigh, Australia) using a mechanical disruption method where the specimen and 5 μL of DNA Extraction Control (Bioline, Eveleigh, Australia) were placed in a supplied 1.5 mL bead-beater with ceramic beads and homogenised with a high-speed benchtop homogeniser (FastPrep-24, MP Biomedicals, Seven Hills, Australia) for 40 s at 6.0 m/s, followed by isolation as per the manufacturer's instructions. The other half of the tick was used for genomic DNA (gDNA) isolation, in accordance with the ISOLATE II Genomic DNA Kit (Bioline, Eveleigh, Australia) with the following changes: tick samples were digested in 180 μL of lysis buffer, 25 μL of proteinase K and 5 μL of DNA Extraction Control (Bioline, Eveleigh, Australia) for 12–16 h in 56 °C on a heat block.

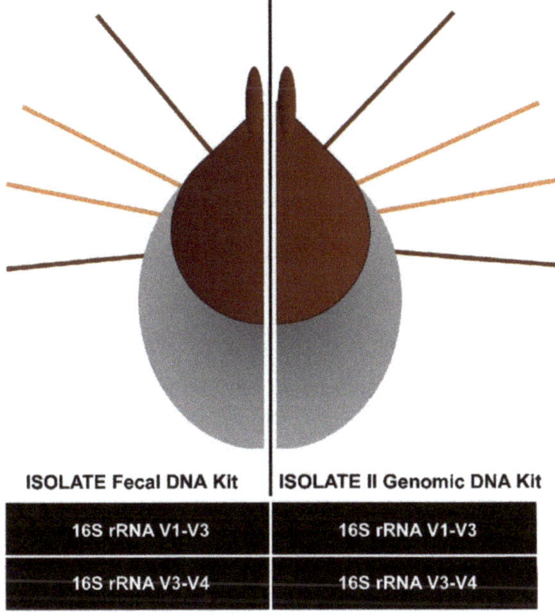

Figure 7. Workflow for the DNA isolation and diversity profiling methods used for the pilot study. Female adult *Ixodes holocyclus* ticks were longitudinally bisected and each half was subject to different DNA isolation protocols—ISOLATE Faecal DNA Kit or ISOLATE II Genomic DNA Kit. The DNA was then subject to sequencing using gene diversity profiling assays, targeting two 16S rRNA hypervariable regions—V1-V3 and V3-V4.

4.3. Morphological Identification and DNA Isolation of Ticks

All ticks examined under a stereo microscope (SMZ-2B, Nikon, Rhodes, Australia) and the species and sex were recorded and identified morphologically with the aid of keys and descriptions [17,18]. Adult ticks were weighed on an analytical balance and the individual weights were recorded.

All tick specimens were surfaced sterilised prior to DNA isolation. Surface sterilisation involved sequential 1 mL washes with a vortex for 1 min in 3% hydrogen peroxide (H_2O_2), two 30 s washes in 70% ethanol (*w/v*), 2 min in phosphate buffered saline (PBS, pH = 7.4) and dried on a Kimwipe (Kimberly-Clark, Ingleburn, Australia) [8].

Nymphs were quadrisected using new, sterile no. 15 scalpel blades and whole tick genomic DNA (gDNA) was extracted in accordance with the ISOLATE II Genomic DNA Kit (Bioline, Eveleigh, Australia), and the modifications made as above. The DNA isolation method was completed as per the kit instructions and total DNA was eluted into 80 µL of elution buffer (Tris buffer, pH = 8.5, preheated to 70 °C). Adult ticks were cut into 1-2mm pieces using a new, sterilised no. 15 or 24 scalpel blade and up to 150 mg tick tissue was used for whole tick gDNA isolation, as above.Eluted whole tick gDNA was stored at −20 °C prior to molecular analysis. As DNA isolation kits contribute bacterial contaminants that can impact microbiome analyses [68], two 'BLANK' DNA extraction control reactions were included.

4.4. Amplification of the Tick Mitochondrially Encoded cox1 Gene

A 604 nucleotide (nt) 5′ fragment of the cytochrome c oxidase subunit I (*cox1*) was amplified using the following primer pairs: LCO1490 (F1) (5′-GGT CAA CAA ATC ATA AAG ATA TTG G-3′) and HCO2198 (R1) (5′-TAA ACT TCA GGG TGA CCA AAA AAT CA-3′) [69] or (F2) (5′-TAC TCT

ACT AAT CAT AAA GAC ATT GG-3′) and S0726 (R2) (5′-CCT CCT CCT GAA GGG TCA AAA AAT GA-3′) [70].

MyTaq™ Red Mix (Bioline, Eveleigh, Australia) was used for *cox*1 amplifications in 30 μL reactions using 2 μL of template DNA. The PCR cycling conditions were as follows: 95 °C for 1 min, 35 cycles of 95 °C for 15 s, 55 °C for 15 s and 72 °C for 10 s followed by 72 °C for 5 min. All reactions had a positive control, and PCR-grade water was as a no template control. DNA from *Rhipicephalus sanguineus* s.l. was used as the positive control for all reactions. All conventional PCR reactions were conducted in an Applied Biosystems Veriti™ Thermal Cycler (Thermo Fisher Scientific, North Ryde, Australia) or a T100™ Thermal Cycler (BioRad, Gladesville, Australia). PCR products were sequenced at Macrogen Ltd. (Seoul, South Korea). The PCR products of 20 nymphs were bi-directionally sequenced, while the remaining 87 were uni-directionally sequenced.

4.5. Tick DNA Sequence Analysis and Phylogeny

Sequences were assembled using CLC Main Workbench 6.8.1 (Qiagen, Vedbæk, Denmark). Phylogenetic analysis of nymph tick DNA and the composition of the nucleotide sequences were determined using MEGA 7.0 [28]. Phylogenetic comparison was made between available *Ixodes* spp. and *Haemaphysalis* spp. tick haplotypes with *Bothriocroton* spp. tick haplotypes as an outgroup from GenBank (National Center for Biotechnology Information, NCBI) using MEGA 7.0 to determine the haplotypic diversity within the Australian tick population [28]. The evolutionary history was inferred using the maximum likelihood (ML) and minimum evolution (ME) method and distances were computed using the Tamura-3 and Kimura-2 methods, respectively, in MEGA 7.0 [28,71,72].

4.6. Detection of Borrelia spp. Spirochaete Amplifying a 16S Ribosomal RNA Gene Fragment

A ~1250-nt fragment of the *16S* ribosomal RNA (rRNA) was amplified in a conventional nested-PCR, using previously published primers [36]. MyTaq™ Red Mix (Bioline, Eveleigh, Australia) was used for *16S* rRNA amplification of *Borrelia* spirochaetes in 30 μL reactions using 10 pmol of each primer (Bor-16F (S0778), 5′-TGC GTC TTA AGC ATG CAA GT-3′/Bor-1360R (S0779), 5′GTA CAA GGC CCG AGA ACG TA-3′ for the first round; Bor-27F (S0780), 5′-CAT GCA AGT CAA ACG GAA TG-3′/Bor-1232R (S0781), 5′-ACT GTT TCG CTT CGC TTT GT-3′ for the second round [36]), template DNA (2 μL of DNA for the primary reaction and 1 μL aliquot of the primary reaction PCR product in the secondary reaction), and PCR-grade water. The nested PCR was run with the cycling conditions as described by Panetta et al. [2]. All reactions contained PCR-grade water as a no template control, but a positive control was omitted to minimise contamination.

4.7. Amplification and Analysis of the Tick Microbial Profile

As a pilot study, the gDNA and faecal DNA isolated using two different methods from the same six female *I. holocyclus* (SC0063-1 to SC0063-6) underwent diversity profiling of the *16S* rRNA gene at the Australian Genome Research Facility (AGRF, Melbourne, Australia). Tick samples (*n* = 24) underwent DNA quality control (QC) screening via PCR and indexing fluorometry at AGRF (Melbourne, Australia) prior to the diversity profiling at the *16S* rRNA primer targets. All tick samples (*n* = 24) passed QC. Two target *16S* rRNA hypervariable regions, V1-V3 and V3-V4 were sequenced on the Illumina MiSeq (300-nt paired end reads) using the following assays: *16S* (V1-V3): 27F (5′-AGA GTT TGA TCM TGG CTC AG-3′) with 519R (5′-GWA TTA CCG CGG CKG CTG-3′) and *16S* (V3-V4) 341F (5′-CCT AYG GGR BGC ASC AG-3′) with 806R (5′-GGA CTA CNN GGG TAT CTA AT-3′). Paired end reads were assembled using PEAR (version 0.9.5) and the primers were identified, trimmed and processed as previously described by Panetta et al. [2].

4.8. Multivariate Statistical Analysis of the Pilot Data for the Single Host Tick Microbiota

Multivariate statistical analyses were used to elucidate patterns of variation within the bacterial composition of the pilot data of *Ixodes holocyclus* (*n* = 6) from a single dog host. The microbiota abundance matrix, OTU taxonomy, and sample factors were imported as metadata for the multivariate analysis in PRIMER v.7 (PRIMER-e, Albany, New Zealand).

The pilot data from the *16S* rRNA V1-V3 (n = 12) and V3-V4 (n = 12) hypervariable regions were analysed separately. Each tick samples were associated with the following sample factors: tick species ID, tick specimen ID, year of collection, host ID, DNA isolation kit used, and MiSeq sequenced *16S* rRNA gene hypervariable region. To clean the data, initially, unassigned OTUs, mitochondria and chloroplasts were excluded from the analyses. The lowest 5% of the data matrix of OTUs and taxonomy abundance was removed manually.

The V1-V3 *16S* rRNA gene diversity profiling assay yielded 1,246,668 raw reads that were filtered into 789,733 high quality reads (min. 3,743; max. 121,393; n = 12). This was clustered into 1642 OTUs and after internal quality control, it was filtered down to 68 OTUs. The V3-V4 *16S* rRNA gene diversity profiling assay yielded 706,769 raw reads that were filtered into 630,336 high quality reads (min. 2149; max. 119,099; n = 12). This was clustered into 313 OTUs and after internal quality control, it was filtered down to 64 OTUs. Internal quality control for both hypervariable regions involved manually removing mitochondria, chloroplasts and unassigned OTUs and the bottom 5% of OTUs.

The cleaned data was then imported as metadata into PRIMER v.7 (PRIMER-e, Albany, New Zealand), and was then standardised (samples by total) and fourth root transformed. Bray-Curtis dissimilarity was utilised to measure the variation in bacterial composition within the ticks. Non-metric multi-dimensional scaling ordination, nMDS [73], was used to view the trends of bacterial community similarity between all samples. The goodness-of-fit from the two-dimensional nMDS plot was measured with Kruskal's stress formula I [73]. Visualisation of the two-dimensional nMDS plots were enriched by annotating the different sample factors (e.g., species of tick, DNA isolation kits) overlaid on to the ordination plots using symbols to assess their possible impacts on the composition of the bacterial communities. Analysis of similarities, ANOSIM [74] (significance level, p = 0.05), was applied to test the null hypothesis of no difference between the bacterial communities.

The Mann-Whitney Test for significance was employed on Prism 8 (GraphPad Software, San Diego, USA) to determine whether the bacterial communities generated from the *16S* rRNA hypervariable regions tested (V1-V3 *versus* V3-V4) were significantly different from each other.

4.9. Multivariate Statistical Analysis of the Adult and Nymph Tick Microbiota from Sydney

Following the pilot study, multivariate statistical analysis was used to determine the patterns of variations within the bacterial composition of the adult and nymph ticks from the Sydney region, and the south coast of NSW as an outgroup. The microbiota abundance matrix, OTU taxonomy, and sample factors were imported as metadata for the multivariate analysis in PRIMER v.7 (PRIMER-e, Albany, New Zealand).

The data from the *16S* rRNA V3-V4 (n = 187) hypervariable regions were analysed separately. Each tick sample was associated with the following sample factors: tick genus, tick species, tick specimen ID, location of collection, region of collection, orientation, coastal proximity (1–5 scale), sex, weight (adult), DNA isolation kit used, and MiSeq hypervariable region. The sample factor orientation was determined to be whether the tick was found towards the north or the south in relation to Sydney. The sample factor coastal proximity was an assigned number from one to five depending on the location of the tick in relation to a coastal body of water (1: 0–4.99 km, 2: 5–9.99 km, 3: 10–14.99 km, 4: 15–19.99 km, 5: ≥ 20 km). A coastal body of water was defined as coastal waters, rivers, lakes, lagoons, undeveloped headlands, marine waters and estuary waters [75].

Of the 203 samples including two 'BLANK' DNA extraction control reactions, 188 passed the in-house QC protocols at AGRF (Melbourne, Australia), and were processed for further sequencing at the *16S* rRNA V3-V4 hypervariable region. The V3-V4 *16S* rRNA gene diversity profiling assays yielded 30,714,537 paired end raw reads (18.49 Gb) that were quality filtered into 20,198,718 (min. 259; max. 217,937; n = 188) high quality reads, excluding singletons, and clustered into 2287 bacterial OTUs. The removal of chloroplasts, mitochondrion, and unassigned OTUs led to 2284 bacterial OTUs.

The removal of OTUs where the total sum was less than 1000 reads led to 137 OTUs being retained. The removal of OTUs found in the 'BLANK' reactions, except for OTU_1 *Candidatus* Midichloria sp. Ixhlol1 and OTU_1948 *Candidatus* Midichloria sp. Ixhlo2, led to 116 bacterial OTUs (12,385,614 paired end reads; $n = 188$) being used for the microbial analysis.

To clean the data, initially, unassigned OTUs, mitochondria and chloroplasts were excluded from the analyses. The total sum of the OTUs from all samples were obtained from the data matrix of OTUs and taxonomy abundance, and if the total sum of the reads within an OTU was < 1000, it was manually removed from analysis. Unassigned OTUs, mitochondria and chloroplasts were excluded from the analysis. OTUs that were present in the 'BLANK' reactions were also removed, except OTU_1 and OTU_1948, which was *Ca.* Midichloria sp. Ixhlol1 and *Ca.* Midichloria sp. Ixhlo2, respectively. OTU_1 and OTU_1948 were present in low numbers in the 'BLANK' DNA extraction control reactions (JS3467: OTU_1 96 reads, OTU_1948 0 reads; JS3374: OTU_1 100 reads, OTU_1948 19 reads), and as they were important in the downstream application for microbial analysis, they were selectively retained.

The cleaned data was then imported into PRIMER v.7 (PRIMER-e, Albany, New Zealand), and the data matrix was standardised (samples by total) and then was fourth root transformed. Bray-Curtis dissimilarity was utilised to measure the variation in bacterial composition within the ticks. Non-metric multi-dimensional scaling ordination, nMDS [73], was used to view the trends of bacterial community similarity between all samples. The goodness-of-fit from the two-dimensional nMDS plot was measured with Kruskal's stress formula I [73]. Visualisation of the two-dimensional nMDS plots were enriched by annotating the different sample factors (e.g., species of tick, DNA isolation kits) overlaid on to the ordination plots using symbols to assess their possible impacts on the composition of the bacterial communities. Analysis of similarities, ANOSIM [74] (significance level, $p = 0.05$), was applied to test the null hypothesis of no difference between the bacterial communities.

Multivariable analyses through nMDS revealed that there were clustering of samples based on the nymph tick species identity (Figure 4a). For the nymph ticks alone, two samples (JS3299 and JS3368) were removed from the analyses and were classified as outliers. Sample JS3299 was removed from the analysis as the number of reads following QC was only 259. Sample JS3368 was removed from the analysis as it did not cluster with other *H. bancrofti* nymphs at the OTU level. At the OTU level, the permutation-based hypothesis testing using analysis of similarities (ANOSIM) histograms of permutations were generated for the nymph ticks to determine whether external factors had an influence on the tick's microbiota. When JS3368 was excluded from the analyses, the ANOSIM histograms revealed that were was not normal distribution at the OTU level ($R = 0.197$, $p = 0.302$), but the data was normally distributed and significant at the family ($R = 0.126$, $p = 0.007$) and order ($R = 0.127$, $p = 0.011$) levels (Figure 8). However, when included in the analyses the data was not normally distributed and was not significant at the OTU ($R = 0.197$, $p = 0.299$), family ($R = 0.140$, $p = 0.308$) or genus ($R = 0.198$, $p = 0.252$) levels (Figure 8). To test it further, the coastal proximity was artificially altered from factor 1: 0–4.99 km to factor 2: 5–9.99 km. This revealed slight non-normal distribution, with the ANOSIM histograms being slightly positively skewed and no significance between coastal proximity and nymph tick's microbiota at the OTU ($R = 0.204$, $p = 0.107$), family ($R = 0.132$, $p = 0.172$) or genus ($R = 0.158$, $p = 0.122$) levels (Figure 8). From this, we believe the tick is a biological outlier, and could have been translocated artificially during the flagging process, or as a larval tick by the host species movement. It was concluded that both samples, JS3299 and JS3368 would be excluded from further nymph tick analyses as they were characterised as outliers.

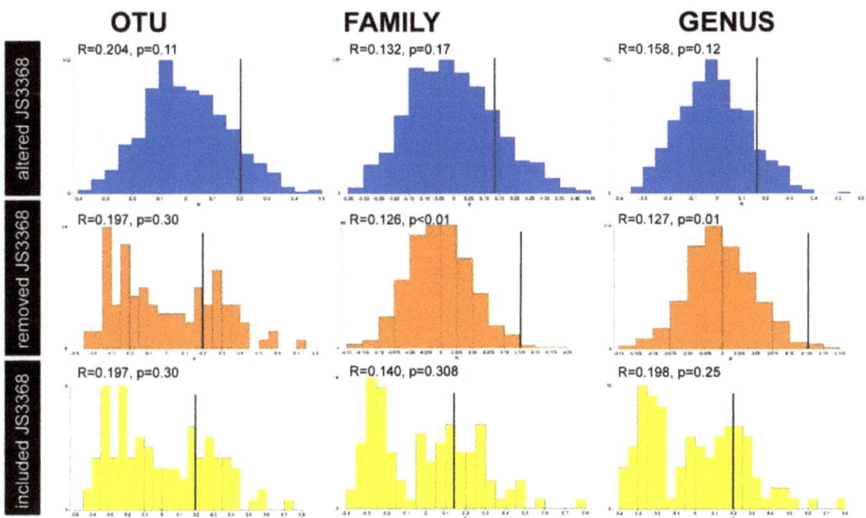

Figure 8. Analysis of similarity (ANOSIM) histograms for the nymph ticks (*Ixodes holocyclus, Ixodes trichosuri, Ixodes tasmani* and *Haemaphysalis bancrofti*) to rationalise the removal of outlier sample JS3368 from further analyses. Histograms of permutated distribution of the test statistic R (up to 999 permutations), with observed R-values and *p*-values noted at different taxonomic levels (OTU, Family, Genus) evaluated at the coastal proximity biotic factor. Outlier sample JS3299 has been removed from all histograms due to low reads. ANOSIM histograms depicting the artificial altering of coastal proximity factor for outlier sample JS3368 from level 1 (0–4.99 km) to level 2 (5–9.99 km) is shown in blue. ANOSIM histograms depicting the removal of outlier sample JS3368 evaluated at the coastal proximity factor is shown in orange. ANOSIM histograms depicting the inclusion of outlier sample JS3368 evaluated at the coastal proximity factor is shown in yellow.

4.10. Availability of Data

Nucleotide sequence data from this study have been deposited to GenBank (National Centre for Biotechnology Information, NCBI), under the accession numbers MT526908–MT527014 for the nymph *cox*1 sequences, and the SRA database under the BioProject ID PRJNA630349 (https://www.ncbi.nlm. nih.gov/sra/PRJNA630349) for the pilot study, and PRJNA631062 (https://www.ncbi.nlm.nih.gov/sra/ PRJNA631062) for the nymph and adult ticks from the North Shore. PRIMER data files and other relevant files are available on LabArchives (https://doi.org/10.25833/x96q-pg77).

5. Conclusions

This study has determined that the predominant nymph tick in the Northern Beaches and North Shore communities is the Australian paralysis tick, *Ixodes holocyclus*. Three other endemic tick species were also recorded: *Ixodes trichosuri, Ixodes tasmani* and *Haemaphysalis bancrofti*. Notably, in veterinary clinics, only *I. holocyclus* was recorded on dogs. We found that external biotic factors (tick species, geographic location of collection, geographic region, north–south orientation) have a significant impact on the bacterial profile within ticks. The microbial analyses revealed that the tick species in the Northern Beaches and North Shore of Sydney, NSW display a core microbiota, which appears to be unique to each species, with overlap with the bacterial genera. The most common endosymbiont in our ticks are *Candidatus* Midichloria sp. Ixholo1 and *Candidatus* Midichloria sp. Ixholo2. Additionally, a novel endosymbiont, *Candidatus* Midichloria sp. OTU_2090, was present in 96.3% and 75.6% *I. holocyclus* nymphs and adults, respectively. *Candidatus* Neoehrlichia arcana and *Candidatus* Neoehrlichia australis was recovered from *I. holocyclus* and one *I. trichosuri* nymph but was absent from *I. holocyclus* adult ticks.

Supplementary Materials: The following are available online at http://www.mdpi.com/2076-0817/9/7/566/s1, Figure S1. Multivariate analyses of pilot study data (V1-V3 vs. V3-V4), Figure S2. Bar graphs for nymph ticks, Table S1. Nymph and adult tick collection information, Table S2. Weight of adult ticks prior to DNA isolation.

Author Contributions: Conceptualization: S.C. and J.Š.; methodology: S.C. and J.Š.; validation: S.C. and J.Š.; formal analysis: S.C.; investigation: S.C.; resources: S.C. and J.Š.; data curation: S.C.; writing—original draft preparation: S.C.; writing—review and editing: S.C. and J.Š.; visualization: S.C.; supervision: J.Š.; project administration: S.C. and J.Š.; funding acquisition: S.C. and J.Š. All authors have read and agreed to the published version of the manuscript.

Funding: S.C. is supported by the University of Sydney Postgraduate Award (UPA) stipend scholarship. The study was in part, supported by the Dugdale Guy Peele Bequest, University of Sydney (J.Š.). The funding bodies had no role in the design, collection, analysis, and interpretation of the data or in writing the manuscript.

Acknowledgments: The authors acknowledge the Sydney Informatics Hub and the University of Sydney's high-performance computing cluster Artemis for providing the high-performance computing resources that have contributed to the research results reported within this paper. The authors wish to acknowledge and thank our previous honours students, Jessica Panuccio and Jessica Panetta, for organising the collection of the adult ticks from the Northern Beaches and South Coast regions in NSW, Australia, including the ticks used in the pilot study. We also wish to extend our gratitude to the staff of the veterinary clinics which participated in the study for the collection of the adult ticks. The authors acknowledge the facilities, and the scientific and technical assistance of Microscopy Australia, formerly the Australian Microscopy and Microanalysis Research Facility, at the Australian Centre for Microscopy & Microanalysis, University of Sydney.

Conflicts of Interest: The authors declare no conflict of interest.

Ethics Statement: Not applicable. Adult tick sample submissions were provided by the veterinary practitioner and approval was not applicable. Nymph tick samples were obtained from the environment.

References

1. Gofton, A.W.; Doggett, S.; Ratchford, A.; Oskam, C.L.; Paparini, A.; Ryan, U.; Irwin, P.; Schneider, B.S. Bacterial profiling reveals novel "*Ca.* Neoehrlichia", *Ehrlichia*, and *Anaplasma* species in Australian human-biting ticks. *PLoS ONE* **2015**, *10*, e0145449. [CrossRef] [PubMed]

2. Panetta, J.L.; Sima, R.; Calvani, N.E.D.; Hajdušek, O.; Chandra, S.; Panuccio, J.; Šlapeta, J. Reptile-associated *Borrelia* species in the goanna tick (*Bothriocroton undatum*) from Sydney, Australia. *Parasits Vectors* **2017**, *10*, 616. [CrossRef] [PubMed]

3. Greay, T.L.; Gofton, A.W.; Paparini, A.; Ryan, U.M.; Oskam, C.L.; Irwin, P.J. Recent insights into the tick microbiome gained through next-generation sequencing. *Parasits Vectors* **2018**, *11*, 12. [CrossRef]

4. Ponnusamy, L.; Gonzalez, A.; Van Treuren, W.; Weiss, S.; Parobek, C.; Juliano, J.; Knight, R.; Roe, R.; Apperson, C.; Meshnick, S. Diversity of *Rickettsiales* in the microbiome of the lone star tick, *Amblyomma americanum*. *Appl. Environ. Microbiol.* **2014**, *80*, 354. [CrossRef]

5. Carpi, G.; Cagnacci, F.; Wittekindt, N.E.; Zhao, F.; Qi, J.; Tomsho, L.P.; Drautz, D.I.; Rizzoli, A.; Schuster, S.C. Metagenomic profile of the bacterial communities associated with *Ixodes ricinus* ticks. *PLoS ONE* **2011**, *6*, e25604. [CrossRef]

6. Ryo, N.; Takashi, A.; Ard, M.N.; Seigo, Y.; Frans, J.; Toshimichi, I.; Chihiro, S. A novel approach, based on BLSOMs (Batch Learning Self-Organizing Maps), to the microbiome analysis of ticks. *ISME J.* **2013**, *7*, 1003–1015.

7. Williams-Newkirk, A.J.; Rowe, L.A.; Mixson-Hayden, T.R.; Dasch, G.A. Characterization of the bacterial communities of life stages of free living lone star ticks (*Amblyomma americanum*). *PLoS ONE* **2014**, *9*, e102130. [CrossRef]

8. Swei, A.; Kwan, J.Y. Tick microbiome and pathogen acquisition altered by host blood meal. *ISME J.* **2017**, *11*, 813–816. [CrossRef]

9. Kaire, G.H. Isolation of tick paralysis toxin from *Ixodes holocyclus*. *Toxicon* **1966**, *4*, 91–97. [CrossRef]

10. Ross, I.C. Tick paralysis in the dog caused by nymphs of *Ixodes holocyclus*. *Aust. Vet. J.* **1932**, *8*, 102–104. [CrossRef]

11. Ross, I.C. An experimental study of tick paralysis in Australia. *Parasitology* **1926**, *18*, 410–429. [CrossRef]

12. Ross, I.C. Tick paralysis: A fatal disease of dogs and other animals in Eastern Australia. *J. Council Sci. Ind. Res.* **1935**, *8*, 8–13.

13. Storer, E.; Sheridan, A.T.; Warren, L.; Wayte, J. Ticks in Australia. *Aust. J. Dermatol.* **2003**, *44*, 83–89. [CrossRef]

14. Brown, A.F.; Hamilton, D.L. Tick bite anaphylaxis in Australia. *J. Accid. Emerg. Med.* **1998**, *15*, 111–113. [CrossRef] [PubMed]

15. Gofton, A.W.; Oskam, C.L.; Lo, N.; Beninati, T.; Wei, H.; McCarl, V.; Murray, D.C.; Paparini, A.; Greay, T.L.; Holmes, A.J.; et al. Inhibition of the endosymbiont "*Candidatus* Midichloria mitochondrii" during 16S rRNA gene profiling reveals potential pathogens in *Ixodes* ticks from Australia. *Parasits Vectors* **2015**, *8*, 345. [CrossRef] [PubMed]

16. Russell, R.C.; Doggett, S.L.; Munro, R.; Ellis, J.; Avery, D.; Hunt, C.; Dickeson, D. Lyme disease: A search for a causative agent in ticks in south-eastern Australia. *Epidemiol. Infect.* **1994**, *112*, 375–384. [CrossRef] [PubMed]

17. Roberts, F.H.S. *Australian Ticks*, 1st ed.; Commonwealth Scientific and Industrial Research Organisation: Melbourne, Australia, 1970; 267p.

18. Barker, S.C.; Walker, A.R. Ticks of Australia: The species that infest domestic animals and humans. *Zootaxa* **2014**, *3816*, 1–144. [CrossRef]

19. Burgdorfer, W. Discovery of the Lyme disease spirochete and its relation to tick vectors. *Yale J. Biol. Med.* **1984**, *57*, 515–520.

20. Burgdorfer, W.; Barbour, A.G.; Hayes, S.F.; Benach, J.L.; Grunwaldt, E.; Davis, J.P. Lyme disease—A tick-borne spirochetosis? *Science* **1982**, *216*, 1317–1319. [CrossRef]

21. Burgdorfer, W.; Barbour, A.G.; Hayes, S.F.; Peter, O.; Aeschlimann, A. Erythema chronicum migrans—A tickborne spirochetosis. *Acta Trop.* **1983**, *40*, 79–83.

22. Burgdorfer, W.; Lane, R.S.; Barbour, A.G.; Gresbrink, R.A.; Anderson, J.R. The western black-legged tick, *Ixodes pacificus*: A vector of *Borrelia burgdorferi*. *Am. J. Trop. Med. Hyg.* **1985**, *34*, 925–930. [CrossRef] [PubMed]

23. Patrican, L.A. Absence of Lyme disease spirochetes in larval progeny of naturally infected *Ixodes scapularis* (Acari: Ixodidae) fed on dogs. *J. Med. Entomol.* **1997**, *34*, 52–55. [CrossRef] [PubMed]

24. Burgdorfer, W.; Hayes, S.F.; Corwin, D. Pathophysiology of the Lyme disease spirochete, *Borrelia burgdorferi*, in Ixodid ticks. *Clin. Infect. Dis.* **1989**, *11*, S1442–S1450. [CrossRef] [PubMed]

25. Jacquet, M.; Genne, D.; Belli, A.; Maluenda, E.; Sarr, A.; Voordouw, M. The abundance of the Lyme disease pathogen *Borrelia afzelii* declines over time in the tick vector *Ixodes ricinus*. *Parasits Vectors* **2017**, *10*, 257. [CrossRef]

26. De Silva, A.M.; Fikrig, E. Growth and migration of *Borrelia burgdorferi* in *Ixodes* ticks during blood feeding. *Am. J. Trop. Med. Hyg.* **1995**, *53*, 397–404. [CrossRef]

27. Lane, R.S.; Piesman, J.; Burgdorfer, W. Lyme borreliosis: Relation of its causative agent to its vectors and hosts in North America and Europe. *Annu. Rev. Entomol.* **1991**, *36*, 587–609. [CrossRef]

28. Kumar, S.; Stecher, G.; Tamura, K. MEGA7: Molecular Evolutionary Genetics Analysis Version 7.0 for bigger datasets. *Mol. Biol. Evol.* **2016**, *33*, 1870–1874. [CrossRef]

29. Kemp, D.H. Identity of *Ixodes holocyclus* and other paralysis ticks in Australia. *Aust. Adv. Vet. Sci.* **1979**, *1979*, 71.

30. Eisen, R.J.; Eisen, L.; Beard, C.B. County-scale distribution of *Ixodes scapularis* and *Ixodes pacificus* (Acari: Ixodidae) in the Continental United States. *J. Med. Entomol.* **2016**, *53*, 349–386. [CrossRef]

31. Piesman, J.; Eisen, L. Prevention of Tick-Borne Diseases. *Annu. Rev. Entomol.* **2008**, *53*, 323–343. [CrossRef]

32. Piesman, J.; Oliver, J.R.; Sinsky, R.J. Growth kinetics of the Lyme disease spirochete (*Borrelia burgdorferi*) in vector ticks (*Ixodes dammini*). *Am. J. Trop. Med. Hyg.* **1990**, *42*, 352–357. [CrossRef] [PubMed]

33. Mather, T.N.; Nicholson, M.C.; Donnelly, E.F.; Matyas, B.T. Entomologic index for human risk of Lyme disease. *Am. J. Epidemiol.* **1996**, *144*, 1066–1069. [CrossRef] [PubMed]

34. Pepin, K.M.; Eisen, R.J.; Mead, P.S.; Piesman, J.; Fish, D.; Hoen, A.G.; Barbour, A.G.; Hamer, S.; Diuk-Wasser, M.A. Geographic variation in the relationship between human Lyme disease incidence and density of infected host-seeking *Ixodes scapularis* nymphs in the Eastern United States. *Am. J. Trop. Med. Hyg.* **2012**, *86*, 1062–1071. [CrossRef] [PubMed]

35. Stafford, K.C., III; Cartter, M.L.; Magnarelli, L.A.; Ertel, S.-H.; Mshar, P.A. Temporal correlations between tick abundance and prevalence of ticks infected with *Borrelia burgdorferi* and increasing incidence of Lyme disease. *J. Clin. Microbiol.* **1998**, *36*, 1240–1244. [CrossRef]

36. Loh, S.-M.; Gofton, A.W.; Lo, N.; Gillett, A.; Ryan, U.M.; Irwin, P.J.; Oskam, C.L. Novel *Borrelia* species detected in echidna ticks, *Bothriocroton concolor*, in Australia. *Parasits Vectors* **2016**, *9*, 339. [CrossRef]

37. Piesman, J.; Stone, B.F. Vector competence of the Australian paralysis tick, *Ixodes holocyclus*, for the Lyme disease spirochete *Borrelia burgdorferi*. *Int. J. Parasitol.* **1991**, *21*, 109–111. [CrossRef]

38. Collignon, P.J.; Lum, G.D.; Robson, J.M.B. Does Lyme disease exist in Australia? *Med. J. Aust.* **2016**, *205*, 413–417. [CrossRef]

39. Irwin, P.J.; Robertson, I.D.; Westman, M.E.; Perkins, M.; Straubinger, R.K. Searching for Lyme borreliosis in Australia: Results of a canine sentinel study. *Parasits Vectors* **2017**, *10*, 114. [CrossRef]

40. Budachetri, K.; Kumar, D.; Crispell, G.; Beck, C.; Dasch, G.; Karim, S. The tick endosymbiont *Candidatus* Midichloria mitochondrii and selenoproteins are essential for the growth of *Rickettsia parkeri* in the Gulf Coast tick vector. *Microbiome* **2018**, *6*, 141. [CrossRef]

41. Douglas, A. Nutritional interactions in insect-microbial symbioses: Aphids and their symbiotic bacteria *Buchnera*. *Annu. Rev. Entomol.* **1998**, *43*, 17–37. [CrossRef]

42. Duron, O.; Morel, O.; Noël, V.; Buysse, M.; Binetruy, F.; Lancelot, R.; Loire, E.; Ménard, C.; Bouchez, O.; Vavre, F.; et al. Tick-bacteria mutualism depends on B vitamin synthesis pathways. *Curr. Biol.* **2018**, *28*, 1896–1902. [CrossRef] [PubMed]

43. Machado-Ferreira, E.; Vizzoni, V.F.; Balsemão-Pires, E.; Moerbeck, L.; Gazeta, G.S.; Piesman, J.; Voloch, C.M.; Soares, C.A. *Coxiella* symbionts are widespread into hard ticks. *Parasitol. Res.* **2016**, *115*, 4691–4699. [CrossRef] [PubMed]

44. Lalzar, I.; Friedmann, Y.; Gottlieb, Y. Tissue tropism and vertical transmission of *Coxiella* in *Rhipicephalus sanguineus* and *Rhipicephalus turanicus* ticks. *Environ. Microbiol.* **2014**, *16*, 3657–3668. [CrossRef] [PubMed]

45. Andreotti, R.; de León, A.A.P.; Dowd, S.E.; Guerrero, F.D.; Bendele, K.G.; Scoles, G.A. Assessment of bacterial diversity in the cattle tick *Rhipicephalus (Boophilus) microplus* through tag-encoded pyrosequencing. *BMC Microbiol.* **2011**, *11*, 6. [CrossRef] [PubMed]

46. Sassera, D.; Beninati, T.; Bandi, C.; Bouman, E.A.P.; Sacchi, L.; Fabbi, M.; Lo, N. 'Candidatus Midichloria mitochondrii', an endosymbiont of the tick *Ixodes ricinus* with a unique intramitochondrial lifestyle. *Int. J. Syst. Evol. Microbiol.* **2006**, *56*, 2535–2540. [CrossRef] [PubMed]

47. Ahantarig, A.; Trinachartvanit, W.; Baimai, V.; Grubhoffer, L. Hard ticks and their bacterial endosymbionts (or would be pathogens). *Folia Microbiol.* **2013**, *58*, 419–428. [CrossRef] [PubMed]

48. Narasimhan, S.; Fikrig, E. Tick microbiome: The force within. *Trends Parasitol.* **2015**, *31*, 315–323. [CrossRef]

49. Chicana, B.; Couper, L.I.; Kwan, J.Y.; Tahiraj, E.; Swei, A. Comparative microbiome profiles of sympatric tick species from the Far-Western United States. *Insects* **2019**, *10*, 353. [CrossRef]

50. Thapa, S.; Zhang, Y.; Allen, M.S. Bacterial microbiomes of *Ixodes scapularis* ticks collected from Massachusetts and Texas, USA. *BMC Microbiol.* **2019**, *19*, 138. [CrossRef]

51. Trout Fryxell, R.T.; DeBruyn, J.M.; Stevenson, B. The microbiome of *Ehrlichia*-infected and uninfected lone star ticks (*Amblyomma americanum*). *PLoS ONE* **2016**, *11*, e0146651.

52. Van Treuren, W.; Ponnusamy, L.; Brinkerhoff, R.J.; Gonzalez, A.; Parobek, C.M.; Juliano, J.J.; Andreadis, T.G.; Falco, R.C.; Beati Ziegler, L.; Hathaway, N.; et al. Variation in the microbiota of *Ixodes* ticks with regard to geography, species, and sex. *Appl. Environ. Microbiol.* **2015**, *81*, 6200–6209. [CrossRef]

53. Beninati, T.; Riegler, M.; Vilcins, I.M.E.; Sacchi, L.; McFadyen, R.; Krockenberger, M.; Bandi, C.; O'Neill, S.L.; Lo, N. Absence of the symbiont *Candidatus* Midichloria mitochondrii in the mitochondria of the tick *Ixodes holocyclus*. *FEMS Microbiol. Lett.* **2009**, *299*, 241–247. [CrossRef] [PubMed]

54. Guizzo, M.G.; Neupane, S.; Kucera, M.; Perner, J.; Frantova, H.; Vaz, I.D.S.; De Oliveira, P.L. Poor unstable midgut microbiome of hard ticks contrasts with abundant and stable monospecific microbiome in ovaries. *Front. Cell. Infect. Microbiol.* **2020**, *10*, 211. [CrossRef] [PubMed]

55. Lewis, D. The detection of Rickettsia-like microorganisms within the ovaries of female *Ixodes ricinus* ticks. *Z. Parasitenkd.* **1979**, *59*, 295–298. [CrossRef]

56. Venere, M.; Fumagalli, M.; Cafiso, A.; Marco, L.; Epis, S.; Plantard, O.; Bardoni, A.; Salvini, R.; Viglio, S.; Bazzocchi, C. *Ixodes ricinus* and its endosymbiont *Midichloria mitochondrii*: A comparative proteomic analysis of salivary glands and ovaries. *PLoS ONE* **2015**, *10*, e0138842. [CrossRef]

57. Lo, N.; Beninati, T.; Sassera, D.; Bouman, E.A.P.; Santagati, S.; Gern, L.; Sambri, V.; Masuzawa, T.; Gray, J.S.; Jaenson, T.G.T.; et al. Widespread distribution and high prevalence of an alpha-proteobacterial symbiont in the tick *Ixodes ricinus*. *Environ. Microbiol.* **2006**, *8*, 1280–1287. [CrossRef]

58. Klubal, R.; Kopecky, J.; Nesvorna, M.; Sparagano, O.A.; Thomayerova, J.; Hubert, J. Prevalence of pathogenic bacteria in *Ixodes ricinus* ticks in Central Bohemia. *Exp. Appl. Acarol.* **2016**, *68*, 127–137. [CrossRef]

59. Gofton, A.W.; Doggett, S.; Ratchford, A.; Ryan, U.; Irwin, P. Phylogenetic characterisation of two novel *Anaplasmataceae* from Australian *Ixodes holocyclus* ticks: 'Candidatus Neoehrlichia australis' and 'Candidatus Neoehrlichia arcana'. *Int. J. Syst. Evol. Microbiol.* **2016**, *66*, 4256–4261. [CrossRef]
60. Silaghi, C.; Beck, R.; Oteo, J.; Pfeffer, M.; Sprong, H. Neoehrlichiosis: An emerging tick-borne zoonosis caused by *Candidatus* Neoehrlichia mikurensis. *Exp. Appl. Acarol.* **2016**, *68*, 279–297. [CrossRef]
61. Rynkiewicz, E.C.; Hemmerich, C.; Rusch, D.B.; Fuqua, C.; Clay, K. Concordance of bacterial communities of two tick species and blood of their shared rodent host. *Mol. Ecol.* **2015**, *24*, 2566–2579. [CrossRef]
62. Van Overbeek, L.; Gassner, F.; van der Plas, C.L.; Kastelein, P.; Nunes-da Rocha, U.; Takken, W. Diversity of *Ixodes ricinus* tick-associated bacterial communities from different forests. *FEMS Microbiol. Ecol.* **2008**, *66*, 72–84. [CrossRef]
63. Sperling, J.L.; Silva-Brandão, K.L.; Brandão, M.M.; Lloyd, V.K.; Dang, S.; Davis, C.S.; Sperling, F.A.H.; Magor, K.E. Comparison of bacterial 16S rRNA variable regions for microbiome surveys of ticks. *Ticks Tick Borne Dis.* **2017**, *8*, 453–461. [CrossRef] [PubMed]
64. Clow, K.M.; Weese, J.S.; Rousseau, J.; Jardine, C.M. Microbiota of field-collected *Ixodes scapularis* and *Dermacentor variabilis* from eastern and southern Ontario, Canada. *Ticks Tick Borne Dis.* **2018**, *9*, 235–244. [CrossRef]
65. Hawlena, H.; Rynkiewicz, E.; Toh, E.; Alfred, A.; Durden, L.A.; Hastriter, M.W.; Nelson, D.E.; Rong, R.; Munro, D.; Dong, Q.; et al. The arthropod, but not the vertebrate host or its environment, dictates bacterial community composition of fleas and ticks. *ISME J.* **2013**, *7*, 221–223. [CrossRef] [PubMed]
66. Rubin, B.E.R.; Sanders, J.G.; Hampton-Marcell, J.; Owens, S.M.; Gilbert, J.A.; Moreau, C.S. DNA extraction protocols cause differences in 16S rRNA amplicon sequencing efficiency but not in community profile composition or structure. *MicrobiologyOpen* **2014**, *3*, 910–921. [CrossRef] [PubMed]
67. Vishnivetskaya, T.A.; Layton, A.C.; Lau, M.C.; Chauhan, A.; Cheng, K.R.; Meyers, A.J.; Murphy, J.R.; Rogers, A.W.; Saarunya, G.S.; Williams, D.E.; et al. Commercial DNA extraction kits impact observed microbial community composition in permafrost samples. *FEMS Microbiol. Ecol.* **2014**, *87*, 217–230. [CrossRef] [PubMed]
68. Salter, S.J.; Cox, M.J.; Turek, E.M.; Calus, S.T.; Cookson, W.O.; Moffatt, M.F.; Turner, P.; Parkhill, J.; Loman, N.J.; Walker, A.W. Reagent and laboratory contamination can critically impact sequence-based microbiome analyses. *BMC Biol.* **2014**, *12*, 87. [CrossRef]
69. Folmer, O.; Black, M.; Hoeh, W.; Lutz, R.; Vrijenhoek, R. DNA primers for amplification of mitochondrial cytochrome c oxidase subunit I from diverse metazoan invertebrates. *Mol. Mar. Biol. Biotechnol.* **1994**, *3*, 294–299.
70. Kushimo, O.M. The Tick Genus *Amblyomma* in Africa: Phylogeny and Mutilocus DNA Barcoding. Master's Thesis, Georgia Southern University, Statesboro, GA, USA, 2013.
71. Tajima, F.; Nei, M. Estimation of evolutionary distance between nucleotide sequences. *Mol. Biol. Evol.* **1984**, *1*, 269–285.
72. Rzhetsky, A.; Nei, M. A simple method for estimating and testing minimum-evolution trees. *Mol. Biol. Evol.* **1992**, *9*, 945–967.
73. Kruskal, J.B.; Wish, M. Multidimensional scaling. In *Quantitative Applications in the Social Sciences*; Uslaner, E.M., Ed.; SAGE Publications Ltd.: Thousand Oaks, CA, USA, 1978; Volume 1, 96p.
74. Clarke, K.R. Non-parametric multivariate analyses of changes in community structure. *Aust. J. Ecol.* **1993**, *18*, 117–143. [CrossRef]
75. NSW Government. Coastal Management Act 2016 No. 20. Available online: https://legislation.nsw.gov.au/#/view/act/2016/20/full (accessed on 3 April 2020).

 *pathogens*

Review

Mosquito-Borne Diseases Emergence/Resurgence and How to Effectively Control It Biologically

Handi Dahmana [1,2] **and Oleg Mediannikov** [1,2,*]

[1] Aix Marseille University, IRD, AP-HM, MEPHI, 13005 Marseille, France; handy92@hotmail.fr
[2] IHU-Méditerranée Infection, 13005 Marseille, France
[*] Correspondence: olegusss1@gmail.com; Tel.: +33-(0)4-13-73-24-01; Fax: +33-(0)4-13-73-24-02

Received: 13 March 2020; Accepted: 21 April 2020; Published: 23 April 2020

Abstract: Deadly pathogens and parasites are transmitted by vectors and the mosquito is considered the most threatening vector in public health, transmitting these pathogens to humans and animals. We are currently witnessing the emergence/resurgence in new regions/populations of the most important mosquito-borne diseases, such as arboviruses and malaria. This resurgence may be the consequence of numerous complex parameters, but the major cause remains the mismanagement of insecticide use and the emergence of resistance. Biological control programmes have rendered promising results but several highly effective techniques, such as genetic manipulation, remain insufficiently considered as a control mechanism. Currently, new strategies based on attractive toxic sugar baits and new agents, such as *Wolbachia* and *Asaia*, are being intensively studied for potential use as alternatives to chemicals. Research into new insecticides, Insect Growth Regulators, and repellent compounds is pressing, and the improvement of biological strategies may provide key solutions to prevent outbreaks, decrease the danger to at-risk populations, and mitigate resistance.

Keywords: mosquito-borne disease; pest control; insecticide resistance; biological control; paratransgenesis; *Wolbachia*; *Asaia*; *Bacillus*

1. Introduction

The significant connection between fauna and flora in the world today is due to many factors, including the highest increase ever experienced in population growth accompanied by the evolution of transport systems. These factors disrupt biogeographic barriers and are followed by the first appearance of species in novel habitats [1,2]. In the Americas, incursions of these species are estimated to cause more than $120 billion in damage every year [3].

Deadly pathogens and parasites may be transmitted by arthropods [4], and the increasing global human and animal populations are threatened by such epidemics and pandemics [5]. Mosquitoes (Diptera: Culicidae) represent the most threatening vector due to their role in the transmission of dangerous pathogens [1]. Through trade and travel, key mosquito species are being introduced into novel habitats [2,6,7].

A number of chemical products formulated to provide a high safety profile are commercially available, but their toxicity to human skin and the nervous system can lead to several serious problems, such as rashes, swelling, and eye irritation [8]. The most important drawback of these products is the incidence of insecticide resistance, which has increased rapidly in recent years [9], and the extremely challenging or downright impossible task of finding and treating all mosquito breeding sites. New approaches and vector-control tools targeting aquatic stages and adults are urgently needed [10].

In this review, we discuss the current state of knowledge about mosquito-borne diseases and the latest figures from these resurgences, highlighting current techniques for their control and their limitations. We then focus on new innovative alternatives currently known but rarely used, others that are not used at all, and those that are still in the test or design phase but are very promising, which we suggest to be considered in the biological control of mosquito-borne diseases.

2. Resurgence of Diseases Transmitted by Mosquitoes

The three main mosquito genera, *Anopheles*, *Aedes*, and *Culex*, transmit the causative agents of numerous important diseases to humans as well as animals [11–14]. In this chapter, we briefly describe the resurgence of essential disease agents transmitted by mosquitoes and their impact on humans and animals.

Malaria is considered the most important parasitic disease of human beings and is currently endemic and transmitted by anopheline mosquitoes in more than 80 countries inhabited by approximately three billion people (Table S1; Figure 1), especially in sub-Saharan Africa, where more than 85% of cases and 90% of deaths occur, mainly in children younger than 5 years old. Malaria continues to cause phenomenal damage to public health (228 million cases worldwide, with 213 million (93%) reported in Africa alone, and severe outbreaks have recently ravaged many areas [15–19].

Wuchereria bancrofti and *Brugia* spp. can be transmitted by numerous mosquito species [13,20–25] (Table S1), and cause various clinical manifestations (25 million men with hydrocele and over 15 million people with lymphoedema) and at least 36 million people continue to have these chronic disease manifestations [26]. However, it is clear that eliminating lymphatic filariasis is not possible without controlling their vectors.

Dengue virus (DENV): Flaviviridae is responsible for dengue disease, caused by four distinct serotypes. Currently, it is the predominant arthropod-borne viral disease affecting humans [27], with 3.6 billion people living in areas at risk of transmission and hundreds of millions of dengue fever cases reported each year [28,29], causing ongoing epidemics in several countries [29,30] (https://www.outbreakobservatory.org/outbreak-thursday) (Table S1; Figure 1)

Zika virus (ZIKV): Flaviviridae also causes ongoing epidemics in several countries in Latin America and the Pacific [30–34] (https://www.who.int/emergencies/diseases/zika/en/) (Figure 1). *Aedes aegypti* is considered to be the primary vector associated with ZIKV outbreaks [35], while *Ae. albopictus* is considered a secondary vector [36]. However, several other species are also involved in the occurrence and transmission of this rapidly spreading virus [34,37,38] (Table S1). Currently, it is considered one of the most serious diseases threatening public health.

Chikungunya virus (CHIKV): Togaviridae is the causal agent of chikungunya fever (CHIKF) (Figure 1), known for producing an antalgic stance gait with severe articular pain [39]. Infected patients evolving to the chronic stage may range from 1.4% to 90% (52% in the American continent) [39]. Numerous outbreaks have recently been reported in several countries [30,40–42].

Yellow fever virus: Flaviviridae [43] is a haemorrhagic and potentially lethal RNA virus that causes outbreaks in several countries, especially in unvaccinated populations [44–48] (Table S1). Its emergence is cyclical, and outbreaks occur approximately 7–10 years apart [49]. In the summer of 2016, 47 countries declared YFV endemic, and 42 countries identified a risk of transmission, with 29 of them in Africa in 2017 [45]. With the highest fatality rate of up to 33.6%, numerous outbreaks continue to be registered [44,50]. Vaccination is safe, affordable, and the most effective way to prevent YF: "70 to 90 million doses are annually produced worldwide" [45].

Annually, the WHO reports approximately 67,000 cases of Japanese encephalitis, 20% to 30% of which are fatal, while 30% to 50% of survivors have significant neurological sequelae [51]. New strains genetically close to strains involved in previous outbreaks continue to be identified [52]. The St. Louis encephalitis virus was the major cause of epidemic encephalitis by an arbovirus in the USA [53]. It is re-emerging, causing numerous cases [54] (Table S1).

Similar to humans, horses are the domesticated animal that is most commonly affected by West Nile virus; 80% of cases are asymptomatic, while neurological signs are the most commonly reported symptom, with 90% of the 20% developing clinical signs, and the mortality rate may reach 30% [55]. Nevertheless, recent outbreaks in humans have been highlighted [56,57].

Different pathogenic blood-borne bacteria are regularly detected in mosquitoes [58,59]. It is not yet clear whether the presence of these bacteria in mosquitoes may be explained by occasional ingestion with blood meals or acquisition from the environment, or whether these bacteria may multiply and eventually be transmitted during blood meals. Different pathogenic alpha-proteobacteria, including *Anaplasma* spp., *Ehrlichia* spp., *Candidatus* Neoehrlichia, *Bartonella* spp., and *Rickettsia* spp., have been identified (xeno-monitoring studies) in adult mosquitoes [59,60]. More interestingly, the agent of febrile rickettsiosis, *Rickettsia felis*, has not only been identified in mosquitoes [58,61] but also shown to be potentially transmitted by *Anopheles* mosquitoes in laboratory experiments [62]. *Francisella tularensis* [63] is also carried by mosquitoes (*Aedes*), which act as a main vector in Sweden and Finland, making it the first reported mosquito-borne bacterium [63].

Several complex factors may explain the expansion of these diseases, such as population growth, globalisation of the economy, international travel (recreational, business, and military), inadequate vector-control efforts, limited access to good healthcare, rapid and unplanned urbanisation of tropical regions coupled with poor sanitary conditions, and a deterioration of public health infrastructures, all of which are related to climate change [64]; but, the major factors remain the mismanagement of insecticide use and the emergence of resistance.

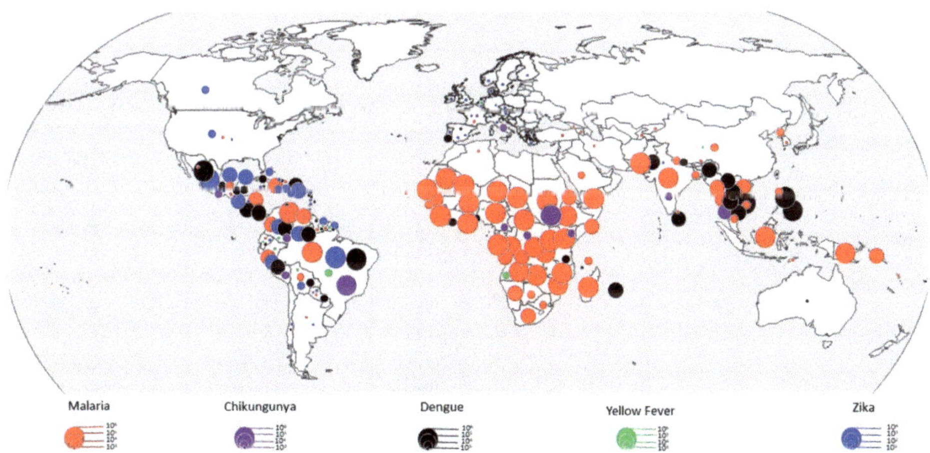

Figure 1. Cartography of significant resurgences of mosquito-borne diseases worldwide (until September 2019). We listed all reported outbreaks and imported and autochthon cases of malaria, dengue fever, yellow fever, chikungunya fever, and Zika fever between 2017 and 2019. This figure clearly shows their resurgence in almost all tropical countries. In many cases they were imported to several northern countries where the competent vector has become established, which may lead to potential local transmission.

3. Actual Insecticide-Based Vector-Control Strategies

The debate regarding dichlorodiphenyltrichloroethane (DDT) use for prevention, especially for malaria control, is polarised because it saved millions of lives worldwide but is unsafe. This has led to the invocation of precautions to enable choices to be made for healthier lives [65]. Some studies have focused on predicting mosquito abundance and assessing aquatic and adult mosquito control strategies [66], but despite the added efforts to develop new insecticides, other new alternative classes are slowly emerging [67,68].

3.1. Indoor Residual Spraying (IRS)

This is a well-developed and effective but potentially underused approach in vector control. It consists of treating the surfaces upon which common mosquitoes rest inside houses with a long-lasting insecticide. The most affected species among the endophilic species will be *Ae. aegypti*, which rests mainly indoors, feeds on humans, and is thus more likely to be reached by IRS than by space sprays [69]. IRS has some limitations and imperfections, such as the need for specialised training, which is time consuming in terms of obtaining public acceptance within a region. It does not prevent people from being bitten but above all, it must be adapted to several factors specific to a region, such as insecticide resistance, which is expensive and takes several years [70]. IRS has had a considerable impact on the mortality of *Ae. aegypti*, and used alone [71] or in combination with larval control [72] contributed to the elimination of *Ae. aegypti* in Guyana and the Cayman Islands, respectively [70]. In 2006, the WHO reaffirmed the importance of IRS for malaria transmission control, which was supported by the President's Malaria Initiative (PMI) in 2012 [73]. Malaria eradication campaigns using IRS in the Mediterranean region seem to have led to the elimination of malaria [74]. New formulations could last between five and eight months [70]. The potential evolution of insecticide resistance in the vector to pyrethroids can be controlled using alternative formulations, such as bendiocarb [75] and other new IRS formulations [67,73,76,77]. Good insecticide management is based on an alternation of formulations to combat the evolution of resistance, which may maintain efficacy over time, especially for location-specific interventions [70].

3.2. Peridomestic Space Spraying

This strategy is attractive because it is highly visible and conveys the message that health authorities use vector-control activities [78]. The risks to humans due to the management of adult mosquitoes are probably negligible [79]. This has no direct impact on immature stages (egg, larvae, or pupae) [80], targeting adult mosquitoes only, and is performed by spraying small droplets of insecticide into the air. It is used mainly in emergency situations to limit the massive production of adult mosquitoes, thus decreasing the risk of existing outbreaks expanding [78].

To perform this intervention, two forms of space sprays are commonly used for control: thermal fog and cold fog, also known as ultralow volume (ULV) sprays. Both can be distributed using a vehicle-mounted or hand-held machine [78]. The insecticide concentration ranges from 2% (pyrethroids) to 95% (organophosphates), depending on the amount of active ingredient in the formulation. The applied volume is dependent on the compound concentration and toxicity to the target species [80]. Aerial spraying of pyrethrin significantly impacts small organisms found in the sprayed zones, which is not the case on large bodies [81].

For dengue control, mosquitoes emerging after treatment can still be vectors because the viruses can be transmitted transovarially. Therefore, their exposure to successive treatments seems necessary and should be done at intervals shorter than the extrinsic incubation period of the virus [82].

A high resurgence of mosquitoes was reported after six days of ULV treatment as a single method in Thailand [81], while good results were observed with a decrease in the incidence of dengue fever after a large emergency vector-control campaign included several space sprays [83].

3.3. Long-Lasting Insecticide-Treated Nets (LLINs)

Designed as a solution to the problems of conventional insecticide-treated nets (ITNs), and based on novel fabric technologies [84], LLINs were developed to resist multiple washes and remain effective for a prolonged time (at least three years). LLINs are considered one of the most successful mosquito control tools, especially for malaria prevention [85]. ITNs with synthetic pyrethroid insecticides either incorporated into or coated around their fibres have resulted in a considerable decline in malaria morbidity and mortality in several countries, especially in sub-Saharan Africa, where over 427 million nets were delivered between 2012 and 2014 [85]. Th annual cost of an LLIN can be as high as US$2.6, while IRS costs about US$4, and standard ITN costs range from US$1.5 to US$6 [85]. The level of use of LLINs varies according to several factors, such as temperature, humidity, season, and, especially, the density of mosquitoes, and access to them plays a major determinant of their use [86,87]. LLINs have contributed to the reduction in malaria over the past 15 years, combined with other new control measures, such as IRS and artemisinin-based combination therapies [88] in children and pregnant women [87,89]. Other important advantages of LLINS include reduced consumption of insecticides and insecticide released into the environment because they do not need retreatment [85]. The efficacy of LLINs is closely related to the molecules used (the choice depends on the presence or absence of its resistance) [90], and their correct use may enhance their efficiency [91]. In a study, the use of LLINs led to a dramatic reduction (97%) in the prevalence of malaria compared to a group of LLIN non-users [92].

3.4. Mosquito Repellents

Mosquitos are mostly attracted to humans by the lactic acid and CO_2 present in our sweat that are detected by chemoreceptors present in their antennae, and repellents mask the human scent [8]. DEET (N,N-diethyl-meta-toluamide) is the most widely used and effective repellent against mosquitoes [93].

Biobased mosquito repellents are pest management tools that are based on safe, biologically based active ingredients derived from plants [94,95], fungi [96], or bacteria [93].

In terms of the effective control of mosquitoes and to ensure human and environmental safety where endemic mosquito resistance and environmental concerns limit the use of products, the use of biobased natural mosquito repellents is preferable to that of chemical repellents [8].

The most effective synthetic repellents are DEET (N,N-diethyl-m-toluamide) and IR3535 (3-(NButyl-N-acetyl)-aminopropionic acid [97]. Several nanoparticles synthesised and successfully impregnated into cotton fabrics in insect-repellent clothing show high efficacy against mosquito larvae and adult populations, which gives them the potential to be used as eco-friendly approaches to control mosquitoes if applied in long-lasting insect-repellent clothing [98,99]. The fact that the use of synthetic repellents causes insecticide resistance in mosquitoes, has a harmful effect on non-target organisms, and threatens the environment has led to widespread discussions around this method of control [97].

The increased involvement of governments and authorities on scientific projects coupled with correct individual action may help to combat the spread of mosquito-borne diseases and limit their devastating transmission.

4. Biological Control

Every year, promising new "eco-friendly" compounds are developed to progressively replace the oldest compounds, which are the most toxic and harmful. The use of biological control programmes, such as genetic modification or biological agents such as predatory fish, bacteria, protozoa, nematodes, and fungi, have rendered some promising results.

4.1. Genetic Modification

The sterile insect technique (SIT) is a species-specific and environmentally benign method for insect population control based on mass rearing, radiation-mediated sterilisation, and the release of a large number of male insects into a given target area, which compete for mates with wild males. A wild female mating with a released sterile male has no or fewer progeny, so the population tends to decline [100–103], which was an improvement on RIDL (release of insects carrying a dominant lethal gene). The lethal dominant gene could be controlled by a female-specific promoter and its expression could be inactivated by antibiotic treatment (tetracycline), allowing the mosquito-colony to be maintained. When male and female separation is required, the antibiotic is removed from the system, causing the death of all females [10]. Some projects cost approximately US$1.1 million [104], and some reports of failure have been published [105]. Mosquito egg production and mass rearing problems were also highlighted [106,107].

Paratransgenic strategies based on genetically modified symbiotic bacteria reintroduced in mosquitoes reveal a very high potential of casually controlling all-important mosquito species, including *Culex*, which is difficult to transform [14]. New studies on RNAi-based bioinsecticides (RNA interference) show promising results [108].

4.2. Fungi

Particular attention has been paid to fungal species belonging to the genera *Lagenidium*, *Coelomomyces*, *Entomophthora*, *Culicinomyces*, *Beauveria*, and *Metarhizium* for their power to reduce mosquito populations, but unfortunately, none of them have been specifically adapted as larvicidal agents against important vector species [109–111], even transgenic ones [112]. Application to surfaces on which mosquitoes land or need to pass through, such as fungus-impregnated cloths around bed nets, attractive bait stations, and adult mosquito traps and PET traps, show promising results, with a 39–50% reduction in survival rates of malaria-carrying mosquitoes and elimination of 95% of *Anopheles arabiensis* mosquitoes in a bait station [113]. One of the most effective fungi studied recently against simultaneously *Ae. albopictus* and *Cx. pipiens* mosquito adults is *Beauveria bassiana*. The production and persistence of its conidia was remarkably high [109].

4.3. Control of Aquatic Stages Using Elephant Mosquito and Fish Predators

The use of fish to control the aquatic stages of mosquitoes was an important tool in the pre-DDT era. These fish were introduced into all potential mosquito-breeding habitats and their use decreased after the introduction of DDT and then was rekindled after the development of resistance and harmful effects [114]. The use of indigenous larvivorous fishes is suggested [115], and a limited number of species are used, primarily *Gambusia affinnis* and *Poecilia reticulata*, although several failures have been reported in the literature [114]. Other aquatic predators may play a role in reducing mosquito populations, especially in rainy periods [116,117], and the combination of multiple predators can reduce mosquito populations [118].

The naturally occurring non-biting *Toxorhynchites* species, which exhibit predatory behaviour during their larval stages, have been explored for their potential use as biological control alternatives to chemical insecticides (the 4th instar larva is the most predaceous) [119,120]. Important progress was made concerning their production for use as biological agents and they demonstrated remarkable effectiveness against numerous mosquito species, such as *Ae. aegypti*, *Ae. albopictus*, and *Cx. quinquefasciatus* [119,121]. In certain situations, they have demonstrated practical potential, but their use continues to be limited by several problems, such as cannibalism during the early instars, temperature (limited by low temperatures), and also the inadequate overlap in the larval habitats between the prey and the predator mosquito [120].

4.4. Protozoan Control

Chilodonella uncinata is a protozoan parasite with many beneficial properties associated with a good microbial pathogen [122]. It causes low to very high (25–100%) mortality in mosquito larvae. It exhibits high virulence and resistance to desiccation and also demonstrates a high reproductive potential when cultured in vitro. Through its mosquito host, *C. uncinata* has the ability to spread in nature by the way of transovarian transmission [122].

5. Bacterial Agents Tested or Used in Control Strategies

Most of the attention of pest control scientists focuses on bacterial agents targeting both aquatic and adult stages. Several studies have shown their efficiency, and their use is recommended by the WHO. Here, we list some bacterial agents currently in use or undergoing tests with promising results.

5.1. Bacillus spp.

Before the discovery of *Bacillus thuringiensis israelensis* (*Bti*) (Figure 2) and *Bacillus sphaericus* (*Bs*) (Figure 3), little attention was paid to bacteria as sources of agents for microbial control of mosquitoes. Around 1500 microorganisms were recently identified as good potential insecticidal agents, and looking for insecticidal activity, metabolites from approximately a thousand microbial isolates were examined [123]. *Bti* formulations are the predominant nonchemical means employed for controlling mosquito larvae [124]. In addition, several studies indicate the highly effective and safe use of individuals or the mixture of *Bti* and *Bs* for mosquito control, and they are considered to be safe to non-target organisms cohabiting with mosquito larvae [125].

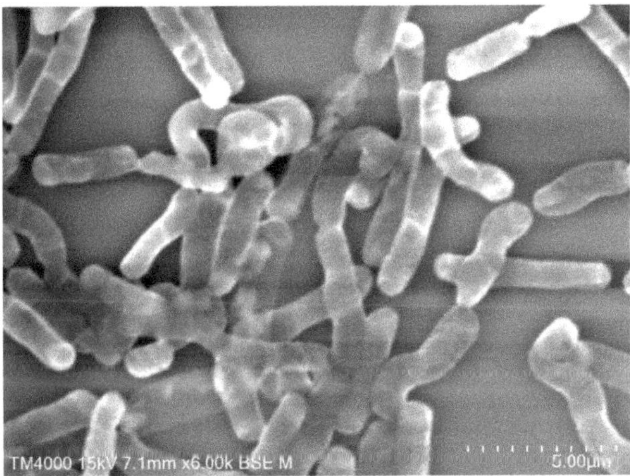

Figure 2. *Bacillus thuringiensis israelensis* (*Bti*), 3 days of culture, 4.9 μm in length (Hitachi TM4000) (personal image).

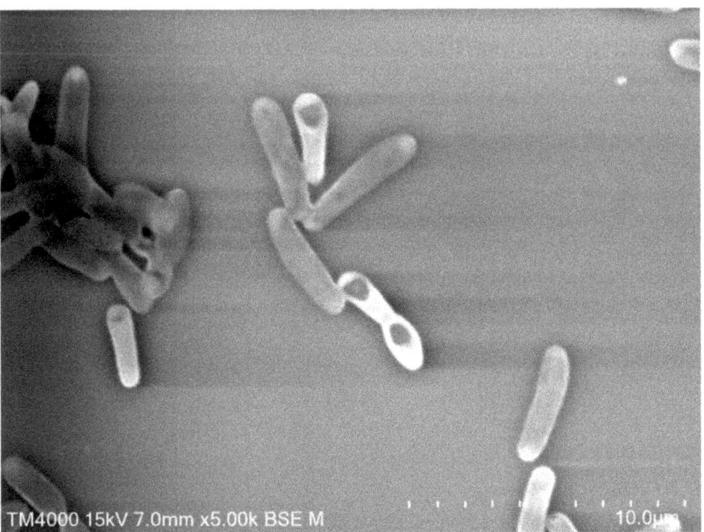

Figure 3. *Lysinibacillus sphaericus* (*Bs*) strain CSURP827, 3 days of culture, 5.02 μm in length (Hitachi TM4000) (personal image).

B. thuringiensis produces three classes of larvicidal proteins: *Cry* (exert intoxication through toxin activation, receptor binding, and pore formation in a suitable larval gut environment), *Cyt* (cytolytic toxicity) when sporulating (parasporal crystals), and *Vip* proteins throughout the vegetative phase (ionic, non-ionic detergents and pore-forming mechanisms of action were suggested), some of which are toxic against a wide range of insect orders, nematodes, and human-cancer cells. This has been widely employed as an effective biopesticide to control pests that are harmful to crops, forests, and humans. *Cyt* toxins possess less toxicity against mosquito larvae than *Cry* toxins [126,127].

Several species of *B. thuringiensis* exhibit a high mortality rate toward all mosquito larval instars, such as *Bti* [128], *B. thuringiensis* var. *krustaki* [129], *B. thuringiensis* var. *jegathesan* (*Btjeg*) [130], *B. thuringiensis* var. *kenyae*, and *B. thuringiensis* var. *entomocidus* [131]. Other species with homology to *Bacillus* show the highest toxicity against dipterans, such as *Clostridium bifermentans* (serovar *malaysia*) [132], *B. circulans* [133], and *B. laterosporus* [134,135]. *Bacillus* spp. remains a massive source of active compounds against pests, which are currently being explored to fill public health needs.

5.2. Insect Growth Regulators (IGRs)

Due to several advantages, such as low toxicity to the environment and selectivity, IGRs present an effective tool to control mosquito populations. They are substances that are analogues or antagonists of hormones and interfere with insect development [136]. There is growing interest in the use of IGRs, such as methoprene and pyriproxyfen, two juvenile hormone agonists belonging to IGR insecticides. They are effective against mosquito larvae and may inhibit the emergence of adults [137]; others include novaluron and diflubenzuron [138] for mosquito control [139]. Numerous recent studies have highlighted that mosquitoes and other pests have developed resistance to commonly used IGRs, such as methoprene and pyriproxyfen [140–142], which reinforces the need to develop new compounds and identify new targets in mosquitoes [143].

5.3. Wolbachia spp.

Mosquito symbiont-associated bacteria may exert a pathogenic effect on their host, interfering with its reproduction and also reducing vector competence [144]. *Wolbachia* are endosymbiotic bacteria that naturally infect approximately 40% of insect species [145,146], and *Wolbachia pipientis* is a unique valid species of the genus [147]. They are present in some major mosquito disease vectors, such as *Cx. quinquefasciatus*, *Ae. albopictus*, and anopheline species, including malaria vectors such as *An. gambiae* and *An. coluzzii* but never *Ae. aegypti* [144,145,148–150]. This maternally transmitted bacterium allowing the invasion of host populations can induce feminisation of males (turning genetic males into females), parthenogenesis (reproduction without males) [144], and cytoplasmic incompatibility, leading to the generation of inviable offspring when a *Wolbachia*-infected male mates with an uninfected female, but not in the contrary case [145]. Successfully used in Myanmar in the 1960s to eradicate *Cx. quinquefasciatus* [151], it is currently also being used to target *Ae. albopictus*, using a triple *Wolbachia*-infected strain [152], and to target *Ae. polynesiensis* (2012) [153]. To date, it has been used in several countries, such as Australia, Brazil, Indonesia, Vietnam, and Colombia. The fear of resistance to the inhibitory effect of *Wolbachia* has been highlighted, but no studies have demonstrated that this scenario is likely to happen, and the creation of *Wolbachia*-superinfected lines, such as *Ae. aegypti* with stable infection, could help to mitigate potential resistance [145,154] and add to their role in reducing vector competence. Studies have reported that *Wolbachia* inhibits the transmission of CHIKV [155], YFV [156], malaria parasites in *An. stephensi* [157] and *An. gambiae* [158], DENV [159], and ZIKV [160]. Recent reviews clearly explain *Wolbachia* as a form of biological control [144,161,162].

Wolbachia-based control constitutes a potentially promising strategy for the control of mosquitoes and their transmitted diseases that urgently needs to be considered and associated with biological control programmes in countries suffering from malaria and arbovirus outbreaks.

5.4. Asaia

To make malaria vectors inefficient, interruption of the cycle within the vector to stop parasite development before the *Anopheles* host becomes infective is a good solution [163]. The simplest approach to this is paratransgenesis, consisting of producing bacterial strains that are able to both live in the midgut of various mosquito species and spread rapidly among wild mosquito populations [164]. Several studies have been performed on the identification and use of competent microorganisms to combat vector-borne diseases [165]. The genus *Asaia*, first discovered in plant nectar, is an excellent candidate [166]; it is localized in many organs of mosquitoes, and can disperse inside the mosquito body through the haemolymph [165,167]. Its distribution in the mosquito population is made possible through several mechanisms (co-feeding, sexual mating, paternal, maternal, and horizontal transmission) [168–170]. *Asaia* bacteria may be genetically modified in order to be recolonised in a new host, resulting in spread within wild populations [166]. Recently, it was isolated and characterised from several *Anopheles* species, which would be beneficial if applied toward achieving paratransgenesis against malaria [165]. Advanced studies recently showed that *Asaia* may activate the mosquito's immune system, leading to a reduction in the development of malaria parasites [171]. In the future, additional assets to which the bacterium may be used in mosquito control may be identified because it seems that *Asaia* plays a key role in the health of the mosquito host, even during its larval stage, allowing the larvae to develop rapidly [172]. Engineering of *Asaia* to produce an antiplasmodial effector causing the mosquito to become refractory to *Plasmodium berghei* is a perfect demonstration of the power of a transgenic microbiota [173], which makes it beneficial to microbial ecology and a potential candidate not only for paratransgenesis but also for general control of mosquitoes and mosquito-borne diseases.

5.5. Spinosyns

Spinosad is a biopesticide derived via fermentation from an actinomycete, *Saccharopolyspora spinosa*, a naturally occurring soil-dwelling bacterium. It contains two insecticidal factors, A ($C_{41}H_{65}NO_{10}$) and D ($C_{42}H_{67}NO_{10}$) [174,175]. It is categorised as a Group 5 insecticide by the Insecticide Resistance Action Committee (IRAC), forming a new class of polyketide-macrolide insecticides that act as nicotinic acetylcholine receptor (nAChR) allosteric modulators. Discovered in the 1980s in an early-stage insecticide screen that included *Ae. aegypti*, it was shown to be highly active against numerous pests in the Lepidoptera, Diptera, Thysanoptera, Coleoptera, Orthoptera, and Hymenoptera orders, and others. Its application to mosquito control is relatively new due to its pesticidal activity after ingestion and cuticle absorption and its highly favourable toxicology profiles in mammals and the environment [176]. It was also recently approved for use as a mosquito larvicide in human drinking water sources and containers [177]. Its applications in natural habitats are too few, but in laboratories it has been demonstrated to be very efficient at preventing and reducing larval development in important medical and veterinary vector species, such as *Ae. aegypti*, *Ae. albopictus* (Figure 4), *Anopheles gambiae* (Figure 5), *An. pseudopunctipennis*, *An. albimanus*, *Cx. pipiens* (Figure 6), *Cx. quinquefasicatus* [175,178], and some anopheline species [179,180].

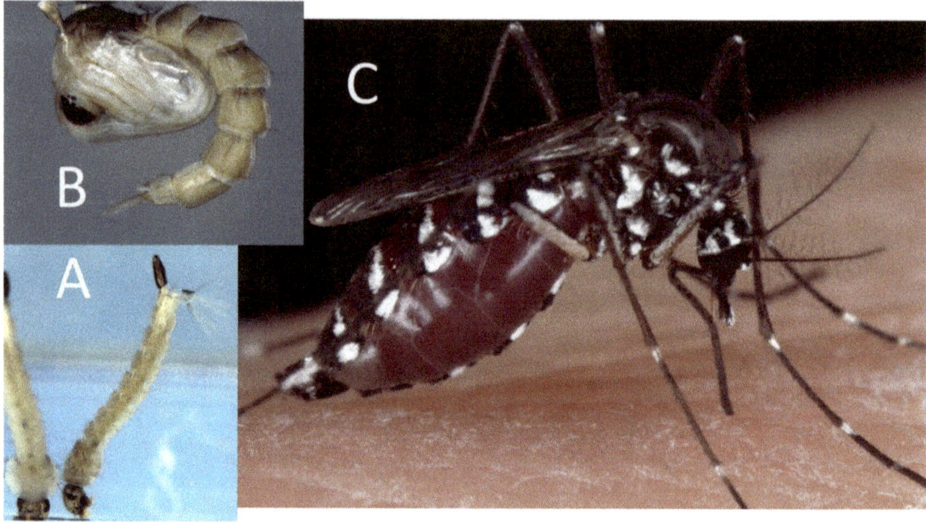

Figure 4. *Aedes albopictus* strain: (**A**) larvae (personal images), (**B**) pupa (personal images), and (**C**) adult (personal images).

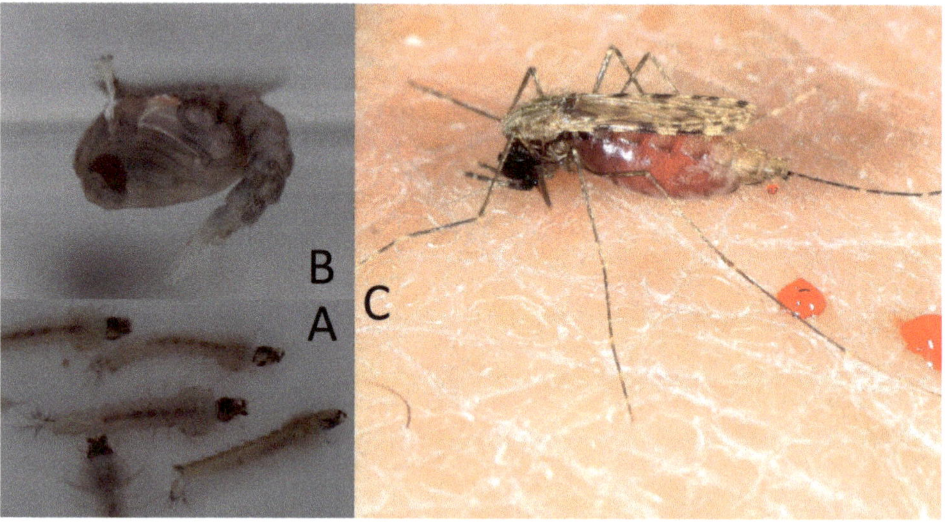

Figure 5. *Anopheles gambiae*: (**A**) larvae (personal images), (**B**) pupa (personal images), and (**C**) adult (personal images).

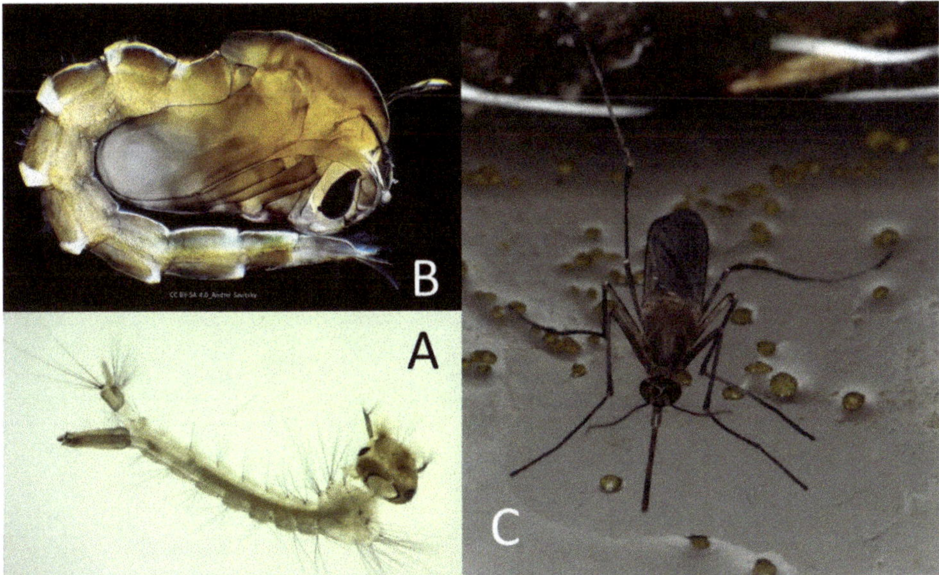

Figure 6. *Culex pipiens*: (**A**) larvae (personal images), (**B**) pupa (CC BY-SA 4.0_Andrei Savitsky), and (**C**) adult (personal images).

As the best solution, biological control requires several components for the design of effective plans for mosquito control, and spinosad, which has no resistance because it was introduced recently into control programmes, will play a very important role.

5.6. Bacterial-Based Feeding Deterrents and Repellents

The fear of the occurrence of possible side effects of DEET, such as toxic encephalopathy, seizures, acute manic psychosis, cardiovascular toxicity, and dermatitis [181], as well as potential resistance that has become a reality with *Ae. aegypti* mosquitoes [182] and *An. gambiae* [183], has led to the use of innovative technologies to create other products free of DEET that are marketed in the form of sprays or creams and include other active ingredients [184], such as picaridin [185] and IR 3535 [186], and a wide range of essential oils, which synergistically use various components and have been reported to provide a higher repellent activity than single isolated components [187]. Recently, a mixture of compounds isolated from *Xenorhabdus budapestensis* (entomopathogenic-associated bacteria) exhibited potent feeding-deterrent activity against three mosquito species considered to be the most important vectors of diseases affecting public health. They belong to the fabclavine class and exhibit a high activity comparable to or better than that of DEET or picaridin in side-by-side assays [93], which supports the attempt to replace toxic molecules by considering bacteria as a very promising source of new alternative molecules for exploitation as mosquito repellents.

6. Biological Insecticide Resistance

In view of their efficacy and safety, the importance of bacterio-insecticides seems to be increasing in insect control activities, which has led researchers to investigate and characterise new bacterial strains with insecticidal properties and identify their active compounds.

6.1. Resistance to Bti

Numerous factors, such as wild proliferation or environmental accumulation, as well as the persistence of human-spread *Bti* in treated larvae breeding sites, may lead to a long exposure time of insects, which may increase the risk of acquiring resistance in target insects and also have a negative impact on non-target insects [188]. A study showed that a high resistance to each individual *Bti* toxin can be obtained under some conditions in the laboratory after only a few generations of selection, and this resistance seems to be lowest for commercial and environmental *Bti*, which might act as a first step in resistance to a complete *Bti* toxin mixture (Table 1). Studies reporting that individuals show resistance to one toxin but not to another suggest that different resistance mechanisms exist [189]. The mechanisms of resistance to *Bti* Cry toxins are widely studied in *Culex* and *Aedes* species [176,190–194], and marginal cross-resistances have been identified [193,195]. To date, resistance of malaria-carrying species to *Bti* has not been found [196].

Table 1. Highlights of field and laboratory insecticide resistances to *Bti* and *Bs*.

Bacteria	Mosquito	Site	Type of Study	Number of Studied Regions	Date	Reference
Bti + *Bs*	*Culex pipiens*–complex	Onondaga County, USA	Field	2	June 2003	[140]
Bti	*Ochlerotatuscataphylla*	Rhône-Alpes, France	Field	4	April 2003	[197]
Bti	*Aedes rusticus*	Rhône-Alpes, France	Field	13	Winters 2005 and 2006	[198]
Bti	*Culex quinquefasciatus*	USA	Laboratory	1	Summer 1990	[191]
Bti	*Aedes aegypti*	USA	Laboratory	1	2011	[192]
Bs	*Culex pipiens*–complex	Utah, USA	Field	3	September 2016	[199]
Bti	*Aedes aegypti*	France	Laboratory	1	2010	[189]

6.2. Resistance to Bs

B. sphaericus (*Bs*) is found in numerous habitats, especially in soils and aquatic habitats. It is known as a producer of a characteristic spherical spore inside the swollen sporangium. Over the past 25 years, scientists had much interest and focused on isolating numerous strains because of their potential use as mosquito larvicides [200]. Several formulations used in biocontrol are highly effective against mosquitoes [125]. Recently renamed *Lysinibacillus sphaericus* (2007) [201], numerous studies have reported various levels of resistance to *Bs* in laboratory and field populations from different countries [176] (Table 1), mostly on *Culex* populations. Different mechanisms [202] have been observed in numerous locations, such as France, China, India, and Brazil [126,203]. If mosquitoes develop resistance to one strain of *Bs*, it appears that they will develop resistance to other *Bs* strains due to the similarity of the binary toxins in most strains; but, they remain susceptible to *Bti* [176]. Resistant strain fitness was found to be heavily impacted, especially fecundity and fertility, which became very low in a study [204,205], although opposite results were achieved in another study [206].

Although it has been tested widely for controlling malaria vectors [207–210], no laboratory or field resistance has been highlighted for *Anopheles* species to date.

6.3. Resistance Management

When it comes to managing the rapid increase in insecticide resistance [9], *Bti* can be used as a powerful tool to mitigate resistance to *Bs* in mosquitoes, although it has been reported that using them in rotation or in a mixture leads to a steady decline in resistance over 30 generations. They can also delay or prevent the emergence of resistance due to the synergistic action between their toxins, and recent formulations have shown greater larvicidal activity and efficacy [176]. Other combinations with botanical pesticides are considered alternatives to mitigating the development of resistance to *Bs* in mosquitoes [211].

The best solution for the management of insecticide resistance is to systematically replace most of the molecules used in chemical control with eco-friendly biological control, but that strategy will depend on the plans conceived, the number of molecules chosen and the associations between the various formulations to prevent and reduce current resistance and avoid the appearance of new resistance.

7. Current Challenges for a Prosperous Future

In view of the current situation and the failures that have been experienced with mosquitos invading new territories associated with devastating outbreaks, new tools, molecules, plans, synergistic associations, and methods of mosquito control are being developed to facilitate strategic objectives, such as protecting at-risk populations, especially in endemic areas; preventing the international spread of mosquitoes and the diseases they carry; and rapidly containing epidemics. Some strategies are in the stage of preliminary testing or in the validation phase and others have recently been introduced into use.

7.1. New Insecticide, IGR, and Repellent Compounds

The most urgent need is to develop new insecticides to fight mosquito-borne diseases due to their crucial efficiency and their economic importance [212]. The Innovative Vector Control Consortium (IVCC) has released new product classes, especially for malaria eradication, and manages international efforts to establish new methods, including producing a new ATSB (attractive targeted sugar bait) product class and programming next-generation IRS projects [213]. Due to their eco-friendly properties and efficiency, entomopathogenic Ascomycete fungi have been suggested for the control of both larval and adult stages of dengue vectors [12,110]. Several other bacteria showing promising results on numerous pests have been suggested to have the same effect on mosquitoes, such as the entomopathogenic nematode-associated bacteria *Xenorhabdus* sp. [93,214]; *Serratia marcescens*, which is often associated with insect infection and shows high insecticidal effects alone [215] or when associated with other insecticides [216]; and entomopathogens [217]. Other bacteria exhibiting toxic effects on mosquitoes, such as *Clostridium bifermentans* [132], may also be considered in control strategies.

Recently, a new compound class, chalcones with JHAN activity, showed impressive insecticide and IGR activity when tested against *Ae. albopictus* larvae and could be useful for the development of environmentally benign IGR insecticides to control mosquitoes [143]. Moreover, the beneficial effects of diterpene and their derivatives as well as their potential use as biological alternatives in dengue fever control has been highlighted [218].

Auto-dissemination is a phenomenon where the dispersal and transfer of active compounds is carried out by contaminated adult mosquitoes to treat undeveloped habitats that are difficult to locate and treat [219]. It can occur through treated materials or dissemination stations, such as modified ovitraps, and can also be combined with other methods, such as SIT [12], which may increase their effectiveness.

7.2. Attractive Toxic Sugar Baits (ATSB)

Bait aims to attract mosquitoes in order to feed them on toxic sugar meals broadly sprayed on plants or placed in bait stations [220,221]. They show the highest efficacy in laboratory and field studies [222] against *Aedes* species, culicines, and sand flies [12].

Whether for indoor or outdoor control, ATBS can reduce mosquito populations through direct mortality caused by feeding them on insecticide-treated bait but also through the spread of mosquito pathogens or non-chemical toxins [223]. Developing mosquito-specific attractants to avoid their effects on non-target species make baits one of the best solutions, and their combination with other strategies, such as genetic ones, will maximise their effectiveness.

7.3. Parasitic Nematodes

Lot of nematodes belonging to numerous orders and families are known to be parasites of insects [224]. Some insect parasitic nematodes that are specific to mosquitoes [225] may be considered alternatives to chemical insecticides [226]. When tested, they were effective against malaria vectors and several other important mosquito species, such as *Ae. aegypti, Ae. albopictus, Cx. quinquefasciatus,* and *An. gambiae* [226–229]. As they are naturally adapted to their host, such nematodes are highly specific to theirs hosts, which they can kill by producing high levels of parasitism. They are free swimming and disseminate easily in the infective stage [225], and species such as *Romanomermis iyengari* are widely suggested to be a component of integrated mosquito control programmes in lymphatic filariasis endemic countries [229].

7.4. Acoustic Larvicides and Traps

These are emerging technologies designed to combat the aquatic stages of mosquitoes by killing them with sound waves resulting in instantaneous mortality or inhibited emergence. They have proven to be effective as a beneficial non-chemical alternative for the treatment of drinking water supplies [230]. This approach has been shown to be highly effective in a range of typical volumes found in peri-domestic water containers [230] without causing resistance within mosquito populations or harming non-target organisms when used properly [231]. Furthermore, even simple and cheap mobile phones can sensitively acquire acoustic data on the species-specific level of adult wingbeat sounds. This makes it possible to simultaneously record the time and location of the encounter between humans and mosquitos, which forms a powerful tool for acoustically mapping mosquito species distribution worldwide [232]. Other innovative acoustic-based tools have been developed to control mosquitoes during rear-and-release operations, such as the low-cost and battery-powered sound-baited gravid *Aedes* trap, which may be an effective replacement for the costly Biogents Sentinel (BGS) trap [233].

7.5. Advanced Genetic Studies

Recently, a new RNAi-based bioinsecticide was developed from D-RNA molecules, which was subsequently tested on *Aedes* larval breeding water [108]. A significant reduction in the viability of the larvae treated with dsRNA was reported while in the surviving larvae and adults, altered morphology and chitin content was observed. In combination with diflubenzuron, this innovative bioinsecticide had insecticidal adjuvant properties [108].

In another study, a considerable reduction in the fertility of *Ae. aegypti* adult males was observed when feeding their larvae double-stranded RNAs (dsRNAs) targeting testis genes. Moreover, several dsRNAs were reported to be inducing males and were remarkably effective in competing for mates. RNAi-mediated knockdown of the female-specific isoform of double-sex was also effective in producing a highly male-biased population of mosquitoes, making it possible to overcome the need to sex-sort insects before release [234].

8. Conclusions

Despite currently deployed methods, epidemics and the spread of mosquito-borne diseases continue as a result of a range of complex reasons, including insecticide resistance, inappropriate design of control programmes, ineffective coverage, missing and poorly trained manpower, as well as a lack of financial resources and infrastructure [12].

Many strategies have been designed for the control of mosquito-borne diseases, each with their strengths and weaknesses. However, approaches such as integrated vector management that adopts receding horizon control strategies, which may consider multiple objectives, seem to provide optimal control solutions that are fast and sustainable but that also offer the most cost-effective control choices [235].

Improving current strategies, such as the sterile insect technique, the release of insects with dominant lethality, or transgenesis, may provide key solutions to preventing outbreaks, decreasing the danger to at-risk populations and mitigating resistance. Meanwhile, promising techniques, such as those discussed in this manuscript, have already proven their effectiveness but remain under-used and require more attention and consideration in vector-control plans.

Supplementary Materials: The following are available online at http://www.mdpi.com/2076-0817/9/4/310/s1, Table S1: Updates concerning important mosquito borne diseases. We listed most of the mosquito-borne diseases, including their actual distribution, transmission, natural occurrence or animal infection, and virulence as well as the existence or absence of treatments or vaccines to date.

Author Contributions: H.D.: Investigation; software; writing—original draft preparation; O.M.: Conceptualisation; methodology; validation; writing—review and editing; supervision. All authors have read and agreed to the published version of the manuscript.

Funding: This research received no external funding.

Acknowledgments: We thank the Editor-in-Chief of the Journal of Pest Science for the invitation to submit a review article. We thank Jean Michel Beranger for providing us with high-quality pictures of insects from our insectary.

Conflicts of Interest: The authors declare no conflict of interest.

References

1. Benelli, G. Research in mosquito control: Current challenges for a brighter future. *Parasitol. Res.* **2015**, *114*, 2801–2805. [CrossRef] [PubMed]
2. Kilpatrick, A.M. Globalization, land use and the invasion of West Nile virus NIH Public Access. *Science* **2011**, *334*, 323–327. [CrossRef] [PubMed]
3. Pimentel, D.; Zuniga, R.; Morrison, D. Update on the environmental and economic costs associated with alien-invasive species in the United States. *Ecol. Econ.* **2005**, *52*, 273–288. [CrossRef]
4. Mehlhorn, H. *Encyclopedia of Parasitology;* Institut für Zoomorphologie, Zellbiologie und Parasitologie: Düsseldorf, Germany, 2016; Volume 1, ISBN 978-3-662-43977-7.
5. Mehlhorn, H.; Al-Rasheid, K.A.S.; Al-Quraishy, S.; Abdel-Ghaffar, F. Research and increase of expertise in arachno-entomology are urgently needed. *Parasitol. Res.* **2012**, *110*, 259–265. [CrossRef]
6. Hubálek, Z.; Halouzka, J. West Nile fever—A reemerging mosquito-borne viral disease in Europe. *Emerg. Infect. Dis.* **1999**, *5*, 643–650. [CrossRef]
7. Vila, M.; Hulme, P.E. *Impact of Biological Invasions on Ecosystem Services*; Springer International Publishing: Berlin/Heidelberg, Germany, 2017; ISBN 978-3-319-45121-3.
8. Shukla, D.; Wijayapala, S.; Vankar, P.S. Effective mosquito repellent from plant based formulation. *Int. J. Mosq. Res.* **2018**, *5*, 19–24.
9. Moyes, C.L.; Vontas, J.; Martins, A.J.; Ng, L.C.; Koou, S.Y.; Dusfour, I.; Raghavendra, K.; Pinto, J.; Corbel, V.; David, J.-P.; et al. Contemporary status of insecticide resistance in the major Aedes vectors of arboviruses infecting humans. *PLoS Negl. Trop. Dis.* **2017**, *11*, e0005625. [CrossRef]
10. Toledo Marrelli, M.; Barretto Bruno WILKE, A.; Toledo Marrelli, M. Genetic Control of Mosquitoes: Population suppression strategies. *Rev. Inst. Med. Trop. São Paulo* **2012**, *54*, 287–292.
11. Benelli, G.; Beier, J.C. Current vector control challenges in the fight against malaria. *Acta Trop.* **2017**, *174*, 91–96. [CrossRef]
12. Achee, N.L.; Grieco, J.P.; Vatandoost, H.; Seixas, G.; Pinto, J.; Ching-Ng, L.; Martins, A.J.; Juntarajumnong, W.; Corbel, V.; Gouagna, C.; et al. Alternative strategies for mosquito-borne arbovirus control. *PLoS Negl. Trop. Dis.* **2019**, *13*, e0006822.
13. Omori, N. A review of the role of mosquitoes in the transmission of Malayan and bancroftian filariasis in Japan. *Bull. World Health Organ.* **1962**, *27*, 585–594. [PubMed]
14. Barretto, A.; Wilke, B.; Toledo Marrelli, M. Paratransgenesis: A promising new strategy for mosquito vector control. *Parasites Vectors* **2015**, *8*, 342.
15. World Malaria Report. 2019. Available online: https://www.who.int/news-room/feature-stories/detail/world-malaria-report-2019 (accessed on 26 March 2020).

16. Rahmah, Z.; Sasmito, S.D.; Siswanto, B.; Sardjono, T.W.; Fitri, L.E. Malaria. *Malays. J. Med. Sci.* **2015**, *22*, 25–32. [PubMed]
17. Feged-Rivadeneira, A.; Ángel, A.; González-Casabianca, F.; Rivera, C. Malaria intensity in Colombia by regions and populations. *PLoS ONE* **2018**, *13*, e0203673. [CrossRef]
18. Amato, R.; Pearson, R.D.; Almagro-Garcia, J.; Amaratunga, C.; Lim, P.; Suon, S.; Sreng, S.; Drury, E.; Stalker, J.; Miotto, O.; et al. Origins of the current outbreak of multidrug-resistant malaria in southeast Asia: A retrospective genetic study. *Lancet Infect. Dis.* **2018**, *18*, 337–345. [CrossRef]
19. Lok, P.; Dijk, S. Malaria outbreak in Burundi reaches epidemic levels with 5.7 million infected this year. *BMJ* **2019**, *366*. [CrossRef]
20. Erickson, S.M.; Thomsen, E.K.; Keven, J.B.; Vincent, N.; Koimbu, G.; Siba, P.M.; Christensen, B.M.; Reimer, L.J. Mosquito-parasite interactions can shape filariasis transmission dynamics and impact elimination programs. *PLoS Negl. Trop. Dis.* **2013**, *7*, e2433. [CrossRef]
21. Gleave, K.; Cook, D.; Taylor, M.J.; Reimer, L.J. Filarial infection influences mosquito behaviour and fecundity. *Sci. Rep.* **2016**, *6*, 36319. [CrossRef]
22. Ughasi, J.; Bekard, H.E.; Coulibaly, M.; Adabie-Gomez, D.; Gyapong, J.; Appawu, M.; Wilson, M.D.; Boakye, D.A. Mansonia africana and Mansonia uniformis are vectors in the transmission of Wuchereria bancrofti lymphatic filariasis in Ghana. *Parasit. Vectors* **2012**, *5*, 89. [CrossRef]
23. Joseph, H.; Moloney, J.; Maiava, F.; McClintock, S.; Lammie, P.; Melrose, W. First evidence of spatial clustering of lymphatic filariasis in an Aedes polynesiensis endemic area. *Acta Trop.* **2011**, *120*, S39–S47. [CrossRef] [PubMed]
24. Southgate, B.A.; Bryan, J.H. Factors affecting transmission of Wuchereria bancrofti by anopheline mosquitoes. 4. Facilitation, limitation, proportionality and their epidemiological significance. *Trans. R Soc. Trop. Med. Hyg.* **1992**, *86*, 523–530. [CrossRef]
25. Wada, Y. Vector mosquitoes of filariasis in Japan. *Trop. Med. Health* **2011**, *39*, 39–45. [PubMed]
26. Lymphatic Filariasis. Available online: https://www.who.int/news-room/fact-sheets/detail/lymphatic-filariasis (accessed on 26 March 2020).
27. Ramos-Castañeda, J.; Barreto Dos Santos, F.; Martínez-Vega, R.; Lio, J.; Galvão De Araujo, M.; Joint, G.; Sarti, E. Dengue in Latin America: Systematic Review of Molecular Epidemiological Trends. *PLoS Negl. Trop. Dis.* **2017**, *11*, e0005224. [CrossRef] [PubMed]
28. Dengue Worldwide Overview. Available online: https://www.ecdc.europa.eu/en/dengue-monthly (accessed on 26 March 2020).
29. Katzelnick, L.C.; Coloma, J.; Harris, E. Dengue: Knowledge gaps, unmet needs, and research priorities. *Lancet Infect. Dis.* **2017**, *17*, e88–e100. [CrossRef]
30. Villamil-Gómez, W.E.; Rodríguez-Morales, A.J.; Uribe-García, A.M.; González-Arismendy, E.; Castellanos, J.E.; Calvo, E.P.; Álvarez-Mon, M.; Musso, D. Zika, dengue, and chikungunya co-infection in a pregnant woman from Colombia. *Int. J. Infect. Dis.* **2016**, *51*, 135–138. [CrossRef]
31. Giron, S.; Franke, F.; Decoppet, A.; Cadiou, B.; Travaglini, T.; Thirion, L.; Durand, G.; Jeannin, C.; L'Ambert, G.; Grard, G.; et al. Vector-borne transmission of Zika virus in Europe, southern France, August 2019. *Euro Surveill.* **2019**, *24*. Available online: https://doi.org/10.2807/1560-7917.ES.2019.24.45.1900655 (accessed on 20 April 2020). [CrossRef]
32. Ruchusatsawat, K.; Wongjaroen, P.; Posanacharoen, A.; Rodriguez-Barraquer, I.; Sangkitporn, S.; Cummings, D.A.; Salje, H. Long-term circulation of Zika virus in Thailand: An observational study. *Lancet Infect. Dis.* **2019**, *19*, 439–446. [CrossRef]
33. Brady, O.J.; Hay, S.I. The first local cases of Zika virus in Europe. *Lancet* **2019**, *394*, 1991–1992. [CrossRef]
34. Ledermann, J.P.; Guillaumot, L.; Yug, L.; Saweyog, S.C.; Tided, M.; Machieng, P.; Pretrick, M.; Marfel, M.; Griggs, A.; Bel, M.; et al. Aedes hensilli as a potential vector of Chikungunya and Zika viruses. *PLoS Negl. Trop. Dis.* **2014**, *8*, e3188. [CrossRef]
35. Diallo, D.; Sall, A.A.; Diagne, C.T.; Faye, O.; Faye, O.; Ba, Y.; Hanley, K.A.; Buenemann, M.; Weaver, S.C.; Diallo, M. Zika virus emergence in mosquitoes in southeastern Senegal, 2011. *PLoS ONE* **2014**, *9*, e109442. [CrossRef]
36. Grard, G.; Caron, M.; Mombo, I.M.; Nkoghe, D.; Mboui Ondo, S.; Jiolle, D.; Fontenille, D.; Paupy, C.; Leroy, E.M. Zika virus in Gabon (Central Africa)—2007: A new threat from Aedes albopictus? *PLoS Negl. Trop. Dis.* **2014**, *8*, e2681. [CrossRef] [PubMed]

37. Song, B.H.; Yun, S.I.; Woolley, M.; Lee, Y.M. Zika virus: History, epidemiology, transmission, and clinical presentation. *J. Neuroimmunol.* **2017**, *308*, 50–64. [CrossRef] [PubMed]
38. Musso, D.; Nilles, E.J.; Cao-Lormeau, V.-M. Rapid spread of emerging Zika virus in the Pacific area. *Clin. Microbiol. Infect.* **2014**, *20*, O595–O596. [CrossRef] [PubMed]
39. Edington, F.; Varjão, D.; Melo, P. Incidence of articular pain and arthritis after chikungunya fever in the Americas: A systematic review of the literature and meta-analysis. *Jt. Bone Spine* **2018**, *85*, 669–678. [CrossRef]
40. Spoto, S.; Riva, E.; Fogolari, M.; Cella, E.; Costantino, S.; Angeletti, S.; Ciccozzi, M. Diffuse maculopapular rash: A family cluster during the last Chikungunya virus epidemic in Italy. *Clin. Case Rep.* **2018**, *6*, 2322–2325. [CrossRef]
41. Rahman, M.M.; Jakaria, S.K.; Sayed, B.; Moniruzzaman, M.; Humayon Kabir, A.K.M.; Mallik, M.U.; Hasan, M.R.; Siddique, A.B.; Hossain, M.A.; Uddin, N.; et al. Clinical and Laboratory Characteristics of an Acute Chikungunya Outbreak in Bangladesh in 2017. *Am. J. Trop. Med. Hyg.* **2018**, *100*, 405–410. [CrossRef]
42. Da Silva Junior, G.B.; Pinto, J.R.; Mota, R.M.S.; da Pires Neto, R.J.; Daher, E.D.F. Risk factors for death among patients with Chikungunya virus infection during the outbreak in northeast Brazil, 2016–2017. *Trans. R. Soc. Trop. Med. Hyg.* **2018**, *113*, 221–226. [CrossRef]
43. De Azevedo Fernandes, N.C.C.; Cunha, M.S.; Guerra, J.M.; Réssio, R.A.; Cirqueira, C.D.S.; Iglezias, S.D.; de Carvalho, J.; Araujo, E.L.L.; Catão-Dias, J.L.; Díaz-Delgado, J. Outbreak of Yellow Fever among Nonhuman Primates, Espirito Santo, Brazil, 2017. *Emerg. Infect. Dis.* **2017**, *23*, 2038–2041. [CrossRef]
44. Possas, C.; Lourenço-de-Oliveira, R.; Tauil, P.L.; de Pinheiro, F.P.; Pissinatti, A.; da Cunha, R.V.; Freire, M.; Martins, R.M.; Homma, A. Yellow fever outbreak in Brazil: The puzzle of rapid viral spread and challenges for immunisation. *Mem. Inst. Oswaldo Cruz* **2018**, *113*, e180278. [CrossRef]
45. Simon, L.V.; Hashmi, M.F.; Torp, K.D. *Yellow Fever*; StatPearls: Tampa/St. Petersburg, FL, USA, 2018.
46. Nwachukwu, W.E.; Yusuff, H.; Nwangwu, U.; Okon, A.; Ogunniyi, A.; Imuetinyan-Clement, J.; Besong, M.; Ayo-Ajayi, P.; Nikau, J.; Baba, A.; et al. The response to re-emergence of yellow fever in Nigeria, 2017. *Int. J. Infect. Dis.* **2020**, *92*, 189–196. [CrossRef]
47. WHO. *Yellow Fever—Nigeria*; WHO: Geneva, Switzerland, 2019.
48. Silva, N.I.O.; Sacchetto, L.; De Rezende, I.M.; Trindade, G.D.S.; Labeaud, A.D.; De Thoisy, B.; Drumond, B.P. Recent sylvatic yellow fever virus transmission in Brazil: The news from an old disease. *Virol. J.* **2020**, *17*, 9. [CrossRef]
49. Auguste, A.J.; Lemey, P.; Bergren, N.A.; Giambalvo, D.; Moncada, M.; Morón, D.; Hernandez, R.; Navarro, J.-C.; Weaver, S.C. Enzootic transmission of yellow fever virus, Venezuela. *Emerg. Infect. Dis.* **2015**, *21*, 99–102. [CrossRef]
50. Selemane, I. Epidemiological monitoring of the last outbreak of yellow fever in Brazil—An outlook from Portugal. *Travel Med. Infect. Dis.* **2019**, *28*, 46–51. [CrossRef]
51. Solomon, T.; Hombach, J.; Jacobson, J.; Hoke, C.; Marfin, A.; Campbell, G.; Ginsburg, A.; Fischer, M.; Tsai, T.; Hills, S.; et al. Estimated Global Incidence of Japanese Encephalitis: A Systematic Review. *Bull. World Health Organ.* **2011**, *89*, 766–774.
52. Fang, Y.; Zhang, Y.; Zhou, Z.B.; Xia, S.; Shi, W.Q.; Xue, J.B.; Li, Y.Y.; Wu, J.T. New strains of Japanese encephalitis virus circulating in Shanghai, China after a ten-year hiatus in local mosquito surveillance. *Parasites Vectors* **2019**, *12*, 22. [CrossRef] [PubMed]
53. Griesemer, S.B.; Kramer, L.D.; Van Slyke, G.A.; Pata, J.D.; Gohara, D.W.; Cameron, C.E.; Ciota, A.T. Mutagen resistance and mutation restriction of St. Louis encephalitis virus. *J. Gen. Virol.* **2017**, *98*, 201–211. [CrossRef] [PubMed]
54. Diaz, A.; Coffey, L.L.; Burkett-Cadena, N.; Day, J.F. Reemergence of St. Louis encephalitis virus in the Americas. *Emerg. Infect. Dis.* **2018**, *24*, 2150–2157. [CrossRef]
55. Castillo-Olivares, J.; Wood, J. West Nile virus infection of horses. *Vet. Res.* **2004**, *35*, 467–483. [CrossRef]
56. Barrett, A.D.T. West Nile in Europe: An increasing public health problem. *J. Travel Med.* **2018**, *25*, tay096. [CrossRef] [PubMed]
57. López-Ruiz, N.; del Montaño-Remacha, M.C.; Durán-Pla, E.; Pérez-Ruiz, M.; Navarro-Marí, J.M.; Salamanca-Rivera, C.; Miranda, B.; Oyonarte-Gómez, S.; Ruiz-Fernández, J. West nile virus outbreak in humans and epidemiological surveillance, West Andalusia, Spain, 2016. *Eurosurveillance* **2018**, *23*, 17-00261. [CrossRef]

58. Zhang, J.; Lu, G.; Li, J.; Kelly, P.; Li, M.; Wang, J.; Huang, K.; Qiu, H.; You, J.; Zhang, R.; et al. Molecular Detection of *Rickettsia felis* and *Rickettsia bellii* in Mosquitoes. *Vector Borne Zoonotic Dis.* **2019**, *19*, 802–809. [CrossRef] [PubMed]

59. Guo, W.P.; Tian, J.H.; Lin, X.D.; Ni, X.B.; Chen, X.P.; Liao, Y.; Yang, S.Y.; Dumler, J.S.; Holmes, E.C.; Zhang, Y.Z. Extensive genetic diversity of *Rickettsiales bacteria* in multiple mosquito species. *Sci. Rep.* **2016**, *6*, 38770. [CrossRef] [PubMed]

60. Krajacich, B.J.; Huestis, D.L.; Dao, A.; Yaro, A.S.; Diallo, M.; Krishna, A.; Xu, J.; Lehmann, T. Investigation of the seasonal microbiome of *Anopheles coluzzii* mosquitoes in Mali. *PLoS ONE* **2018**, *13*, e0194899. [CrossRef] [PubMed]

61. Socolovschi, C.; Pagés, F.; Raoult, D. Rickettsia felis in aedes albopictus mosquitoes, libreville, gabon. *Emerg. Infect. Dis.* **2012**, *18*, 1688–1689. [CrossRef] [PubMed]

62. Dieme, C.; Bechah, Y.; Socolovschi, C.; Audoly, G.; Berenger, J.M.; Faye, O.; Raoult, D.; Parola, P. Transmission potential of rickettsia felis infection by *Anopheles gambiae* mosquitoes. *Proc. Natl. Acad. Sci. USA* **2015**, *112*, 8088–8093. [CrossRef]

63. Eliasson, H.; Broman, T.; Forsman, M.; Bäck, E. Tularemia: Current Epidemiology and Disease Management. *Infect. Dis. Clin. North Am.* **2006**, *20*, 289–311. [CrossRef]

64. San Martín, J.L.; Brathwaite, O.; Zambrano, B.; Solórzano, J.O.; Bouckenooghe, A.; Dayan, G.H.; Guzmán, M.G. The epidemiology of dengue in the americas over the last three decades: A worrisome reality. *Am. J. Trop. Med. Hyg.* **2010**, *82*, 128–135. [CrossRef]

65. Bouwman, H.; van den Berg, H.; Kylin, H. DDT and Malaria Prevention: Addressing the Paradox. *Environ. Health Perspect.* **2011**, *119*, 744–747. [CrossRef]

66. Cailly, P.; Tran, A.; Balenghien, T.; L'Ambert, G.; Toty, C.; Ezanno, P. A climate-driven abundance model to assess mosquito control strategies. *Ecol. Model.* **2012**, *227*, 7–17. [CrossRef]

67. Hemingway, J.; Beaty, B.J.; Rowland, M.; Scott, T.W.; Sharp, B.L. The Innovative Vector Control Consortium: Improved control of mosquito-borne diseases. *Trends Parasitol.* **2006**, *22*, 308–312. [CrossRef]

68. Oxborough, R.M.; Kitau, J.; Jones, R.; Mosha, F.W.; Rowland, M.W. Experimental hut and bioassay evaluation of the residual activity of a polymer-enhanced suspension concentrate (SC-PE) formulation of deltamethrin for IRS use in the control of *Anopheles arabiensis*. *Parasites Vectors* **2014**, *7*, 454. [CrossRef] [PubMed]

69. Dzul-Manzanilla, F.; Ibarra-López, J.; Marín, W.B.; Martini-Jaimes, A.; Leyva, J.T.; Correa-Morales, F.; Huerta, H.; Manrique-Saide, P.; Vazquez-Prokopec, G.M.; Day, J. Indoor resting behavior of *Aedes aegypti* (Diptera: Culicidae) in Acapulco, Mexico. *J. Med. Entomol.* **2018**, *54*, 501–504.

70. Hladish, T.J.; Pearson, C.A.B.; Rojas, D.P.; Gomez-Dantes, H.; Halloran, M.E.; Vazquez-Prokopec, G.M.; Longini, I.M. Forecasting the effectiveness of indoor residual spraying for reducing dengue burden. *PLoS Negl. Trop. Dis.* **2018**, *12*, e0006570. [CrossRef] [PubMed]

71. Giglioli, G. An Investigation of the House-Frequenting Habits of Mosquitoes of the British Guiana Coastland in Relation to the Use of DDT 1. *Am. J. Trop. Med. Hyg.* **1948**, *1*, 43–70. [CrossRef]

72. Nathan, M.B.; Giglioli, M.E. Eradication of *Aedes aegypti* on Cayman Brac and Little Cayman, West Indies, with Abate (Temephos) in 1970–1971. *Bull. Pan Am. Health Organ.* **1982**, *16*, 28–39.

73. Oxborough, R.M.; Kitau, J.; Jones, R.; Feston, E.; Matowo, J.; Mosha, F.W.; Rowland, M.W. Long-lasting control of Anopheles arabiensis by a single spray application of micro-encapsulated pirimiphos-methyl (Actellic® 300 CS). *Malar. J.* **2014**, *13*, 37. [CrossRef]

74. WHO. *Pesticides and Their Application: For the Control of Vectors and Pests of Public Health Importance*, 6th ed.; WHO: Geneva, Switzerland, 2006; Available online: https://apps.who.int/iris/handle/10665/69223 (accessed on 20 April 2020).

75. Vazquez-Prokopec, G.M.; Medina-Barreiro, A.; Che-Mendoza, A.; Dzul-Manzanilla, F.; Correa-Morales, F.; Guillermo-May, G.; Bibiano-Marín, W.; Uc-Puc, V.; Geded-Moreno, E.; Vadillo-Sánchez, J.; et al. Deltamethrin resistance in Aedes aegypti results in treatment failure in Merida, Mexico. *PLoS Negl. Trop. Dis.* **2017**, *11*, e0005656. [CrossRef]

76. Uragayala, S.; Kamaraju, R.; Tiwari, S.N.; Sreedharan, S.; Ghosh, S.K.; Valecha, N. Village-scale (Phase III) evaluation of the efficacy and residual activity of SumiShield ® 50 WG (Clothianidin 50%, w/w) for indoor spraying for the control of pyrethroid-resistant *Anopheles culicifacies* Giles in Karnataka state, India. *Trop. Med. Int. Heal.* **2018**, *23*, 605–615. [CrossRef]

77. Zaim, M.; Guillet, P. Alternative insecticides: An urgent need. *Trends Parasitol.* **2002**, *18*, 161–163. [CrossRef]

78. Esu, E.; Lenhart, A.; Smith, L.; Horstick, O. Effectiveness of peridomestic space spraying with insecticide on dengue transmission; Systematic review. *Trop. Med. Int. Heal.* **2010**, *15*, 619–631. [CrossRef]

79. Peterson, R.K.D.; Macedo, P.A.; Davis, R.S. A human-health risk assessment for West Nile virus and insecticides used in mosquito management. *Environ. Health Perspect.* **2006**, *114*, 366–372. [CrossRef] [PubMed]

80. Bonds, J.A.S. Ultra-low-volume space sprays in mosquito control: A critical review. *Med. Vet. Entomol.* **2012**, *26*, 121–130. [CrossRef] [PubMed]

81. Boyce, W.M.; Lawler, S.P.; Schultz, J.M.; McCauley, S.J.; Kimsey, L.S.; Niemela, M.K.; Nielsen, C.F.; Reisen, W.K. Nontarget effects of the mosquito adulticide pyrethrin applied aerially during a West Nile virus outbreak in an urban California environment. *J. Am. Mosq. Control Assoc.* **2007**, *23*, 335–339. [CrossRef]

82. Nathan, M.; Reiter, P.; WHO. *Guidelines for Assessing the Efficacy of Insecticidal Space Sprays for Control of the Dengue Vector: Aedes aegypti*; WHO: Geneva, Switzerland, 2001; Available online: https://apps.who.int/iris/handle/10665/67047 (accessed on 20 April 2020).

83. Teng, H.-J.; Chen, T.-J.; Tsai, S.-F.; Lin, C.-P.; Chiou, H.-Y.; Lin, M.-C.; Yang, S.-Y.; Lee, Y.-W.; Kang, C.-C.; Hsu, H.-C.; et al. Emergency vector control in a DENV-2 outbreak in 2002 in Pingtung City, Pingtung County, Taiwan. *Jpn. J. Infect. Dis.* **2007**, *60*, 271–279. [PubMed]

84. Guillet, P.; Alnwick, D.; Cham, M.K.; Neira, M.; Zaim, M.; Heymann, D.; Mukelabai, K. Long-lasting treated mosquito nets: A breakthrough in malaria prevention. *Bull. World Health Organ.* **2001**, *79*, 998. [PubMed]

85. Yang, G.G.; Kim, D.; Pham, A.; Paul, C.J. A meta-regression analysis of the effectiveness of mosquito nets for malaria control: The value of long-lasting insecticide nets. *Int. J. Environ. Res. Public Health* **2018**, *15*, 1–12. [CrossRef]

86. Rowland, M.; Webster, J.; Saleh, P.; Chandramohan, D.; Freeman, T.; Pearcy, B.; Durrani, N.; Rab, A.; Mohammed, N. Prevention of malaria in Afghanistan through social marketing of insecticide-treated nets: Evaluation of coverage and effectiveness by cross sectional surveys and passive surveillance. *Trop. Med. Int. Heal.* **2002**, *7*, 813–822. [CrossRef]

87. Arroz, J.A.H.; Candrinho, B.; Mendis, C.; Varela, P.; Pinto, J.; Do, M.; Martins, R.O. Effectiveness of a new long-lasting insecticidal nets delivery model in two rural districts of Mozambique: A before-after study. *Malar J.* **2018**, *17*, 66. [CrossRef]

88. Girond, F.; Madec, Y.; Kesteman, T.; Randrianarivelojosia, M.; Randremanana, R.; Randriamampionona, L.; Randrianasolo, L.; Ratsitorahina, M.; Herbreteau, V.; Hedje, J.; et al. Evaluating effectiveness of mass and continuous long-lasting insecticidal net distributions over time in Madagascar: A sentinel surveillance based epidemiological study. *EClinicalMedicine* **2018**, *1*, 62–69. [CrossRef]

89. Hounkonnou, C.; Djènontin, A.; Egbinola, S.; Houngbegnon, P.; Bouraima, A.; Soares, C.; Fievet, N.; Accrombessi, M.; Yovo, E.; Briand, V.; et al. Impact of the use and efficacy of long lasting insecticidal net on malaria infection during the first trimester of pregnancy—A pre-conceptional cohort study in southern Benin. *BMC Public Health* **2018**, *18*, 683. [CrossRef]

90. Trape, J.F.; Tall, A.; Diagne, N.; Ndiath, O.; Ly, A.B.; Faye, J.; Dieye-Ba, F.; Roucher, C.; Bouganali, C.; Badiane, A.; et al. Malaria morbidity and pyrethroid resistance after the introduction of insecticide-treated bednets and artemisinin-based combination therapies: A longitudinal study. *Lancet Infect. Dis.* **2011**, *11*, 925–932. [CrossRef]

91. Sougoufara, S.; Thiaw, O.; Cailleau, A.; Diagne, N.; Harry, M.; Bouganali, C.; Sembène, P.M.; Doucoure, S.; Sokhna, C. The Impact of Periodic Distribution Campaigns of Long-Lasting Insecticidal-Treated Bed Nets on Malaria Vector Dynamics and Human Exposure in Dielmo, Senegal. *Am. J. Trop. Med. Hyg.* **2018**, *98*, 1343–1352. [CrossRef]

92. Soleimani-Ahmadi, M.; Vatandoost, H.; Shaeghi, M.; Raeisi, A.; Abedi, F.; Eshraghian, M.R.; Madani, A.; Safari, R.; Oshaghi, M.A.; Abtahi, M.; et al. Field evaluation of permethrin long-lasting insecticide treated nets (Olyset®) for malaria control in an endemic area, southeast of Iran. *Acta Trop.* **2012**, *123*, 146–153. [CrossRef]

93. Kajla, M.K.; Barrett-Wilt, G.A.; Paskewitz, S.M. Bacteria: A novel source for potent mosquito feeding-deterrents. *Sci. Adv.* **2019**, *5*, eaau6141. [CrossRef] [PubMed]

94. Nerio, L.S.; Olivero-Verbel, J.; Stashenko, E. Repellent activity of essential oils: A review. *Bioresour. Technol.* **2010**, *101*, 372–378. [CrossRef]

95. Trongtokit, Y.; Rongsriyam, Y.; Komalamisra, N.; Apiwathnasorn, C. Comparative repellency of 38 essential oils against mosquito bites. *Phyther. Res.* **2005**, *19*, 303–309. [CrossRef] [PubMed]

96. Daisy, B.H.; Strobel, G.A.; Castillo, U.; Ezra, D.; Sears, J.; Weaver, D.K.; Runyon, J.B. Naphthalene, an insect repellent, is produced by *Muscodor vitigenus*, a novel endophytic fungus. *Microbiology* **2002**, *148*, 3737–3741. [CrossRef]

97. Naseem, S.; Munir, T.; Faheem Malik, M. Mosquito management: A review. *J. Entomol. Zool. Stud.* **2016**, *4*, 73–79.

98. Balaji, A.P.B.; Ashu, A.; Manigandan, S.; Sastry, T.P.; Mukherjee, A.; Chandrasekaran, N. Polymeric nanoencapsulation of insect repellent: Evaluation of its bioefficacy on *Culex quinquefasciatus* mosquito population and effective impregnation onto cotton fabrics for insect repellent clothing. *J. King Saud Univ. Sci.* **2017**, *29*, 517–527. [CrossRef]

99. Soni, N.; Prakash, S. Green nanoparticles for mosquito control. *Sci. World J.* **2014**, *2014*, 496362. [CrossRef]

100. Knipling, E.F. Possibilities of Insect Control or Eradication Through the Use of Sexually Sterile Males1. *J. Econ. Entomol.* **1955**, *48*, 459–462. [CrossRef]

101. Phuc, H.K.; Andreasen, M.H.; Burton, R.S.; Vass, C.; Epton, M.J.; Pape, G.; Fu, G.; Condon, K.C.; Scaife, S.; Donnelly, C.A.; et al. Late-acting dominant lethal genetic systems and mosquito control. *BMC Biol.* **2007**, *5*, 11. [CrossRef]

102. Reiter, P. Oviposition, Dispersal, and Survival in *Aedes aegypti*: Implications for the Efficacy of Control Strategies. *Vector Borne Zoonotic Dis.* **2007**, *7*, 261–273. [CrossRef]

103. Onyekwere, J.; Nnamonu, E.; Bede, E.; Okoye, C.; Okoro, J.; Nnamonu, E.; Bede, E.; Okoye, I.C.; Onyekwere, J.; Nnamonu, E.; et al. Application of genetically modified mosquitoes (Anopheles species) in the control of malaria transmission. *Asian J. Biotechnol. Genet. Eng.* **2018**, *1*, 1–16.

104. Meghani, Z.; Boëte, C. Genetically engineered mosquitoes, Zika and other arboviruses, community engagement, costs, and patents: Ethical issues. *PLoS Negl. Trop. Dis.* **2018**, *12*, e0006501. [CrossRef]

105. GMWATCH. New Documents Show Oxitec's GM Mosquitoes Ineffective and Risky. Available online: https://www.gmwatch.org/en/news/latest-news/17828-new-documents-show-oxitec-s-gm-mosquitoes-ineffective-and-risky (accessed on 6 April 2019).

106. Yamada, H.; Kraupa, C.; Lienhard, C.; Parker, A.G.; Maiga, H.; de Oliveira Carvalho, D.; Zheng, M.; Wallner, T.; Bouyer, J. Mosquito mass rearing: Who's eating the eggs? *Parasite* **2019**, *26*, 75. [CrossRef] [PubMed]

107. Mukherjee, S.; Blaustein, L. Effects of predator type and alternative prey on mosquito egg raft predation and destruction. *Hydrobiologia* **2019**, *846*, 215–221. [CrossRef]

108. Lopez, S.B.G.; Guimarães-Ribeiro, V.; Rodriguez, J.V.G.; Dorand, F.A.P.S.; Salles, T.S.; Sá-Guimarães, T.E.; Alvarenga, E.S.L.; Melo, A.C.A.; Almeida, R.V.; Moreira, M.F. RNAi-based bioinsecticide for Aedes mosquito control. *Sci. Rep.* **2019**, *9*, 1–13. [CrossRef]

109. Lee, J.Y.; Woo, R.M.; Choi, C.J.; Shin, T.Y.; Gwak, W.S.; Woo, S.D. *Beauveria bassiana* for the simultaneous control of *Aedes albopictus* and *Culex pipiens* mosquito adults shows high conidia persistence and productivity. *AMB Express* **2019**, *9*, 206. [CrossRef]

110. Scholte, E.-J.; Knols, B.G.J.; Samson, R.A.; Takken, W. Entomopathogenic fungi for mosquito control: A review. *J. Insect Sci.* **2004**, *4*, 19. [CrossRef]

111. Noskov, Y.A.; Polenogova, O.V.; Yaroslavtseva, O.N.; Belevich, O.E.; Yurchenko, Y.A.; Chertkova, E.A.; Kryukova, N.A.; Kryukov, V.Y.; Glupov, V.V. Combined effect of the entomopathogenic fungus *Metarhizium robertsii* and avermectins on the survival and immune response of *Aedes aegypti* larvae. *PeerJ* **2019**, *7*, e7931. [CrossRef] [PubMed]

112. Lovett, B.; Bilgo, E.; Diabate, A.; St Leger, R. A review of progress toward field application of transgenic mosquitocidal entomopathogenic fungi. *Pest Manag. Sci.* **2019**, *75*, 2316–2324. [CrossRef] [PubMed]

113. Silva, L.E.I.; Paula, A.R.; Ribeiro, A.; Butt, T.M.; Silva, C.P.; Samuels, R.I. A new method of deploying entomopathogenic fungi to control adult *Aedes aegypti* mosquitoes. *J. Appl. Entomol.* **2018**, *142*, 59–66. [CrossRef]

114. Louca, V.; Lucas, M.C.; Green, C.; Majambere, S.; Fillinger, U.; Lindsay, S.W. Role of fish as predators of mosquito larvae on the floodplain of the Gambia River. *J. Med. Entomol.* **2009**, *46*, 546–556. [CrossRef] [PubMed]

115. Aditya, G.; Pal, S.; Saha, N.; Saha, G. Efficacy of indigenous larvivorous fishes against *Culex quinquefasciatus* in the presence of alternative prey: Implications for biological control. *J. Vector Borne Dis.* **2012**, *49*, 217–225.

116. Sareein, N.; Phalaraksh, C.; Rahong, P.; Techakijvej, C.; Seok, S.; Bae, Y.J. Relationships between predatory aquatic insects and mosquito larvae in residential areas in northern Thailand. *J. Vector Ecol.* **2019**, *44*, 223–232. [CrossRef]

117. Früh, L.; Kampen, H.; Schaub, G.A.; Werner, D. Predation on the invasive mosquito *Aedes japonicus* (Diptera: Culicidae) by native copepod species in Germany. *J. Vector Ecol.* **2019**, *44*, 241–247. [CrossRef]

118. Cuthbert, R.N.; Callaghan, A.; Sentis, A.; Dalal, A.; Dick, J.T.A. Additive multiple predator effects can reduce mosquito populations. *Ecol. Entomol.* **2019**, *45*, 243–250. [CrossRef]

119. Digma, J.R.; Sumalde, A.C.; Salibay, C.C. Laboratory evaluation of predation of *Toxorhynchites amboinensis* (Diptera:Culicidae) on three mosquito vectors of arboviruses in the Philippines. *Biol. Control* **2019**, *137*, 104009. [CrossRef]

120. Focks, D.A. Toxorhynchites as biocontrol agents. *J. Am. Mosq. Control Assoc.* **2007**, *23*, 118–127. [CrossRef]

121. Schiller, A.; Allen, M.; Coffey, J.; Fike, A.; Carballo, F. Updated Methods for the Production of *Toxorhynchites rutilus* septentrionalis (Diptera, Culicidae) for Use as Biocontrol Agent Against Container Breeding Pest Mosquitoes in Harris County, Texas. *J. Insect Sci.* **2019**, *19*, 8. [CrossRef] [PubMed]

122. Das, B.P. Chilodonella uncinate—A protozoa pathogenic to mosquito larvae. *Curr. Sci.* **2003**, *85*, 483–489.

123. Dhanasekaran, D.; Thangaraj, R. Microbial secondary metabolites are an alternative approaches against insect vector to prevent zoonotic diseases. *Asian Pac. J. Trop. Dis.* **2014**, *4*, 253–261. [CrossRef]

124. Lacey, L.A. Bacillus thuringiensis serovariety israelensis and Bacillus sphaericus for mosquito control. *J. Am. Mosq. Control Assoc.* **2007**, *23*, 93–109. [CrossRef]

125. Derua, Y.A.; Kahindi, S.C.; Mosha, F.W.; Kweka, E.J.; Atieli, H.E.; Wang, X.; Zhou, G.; Lee, M.; Githeko, A.K.; Yan, G. Microbial larvicides for mosquito control: Impact of long lasting formulations of *Bacillus thuringiensis* var. *Israelensis* and *Bacillus sphaericus* on non target organisms in western Kenya highlands. *Ecol. Evol.* **2018**, *8*, 7563–7573. [CrossRef]

126. Zhang, Q.; Hua, G.; Adang, M.J. Effects and mechanisms of Bacillus thuringiensis crystal toxins for mosquito larvae. *Insect Sci.* **2017**, *24*, 714–729. [CrossRef]

127. Palma, L.; Muñoz, D.; Berry, C.; Murillo, J.; Caballero, P. Bacillus thuringiensis toxins: An overview of their biocidal activity. *Toxins* **2014**, *6*, 3296–3325. [CrossRef]

128. Chee Dhang, C.; Han Lim, L.; Wasi Ahmad, N.; Benjamin, S.; Koon Weng, L.; Abdul Rahim, D.; Syafinaz Safian, E.; Sofian-Azirun, M. Field effectiveness of *Bacillus thuringiensis israelensis* (*Bti*) against *Aedes* (*Stegomyia*) *aegypti* (Linnaeus) in ornamental ceramic containers with common aquatic plants. *Trop. Biomed.* **2009**, *26*, 100–105.

129. Saliha, B.; Wafa, H.; Laid, O.M. Effect of Bacillus thuringiensis var krustaki on the mortality and development of *Culex pipiens* (Diptera; Cullicidae). *Int. J. Mosq. Res.* **2017**, *4*, 20–23.

130. Delécluse, A.; Rosso, M.L.; Ragni, A. Cloning and expression of a novel toxin gene from *Bacillus thuringiensis* subsp. jegathesan encoding a highly mosquitocidal protein. *Appl. Environ. Microbiol.* **1995**, *61*, 4230–4235. [CrossRef]

131. López-Meza, J.; Federici, B.A.; Poehner, W.J.; Martinez-Castillo, A.; Ibarra, J.E. Highly mosquitocidal isolates of Bacillus thuringiensis subspecies kenyae and entomocidus from Mexico. *Biochem. Syst. Ecol.* **1995**, *23*, 461–468. [CrossRef]

132. De Barjac, H.; Sebald, M.; Charles, J.F.; Cheong, W.H.; Lee, H.L. Clostridium bifermentans serovar malaysia, a new anaerobic bacterium pathogen to mosquito and blackfly larvae. *C. R. Acad. Sci. Iii.* **1990**, *310*, 383–387. [PubMed]

133. Darriet, F.; Hougard, J.-M. An isolate of Bacillus circulans toxic to mosquito larvae. *J. Am. Mosq. Control Assoc.* **2002**, *18*, 65–67. [PubMed]

134. Favret, M.E.; Yousten, A.A. Insecticidal activity of Bacillus laterosporus. *J. Invertebr. Pathol.* **1985**, *45*, 195–203. [CrossRef]

135. Orlova, M.V.; Smirnova, T.A.; Ganushkina, L.A.; Yacubovich, V.Y.; Azizbekyan, R.R. Insecticidal activity of Bacillus laterosporus. *Appl. Environ. Microbiol.* **1998**, *64*, 2723–2725. [CrossRef]

136. Pener, M.P. An Overview of Insect Growth Disruptors; Applied Aspects. *Adv. Insect Phys.* **2012**, *43*, 1–162.

137. Yapabandara, A.M.G.M.; Curtis, C.F. Laboratory and field comparisons of pyriproxyfen, polystyrene beads and other larvicidal methods against malaria vectors in Sri Lanka. *Acta Trop.* **2002**, *81*, 211–223. [CrossRef]

138. Ansari, M.A.; Razdan, R.K.; Sreehari, U. Laboratory and field evaluation of Hilmilin against mosquitoes. *J. Am. Mosq. Control Assoc.* **2005**, *21*, 432–436. [CrossRef]

139. Raghavendra, K.; Barik, T.K.; Reddy, B.P.N.; Sharma, P.; Dash, A.P. Malaria vector control: From past to future. *Parasitol. Res.* **2011**, *108*, 757–779. [CrossRef]

140. Paul, A.; Harrington, L.C.; Zhang, L.; Scott, J.G. Insecticide resistance in *Culex pipiens* from New York. *J. Am. Mosq. Control Assoc.* **2005**, *21*, 305–309. [CrossRef]

141. De Silva, J.J.; Mendes, J. Susceptibility of *Aedes aegypti* (L) to the insect growth regulators diflubenzuron and methoprene in Uberlândia, State of Minas Gerais. *Rev. Soc. Bras. Med. Trop.* **2007**, *40*, 612–616. [CrossRef]

142. Dennehy, T.J.; Degain, B.A.; Harpold, V.S.; Zaborac, M.; Morin, S.; Fabrick, J.A.; Nichols, R.L.; Brown, J.K.; Byrne, F.J.; Li, X. Extraordinary resistance to insecticides reveals exotic Q biotype of *Bemisia tabaci* in the New World. *J. Econ. Entomol.* **2010**, *103*, 2174–2186. [CrossRef] [PubMed]

143. Lee, S.-H.; Young CHOI, J.; Ram LEE, B.; Fang, Y.; Hoon KIM, J.; Hwan Park, D.; Gu Park, M.; Mi Woo, R.; Jin Kim, W.; Ho, Y.J.; et al. Insect growth regulatory and larvicidal activity of chalcones against *Aedes albopictus*. *Entomol. Rep.* **2018**, *48*, 55–59. [CrossRef]

144. Niang, E.H.A.; Bassene, H.; Fenollar, F.; Mediannikov, O. Biological Control of Mosquito-Borne Diseases: The Potential of Wolbachia -Based Interventions in an IVM Framework. *J. Trop. Med.* **2018**, *2018*, 1–15. [CrossRef] [PubMed]

145. Benelli, G.; Jeffries, C.L.; Walker, T. Biological Control of Mosquito Vectors: Past, Present, and Future. *Insects* **2016**, *7*, 52. [CrossRef]

146. Zug, R.; Hammerstein, P. Still a Host of Hosts for Wolbachia: Analysis of Recent Data Suggests That 40% of Terrestrial Arthropod Species Are Infected. *PLoS ONE* **2012**, *7*, 38544. [CrossRef]

147. Skerman, V.B.D.; McGowan, V.F.; Sneath, P.H.A.; Peter, H.A. *Approved Lists of Bacterial Names*; Skerman, V.B.D., McGowan, V.F., Sneath, P.H.A., Eds.; American Society for Microbiology: Washington, DC, USA, 1989; ISBN 9781555810146.

148. Ogunbiyi, T.S.; Eromon, P.; Oluniyi, P.; Ayoade, F.; Oloche, O.; Oguzie, J.U.; Folarin, O.; Happi, C.; Komolafe, I. First Report of Wolbachia from Field Populations of Culex Mosquitoes in South-Western Nigeria. *Afr. Zool.* **2019**, *54*, 181–185. [CrossRef]

149. Balaji, S.; Jayachandran, S.; Prabagaran, S.R. Evidence for the natural occurrence of Wolbachia in *Aedes aegypti* mosquitoes. *Fems Microbiol. Lett.* **2019**, *366*. Available online: https://doi.org/10.1093/femsle/fnz055 (accessed on 20 April 2020). [CrossRef]

150. Mohanty, I.; Rath, A.; Swain, S.P.; Pradhan, N.; Hazra, R.K. Wolbachia Population in Vectors and Non-vectors: A Sustainable Approach Towards Dengue Control. *Curr. Microbiol.* **2019**, *76*, 133–143. [CrossRef]

151. Laven, H. Eradication of *Culex pipiens* fatigans through Cytoplasmic Incompatibility. *Nature* **1967**, *216*, 383–384. [CrossRef]

152. Zhang, D.; Zheng, X.; Xi, Z.; Bourtzis, K.; Gilles, J.R.L. Combining the Sterile Insect Technique with the Incompatible Insect Technique: I-Impact of Wolbachia Infection on the Fitness of Triple- and Double-Infected Strains of *Aedes albopictus*. *PLoS ONE* **2015**, *10*, e0121126. [CrossRef] [PubMed]

153. O'Connor, L.; Plichart, C.; Sang, A.C.; Brelsfoard, C.L.; Bossin, H.C.; Dobson, S.L. Open Release of Male Mosquitoes Infected with a Wolbachia Biopesticide: Field Performance and Infection Containment. *PLoS Negl. Trop. Dis.* **2012**, *6*, e1797. [CrossRef]

154. Joubert, D.A.; Walker, T.; Carrington, L.B.; De Bruyne, J.T.; Kien, D.H.T.; Hoang, N.L.T.; Chau, N.V.V.; Iturbe-Ormaetxe, I.; Simmons, C.P.; O'Neill, S.L. Establishment of a Wolbachia Superinfection in *Aedes aegypti* Mosquitoes as a Potential Approach for Future Resistance Management. *PLoS Pathog.* **2016**, *12*, e1005434. [CrossRef] [PubMed]

155. Aliota, M.T.; Walker, E.C.; Yepes, A.U.; Velez, I.D.; Christensen, B.M.; Osorio, J.E. The wMel Strain of Wolbachia Reduces Transmission of Chikungunya Virus in *Aedes aegypti*. *PLoS Negl. Trop. Dis.* **2016**, *10*, e0004677. [CrossRef]

156. Van den Hurk, A.F.; Hall-Mendelin, S.; Pyke, A.T.; Frentiu, F.D.; McElroy, K.; Day, A.; Higgs, S.; O'Neill, S.L. Impact of Wolbachia on infection with chikungunya and yellow fever viruses in the mosquito vector *Aedes aegypti*. *PLoS Negl. Trop. Dis.* **2012**, *6*, e1892. [CrossRef] [PubMed]

157. Bian, G.; Joshi, D.; Dong, Y.; Lu, P.; Zhou, G.; Pan, X.; Xu, Y.; Dimopoulos, G.; Xi, Z. Wolbachia Invades *Anopheles stephensi* Populations and Induces Refractoriness to Plasmodium Infection. *Science* **2013**, *340*, 748–751. [CrossRef]

158. Hughes, G.L.; Koga, R.; Xue, P.; Fukatsu, T.; Rasgon, J.L. Wolbachia Infections Are Virulent and Inhibit the Human Malaria Parasite Plasmodium Falciparum in Anopheles Gambiae. *PLoS Pathog.* **2011**, *7*, e1002043. [CrossRef]

159. Ford, S.A.; Allen, S.L.; Ohm, J.R.; Sigle, L.T.; Sebastian, A.; Albert, I.; Chenoweth, S.F.; McGraw, E.A. Selection on *Aedes aegypti* alters Wolbachia-mediated dengue virus blocking and fitness. *Nat. Microbiol.* **2019**, *4*, 1832–1839. [CrossRef]

160. Dutra, H.L.C.; Rocha, M.N.; Dias, F.B.S.; Mansur, S.B.; Caragata, E.P.; Moreira, L.A. Wolbachia Blocks Currently Circulating Zika Virus Isolates in Brazilian *Aedes aegypti* Mosquitoes. *Cell Host Microbe* **2016**, *19*, 744–771. [CrossRef]

161. Mariño, Y.; Verle Rodrigues, J.; Bayman, P. Wolbachia Affects Reproduction and Population Dynamics of the Coffee Berry Borer (*Hypothenemus hampei*): Implications for Biological Control. *Insects* **2017**, *8*, 8. [CrossRef]

162. Chegeni, T.N.; Fakhar, M. Promising Role of Wolbachia as Anti-parasitic Drug Target and Eco-Friendly Biocontrol Agent. *Recent Pat. Antiinfect. Drug Discov.* **2019**, *14*, 69–79. [CrossRef] [PubMed]

163. Bongio, N.J.; Lampe, D.J. Inhibition of *Plasmodium berghei* Development in Mosquitoes by Effector Proteins Secreted from *Asaia* sp. Bacteria Using a Novel Native Secretion Signal. *PLoS ONE* **2015**, *10*. Available online: https://doi.org/10.1371/journal.pone.0143541 (accessed on 20 April 2020). [CrossRef] [PubMed]

164. Wang, S.; Jacobs-Lorena, M. Genetic approaches to interfere with malaria transmission by vector mosquitoes. *Trends Biotechnol.* **2013**, *31*, 185–193. [CrossRef] [PubMed]

165. Rami, A.; Raz, A.; Zakeri, S.; Dinparast Djadid, N. Isolation and identification of *Asaia* sp. in *Anopheles* spp. mosquitoes collected from Iranian malaria settings: Steps toward applying paratransgenic tools against malaria. *Parasit Vectors* **2018**, *11*, 367. [CrossRef] [PubMed]

166. Favia, G.; Ricci, I.; Marzorati, M.; Negri, I.; Alma, A.; Sacchi, L.; Bandi, C.; Daffonchio, D. Bacteria of the Genus Asaia: A Potential Paratransgenic Weapon Against Malaria. In *Transgenesis and the Management of Vector-Borne Disease*; Springer: New York, NY, USA, 2008; pp. 49–59.

167. Favia, G.; Ricci, I.; Damiani, C.; Raddadi, N.; Crotti, E.; Marzorati, M.; Rizzi, A.; Urso, R.; Brusetti, L.; Borin, S.; et al. Bacteria of the genus Asaia stably associate with *Anopheles stephensi*, an Asian malarial mosquito vector. *Proc. Natl. Acad. Sci. USA* **2007**, *104*, 9047–9051. [CrossRef] [PubMed]

168. Damiani, C.; Ricci, I.; Crotti, E.; Rossi, P.; Rizzi, A.; Scuppa, P.; Esposito, F.; Bandi, C.; Daffonchio, D.; Favia, G. Paternal transmission of symbiotic bacteria in malaria vectors. *Curr. Biol.* **2008**, *18*, R1087–R1088. [CrossRef]

169. Nugapola, N.W.N.P.; De Silva, W.A.P.P.; Karunaratne, S.H.P.P. Distribution and phylogeny of Wolbachia strains in wild mosquito populations in Sri Lanka. *Parasites Vectors* **2017**, *10*, 230. [CrossRef]

170. Mamlouk, D.; Gullo, M. Acetic Acid Bacteria: Physiology and Carbon Sources Oxidation. *Indian J. Microbiol.* **2013**, *53*, 377. [CrossRef]

171. Cappelli, A.; Damiani, C.; Mancini, M.V.; Valzano, M.; Rossi, P.; Serrao, A.; Ricci, I.; Favia, G. Asaia Activates Immune Genes in Mosquito Eliciting an Anti-Plasmodium Response: Implications in Malaria Control. *Front. Genet.* **2019**, *10*, 836. [CrossRef]

172. Mitraka, E.; Stathopoulos, S.; Siden-Kiamos, I.; Christophides, G.K.; Louis, C. Asaia accelerates larval development of *Anopheles gambiae*. *Pathog. Glob. Health* **2013**, *107*, 305. [CrossRef]

173. Shane, J.L.; Grogan, C.L.; Cwalina, C.; Lampe, D.J. Blood meal-induced inhibition of vector-borne disease by transgenic microbiota. *Nat. Commun.* **2018**, *9*, 1–10. [CrossRef] [PubMed]

174. Mertz, F.P.; YAO, R.C. *Saccharopolyspora spinosa* sp. nov. Isolated from Soil Collected in a Sugar Mill Rum Still. *Int. J. Syst. Bacteriol.* **1990**, *40*, 34–39. [CrossRef]

175. Hertlein, M.B.; Mavrotas, C.; Jousseaume, C.; Lysandrou, M.; Thompson, G.D.; Jany, W.; Ritchie, S.A. A Review of Spinosad as a Natural Product for Larval Mosquito Control. *J. Am. Mosq. Control Assoc.* **2010**, *26*, 67–87. [CrossRef]

176. Su, T. *Resistance and Its Management to Microbial and Insect Growth Regulator Larvicides in Mosquitoes*; InTech Europe: Rijeka, Croatia, 2016; pp. 135–154. [CrossRef]

177. WHO. *Spinosad DT in Drinking-Water: Use for Vector Control in Drinking-Water Sources and Containers*; WHO: Geneva, Switzerland, 2010; Available online: https://www.who.int/water_sanitation_health/dwq/chemicals/spinosadbg.pdf (accessed on 20 April 2020).

178. Marina, C.F.; Bond, J.G.; Muñoz, J.; Valle, J.; Chirino, N.; Williams, T. Spinosad: A biorational mosquito larvicide for use in car tires in southern Mexico. *Parasit. Vectors* **2012**, *5*, 95. [CrossRef] [PubMed]

179. Marina, C.F.; Bond, J.; Muñoz, J.; Valle, J.; Novelo-Gutiérrez, R.; Williams, T. Efficacy and non-target impact of spinosad, Bti and temephos larvicides for control of *Anopheles* spp. in an endemic malaria region of southern Mexico. *Parasit. Vectors* **2014**, *7*, 55. [CrossRef] [PubMed]

180. Prabhu, K.; Murugan, K.; Nareshkumar, A.; Bragadeeswaran, S. Larvicidal and pupicidal activity of spinosad against the malarial vector *Anopheles stephensi*. *Asian Pac. J. Trop. Med.* **2011**, *4*, 610–613. [CrossRef]

181. Qiu, H.; Jun, H.W.; Mccall, J.W. Pharmacokinetics, formulation, and safety of insect repellent N,N-Diethyl-3-methylbenzamide (deet): A review. *J. Am. Mosq. Control Assoc.* **1998**, *14*, 12–27.

182. Stanczyk, N.M.; Brookfield, J.F.Y.; Field, L.M.; Logan, J.G. Aedes aegypti Mosquitoes Exhibit Decreased Repellency by DEET following Previous Exposure. *PLoS ONE* **2013**, *8*, e54438. [CrossRef]

183. Deletre, E.; Martin, T.; Duménil, C.; Chandre, F. Insecticide resistance modifies mosquito response to DEET and natural repellents. *Parasit. Vectors* **2019**, *12*, 89. [CrossRef] [PubMed]

184. Debboun, M.; Frances, S.P.; Strickman, D. *Insect Repellents Handbook*; CRC Press: Boca Raton, FL, USA, 2014; ISBN 9781466553552.

185. Van Roey, K.; Sokny, M.; Denis, L.; Van den Broeck, N.; Heng, S.; Siv, S.; Sluydts, V.; Sochantha, T.; Coosemans, M.; Durnez, L. Field evaluation of picaridin repellents reveals differences in repellent sensitivity between Southeast Asian vectors of malaria and arboviruses. *PLoS Negl. Trop. Dis.* **2014**, *8*, e3326. [CrossRef]

186. Carroll, S.P. Prolonged efficacy of IR3535 repellents against mosquitoes and blacklegged ticks in North America. *J. Med. Entomol.* **2008**, *45*, 706–714. [CrossRef] [PubMed]

187. Lee, M.Y. Essential Oils as Repellents against Arthropods. *Biomed Res. Int.* **2018**, *2018*, 6860271. [CrossRef] [PubMed]

188. Tilquin, M.; Paris, M.; Reynaud, S.; Despres, L.; Ravanel, P.; Geremia, R.A.; Gury, J. Long lasting persistence of *Bacillus thuringiensis* Subsp. israelensis (Bti) in mosquito natural habitats. *PLoS ONE* **2008**, *3*, e3432. [CrossRef] [PubMed]

189. Paris, M.; Tetreau, G.; Laurent, F.; Lelu, M.; Despres, L.; David, J.-P. Persistence of *Bacillus thuringiensis* israelensis (Bti) in the environment induces resistance to multiple Bti toxins in mosquitoes. *Pest Manag. Sci.* **2011**, *67*, 122–128. [CrossRef] [PubMed]

190. Bravo, A.; Soberón, M. How to cope with insect resistance to Bt toxins? *Trends Biotechnol.* **2008**, *26*, 573–579. [CrossRef]

191. Georghiou, G.P.; Wirth, M.C. Influence of Exposure to Single versus Multiple Toxins of *Bacillus thuringiensis* subsp. israelensis on Development of Resistance in the Mosquito *Culex quinquefasciatus* (Diptera: Culicidae). *Appl. Environ. Microbiol.* **1997**, *63*, 1095–1101. [CrossRef] [PubMed]

192. Cadavid-Restrepo, G.; Sahaza, J.; Orduz, S. Treatment of an *Aedes aegypti* colony with the Cry11Aa toxin for 54 generations results in the development of resistance. *Mem. Inst. Oswaldo Cruz* **2012**, *107*, 74–79. [CrossRef]

193. Stalinski, R.; Tetreau, G.; Gaude, T.; Després, L. Pre-selecting resistance against individual Bti Cry toxins facilitates the development of resistance to the Bti toxins cocktail. *J. Invertebr. Pathol.* **2014**, *119*, 50–53. [CrossRef]

194. Paris, M.; Melodelima, C.; Coissac, E.; Tetreau, G.; Reynaud, S.; David, J.-P.; Despres, L. Transcription profiling of resistance to Bti toxins in the mosquito *Aedes aegypti* using next-generation sequencing. *J. Invertebr. Pathol.* **2012**, *109*, 201–208. [CrossRef]

195. Cheong, H.; Dhesi, R.K.; Gill, S.S. Marginal cross-resistance to mosquitocidal Bacillus thuringiensis strains in Cry11A-resistant larvae: Presence of Cry11A-like toxins in these strains. *Fems Microbiol. Lett.* **2006**, *153*, 419–424. [CrossRef]

196. Demissew, A.; Balkew, M.; Girma, M. Larvicidal activities of chinaberry, neem and *Bacillus thuringiensis* israelensis (Bti) to an insecticide resistant population of *Anopheles arabiensis* from Tolay, Southwest Ethiopia. *Asian Pac. J. Trop. Biomed.* **2016**, *6*, 554–561. [CrossRef]

197. Boyer, S.; Tilquin, M.; Ravanel, P. differential sensitivity to *Bacillus thuringiensis* var. israelensis and temephos in field mosquito populations of *Ochlerotatus cataphylla* (Diptera: Culicidae): Toward resistance? *Environ. Toxicol. Chem.* **2007**, *26*, 157. [CrossRef] [PubMed]

198. Paris, M.; Boyer, S.; Bonin, A.; Collado, A.; David, J.-P.; Despres, L. Genome scan in the mosquito *Aedes rusticus*: Population structure and detection of positive selection after insecticide treatment. *Mol. Ecol.* **2010**, *19*, 325–337. [CrossRef] [PubMed]

199. Su, T.; Thieme, J.; White, G.S.; Lura, T.; Mayerle, N.; Faraji, A.; Cheng, M.L.; Brown, M.Q. High Resistance to *Bacillus sphaericus* and Susceptibility to Other Common Pesticides in *Culex pipiens* (Diptera: Culicidae) from Salt Lake City, UT. *J. Med. Entomol.* **2019**, *56*, 506–513. [CrossRef]
200. Park, H.-W.; Bideshi, D.K.; Federici, B.A. Properties and applied use of the mosquitocidal bacterium, *Bacillus sphaericus*. *J. Asia. Pac. Entomol.* **2010**, *13*, 159–168. [CrossRef]
201. Ahmed, I.; Yokota, A.; Yamazoe, A.; Fujiwara, T. Proposal of *Lysinibacillus boronitolerans* gen. nov. sp. nov., and transfer of *Bacillus fusiformis* to *Lysinibacillus fusiformis* comb. nov. and *Bacillus sphaericus* to *Lysinibacillus sphaericus* comb. nov. *Int. J. Syst. Evol. Microbiol.* **2007**, *57*, 1117–1125. [CrossRef]
202. Nielsen-Leroux, C.; Pasquier, F.; Charles, J.F.; Sinègre, G.; Gaven, B.; Pasteur, N. Resistance to *Bacillus sphaericus* involves different mechanisms in *Culex pipiens* (Diptera:Culicidae) larvae. *J. Med. Entomol.* **1997**, *34*, 321–327. [CrossRef]
203. Rao, D.R.; Mani, T.R.; Rajendran, R.; Joseph, A.S.; Gajanana, A.; Reuben, R. Development of a high level of resistance to *Bacillus sphaericus* in a field population of *Culex quinquefasciatus* from Kochi, India. *J. Am. Mosq. Control Assoc.* **1995**, *11*, 1–5.
204. Rodcharoen, J.; Mulla, M.S. Biological Fitness of *Culex quinquefasciatus* (Diptera: Culicidae) Susceptible and Resistant to *Bacillus sphaericus*. *J. Med. Entomol.* **1997**, *34*, 5–10. [CrossRef]
205. De Oliveira, C.M.F.; Filho, F.C.; Beltràn, J.E.N.; Silva-Filha, M.H.; Regis, L. Biological fitness of a *Culex quinquefasciatus* population and its resistance to *Bacillus sphaericus*. *J. Am. Mosq. Control Assoc.* **2003**, *19*, 125–129.
206. Amorim, L.B.; de Barros, R.A.; de Melo Chalegre, K.D.; de Oliveira, C.M.F.; Narcisa Regis, L.; Silva-Filha, M.H.N.L. Stability of *Culex quinquefasciatus* resistance to *Bacillus sphaericus* evaluated by molecular tools. *Insect Biochem. Mol. Biol.* **2010**, *40*, 311–316. [CrossRef] [PubMed]
207. Rojas, J.E.; Mazzarri, M.; Sojo, M.; García, A.G.Y. Effectiveness of *Bacillus sphaericus* strain 2362 on larvae of Anopheles nuñeztovari. *Investig. Clin.* **2001**, *42*, 131–146. [PubMed]
208. Nicolas, L.; Darriet, F.; Hougard, J.M. Efficacy of *Bacillus sphaericus* 2362 against larvae of *Anopheles gambiae* under laboratory and field conditions in West Africa. *Med. Vet. Entomol.* **1987**, *1*, 157–162. [CrossRef] [PubMed]
209. Derua, Y.A.; Kahindi, S.C.; Mosha, F.W.; Kweka, E.J.; Atieli, H.E.; Zhou, G.; Lee, M.-C.; Githeko, A.K.; Yan, G. Susceptibility of *Anopheles gambiae* complex mosquitoes to microbial larvicides in diverse ecological settings in western Kenya. *Med. Vet. Entomol.* **2019**, *33*, 220–227. [CrossRef]
210. Skovmand, O.; Bauduin, S. Efficacy of a granular formulation of *Bacillus sphaericus* against *Culex quinquefasciatus* and *Anopheles gambiae* in West African countries. *J. Vector Ecol.* **1997**, *22*, 43–51.
211. Poopathi, S.; Mani, T.R.; Rao, D.R.; Kabilan, L. Evaluation of Synergistic Interaction between *Bacillus sphaericus* and a Neem-based Biopesticide on Bsph-Susceptible *Culex quinquefasciatus* Say Larvae. *Int. J. Trop. Insect Sci.* **2011**, *22*, 303–306. [CrossRef]
212. Raoult, D.; Abat, C. Developing new insecticides to prevent chaos: The real future threat. *Lancet Infect. Dis.* **2017**, *17*, 804–805. [CrossRef]
213. IVCC Annual Report 2017–2018. 2018. Available online: http://www.ivcc.com/about/governance/annual-reports (accessed on 20 April 2020).
214. Sergeant, M.; Baxter, L.; Jarrett, P.; Shaw, E.; Ousley, M.; Winstanley, C.; Morgan, J.A.W. Identification, typing, and insecticidal activity of Xenorhabdus isolates from entomopathogenic nematodes in United Kingdom soil and characterization of the xpt toxin loci. *Appl. Environ. Microbiol.* **2006**, *72*, 5895–5907. [CrossRef]
215. Pineda-Castellanos, M.L.; Rodríguez-Segura, Z.; Villalobos, F.J.; Hernández, L.; Lina, L.; Nuñez-Valdez, M.E. Pathogenicity of Isolates of Serratia Marcescens towards Larvae of the Scarab Phyllophaga Blanchardi (Coleoptera). *Pathogens* **2015**, *4*, 210–228. [CrossRef]
216. Niu, H.; Wang, N.; Liu, B.; Xiao, L.; Wang, L.; Guo, H. Synergistic and additive interactions of *Serratia marcescens* S-JS1 to the chemical insecticides for controlling *Nilaparvata lugens* (Hemiptera: Delphacidae). *J. Econ. Entomol.* **2018**, *111*, 823–828. [CrossRef]
217. Wei, G.; Lai, Y.; Wang, G.; Chen, H.; Li, F.; Wang, S. Insect pathogenic fungus interacts with the gut microbiota to accelerate mosquito mortality. *Proc. Natl. Acad. Sci. USA* **2017**, *114*, 5994–5999. [CrossRef] [PubMed]
218. Islam, M.T.; Mubarak, M.S. Diterpenes and their derivatives as promising agents against dengue virus and dengue vectors: A literature-based review. *Phyther. Res.* **2019**. Available online: https://doi.org/10.1002/ptr.6562 (accessed on 20 April 2020).

219. Itoh, T.; Kawada, H.; Abe, A.; Eshita, Y.; Rongsriyam, Y.; Igarashi, A. Utilization of bloodfed females of *Aedes aegypti* as a vehicle for the transfer of the insect growth regulator pyriproxyfen to larval habitats. *J. Am. Mosq. Control Assoc.* **1994**, *10*, 344–347. [PubMed]

220. Revay, E.E.; Müller, G.C.; Qualls, W.A.; Kline, D.L.; Naranjo, D.P.; Arheart, K.L.; Kravchenko, V.D.; Yefremova, Z.; Hausmann, A.; Beier, J.C.; et al. Control of *Aedes albopictus* with attractive toxic sugar baits (ATSB) and potential impact on non-target organisms in St. Augustine, Florida. *Parasitol. Res.* **2014**, *113*, 73–79. [CrossRef]

221. Naranjo, D.P.; Qualls, W.A.; Müller, G.C.; Samson, D.M.; Roque, D.; Alimi, T.; Arheart, K.; Beier, J.C.; Xue, R.-D. Evaluation of boric acid sugar baits against *Aedes albopictus* (Diptera: Culicidae) in tropical environments. *Parasitol. Res.* **2013**, *112*, 1583–1587. [CrossRef]

222. Müller, G.C.; Beier, J.C.; Traore, S.F.; Toure, M.B.; Traore, M.M.; Bah, S.; Doumbia, S.; Schlein, Y. Successful field trial of attractive toxic sugar bait (ATSB) plant-spraying methods against malaria vectors in the Anopheles gambiae complex in Mali, West Africa. *Malar. J.* **2010**, *9*, 210. [CrossRef] [PubMed]

223. Allan, S.A. Susceptibility of adult mosquitoes to insecticides in aqueous sucrose baits. *J. Vector Ecol.* **2011**, *36*, 59–67. [CrossRef]

224. George, O. *Poinar Nematodes for Biological Control of Insects*; CRC Press: Boca Raton, FL, USA, 2018. [CrossRef]

225. Petersen, J.J. Role of mermithid nematodes in biological control of mosquitoes. *Exp. Parasitol.* **1973**, *33*, 239–247. [CrossRef]

226. Abagli, A.Z.; Alavo, T.B.; Platzer, E.G. Efficacy of the insect parasitic nematode, *Romanomermis iyengari*, for malaria vector control in Benin West Africa. *Malar. J.* **2012**, *11*, P5. [CrossRef]

227. Paily, K.P.; Balaraman, K. Susceptibility of ten species of mosquito larvae to the parasitic nematode *Romanomermis iyengari* and its development. *Med. Vet. Entomol.* **2000**, *14*, 426–429. [CrossRef]

228. Abagli, A.Z.; Alavo, T.B.C.; Perez-Pacheco, R.; Platzer, E.G. Efficacy of the mermithid nematode, *Romanomermis iyengari*, for the biocontrol of *Anopheles gambiae*, the major malaria vector in sub-Saharan Africa. *Parasit. Vectors* **2019**, *12*, 253. [CrossRef]

229. Abagli, A.Z.; Alavo. Biocontrol of *Culex quinquefasciatus* using the insect parasitic nematode, Romanomermis iyengari *(Nematoda: Mermithidae)*. *Trop. Biomed.* **2019**, *36*, 1003–1013. Available online: http://msptm.org/files/Vol36No4/1003-1013-Alavo-TBC.pdf (accessed on 20 April 2020).

230. Britch, S.C.; Nyberg, H.; Aldridge, R.L.; Swan, T.; Linthicum, K.J. Acoustic Control of Mosquito Larvae in Artificial Drinking Water Containers. *J. Am. Mosq. Control Assoc.* **2016**, *32*, 341–344. [CrossRef]

231. Fredregill, C.L.; Motl, G.C.; Dennett, J.A.; Bueno, R.; Debboun, M. Efficacy of Two Larvasonic[TM] Units Against *Culex* Larvae and Effects on Common Aquatic Nontarget Organisms in Harris County, Texas [1]. *J. Am. Mosq. Control Assoc.* **2015**, *31*, 366–370. [CrossRef]

232. Mukundarajan, H.; Hol, F.J.H.; Castillo, E.A.; Newby, C.; Prakash, M. Using mobile phones as acoustic sensors for high-throughput mosquito surveillance. *Elife* **2017**, *6*, e27854. [CrossRef] [PubMed]

233. Johnson, B.J.; Rohde, B.B.; Zeak, N.; Staunton, K.M.; Prachar, T.; Ritchie, S.A. A low-cost, battery-powered acoustic trap for surveilling male *Aedes aegypti* during rear-and-release operations. *PLoS ONE* **2018**, *13*, e0201709. [CrossRef] [PubMed]

234. Whyard, S.; Erdelyan, C.N.G.; Partridge, A.L.; Singh, A.D.; Beebe, N.W.; Capina, R. Silencing the buzz: A new approach to population suppression of mosquitoes by feeding larvae double-stranded RNAs. *Parasit. Vectors* **2015**, *8*, 96. [CrossRef] [PubMed]

235. Jesus, T.; Wanner, E.; Cardoso, R. A receding horizon control approach for integrated vector management of *Aedes aegypti* using chemical and biological control: A mono and a multiobjective approach. *Math. Methods Appl. Sci.* **2019**. Available online: https://doi.org/10.1002/mma.6115 (accessed on 20 April 2020).

*pathogens*

Article

Highly Sensitive Virome Characterization of *Aedes aegypti* and *Culex pipiens* Complex from Central Europe and the Caribbean Reveals Potential for Interspecies Viral Transmission

Jakob Thannesberger [1], Nicolas Rascovan [2], Anna Eisenmann [1], Ingeborg Klymiuk [3], Carina Zittra [4,5], Hans-Peter Fuehrer [4], Thea Scantlebury-Manning [6], Marquita Gittens-St.Hilaire [7], Shane Austin [6], Robert Clive Landis [8] and Christoph Steininger [1,*]

[1] Division of Infectious Diseases, Department of Medicine 1, Medical University of Vienna, 1090 Vienna, Austria; jakob.thannesberger@meduniwien.ac.at (J.T.); anna.eisenmann@gmx.at (A.E.)
[2] Department of Genomes & Genetics, Institut Pasteur, 75015 Paris, France; nicorasco@gmail.com
[3] Center for Medical Research, Core Facility Molecular Biology, Medical University of Graz, 8036 Graz, Austria; ingeborg.klymiuk@medunigraz.at
[4] Institute of Parasitology, University of Veterinary Medicine, 1210 Vienna, Austria; carina.zittra@univie.ac.at (C.Z.); Hans-Peter.Fuehrer@vetmeduni.ac.at (H.-P.F.)
[5] Unit Limnology, Department of Functional and Evolutionary Ecology, University of Vienna, 1010 Vienna, Austria
[6] Department of Biological and Chemical Sciences, Faculty of Science and Technology, Cave Hill Campus, The University of the West Indies, Bridgetown BB11000, Barbados; thea.scantlebury-manning@cavehill.uwi.edu (T.S.-M.); shane.austin@cavehill.uwi.edu (S.A.)
[7] Faculty of Medical Sciences, University of the West Indies, Queen Elizabeth Hospital, St. Michael BB14004, Barbados; marquita.gittens@cavehill.uwi.edu
[8] Edmund Cohen Laboratory for Vascular Research, George Alleyne Chronic Disease Research Centre, The University of the West Indies, Bridgetown BB11115, Barbados; clive.landis@cavehill.uwi.edu
*   Correspondence: christoph.steininger@meduniwien.ac.at; Tel.: +43-140-4004-4400; Fax: +43-140-4004-4180

Received: 16 July 2020; Accepted: 19 August 2020; Published: 21 August 2020

Abstract: Mosquitoes are the most important vectors for arthropod-borne viral diseases. Mixed viral infections of mosquitoes allow genetic recombination or reassortment of diverse viruses, turning mosquitoes into potential virologic mixing bowls. In this study, we field-collected mosquitoes of different species (*Aedes aegypti* and *Culex pipiens complex*), from different geographic locations and environments (central Europe and the Caribbean) for highly sensitive next-generation sequencing-based virome characterization. We found a rich virus community associated with a great diversity of host species. Among those, we detected a large diversity of novel virus sequences that we could predominately assign to circular Rep-encoding single-stranded (CRESS) DNA viruses, including the full-length genome of a yet undescribed *Gemykrogvirus* species. Moreover, we report for the first time the detection of a potentially zoonotic CRESS-DNA virus (*Cyclovirus VN*) in mosquito vectors. This study expands the knowledge on virus diversity in medically important mosquito vectors, especially for CRESS-DNA viruses that have previously been shown to easily recombine and jump the species barrier.

Keywords: mosquito virome; CRESS-DNA viruses; CyCV-VN; insect-specific viruses; ISV; BatCV

---

1. Introduction

Mosquitoes represent the most medically important group of arthropod vectors. In particular, mosquitoes of the *Aedes* and *Culex* genera are well adapted to human environments, making them the most important transmission vectors for arthropod-borne viruses (arboviruses) such as Dengue virus, Zika virus, or West Nile virus [1]. Nevertheless, arboviruses sensu stricto are not the only viruses that replicate in mosquito vectors [2–5].

Mosquitoes may be infected by a range of insect-specific viruses (ISVs) that replicate only in arthropod cells, unlike arboviruses that also replicate in vertebrate cells. Arboviruses are closely related to ISVs in terms of biological properties and phylogenetic distance and may have arisen from ISVs that occasionally gained the capacity to infect secondary hosts [6–8]. Most RNA ISVs are classified in the families of *Flaviviridae*, *Togaviridae*, the order of *Bunyavirales*, and, recently, *Mesoniviridae* [6]. Despite being confined to the arthropod host, ISVs are medically relevant as they may alter arbovirus replication and transmission [6,9–12]. For example, coinfection with ISVs has been shown to attenuate arbovirus replication in *Aedes* cell lines in vitro, including West Nile virus, Japanese encephalitis virus, and Zika virus [13,14]. In vivo studies have further confirmed that dissemination of the West Nile virus is attenuated in naturally ISV-infected mosquito populations [15].

After identification of the first ISV (cell-fusing agent virus) in 1975, it took almost 25 years for the second ISV to be discovered (Kamiti river virus) and seven more years for the third (Culex flavivirus) [6]. While advances in metagenomic sequencing are constantly accelerating the detection rate of novel ISVs, the vast diversity of ISVs has yet to be fully explored and understood [6]. Comprehensive knowledge of ISV diversity and abundance is pivotal to understanding arbovirus evolution and the interaction of ISVs with medically important mosquito vectors. Thereby, it may also help to identify ways to use ISVs in arbovirus control efforts.

In addition to arboviruses and ISVs, mosquitoes may temporarily host additional virus species that do not replicate in the arthropod hosts but become transiently associated with the mosquito vector through dynamic interaction with virus sources of the prevailing environment. For example, depending on species and sex, mosquitoes can be nourished on the blood of vertebrate hosts (female mosquitoes only) or on the nectar of pollinating plants (both female and male). The diversity of the mosquito virome is, therefore, in constant flux, depending on the surrounding environment [4]. Viruses ingested from other host organisms may, therefore, constitute an abundant and diverse fraction of the mosquito virome, even if they may not be capable of replicating within the mosquito [16].

The presence of a range of viruses from different hosts may drive interspecies recombination or reassortment within mosquitoes and promote the emergence of novel virus species [17]. The likelihood of a recombination event is increased when phylogenetically closely related viruses are present in the same biological reservoir, and this may be especially true of viruses with low-complexity genomes such as circular Rep-encoding ssDNA viruses (CRESS-DNA viruses) [18]. CRESS-DNA viruses undergo rapid evolution and are prone to recombination events due to their highly conserved genome organization throughout all known lineages. Genomic analysis of the recently identified CRESS-DNA virus Porcine circovirus 3 revealed that the virus genome was most likely formed by recombination of avian and mammalian (especially bat) Circovirus strains, which subsequently led to a species jump into pigs and an attendant economic loss of livestock production [19]. Mixed CRESS-DNA virus infections of mosquito vectors might, therefore, drive virus recombination events and subsequent formation of novel viruses with an altered host spectrum. CRESS-DNA viruses are highly abundant in many ecosystems and have recently gained attention as potential human pathogenic viruses [18].

The replication of arboviruses and ISVs within the same biological compartment and the transfer of viruses between ecological niches provide the prerequisite for the emergence of novel mutant viruses with pathogenic potential. Metagenomic analysis of the mosquito virome can promote unbiased detection of viruses that might not have been associated with mosquito vectors before, including zoonotic viruses. In this study, we aim to analyze the metagenomic virome diversity of medically important mosquito vectors. We performed highly sensitive virus metagenome sequencing of field-collected *Aedes aegypti* (*Ae.ae.*) and *Culex pipiens* complex (*C.pip.cl.*) mosquitoes from the Caribbean (Barbados; *Ae. aegypti* and *Cx. pipiens* complex) and central Europe (Austria; *Cx. pipiens* complex) to conduct comparative virome analysis of the two ecologically divergent areas. We characterized phylogenetic virome patterns and the spectrum of virus-associated hosts. Viral hits were categorized by concordance with database entries, hence revealing any novel viruses. Among those, we describe the genome organization of a novel virus of the genus of *Gemykrogvirus* and, for the first time, identify a potentially zoonotic Cyclovirus (*CyCV-VN*) in mosquito vectors.

2. Material and Methods

2.1. Mosquito Collection and Taxonomic Identification

Mosquito collection was performed as previously described, using carbon dioxide-equipped BG sentinel traps (Biogents AG, Regensburg, Germany) for 24-h time periods [20]. To compare sites of divergent ecological preconditions, mosquitoes were collected in Austria and Barbados. Austria is located in the temperate humid region of central Europe. Barbados, a Caribbean island of the West Indies, is within the tropical humid region.

In Austria, traps were set up at two trapping sites at the municipal area of Vienna (Supplementary Table S1). Locations were characterized based on satellite images (Google® Inc, Mountain View, CA, USA) and land cover maps provided by the Austrian Environment Agency (CORINE Land cover map Austria; Supplementary Figure S4) [21]. Calculations on absolute and relative coverage were done in ImageJ [22]. Mosquito collection in Barbados was performed during October and November 2016 over eight consecutive days. Twelve collection sites, in immediate proximity to human housing, were chosen. A detailed description of the collection spots has been previously published [20]. *Culex pipiens* complex mosquitoes yielded at collection spot Bbd03 (13.2693939208984, −59.6246032714843) were used for this study. Trapped mosquitoes were shock-frozen at −80°C within a maximum time of 60 min after trap disassembly and transferred to the research laboratory of the University of Veterinary Medicine, Vienna. Female mosquitoes were specified (morphologically) using the key of Becker and (single legs) genetically verified with molecular barcoding, as previously reported [23,24]. None of the individuals included in subsequent analysis showed signs of recent blood meal intake.

2.2. Sample Preparation—Virus Purification and Enrichment Protocol (VIPEP)

Mosquito individuals of each spot were pooled in numbers of 50. Mechanical homogenization was performed using a TissueLyser II (Qiagen, Venlo, The Netherlands) with five 2.8 mm ceramic beads (Peqlab, Erlangen, Germany) and 1 mL of Dulbecco's phosphate-buffered saline (DPBS) buffer at a frequency of 30 strokes per second for a duration of 4 × 30 s interspersed by a 30 s pause. Virus purification and enrichment protocol (VIPEP) was performed on mosquito homogenates in combination with an additional SpeedVac (Thermo Fisher Scientific, Waltham, MA, USA) concentration step, as previously described [25]. In short, virus particles were enriched by a nine-step procedure that included initial resuspension of the homogenates in DPBS buffer pH7 (Dulbecco's PBS, no calcium, no magnesium, Thermo Fisher Scientific), two centrifugation steps for 5 min at 2500× *g* and 15 min at 4800× *g*, filtration through a 0.45 µM syringe filter, ultrafiltration using 50 kDa molecular weight cut-off filtration units (Amicon Ultra-15 50K, Merck Millipore, Cork, Ireland), DNase I digestion (Qiagen), DNA and RNA preparation using a QIAamp UCP Micro Kit (Qiagen), blocking of ribosomal RNA sequences using a set of 5 specific oligonucleotides, cDNA synthesis using nonribosomal hexanucleotides together with the Super Script IV enzyme (Thermo Fisher Scientific), and final amplification of the total nucleic acids using a Repli-g kit (Qiagen).

2.3. Illumina Library Preparation and Sequencing

MDA-amplified double-stranded cDNA and genomic DNA were quantified with a Qubit dsDNA High Sensitivity Kit on a Qubit 4 fluorometer (Thermo Fisher Scientific) according to the manufacturer's instructions. Shotgun library preparation was done with a NEBNext Ultra II DNA Library Prep Kit for Illumina in combination with the Index Primers Set 1 and 2 (New England BioLabs, Frankfurt, Germany) according to the manufacturer's instructions. Briefly, up to 30 ng dsDNA were fragmented by ultrasonication in a Bioruptor Pico sonication system (Diagenode, Liege, Belgium) in a total volume of 55 µL 1× TE for 5 cycles of 15 s on and 30 s off. Then, 50 µL of fragmented DNA was used for end repair and adapter ligation reactions, according to the manufacturer's instructions, for a DNA input of less than 100 ng. Size selection and purification were performed according to instructions for 500 to 700 bp insert size. Subsequent PCR amplification was performed with 12 cycles, and libraries eluted after amplification and purification in 33 µL 1× TE buffer (pH 8.0). For quality control, libraries were analyzed with a DNA High Sensitivity Kit on a 2100 Bioanalyzer system (Agilent Technologies, Santa Clara, CA, USA) and quantified on a Quantus™ fluorometer (Promega, Walldorf, Germany). Following library preparation, equimolar pools were sequenced on an Illumina MiSeq desktop sequencer (Illumina, San Diego, CA, USA). Libraries were diluted to 8 pM and run without PhiX control for 600 cycles with version three chemistry according to the manufacturer's instructions. FASTQ Files were used for data analysis.

2.4. Bioinformatic Analysis

Raw sequencing data were quality trimmed and adaptors removed using AdapterRemoval v2.2.0 (https://adapterremoval.readthedocs.io), keeping only sequences longer than 30 bp, with a maximum error rate of 3 and trimming ambiguous bases in 3′ and 5′ ends. Trimmed datasets were de novo assembled using metaSPAdes v3.12.0 (https://cab.spbu.ru/spades) with default parameters. Contigs larger than 500 bp were used for downstream analyses. Open reading frames (ORF) were predicted on the contig sequences using MetaGeneMark v3.38. BLASTN analysis (Blast+ v2.5.0) (http://exon.gatech.edu/meta_gmhmmp.cgi) against the NCBI nt database was used as a first approach to determine contig identity, employing a minimum e-value threshold of 1×10^{-5} and keeping the first 25 hits. A BLASTP analysis of the predicted ORF sequences (in amino acids) against the NCBI nr database (*e*-value $< 1 \times 10^{-5}$, 25 first hits) was used as a second approach. Depending on the case (see the Results section), taxonomic annotation of contigs was resolved by either importing BLAST results to MEGAN v6.10.13, which uses the lowest common ancestor algorithm for classification, or by considering the best BLAST hit. The relative abundance of each contig in each sample was calculated using the RPKM metric (reads per kilobase per million mapped reads), which normalizes by contig lengths and sequencing depth of samples. For this, reads were remapped on contigs using bwa v0.7.10-r789 with the *"aln"* method, while coverage and depth were calculated on the resulting bam files using custom scripts and bedtools v2.26.0 (https://bedtools.readthedocs.io). In addition, the abundance of each individual taxa (based on NCBI taxID) was also estimated using the RPKM method, utilizing the average of RPKM of all contigs assigned to the same taxID.

Vector sequences and hits that were also detected in the negative control were excluded from further analysis. To simplify the taxonomic analysis, best BlastN hits to identical NCBI taxonomy ID were clustered. If more than one contig matched the same best hit, mean pairwise sequence identity (% ID) and mean coverage of query sequence (% subject coverage) were used for the analysis. Alternatively, when contigs were found to hit different sequences but with the same NCBI taxonomy ID, mean % ID and mean % query coverage were calculated for each subject sequence.

2.5. Phylogenetic Analysis

Predicted Rep and Cap protein sequences of the novel virus were used for phylogenetic analysis. Amino acid sequences of Rep and Cap proteins of (i) proposed type species of the nine phylogenetic genera of the family of *Genomoviridae*, according to Zhao et al., and (ii) ten other *Genomoviridae* species identified by BlastN were retrieved from the NCBI database [26]. Additionally, the homologous protein sequence of *Gemycircularvirus* type species was included as the outlier. Sequence alignments were created in T-Coffee using the PSI-Coffee protein alignment algorithm, including protein structure information [27]. Alignments of Rep and Cap protein sequences were trimmed to conserve only regions covered by the contig sequence, and trimmed alignments were concatenated for phylogenetic tree construction.

The phylogenetic tree was generated using the maximum likelihood algorithm based on the LG model including discrete Gamma distribution (+G) and by assuming that a certain fraction of sites are evolutionarily invariable (+I). This model was identified as the best fitting model with the lowest BIC (Bayesian information criterion) in MEGA10 [28,29]. The bootstrap consensus tree was inferred from 100 repetitions. Initial trees for the heuristic search were obtained automatically by applying neighbor-ioin (NJ) and BioNJ algorithms to a matrix of pairwise distances estimated using a JTT model and then selecting the topology with a superior log likelihood value.

2.6. Virus-Specific Polymerase Chain Reactions (PCRs)

PCR reactions were performed for verification of metagenomic detection of Bat circovirus POA/2012/II (BatCv) and Cyclovirus VN isolate hcf1 (CyCV-VN). Primers were designed in Primer3 (http://primer3.ut.ee) software, and primer sequences were tested for specificity by Primer Blast analysis (https://www.ncbi.nlm.nih.gov/tools/primer-blast/). Amplicon length was chosen to be less than 200 nt (132 bp for BatCV and 160 bp for CyCV-VN), as required for testing shared sequencing libraries. The following primer pairs were chosen: BatCV_fw (5′- ATCCAGCCGTAGAAGTCGTC-3′) and BatCV_rv (5′-CGGAAAATCAAAGCGTGCAC-3′), CyCV_fw (5′- TGAAGGAGGAGAGACATGCC-3′) and CyCV_rv (5′- TGTTCCAGTCGATCCCCAAA-3′). The PCR mixture contained 1× iTaq PCR buffer (Bio-Rad, Hercules, CA, USA), 200 mM of each deoxynucleotide triphosphate (dNTP) mix, 2 mM MgCl$_2$, 5 mM of each forward and reverse primer, and 0.4 U iTaq DNA polymerase (Bio-Rad) in a 20 µL reaction volume. Additionally, 2 µL of VIPEP enriched Ae.ae.BRB or C.pip.cl.BRB was used as a template and 2 µL of nucleic acid-free water as a negative control. PCR reactions included initial denaturation in a thermal cycler at 95 °C for 15 min, followed by 35 cycles of denaturation at 94 °C for 60 s, annealing at 59 °C (BatCV) or 54 °C (CyCV-VN) for 60 s and extension at 72 °C for 2 min, followed by a final extension at 72 °C for 10 min. Amplicons were visualized by electrophoresis on 2% agarose gel.

Verification of human Torquet Teno virus (TTV) was done by real-time PCR (RT-PCR), as previously described by Maggi et al. [30]. Briefly, PCR reactions were done in a total volume of 20 µL that contained 10 µL iTaq Universal Supermix (Bio-Rad), 300 nM concentration of primers (AMTS and AMTAS), 200 nM TaqMan probe (AMTPTU), and 2 µL of template DNA. Mixtures were prepared in 96-well optical microtiter plates (Thermo Fisher Scientific) and amplified on a StepOnePlus real-time PCR system (Thermo Fisher Scientific) by using the following cycling parameters: denaturation for 90 s at 95 °C, followed by 40 cycles of denaturation for 15 s at 95 °C, and annealing and extension for 60 s at 60 °C. A plasmid carrying the amplicon insert was used as positive control, and nucleic acid-free water was used as negative control. All samples were tested in duplicates.

3. Results

3.1. Highly Sensitive Virome Characterization

For this study, we used four pools of field-collected mosquitoes from Austria and Barbados (Supplementary Table S1). Highly specific purification of virus nucleic acids was performed for each pool separately, including a reverse transcriptase step to detect DNA and RNA viruses. Virus metagenome sequencing on the Illumina MiSeq platform generated 2.00×10^6 to 1.41×10^7 raw sequencing read pairs per sample; read pairs were consequently trimmed and assembled to contigs. Taxonomic annotation matched 66% to 83% of contigs longer than 500 bp to NCBI database entries (Supplementary Table S1). Efficient presequencing elimination of cells, cell-derived debris, and nonviral sequences yielded highly sensitive sequencing of viral genomes. Hits to plasmid sequences and sequences matching hits that were detected in no-template control samples were removed from the dataset. Eukaryotic and prokaryotic sequences accounted only for a minor fraction (3–12%) of total assigned sequences, and the majority of reads could be assigned to viral genomes (88–97%) (Supplementary Figure S1). Viral contigs that were assigned to broad categories, such as uncultured or unclassified viruses, were excluded from further analysis. We found that mosquitoes hosted a highly diverse set of viruses that differed greatly between the pools. The overall richness was high, with 103 different viral taxa identified at the species level, of which 86 were detected in a single sample (Figure 1). We further assessed virome richness at the family taxonomic level. We found that the mosquito virome in all four pools was predominantly constituted by families from the class of circular CRESS-DNA viruses (Supplementary Figure S2), which were also the most abundant fraction in all pools (Supplementary Figure S3).

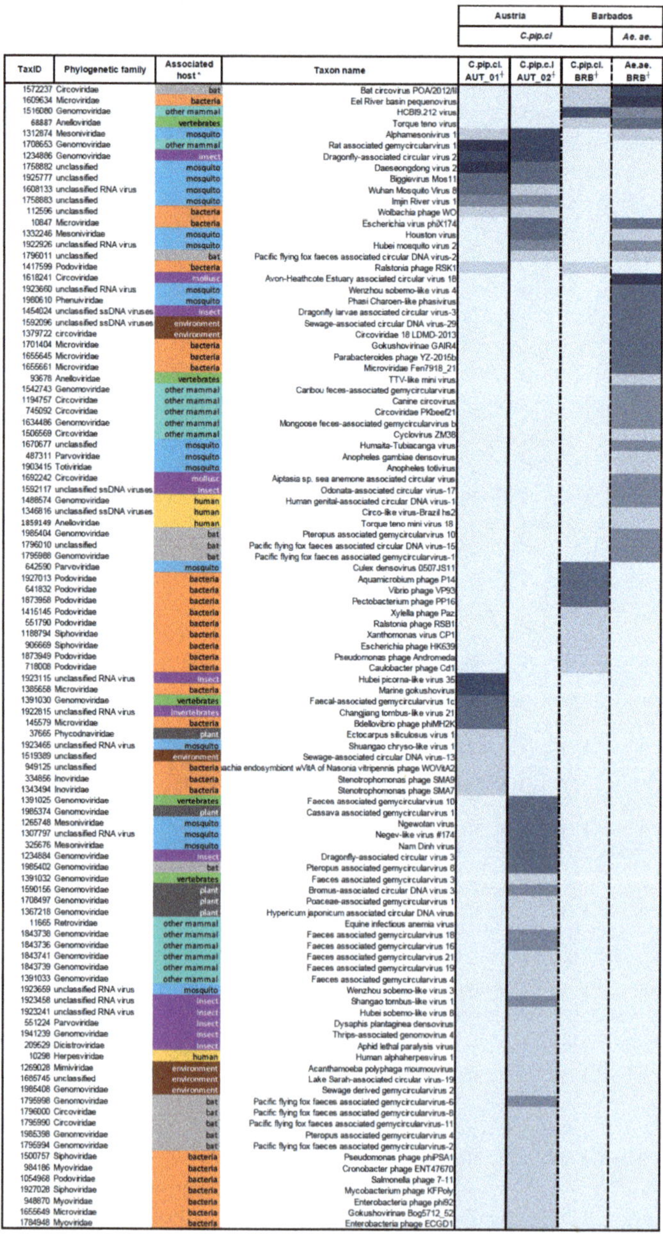

The four abundance columns (shades of blue) are, under Austria / *C.pip.cl*: C.pip.cl. AUT_01[†] and C.pip.c.l AUT_02[†]; under Barbados: C.pip.cl. BRB[†] (*C.pip.cl*) and Ae.ae. BRB[†] (*Ae. ae.*).

TaxID	Phylogenetic family	Associated host*	Taxon name
1572237	Circoviridae	bat	Bat circovirus POA/2012/II
1609634	Microviridae	bacteria	Eel River basin pequenovirus
1516080	Genomoviridae	other mammal	HCBI9.212 virus
68887	Anelloviridae	vertebrates	Torque teno virus
1312874	Mesoniviridae	mosquito	Alphamesonivirus 1
1708653	Genomoviridae	other mammal	Rat associated gemycircularvirus 1
1234886	Genomoviridae	insect	Dragonfly-associated circular virus 2
1758882	unclassified	mosquito	Daeseongdong virus 2
1925777	unclassified	mosquito	Biggievirus Mos11
1608133	unclassified RNA virus	mosquito	Wuhan Mosquito Virus 8
1758883	unclassified	mosquito	Imjin River virus 1
112596	unclassified	bacteria	Wolbachia phage WO
10847	Microviridae	bacteria	Escherichia virus phiX174
1332246	Mesoniviridae	mosquito	Houston virus
1922926	unclassified RNA virus	mosquito	Hubei mosquito virus 2
1796011	unclassified	bat	Pacific flying fox faeces associated circular DNA virus-2
1417599	Podoviridae	bacteria	Ralstonia phage RSK1
1618241	Circoviridae	mollusc	Avon-Heathcote Estuary associated circular virus 18
1923660	unclassified RNA virus	mosquito	Wenzhou sobemo-like virus 4
1980610	Phenuiviridae	mosquito	Phasi Charoen-like phasivirus
1454024	unclassified ssDNA viruses	insect	Dragonfly larvae associated circular virus-3
1592096	unclassified ssDNA viruses	environment	Sewage-associated circular DNA virus-29
1379722	circoviridae	environment	Circoviridae 18 LDMD-2013
1701404	Microviridae	bacteria	Gokushovirinae GAIR4
1655645	Microviridae	bacteria	Parabacteroides phage YZ-2015b
1655661	Microviridae	bacteria	Microviridae Fen7918_21
93678	Anelloviridae	vertebrates	TTV-like mini virus
1542743	Genomoviridae	other mammal	Caribou feces-associated gemycircularvirus
1194757	Circoviridae	other mammal	Canine circovirus
745092	Circoviridae	other mammal	Circoviridae PKbeef21
1634486	Genomoviridae	other mammal	Mongoose feces-associated gemycircularvirus b
1506569	Circoviridae	other mammal	Cyclovirus ZM38
1670677	unclassified	mosquito	Humaita-Tubiacanga virus
487311	Parvoviridae	mosquito	Anopheles gambiae densovirus
1903415	Totiviridae	mosquito	Anopheles totivirus
1692242	Circoviridae	mollusc	Aiptasia sp. sea anemone associated circular virus
1592117	unclassified ssDNA viruses	insect	Odonata-associated circular virus-17
1488574	Genomoviridae	human	Human genital-associated circular DNA virus-1
1346816	unclassified ssDNA viruses	human	Circo-like virus-Brazil hs2
1859149	Anelloviridae	human	Torque teno mini virus 18
1985404	Genomoviridae	bat	Pteropus associated gemycircularvirus 10
1796010	unclassified	bat	Pacific flying fox faeces associated circular DNA virus-15
1795988	Genomoviridae	bat	Pacific flying fox faeces associated gemycircularvirus-1
642590	Parvoviridae	mosquito	Culex densovirus 0507JS11
1927013	Podoviridae	bacteria	Aquamicrobium phage P14
641832	Podoviridae	bacteria	Vibrio phage VP93
1873958	Podoviridae	bacteria	Pectobacterium phage PP16
1415145	Podoviridae	bacteria	Xylella phage Paz
551790	Podoviridae	bacteria	Ralstonia phage RSB1
1188794	Siphoviridae	bacteria	Xanthomonas virus CP1
906669	Siphoviridae	bacteria	Escherichia phage HK639
1873949	Podoviridae	bacteria	Pseudomonas phage Andromeda
718008	Podoviridae	bacteria	Caulobacter phage Cd1
1923115	unclassified RNA virus	insect	Hubei picorna-like virus 35
1385658	Microviridae	bacteria	Marine gokushovirus
1391030	Genomoviridae	vertebrates	Faecal-associated gemycircularvirus 1c
1922815	unclassified RNA virus	invertebrates	Changjiang tombus-like virus 21
145579	Microviridae	bacteria	Bdellovibrio phage phiMH2K
37665	Phycodnaviridae	plant	Ectocarpus siliculosus virus 1
1923465	unclassified RNA virus	mosquito	Shuangao chryso-like virus 1
1519389	unclassified	environment	Sewage-associated circular DNA virus-13
949125	unclassified	bacteria	Wolbachia endosymbiont wVitA of Nasonia vitripennis phage WOVitA2
334856	Inoviridae	bacteria	Stenotrophomonas phage SMA9
1343494	Inoviridae	bacteria	Stenotrophomonas phage SMA7
1391025	Genomoviridae	vertebrates	Faeces associated gemycircularvirus 10
1985374	Genomoviridae	plant	Cassava associated gemycircularvirus 1
1265748	Mesoniviridae	mosquito	Ngewotan virus
1307797	unclassified RNA virus	mosquito	Negev-like virus #174
325676	Mesoniviridae	mosquito	Nam Dinh virus
1234884	Genomoviridae	insect	Dragonfly-associated circular virus 3
1985402	Genomoviridae	bat	Pteropus associated gemycircularvirus 8
1391032	Genomoviridae	vertebrates	Faeces associated gemycircularvirus 3
1590156	Genomoviridae	plant	Bromus-associated circular DNA virus 3
1708497	Genomoviridae	plant	Poaceae-associated gemycircularvirus 1
1367218	Genomoviridae	plant	Hypericum japonicum associated circular DNA virus
11665	Retroviridae	other mammal	Equine infectious anemia virus
1843738	Genomoviridae	other mammal	Faeces associated gemycircularvirus 18
1843736	Genomoviridae	other mammal	Faeces associated gemycircularvirus 16
1843741	Genomoviridae	other mammal	Faeces associated gemycircularvirus 21
1843739	Genomoviridae	other mammal	Faeces associated gemycircularvirus 19
1391033	Genomoviridae	other mammal	Faeces associated gemycircularvirus 4
1923659	unclassified RNA virus	mosquito	Wenzhou sobemo-like virus 3
1923458	unclassified RNA virus	insect	Shangao tombus-like virus 1
1923241	unclassified RNA virus	insect	Hubei sobemo-like virus 8
551224	Parvoviridae	insect	Dysaphis plantaginea densovirus
1941239	Genomoviridae	insect	Thrips-associated genomovirus 4
209529	Dicistroviridae	insect	Aphid lethal paralysis virus
10298	Herpesviridae	human	Human alphaherpesvirus 1
1269028	Mimiviridae	environment	Acanthamoeba polyphaga moumouvirus
1685745	unclassified	environment	Lake Sarah-associated circular virus-19
1985408	Genomoviridae	environment	Sewage derived gemycircularvirus 2
1795998	Genomoviridae	bat	Pacific flying fox faeces associated gemycircularvirus-6
1796000	Circoviridae	bat	Pacific flying fox faeces associated gemycircularvirus-8
1795990	Circoviridae	bat	Pacific flying fox faeces associated gemycircularvirus-11
1985398	Genomoviridae	bat	Pteropus associated gemycircularvirus 4
1795994	Genomoviridae	bat	Pacific flying fox faeces associated gemycircularvirus-2
1500757	Siphoviridae	bacteria	Pseudomonas phage phiPSA1
984186	Myoviridae	bacteria	Cronobacter phage ENT47670
1054968	Podoviridae	bacteria	Salmonella phage 7-11
1927028	Siphoviridae	bacteria	Mycobacterium phage KFPoly
948870	Myoviridae	bacteria	Enterobacteria phage ph92
1655649	Microviridae	bacteria	Gokushovirinae Bog5712_52
1784948	Myoviridae	bacteria	Enterobacteria phage ECGD1

Figure 1. Metagenomic virome richness; * host organisms grouped by color, † shades of blue indicate relative abundance in RPKM.

3.2. Influence of the Environment on the Mosquito Virome

Mosquito pools from similar environments were found to share mutual virus hits, referred to as ecosystem signature viruses. *C.pip.cl.* mosquito pools from Austria (AUT_01 and AUT_02) shared eight viral hits while mosquitoes from Barbados (Ae.ae.BRB and C.pip.cl.BRB) shared a set of four mutual viral hits (Figure 1). We did not find any evidence that mosquitoes of the same species complex from Austria and Barbados carried common viral infections. Moreover, the virome of *Cx. pipiens* complex mosquitoes from Barbados was more similar to *Ae.aegypti* mosquitoes from the same ecosystem (four mutual viral taxa) than compared to the pools of *Cx. pipiens* complex mosquitoes from Austria (one mutual viral taxon with C.pip.cl.AUT_01). However, these findings are limited by the small number of pools tested.

We then analyzed the impact of the immediate surrounding environment within the range of the average flight distance of *Cx. pipiens* complex mosquitoes from the collection spots. Governmental land coverage maps were used to characterize the immediate surroundings (Supplementary Figure S4) [20]. Although the number of sites was relatively limited, we did not observe a higher virome richness in more diverse environments. For example, the area surrounding collection spot AUT_01 was considerably more diverse (comprising seven different categories of land coverage) and included a higher proportion of natural-state areas (58.1%) such as waterbodies, urban green lands, and recreational areas than found at spot AUT_02 (37.1%) (Supplementary Table S2). However, the relative abundance of viral reads (97%) and the overall richness of viral taxa (58) was higher in pool AUT_02 than in pool AUT_01 (Supplementary Figure S1).

3.3. The Mosquito Virome is Comprised of Viruses from A Wide Diversity of Hosts

To further characterize the mosquito virome, we analyzed virus hits for their associated host species annotated in the NCBI database. We identified viruses across a wide range of hosts in both *Ae. aegypti* and *Cx. pipiens* complex mosquitoes. Besides viruses of vertebrate hosts, such as human, bat, and other mammal viruses, we detected viruses associated with invertebrate hosts, plant viruses, environmental viruses, and bacteriophages. When viral taxa were clustered according to related host groups, we observed that taxonomic richness was highest for mosquito-associated viruses and bacteriophages (Figure 2A). Within the group of vertebrate hosts, viruses associated with mammals and, in particular, to bats prevailed at high diversity. These two groups were among the most abundant throughout all four samples (Figure 2B). However, this finding might be biased by the fact that the viral microbiome of bats has been sampled more extensively than that of other wild mammal species.

3.4. Mosquito-Specific Viruses form the Core Group of the Mosquito Virome

We separated metagenomic sequence assignments to their goodness-of-assignment fit by plotting pairwise sequence identity versus query coverage in a two-dimensional scatterplot. Viral hits to broader categories were included in this analysis to gain further information on sequence similarity, regardless of classification. To test this method of analysis, we used a 100-nucleotide model sequence from an NCBI database virus genome sequence (LK931484.1) and introduced variable numbers of random point mutations between 10% to 50% of total length (Supplementary Table S3). Assignment of sequences with mutation rates greater than 30% resulted in hits, which were taxonomically unrelated to the primary sequence (e.g., plants, fish, bacteria). Sequences with 25–30% of mutated nucleotides were at the borderline of being correctly annotated (Supplementary Table S4). Assignments to taxonomically unrelated taxa were based on BLAST alignments with high pairwise sequence identity, though covering only a minor fraction of query sequence. When plotted, biologically implausible hits clustered clearly apart from correct assignments, separated by a query coverage of less than 60%. Hits to taxonomically related taxa (i.e., virus sequences) fell within close proximity to the group of correct assignments but with lower pairwise sequence identity (Supplementary Figure S5).

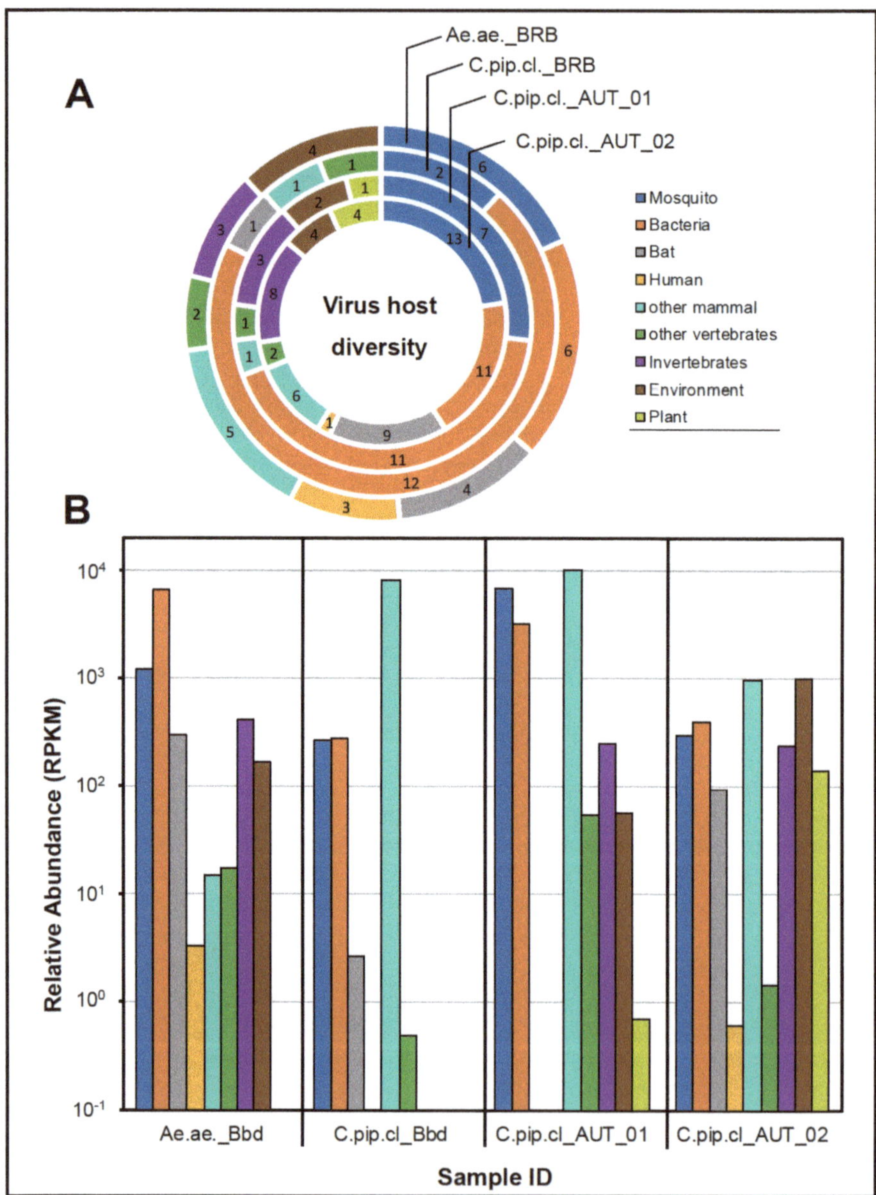

Figure 2. All obtained hits to viral taxa grouped by the associated host organism (NCBI). (**A**) Richness in number of hits to indicated host group for each mosquito pool; (**B**) accumulative relative abundance (RPKM) of viral hits grouped by host organism.

We used this method of analysis to identify those hits of the mosquito virome that closely resembled assigned virus taxa (Figure 3). Among this group of high-likelihood hits (>60% query coverage) mosquito-associated viruses were the predominant virus group besides hits to bacteriophages, bat-associated viruses, and a few vertebrate viruses. Of six high-likelihood hits to bacteriophages, two resembled *Wolbachia-*

infecting phage species. Moreover, we found that ecosystem signature viruses were predominantly assigned with a high degree of confidence.

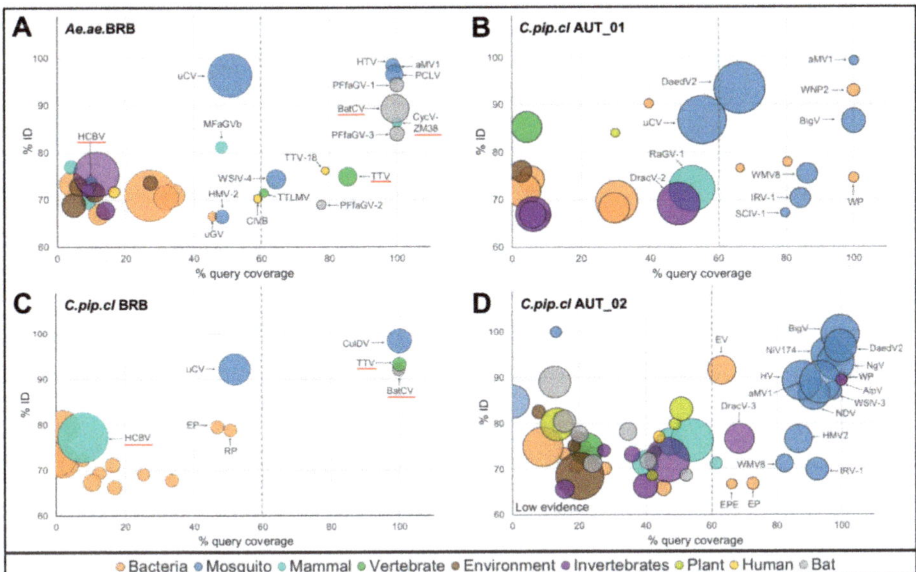

Figure 3. Goodness-of-assignment fit for viral metagenomes of (**A**) *Aedes aegypti* mosquitoes, Barbados, (**B**) *Culex pipiens* complex mosquitoes, Austria pool 01, (**C**) *Culex pipiens* complex mosquitoes, Barbados, (**D**) *Culex pipiens* complex mosquitoes, Austria pool 02. Bubbles represent metagenomic viral hits separated by % coverage of query sequence (*x*-axis) and % pairwise sequence identity (*y*-axis), size of bubbles represents relative abundance (RPKM), color indicates associated virus host; underlined sequences have been verified by PCR assay. aMV, alphamesoniviurs 1; AlpV, aphid lethal paralysis virus; BatCV, bat circovirus POA/2012/II; BigV, biggievirus Mos11; ClVB, circo-like virus, Brazil hs2; CulDV, *Culex* densovirus 0507JS11; CycV-ZM38, cyclovirus ZM38; DaedV2, Daeseongdong virus 2; DracV-2, dragonfly-associated circular virus 2; DracV-3, dragonfly-associated circular virus 3; EP, *Escherichia* phage HK639; EV, *Escherichia* virus phiX174; HMV-2, Hubei mosquito virus 2; HCBV, HCBI9.212 virus; HTV, Humaita–Tubiacanga virus; HV, Houston virus; IRV-1, Imjin River virus 1; MFaGVb, mongoose feces-associated gemycircularvirus b; NDV, Nam Dinh virus; NgV, Ngewotan virus; NlV174, Negev-like virus #174; PCLV, Phasi Charoen-like virus; PFfaGV-1, Pacific flying fox feces-associated gemycircularvirus-1; PFfaGV-2, Pacific flying fox feces-associated gemycircularvirus-2; PFfaGV-3, Pacific flying fox feces-associated gemycircularvirus-3; RaGV-1, rat-associated gemycircularvirus 1; RP, *Ralstonia* phage RSK1; SClV-1, Shuangao chryso-like virus 1; TTV, Torque teno virus; TTV-18, Torque teno virus 18; TTLMV, TTV- like mini virus; uCV, uncultured circovirus; uGV, uncultured Gokushovirinae; WP, *Wolbachia* phage WO; WN2P, *Wolbachia* endosymbiont wVitA of *Nasonia vitripennis* phage 2; WMV8, Wuhan mosquito virus 8; WSlV-3, Wenzhou sobemo-like virus 3; WSlV-4, Wenzhou sobemo-like virus 4.

For the two Austrian mosquito pools, six out of eight ecosystem signature viruses were assigned with query coverage >60 (Figure 3B,D). This group of viruses was mainly formed by mosquito-specific viruses (5 out of 6) and two *Wolbachia*-infecting bacteriophages. Daesongdong virus 2 (DaedV2) was previously identified in *C.pip.cl.* mosquitoes from South Korea, while different strains of Biggievirus Mos11 (BigV) have been identified in *C.pip.cl.* mosquitoes from the US, Italy, and India (GI KX924639, MF281708, MF281709, MH603566). *Imjin River virus 1* (*IRV1*), an ssRNA virus taxonomically related to *Wuhan mosquito virus 8* (WMV8), has been previously identified in virus metagenomes of *Cx.*

bitaeniorhynchus from South Korea [31]. For mosquitoes from Barbados, we identified two mutual viruses that lay within the range of >60% query coverage (vertebrate-infecting *Torquet Teno virus* (*TTV*) and *Bat circovirus POA/2012/II*). Notably, the fraction of CRESS-DNA viruses in the group of high-confidence hits, across all samples, was less than in the overall metagenomes (34.9% vs. 55.2%; Supplementary Table S5).

3.5. Validation of Metagenomic Results

We validated metagenomic sequencing results by using molecular detection methods. To verify single surrogate CRESS-DNA hits with a high goodness-of-assignment fit from the Ae.ae.BRB virome (Bat circovirus POA/2012/II, Cyclovirus ZM38, and Torquet Teno virus), we designed specific PCR assays on the respective metagenomic sequences. All four PCR assays yielded uniformly positive results, and sequencing of PCR amplicons verified NGS-derived metagenomic sequences (Supplementary Figure S6).

CycV-ZM38 resembles a strain of *Cyclovirus VN* (*CyCV-VN*) virus species. Sequence analysis of the contig sequence annotated as CycV-ZM38 revealed a pairwise sequence identity to CyCV-VN of 78%, including a 141 nt sequence that is identical to the last 141 nt (1855–1995) of the contig, indicative of the circular genomic structure. The contig sequence displayed a conserved genome organization, with two ORFs resembling a capsid protein (cap) coding gene and a replicase (Rep) coding gene in BlastP analysis. According to the species demarcation criteria for circoviruses, being <75% of genome nucleotide identity, the metagenomic sequence resembles a closely related, novel variant of CyCV-VN.

3.6. Novel Viruses Were Mostly CRESS-DNA Viruses

Hits with 0–60% query coverage matched database sequences only at short stretches, thereby resembling widely novel sequences. Of these, 64.9% were assigned as CRESS-DNA viruses (Supplementary Table S5). Compared to the total dataset, hits with less than 60% of query coverage were significantly enriched in CRESS-DNA viruses (chi-square test, R = 10.6, p = 0.001).

High-fidelity virus assignments were associated with hosts that fit into the biological context (i.e., mosquitoes, bats, symbiotic bacteria). Hits of lower goodness-of-assignment fit corresponded to hosts that were inappropriate to the ecosystem of mosquito collection (e.g., caribou feces-associated gemycircularvirus). Moreover, environmental viruses, plant viruses, and the majority of bacteriophage sequences were clustered in this field of low goodness-of-assignment fit.

As seen for the model sequence in Supplementary Figure S5, mutation rates exceeding 30% in a random database sequence mislead BLASTN sequence annotation to taxonomically unrelated taxa. However, this model-simulated accumulation of random point mutations assumed a completely nonconserved genome sequence. This cannot be assumed for viral genomes that include protein-coding regions or other conserved genome segments. Hits with low goodness-of-assignment fit are particularly susceptible to a change in their annotation alongside the expansion of databases, as seen for the HCBI9.212 virus. At the time of metagenome analysis for this study (10/2018), two contig sequences from the Ae.ae.BRB and C.pip.clBRB metagenomes were assigned as the HCBI9.212 virus (query coverage 10% and 7.8% and pairwise similarity 73% and 77%, respectively). Since then, the NCBI database has been constantly expanded by novel annotations, assigning both contigs, at the time of writing (02/2020), as apis mellifera genomovirus 2, with increased goodness-of-assignment fit (Supplementary Figure S7). However, further investigation of the contig sequence revealed that the short aligning sequence was located at the capsid protein gene sequence, highly conserved in the family of *Genomoviridae*. The full-length contig sequence *k141_1014* (2222 nucleotide) resembled the circular genome of a novel CRESS-DNA virus of the family of *Genomoviridae*, tentatively named *mosquito-associated virus Barbados* (*MaVBRB*; Figure 4A). Comparing the *k141_1014* sequence to the C.pip.cl.BRB-derived metagenomic sequence assigned as HCBI9.212, we found that the underlying contig sequence shared 99.5% pairwise sequence identity to the MaVBRB genomic sequence from the *Ae.ae.* pool. The sequence included two inversely oriented open reading frames coding for a replication-associated protein (Rep) and a capsid protein (Cap) and presented the structural features of two palindromic sequences flanking a putative origin of replication. In the phylogenetic tree constructed using Rep amino acid sequences from viruses of the family of *Genomoviridae*, MaVBRB clusters were within the genus of *Gemykrogvirus* (Figure 4B). Covering the total sequence with 5 abutting-primer PCRs, we confirmed the NGS-derived genomic virus sequence (Supplementary Figure S8).

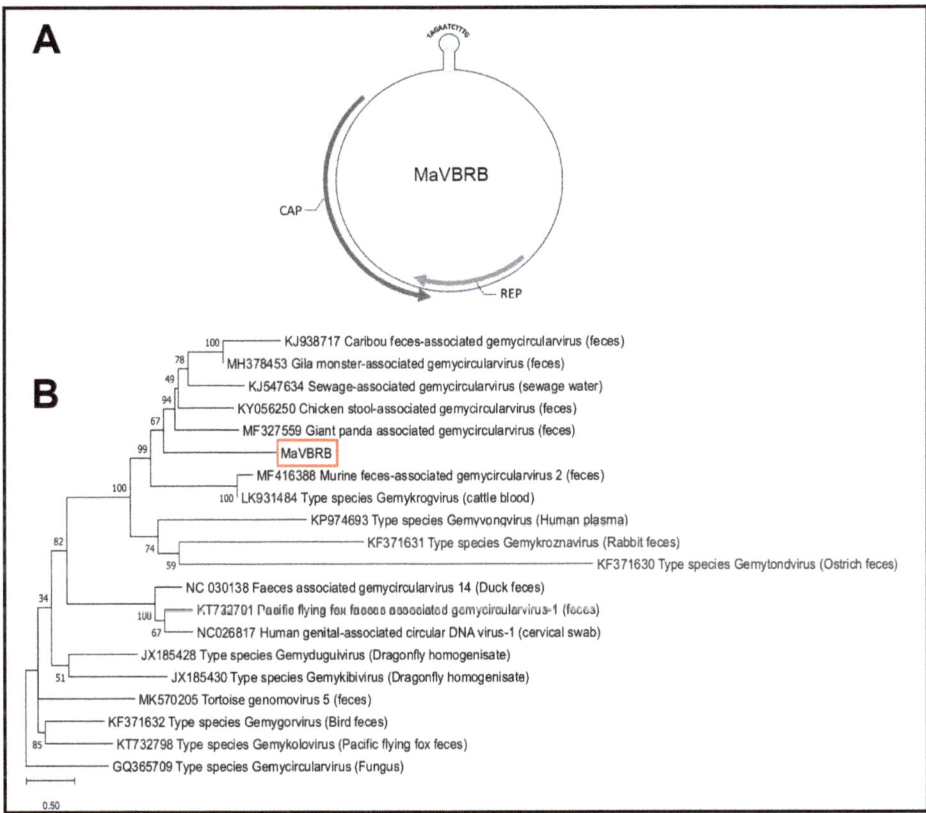

Figure 4. (**A**) Genome organization of novel *mosquito-associated virus Barbados* (*MaVBRB*) consensus sequence identified in Ae.ae.BRB and C.pip.cl.BRB mosquito pools. (**B**) Phylogenetic tree of members of the family of *Genomoviridae* using replication-associated protein (Rep) and capsid protein (Cap) sequences, tree branches indicate sequence GI and primary source of sequence isolation; *MaVBRB* clusters in the genus of *Gemykrogvirus* type species.

4. Discussion

Mosquitoes are the most important vectors for the global transmission of arthropod-borne diseases that account for more than 700,000 deaths annually [32]. Studying the viral microbiome of medically important mosquito vectors can help us to better understand virus dynamics and epidemiology of known and novel human pathogenic viruses.

In this study, we analyzed and compared virome patterns of two medically relevant mosquito species (*Cx. pipiens* complex. and *Ae. aegypti)* trapped at distinct ecosystems (Austria and Barbados). We assembled a large diversity of viral sequences matching database entries of viruses that were previously associated with a range of hosts. Separating hits by their goodness-of-assignment fit, we provide evidence that mosquito vectors host a "core virome" that is closely related to previously identified virus taxa. Strikingly, a large fraction of this set of viruses was shared among mosquito pools from the same ecosystem, even though collection points were separated by several kilometers in distance. Ecosystem signature viruses of the Austrian pools predominantly resembled close taxonomic relatives of mosquito-specific viruses that have been identified previously in *Culex* mosquitoes from different parts of the world [31,33]. Hence, our data support the previous host association of DaedV2 and BigV to *C.pip.cl.* and emphasize the global abundance of these mosquito-specific viruses. However, despite the limiting number of samples, we observed more similarity in viromes from closely related environments than in the same mosquito species from different environments. We, therefore, hypothesize that the local ecosystem may play an important role (arguably, more important than the host species) in shaping the viral communities in mosquitoes as individuals of different species are exposed to common sources of virus infection.

Moreover, we found evidence that different *Wolbachia*-infecting bacteriophage strains are part of the viral microbiome of sampled Austrian *C.pip.cl.* mosquitoes. *Wolbachia* are intracellular alphaproteobacteria, often living in endosymbiosis with arthropods. Infection with endosymbiotic *Wolbachia* has been shown to alter the reproductive behavior of its arthropod host [34]. Therefore, Wolbachia-infecting prophage WO has been recently proposed as a beneficial tool for vector control efforts [35]. This is the first report of *Wolbachia-infecting prophage WO* (Tax ID 112596) and *Wolbachia phage WOVitA2* (Tax ID 949125) in field-collected mosquitoes from central Europe. However, we did not recover full genomic sequences as corresponding contigs covered just parts of the bacteriophage genomes that might represent conserved sections of this taxonomic group.

Analyzing the viral metagenome on a taxonomic level, we found that CRESS-DNA viruses account for a disproportionally prominent fraction of the taxonomic richness and form the most abundant taxonomic group in the mosquito virome. The taxonomic clade of Rep-encoding ssDNA viruses (CRESS-DNA viruses) is highly diverse and abundant. CRESS-DNA viruses share a similar genome organization, all encoding a highly conserved replication-associated protein (Rep) and at least one further, less conserved capsid (Cap) protein. In many virus metagenomic studies, multiple displacement amplification (MDA) is used as a standard method for unspecific sequence amplification, whereby small circular ssDNA sequences such as CRESS-DNA viral genomes are preferentially amplified. Hence, it has to be taken into account that relative virus representation might thereby be biased [36]. However, the usage of MDA has uncovered the vast abundance of CRESS-DNA viruses prevailing almost ubiquitously throughout most ecosystems, including the human body.

Among CRESS-DNA viruses of vertebrate hosts, bat viruses accounted for a disproportionally high fraction of the *Ae. ae.* virome. Bat circovirus POA/2012/II (BatCv POA/II) is a CRESS-DNA virus of the family of *Circoviridae* that was initially detected in bat feces collected in Southern Brazil [37]. Besides this first metagenomic identification of BatCv POA/II, there have been no other detections to differentiate whether the virus effectively replicates in bats or if the virus simply passes through the digestive tract of these insectivorous animals, as previously suggested [37]. Hence, identification of BatCV POA/II in two mosquito pools of different species indicates that the virus might primarily infect mosquitoes rather than bats. Likewise, this observation was discussed for other families of CRESS-DNA viruses, as two out of three phylogenetic clusters of *Cycloviruses* may only infect arthropods while they are currently being associated with diverse vertebrate hosts [38]. All four bat-associated viruses of the *Ae. ae.* virome were exclusively identified by metagenomic sequencing of bat feces. Final host assignment can only be done by further virus isolation experiments. However, the current findings suggest that a spillover of CRESS-DNA viruses does not only occur from vertebrate hosts downwards onto mosquitoes but also from mosquitoes upwards. Thereby, mosquitoes act as mixing vessels for CRESS-DNA viruses from different environmental sources, which may promote recombination events leading to novel virus variants.

CRESS-DNA viruses evolve rapidly, with comparable evolution rates to RNA viruses [39]. High mutation rates and the possibility for genome recombination are consistent with a high speed of CRESS-DNA virus evolution [18,26,40,41]. By bringing together viruses from different hosts, mosquito vectors might provide ideal preconditions for recombination events of highly abundant CRESS-DNA viruses. By mapping hits by their pairwise sequence identity and query coverage, we were able to identify metagenomic sequences that most likely match known viruses from sequences that represent novel virus sequences. We provide evidence that the group of novel viruses in the mosquito virome is predominantly formed by unknown CRESS-DNA viruses. Of those, we picked one hit to characterize the viral genome of a novel *Gemykrogvirus*, tentatively named MaVBRB. MaVBRB forms a phylogenetic cluster with a later identified AmGV-2, infecting honeybees. Most of the other members of the genus of *Gemykrogviruses* have previously been detected in the feces of various animal species [39]. Since virus–host specificity is commonly conserved among viral families, the identification of MaVBRB and AmGV-2 might provide an important lead for prospective host assignment of *Gemykrogviruses*. It is still a matter of speculation what the true hosts of *Gemykrogviruses* are. One possibility is that *Gemykrogviruses* actually infect plants, which would explain their detection in the feces of herbivorous animals and in mosquitoes that feed on pollen.

The host range of eukaryotic CRESS-DNA viruses is highly diverse and includes plants and various invertebrate and vertebrate species, including humans [26,42]. Indeed, human Torquet Teno virus (TTV), a CRESS-DNA virus from the *Anelloviridae* family, accounts for the most abundant virus infection of humans, with estimated worldwide infection rates of 70–100% [43–45]. In this study, we identified TTV genomic sequences in two separate mosquito pools collected in Barbados and verified the metagenomic finding by qPCR (specific for human TTV species). Finding human-infecting TTV in field-caught mosquito populations suggests a spillover of CRESS-DNA viruses to mosquitoes, most likely transmitted by feeding on the blood of infected vertebrate hosts. It may be assumed that the detected mosquito virome diversity is driven by viruses of different sources that passage through the individual mosquito's digestive tract. CO2-baited trapping selects for unfed mosquitoes, and none of the included mosquito individuals showed morphologic signs of recent blood meal intake. However, it cannot be totally excluded that CRESS-DNA virus detection might have been influenced by the remainder of vertebrate blood within the individuals.

Moreover, single CRESS-DNA viruses have been identified in serum and brain biopsies of multiple sclerosis patients, in the pericardial fluid of pericarditis patients, and the cerebrospinal fluid of encephalitis patients [46–48]. CyCV-VN was initially identified in cerebrospinal fluid (CSF) from patients with suspected central nervous system infection from Vietnam, as well as in feces from humans and pigs from the same area [48]. The pathogenic role of CyCV-VN remains unclear as screening of >600 CSF patient samples from Vietnam, Cambodia, Nepal, and the Netherlands failed to detect the virus [49]. However, in later studies, CyCV-VN was detected in stool samples of healthy children from Madagascar, pig feces from Cameroon, and shrew enteric samples from Zambia [50,51]. A recent study reported a 43% plasma prevalence of CyCV-VN in healthy blood donors from Madagascar [52]. In this study, we identified a novel genetic variant of CyCV-VN. To our knowledge, this is the first study to detect CyCV-VN in mosquito vectors, possibly suggesting transmission via *Aedes aegypti*. However, further studies are needed to confirm these findings in larger cohorts and to investigate mosquito vector competence for CyCV-VN transmission.

In conclusion, this study comprehensively characterized the viral metagenome of field-collected *C.pip.cl.* and *Ae. ae.* mosquitoes from two distinctive geographic locations. Among all phylogenetic virus clades, CRESS-DNA viruses constituted a diverse and highly abundant portion of the mosquito virome. We present evidence that mosquito vectors are important hubs for CRESS-DNA virus transmission between different environmental sources. Thus, mosquitoes contain a large pool of novel CRESS-DNA viruses from which we identified a novel variant of a human-infecting CRESS-DNA virus, currently not linked to vector-related transmission.

Supplementary Materials: The following are available online at http://www.mdpi.com/2076-0817/9/9/686/s1, Figure S1: Fraction of reads mapping on contigs classified as viruses, prokaryotes and eukaryotes, Figure S2: Virome composition on taxonomic family level as number of hits to indicated clade, Figure S3: Relative abundance of viral hits on taxonomic family level in RPKM, Figure S4: Land coverage of Austrian collection spots AUT_1 and AUT_2, Figure S5: Goodness of alignment- fit of best BLAST N hits for for mutated model sequences, Figure S6: Verification of metagenomic hits, Figure S7: Goodness of assignment- fit for best BLAST N hit of contig sequence k141_1014 from *Aedes aegypti* (*Ae.ae.*) metagenome and contig sequence NODE_38 from *Culex pipiens* complex (*C.pip.cl.*) Barbados (BRB) using NCBI database version 2018 and updated database version 2020; Figure S8: Verification of metagenomic assembled *mosquito associated virus Barbados* (*MaVBRB*) genome sequences by abutting primer PCR reactions; Table S1: Sequencing results, sequence separation and quality trimming, Table S2: Land coverage, AUT_01 and AUT_02, Table S3: Model sequences of 100 nucleotide length from primary NCBI entry LK931484.1 with indicated number of random point mutations, Table S4: Best BLAST N hits of model sequence to NCBI database, Table S5: Number of CRESS-DNA viruses hits below and above 60% query coverage.

Author Contributions: Conceptualization C.S.; field work C.S., J.T., C.Z., H.-P.F., T.S.-M., M.G.-S., S.A.; methodology J.T., A.E., C.Z., H.-P.F.; bioinformatic analysis N.R., J.T.; next generation sequencing I.K.; writing original draft preparation J.T.; writing review and editing J.T., C.S., R.C.L., N.R., H.-P.F. and C.Z.; funding acquisition C.S., R.C.L. All authors have read and agreed to the published version of the manuscript.

Funding: This research received no external funding.

Conflicts of Interest: The authors declare no conflict of interest.

References

1. Aspöck, H. Medical Entomology in the 21 st Century: Retrospect and Challenges. *Nov. Acta Leopold.* **2016**, *411*, 241–258.

2. Shi, C.; Liu, Y.; Hu, X.; Xiong, J.; Zhang, B.; Yuan, Z. A metagenomic survey of viral abundance and diversity in mosquitoes from hubei province. *PLoS ONE* **2015**, *10*, e0129845. [CrossRef] [PubMed]

3. Sadeghi, M.; Altan, E.; Deng, X.; Barker, C.M.; Fang, Y.; Coffey, L.L.; Delwart, E. Virome of > 12 thousand Culex mosquitoes from throughout California. *Virology* **2018**, *523*, 74–88. [CrossRef] [PubMed]

4. Shi, C.; Beller, L.; Deboutte, W.; Yinda, K.C.; Delang, L.; Vega-Rúa, A.; Failloux, A.B.; Matthijnssens, J. Stable distinct core eukaryotic viromes in different mosquito species from Guadeloupe, using single mosquito viral metagenomics. *Microbiome* **2019**, *7*, 121. [CrossRef] [PubMed]

5. Hameed, M.; Liu, K.; Anwar, M.N.; Wahaab, A.; Li, C.; Di, D.; Wang, X.; Khan, S.; Xu, J.; Li, B.; et al. A viral metagenomic analysis reveals rich viral abundance and diversity in mosquitoes from pig farms. *Transbound. Emerg. Dis.* **2020**, *67*, 328–343. [CrossRef]

6. Bolling, B.G.; Weaver, S.C.; Tesh, R.B.; Vasilakis, N. Insect-specific virus discovery: Significance for the arbovirus community. *Viruses* **2015**, *7*, 4911–4928. [CrossRef]

7. Shi, M.; Lin, X.D.; Tian, J.H.; Chen, L.J.; Chen, X.; Li, C.X.; Qin, X.C.; Li, J.; Cao, J.P.; Eden, J.S.; et al. Redefining the invertebrate RNA virosphere. *Nature* **2016**, *540*, 539–543. [CrossRef]

8. Junglen, S.; Drosten, C. Virus discovery and recent insights into virus diversity in arthropods. *Curr. Opin. Microbiol.* **2013**, *16*, 507–513. [CrossRef]

9. Cook, S.; Moureau, G.; Kitchen, A.; Gould, E.A.; de Lamballerie, X.; Holmes, E.C.; Harbach, R.E. Molecular evolution of the insect-specific flaviviruses. *J. Gen. Virol.* **2012**, *93*, 223–234. [CrossRef]

10. Huhtamo, E.; Cook, S.; Moureau, G.; Uzcátegui, N.Y.; Sironen, T.; Kuivanen, S.; Putkuri, N.; Kurkela, S.; Harbach, R.E.; Firth, A.E.; et al. Novel flaviviruses from mosquitoes: Mosquito-specific evolutionary lineages within the phylogenetic group of mosquito-borne flaviviruses. *Virology* **2014**, 320–329. [CrossRef]

11. Zakrzewski, M.; Rašić, G.; Darbro, J.; Krause, L.; Poo, Y.S.; Filipović, I.; Parry, R.; Asgari, S.; Devine, G.; Suhrbier, A. Mapping the virome in wild-caught Aedes aegypti from Cairns and Bangkok. *Sci. Rep.* **2018**, *8*, 4690. [CrossRef] [PubMed]

12. Faizah, A.N.; Kobayashi, D.; Isawa, H.; Amoa-Bosompem, M.; Murota, K.; Higa, Y.; Futami, K.; Shimada, S.; Kim, K.S.; Itokawa, K.; et al. Deciphering the Virome of Culex vishnui Subgroup Mosquitoes, the Major Vectors of Japanese Encephalitis, in Japan. *Viruses* **2020**, *12*, 264. [CrossRef] [PubMed]

13. Kenney, J.L.; Solberg, O.D.; Langevin, S.A.; Brault, A.C. Characterization of a novel insect-specific flavivirus from Brazil: Potential for inhibition of infection of arthropod cells with medically important flaviviruses. *J. Gen. Virol.* **2014**, *95*, 2796–2808. [CrossRef] [PubMed]

14. Schultz, M.J.; Frydman, H.M.; Connor, J.H. Dual Insect specific virus infection limits Arbovirus replication in Aedes mosquito cells. *Virology* **2018**, *518*, 406–413. [CrossRef]

15. Bolling, B.G.; Olea-Popelka, F.J.; Eisen, L.; Moore, C.G.; Blair, C.D. Transmission dynamics of an insect-specific flavivirus in a naturally infected Culex pipiens laboratory colony and effects of co-infection on vector competence for West Nile virus. *Virology* **2012**, *427*, 90–97. [CrossRef]

16. Ng, T.F.F.; Willner, D.L.; Lim, Y.W.; Schmieder, R.; Chau, B.; Nilsson, C.; Anthony, S.; Ruan, Y.; Rohwer, F.; Breitbart, M. Broad surveys of DNA viral diversity obtained through viral metagenomics of mosquitoes. *PLoS ONE* **2011**, *6*, e20579. [CrossRef]

17. Davidson, I.; Silva, R.F. Creation of diversity in the animal virus world by inter-species and intra-species recombinations: Lessons learned from poultry viruses. *Virus Genes* **2008**, *36*, 1–9. [CrossRef]

18. Malathi, V.G.; Renuka Devi, P. ssDNA viruses: Key players in global virome. *VirusDisease* **2019**, *30*, 3–12. [CrossRef]

19. Franzo, G.; Segales, J.; Tucciarone, C.M.; Cecchinato, M.; Drigo, M. The analysis of genome composition and codon bias reveals distinctive patterns between avian and mammalian circoviruses which suggest a potential recombinant origin for Porcine circovirus 3. *PLoS ONE* **2018**, *13*, e0199950. [CrossRef]

20. Thannesberger, J.; Rascovan, N.; Eisenmann, A.; Klymiuk, I.; Zittra, C.; Fuehrer, H.P.; Scantlebury-Manning, T.; Gittens-St.Hilaire, M.; Austin, S.; Landis, R.C. Co-circulation of novel Zika virus strains in Aedes mosquitoes from Barbados. **2019**, Unpublished work.

21. CORINE Land Cover Map Austria. Available online: https://www.data.gv.at/anwendungen/web-applikation-zu-corine-land-cover-oesterreich (accessed on 12 February 2019).

22. Schneider, C.A.; Rasband, W.S.; Eliceiri, K.W. NIH Image to ImageJ: 25 years of image analysis. *Nat. Methods* **2012**, *9*, 671–675. [CrossRef]

23. Becker, N.; Petric, D.; Zgomba, M.; Boase, C.; Madon, M.; Dahl, C.; Kaiser, A. *Mosquitoes and Their Control*; Springer: Berlin/Heidelberg, Germany, 2010; ISBN 978-3-540-92873-7.

24. Werblow, A.; Flechl, E.; Klimpel, S.; Zittra, C.; Lebl, K.; Kieser, K.; Laciny, A.; Silbermayr, K.; Melaun, C.; Fuehrer, H.P. Direct PCR of indigenous and invasive mosquito species: A time- and cost-effective technique of mosquito barcoding. *Med. Vet. Entomol.* **2016**, *30*, 8–13. [CrossRef] [PubMed]

25. Thannesberger, J.; Hellinger, H.-J.; Klymiuk, I.; Kastner, M.-T.; Rieder, F.J.J.; Schneider, M.; Fister, S.; Lion, T.; Kosulin, K.; Laengle, J.; et al. Viruses comprise an extensive pool of mobile genetic elements in eukaryote cell cultures and human clinical samples. *FASEB J.* **2017**, *31*, 1987–2000. [CrossRef] [PubMed]

26. Zhao, L.; Rosario, K.; Breitbart, M.; Duffy, S. *Eukaryotic Circular Rep-Encoding Single-Stranded DNA (CRESS DNA) Viruses: Ubiquitous Viruses With Small Genomes and a Diverse Host Range*, 1st ed.; Elsevier Inc: Amsterdam, The Netherlands, 2019.

27. Notredame, C.; Higgins, D.G.; Heringa, J. T-coffee: A novel method for fast and accurate multiple sequence alignment 1 1Edited by J. Thornton. *J. Mol. Biol.* **2000**, *302*, 205–217. [CrossRef] [PubMed]

28. Le, S.Q.; Gascuel, O. An Improved General Amino Acid Replacement Matrix. *Mol. Biol. Evol.* **2008**, *25*, 1307–1320. [CrossRef]

29. Kumar, S.; Stecher, G.; Tamura, K. MEGA7: Molecular Evolutionary Genetics Analysis Version 7.0 for Bigger Datasets. *Mol. Biol. Evol.* **2016**, *33*, 1870–1874. [CrossRef]

30. Maggi, F.; Pifferi, M.; Fornai, C.; Tempestini, E.; Vatteroni, M.; Marchi, S.; Pietrobelli, A.; Andreoli, E.; Presciuttini, S. TT Virus in the Nasal Secretions of Children with Acute Respiratory Diseases: Relations to Viremia and Disease Severity. *J. Virol.* **2003**, *77*, 2418–2425. [CrossRef]

31. Hang, J.; Klein, T.A.; Kim, H.; Yang, Y.; Jima, D.D.; Richardson, J.H.; Jarman, G. Genome Sequences of Five Arboviruses in Field-Captured Mosquitoes in a Unique Rural Environment of South Korea. *Genome Announc* **2016**, *4*. [CrossRef]

32. WHO. WHO—Vector Born Diseases. Available online: https://www.who.int/news-room/fact-sheets/detail/vector-borne-diseases (accessed on 3 February 2020).

33. Pettersson, J.H.O.; Shi, M.; Eden, J.S.; Holmes, E.C.; Hesson, J.C. Meta-transcriptomic comparison of the RNA Viromes of the mosquito vectors culex pipiens and culex torrentium in Northern Europe. *Viruses* **2019**, *11*, 1033. [CrossRef]

34. Reveillaud, J.; Bordenstein, S.R.; Cruaud, C.; Shaiber, A.; Esen, Ö.C.; Weill, M.; Makoundou, P.; Lolans, K.; Watson, A.R.; Rakotoarivony, I.; et al. The Wolbachia mobilome in Culex pipiens includes a putative plasmid. *Nat. Commun.* **2019**, *10*, 1051. [CrossRef] [PubMed]

35. Kent, B.N.; Bordenstein, S.R. Phage WO of Wolbachia: Lambda of the endosymbiont world Bethany. *Trends Microbiol.* **2010**, *18*, 173–181. [CrossRef] [PubMed]

36. Kim, K.H.; Bae, J.W. Amplification methods bias metagenomic libraries of uncultured single-stranded and double-stranded DNA viruses. *Appl. Environ. Microbiol.* **2011**, *77*, 7663–7668. [CrossRef] [PubMed]

37. De Sales Lima, F.E.; Cibulski, S.P.; Dos Santos, H.F.; Teixeira, T.F.; Varela, A.P.M.; Roehe, P.M.; Delwart, E.; Franco, A.C. Genomic characterization of novel circular ssDNA viruses from insectivorous bats in Southern Brazil. *PLoS ONE* **2015**, *10*, e0118070.

38. Dennis, T.P.W.; Flynn, P.J.; De Souza, M.; Singer, J.B.; Gifford, R.J. Insights into Circovirus Host Range from the Genomic Fossil Record. *J. Virol.* **2018**, *92*, JVI.00145-18. [CrossRef] [PubMed]

39. Rosario, K.; Breitbart, M.; Harrach, B.; Segalés, J.; Delwart, E.; Biagini, P.; Varsani, A. Revisiting the taxonomy of the family Circoviridae: Establishment of the genus Cyclovirus and removal of the genus Gyrovirus. *Arch. Virol.* **2017**, *162*, 1447–1463. [CrossRef]

40. Firth, C.; Charleston, M.A.; Duffy, S.; Shapiro, B.; Holmes, E.C. Insights into the Evolutionary History of an Emerging Livestock Pathogen: Porcine Circovirus 2. *J. Virol.* **2009**, *83*, 12813–12821. [CrossRef]

41. Duffy, S.; Shackelton, L.A.; Holmes, E.C. Rates of evolutionary change in viruses: Patterns and determinants. *Nat. Rev. Genet.* **2008**, *9*, 267–276. [CrossRef]

42. Rosario, K.; Duffy, S.; Breitbart, M. A field guide to eukaryotic circular single-stranded DNA viruses: Insights gained from metagenomics. *Arch. Virol.* **2012**, *157*, 1851–1871. [CrossRef]

43. Maggi, F.; Bendinelli, M. Human anelloviruses and the central nervous system. *Rev. Med. Virol.* **2010**, *19*, 57–64. [CrossRef]

44. Vasilyev, E.V.; Trofimov, D.Y.; Tonevitsky, A.G.; Ilinsky, V.V.; Korostin, D.O.; Rebrikov, D.V. Torque Teno Virus (TTV) distribution in healthy Russian population. *Virol. J.* **2009**, *6*, 134. [CrossRef]

45. AbuOdeh, R.; Al-Mawlawi, N.; Al-Qahtani, A.A.; Bohol, M.F.F.; Al-Ahdal, M.N.; Hasan, H.A.; AbuOdeh, L.; Nasrallah, G.K. Detection and genotyping of torque teno virus (TTV) in healthy blood donors and patients infected with HBV or HCV in Qatar. *J. Med. Virol.* **2015**, *87*, 1184–1191. [CrossRef]

46. Lamberto, I.; Gunst, K.; Müller, H.; Zur Hausen, H.; de Villiers, E.-M. Mycovirus-like DNA virus sequences from cattle serum and human brain and serum samples from multiple sclerosis patients. *Genome Announc.* **2014**, *2*, 00848-14. [CrossRef]

47. Halary, S.; Duraisamy, R.; Fancello, L.; Monteil-Bouchard, S.; Jardot, P.; Biagini, P.; Gouriet, F.; Raoult, D.; Desnues, C. Novel single-stranded DNA circular viruses in pericardial fluid of patient with recurrent pericarditis. *Emerg. Infect. Dis.* **2016**, *22*, 1839–1841. [CrossRef] [PubMed]

48. Van Tan, L.; Van Doorn, H.R.; Trung, D.; Hong, T.; Phuong, T.; Vries, M. De Identification of a New Cyclovirus in Cerebrospinal Fluid of Patients with Acute Central Nervus System Infections. *mBio* **2013**, *4*, e00231-13. [CrossRef] [PubMed]

49. Van Tan, L.; De Jong, M.D.; Van Kinh, N.; Trung, N.V.; Taylor, W.; Wertheim, H.F.L.; Van Der Ende, A.; Van Der Hoek, L.; Canuti, M.; Crusat, M.; et al. Limited geographic distribution of the novel cyclovirus CyCV-VN. *Sci. Rep.* **2014**, *4*, 7–10. [CrossRef]

50. Sasaki, M.; Orba, Y.; Ueno, K.; Ishii, A.; Moonga, L.; Hangombe, B.M.; Mweene, A.S.; Ito, K.; Sawa, H. Metagenomic analysis of the shrew enteric virome reveals novel viruses related to human stool-associated viruses. *J. Gen. Virol.* **2015**, *96*, 440–452. [CrossRef]

51. Garigliany, M.M.; Hagen, R.M.; Frickmann, H.; May, J.; Schwarz, N.G.; Perse, A.; Jöst, H.; Börstler, J.; Shahhosseini, N.; Desmecht, D.; et al. Cyclovirus CyCV-VN species distribution is not limited to Vietnam and extends to Africa. *Sci. Rep.* **2014**, *4*, 7552. [CrossRef] [PubMed]

52. Sauvage, V.; Gomez, J.; Barray, A.; Vandenbogaert, M.; Boizeau, L.; Tagny, C.T.; Rakoto, O.; Bizimana, P.; Guitteye, H.; Ciré, B.B.; et al. High prevalence of cyclovirus Vietnam (CyCV-VN) in plasma samples from Madagascan healthy blood donors. *Infect. Genet. Evol.* **2018**, *66*, 9–12. [CrossRef] [PubMed]

Article

Mosquitoes (Diptera: Culicidae) in the Dark—Highlighting the Importance of Genetically Identifying Mosquito Populations in Subterranean Environments of Central Europe

Carina Zittra [1], Simon Vitecek [2,3], Joana Teixeira [4], Dieter Weber [4], Bernadette Schindelegger [2], Francis Schaffner [5] and Alexander M. Weigand [4,*]

1 Unit Limnology, Department of Functional and Evolutionary Ecology, University of Vienna, 1090 Vienna, Austria; carina.zittra@univie.ac.at

2 WasserCluster Lunz—Biologische Station, 3293 Lunz am See, Austria; simon.vitecek@wcl.ac.at (S.V.); bernadette.schindelegger@wcl.ac.at (B.S.)

3 Institute of Hydrobiology and Aquatic Ecosystem Management, University of Natural Resources and Life Sciences, Vienna, Gregor-Mendel-Strasse 33, 1180 Vienna, Austria

4 Zoology Department, Musée National d'Histoire Naturelle de Luxembourg (MNHNL), 2160 Luxembourg, Luxembourg; joanamtl@gmail.com (J.T.); dieter.weber124@gmx.de (D.W.)

5 Francis Schaffner Consultancy, 4125 Riehen, Switzerland; francis.schaffner@uzh.ch

* Correspondence: alexander.weigand@mnhn.lu; Tel.: +352-462-240-212

Citation: Zittra, C.; Vitecek, S.; Teixeira, J.; Weber, D.; Schindelegger, B.; Schaffner, F.; Weigand, A.M. Mosquitoes (Diptera: Culicidae) in the Dark—Highlighting the Importance of Genetically Identifying Mosquito Populations in Subterranean Environments of Central Europe. *Pathogens* **2021**, *10*, 1090. https://doi.org/10.3390/pathogens10091090

Academic Editor: Vito Colella

Received: 16 June 2021
Accepted: 24 August 2021
Published: 26 August 2021

Publisher's Note: MDPI stays neutral with regard to jurisdictional claims in published maps and institutional affiliations.

Abstract: The common house mosquito, *Culex pipiens* s. l. is part of the morphologically hardly or non-distinguishable *Culex pipiens* complex. Upcoming molecular methods allowed us to identify members of mosquito populations that are characterized by differences in behavior, physiology, host and habitat preferences and thereof resulting in varying pathogen load and vector potential to deal with. In the last years, urban and surrounding periurban areas were of special interest due to the higher transmission risk of pathogens of medical and veterinary importance. Recently, surveys of underground habitats were performed to fully evaluate the spatial distribution of rare members of the *Cx. pipiens* complex in Europe. Subterranean environments and their contribution to mosquito-borne pathogen transmission are virtually unknown. Herein, we review the underground community structures of this species complex in Europe, add new data to Germany and provide the first reports of the *Cx. pipiens* complex and usually rarely found mosquito taxa in underground areas of Luxembourg. Furthermore, we report the first finding of *Culiseta glaphyroptera* in Luxembourg. Our results highlight the need for molecular specimen identifications to correctly and most comprehensively characterize subterranean mosquito community structures.

Keywords: *Culex pipiens* s. l.; *Culex torrentium*; *Culiseta glaphyroptera*; caves; subterranean environment; Luxembourg; Germany

1. Introduction

The globally distributed *Culex pipiens* complex (or *Culex pipiens* assemblage *sensu* Harbach, 2012) consists of several taxa: *Cx. quinquefasciatus*, *Cx. pipiens pallens*, *Cx. australicus* and the nominate taxon *Culex pipiens* in Europe [1]. The latter taxon includes two behaviorally and genetically distinct forms ('f.'), *Cx. pipiens* f. *pipiens* and *Cx. pipiens* f. *molestus*, that do not differ morphologically and are able to hybridize in areas of coexistence [2]. Members of the complex are of medical importance as they are primary vectors of several pathogens, including the West Nile virus, the widespread cause of arboviral neurological disease, and are often the most abundant mosquitoes in urbanized areas [3]. Emergence, distribution and transmission of the West Nile virus and other mosquito-borne pathogens are regulated through potential vector communities that link suitable reservoirs and susceptible hosts [4,5]. Differences in vectorial capacity and vector competence between the

Cx. pipiens complex members due to specific ecology, physiology, and behavior, therefore, have direct reverberations on animal and human health [6].

Forms of *Cx. pipiens* are reported to differ, in which eurygamous *Cx. pipiens* f. *pipiens* requires more space for mating than stenogamous *Cx. pipiens* f. *molestus*, which does not mate in swarms. These observations led to the general assumption that *Cx. pipiens* f. *pipiens* is restricted to epigean (above-ground) sites, while *Cx. pipiens* f. *molestus* is considered to inhabit mostly hypogean (underground, subterranean) sites, especially such close to human settlements [7]. Further, different host and habitat preferences of both forms and their hybrid offspring potentially lead to distinct roles in host-vector-pathogen dynamics: In contrast to *Cx. pipiens* f. *molestus*, *Cx. pipiens* f. *pipiens* is reported to prefer avian hosts, while the host and habitat preference of their cross-bred offspring is not sufficiently known. Additionally, they have contrasting strategies to survive winter, where female *Cx. pipiens* f. *molestus* remain active but female *Cx. pipiens* f. *pipiens* overwinter undergoing diapause in shelters associated with human settlements like cellars or attics [8].

Different populations of the *Cx. pipiens* complex were discovered earlier to inhabit fully enclosed sites, but also crevices connected to above-ground habitats or open-air habitats [9]). However, the allozyme loci used then to differentiate between taxa precluded identification of *Cx. pipiens* forms [10,11]. Once standardized and replicable molecular methods [12,13] allowed reliable differentiation of the *Cx. pipiens* complex members, this taxon was examined in Europe in epigean sites [2,14]. In contrast, the composition and seasonality of the *Cx. pipiens* complex in relation to abiotic parameters of subterranean resting and hibernation habitats were often neglected.

Subterranean sites such as natural caves, mining galleries, tunnels, and culverts are resting and hibernation shelters for several subtroglophilous insects of the order Diptera, not forming permanent subterranean populations but seasonally inhabiting underground habitats [15–17]. Mosquitoes of the genus *Culex* are known to use non-urban subterranean habitats as hibernation and resting sites. The purported adaptation to urban environments, therefore, has been discussed for decades [9,17]. Amongst the earliest reports is that of Legendre [9], who observed the larval development of *Cx. pipiens* s. l. in well-connected underground cave systems with strong exchange with above-ground habitats if water and air temperatures were suitable and nutrients available [9]. Subterranean sites provide stable and adequate conditions (in terms of humidity and temperature) and are visited actively by a range of mosquito taxa, probably to avoid unfavorable conditions such as dryness and cold temperatures or to reduce predation pressure [17,18]. This possibly is a behavioral adaptation but seems to be a common trait in mosquitoes, and while above-ground distribution patterns of the *Cx. pipiens* forms, their offspring and *Cx. torrentium* were recently examined in Germany and Austria [2,14,19], the hypogean distribution of these taxa remains obscure at greater scales. Life in caves in Central Europe is well-known, particularly in Germany and Luxembourg [15,16,20–27], but mosquitoes collected or spotted in caves were often mostly identified to family-level ([16,28], long-term collection data for [29]). With regard to members of the *Cx. pipiens* complex—a taxon reported as widely distributed and highly abundant in underground habitats [30]—the general practice of reporting combined occurrence data for *Cx. pipiens* s. l. and *Cx. torrentium* as "Culicidae" or even "*Culex pipiens*" is not ideal given their potentially different distribution patterns and epidemiological relevance [31].

Within the *Cx. pipiens* complex, *Cx. pipiens* f. *molestus* is generally accepted as the hypogean counterpart of *Cx. pipiens* f. *pipiens* [10,32], but both forms are known to occur in sympatry above-ground [2]. Yet, it appears that they can become strongly isolated, as observed in the London underground railway system where a dominance of f. *molestus* was found [10]. Generalizations and expected distribution patterns extrapolated from these records were, however, not confirmed as members of the *Cx. pipiens* complex were found in sympatry in subterranean habitats of both urban and rural areas in Austria, Germany, and Hungary [18,29,33]

At present, mosquito community composition, including members of the *Cx. pipiens* complex, in the subterranean realm, has been rarely studied using molecular tools, and baseline data [18,29,33] are slowly emerging in Europe. Better knowledge about the distribution and composition of mosquitoes and the *Cx. pipiens* complex in particular, is crucial to assess vector-borne pathogen dynamics in rural habitats and to estimate the potential impact of to date neglected subterranean sites on public health. In this contribution, we review available literature and present new data to provide the first synopsis on mosquitoes of the *Cx. pipiens* complex in underground environments and provide a summary of Culicidae in subterranean habitats.

2. Results

In Germany, 151 specimens belonging to the *Cx. pipiens* complex were molecularly analyzed (Table 1). A total of 56% were found in the transition zone, and most of the specimens were collected in autumn (Supplementary Table S1). *Culex pipiens* f. *pipiens* was most abundant with 99 specimens including two males, found exclusively in the transition zone. *Culex pipiens* f. *molestus* was rarely represented by two specimens found in autumn in the Westerwald and the Swabian Jura again in the transition zone. Hybrids were represented by four specimens collected in spring and autumn at the Swabian Jura. A total of 46 specimens were identified as *Cx. torrentium*.

Table 1. The number of individuals and percentage of the *Culex pipiens* complex taxa and *Culex torrentium* sampled in Germany and Luxembourg (this study) in comparison to Austria [33]. Provided are the total number of genetically analyzed specimens per species and their frequencies in the total regional datasets.

Country	*Cx. p.* f. *pipiens*	*Cx. p.* f. *molestus*	*Cx. p.* f. *pipiens* X f. *molestus*	*Cx. torrentium*	Total (*n*)
Germany	99 (66%)	2 (1%)	4 (3%)	46 (30%)	151
Luxembourg	119 (75%)	2 (1%)	0	38 (24%)	159
Austria	44 (34%)	3 (2%)	13 (11%)	69 (53%)	126

In Luxembourg, 159 mosquitoes belonging to the *Cx. pipiens* complex were found in subterranean areas in Luxembourg. *Culex pipiens* f. *pipiens* was the most abundant taxon with 119 specimens (75%), followed by *Cx. torrentium* (38 specimens; 24%) and *Cx. pipiens* f. *molestus* (2 specimens; 1%). Hybrids of *Cx. pipiens* f. *pipiens* and *Cx. pipiens* f. *molestus* were not detected in Luxembourg samples.

We first collected *Culiseta annulata* and *Cs. glaphyroptera* in Luxembourg caves, representing at the same time the first record of *Cs. glaphyroptera* in Luxembourg (Table 2). Studies of subterranean mosquito populations in Europe are available from Austria [30,33,34], Germany [23–29,33], Croatia [35], Czech Republic [34,36,37], France [38], Hungary [18,34], Italy [39], Luxembourg [16], Poland [40], Norway [17], Slovakia [41–43], and Sweden [44] (Table 2). Several mosquito taxa are reported from subterranean habitats at larger geographical scales (i.e., spanning more than three European countries), including *Anopheles maculipennis* s. l., *Cs. alaskaensis*, *Cs. annulata*, *Cs. glaphyroptera*, and members of the *Cx. pipiens* complex. The highest *Culex* diversity in subterranean habitats was found in Austria and Germany, comprising seven and five taxa, respectively. Additionally, unregular occurrences of other mosquito taxa in subterranean habitats are reported: *Aedes cinereus/geminus*, *Ae. cataphylla*, *Ae. rossicus*, *An. messae*, *An. claviger*, *Cx. hortensis*, *Cx. modestus*, *Cx. territans*, and *Uranotaenia unguiculata* (but not *An. marteri* that was previously reported from subterranean sites in Hesse, Germany [29], due to a database error and should be considered reports of *An. maculipennis* s. l.).

Table 2. Mosquito species detected in artificial or natural subterranean shelters.

Taxon	AT *	CZ	DE *	FR	HR	HU	IT	LU *	NO	PL	SE	SK
Aedes cinereus/geminus			x									
Ae. cataphylla			x									
Ae. communis			x									
Ae. geniculatus	x											
Ae. rossicus			x									
Anopheles maculipennis s. l.	x	x	x	x		x						
An. messeae		x				x						
An. claviger			x									
Culiseta alaskaensis		x	x						x	x	x	x
Cs. annulata	x	x	x	x		x		x [1]	x			
Cs. glaphyroptera		x	x					x [1]				x
Culex hortensis	x			x		x						
Cx. modestus	x		x									
Culex sp.					x							
Cx. pipiens s. l.		x		x			x		x			
Cx. p. f. *molestus*	x		x [1]			x		x [1]				
Cx. p. f. *pipiens*	x		x [1]			x		x [1]				
Cx. p. f. *pipiens* X *molestus*	x		x [1]									
Cx. territans	x					x				x		
Cx. torrentium	x		x [1]					x [1]				
Uranotaenia unguiculata	x	x				x						

Countries in which species identification was supported by the usage of reliable molecular tools are indicated with *, original data compiled in this study are indicated with [1], AT = Austria, CZ = Czechia, DE = Germany, FR = France, HR = Croatia, HU = Hungary, IT = Italy, LU = Luxembourg, NO = Norway, PL = Poland, SE = Sweden, SK = Slovakia.

3. Discussion

It appears that molecular analysis of hardly or un-identifiable *Culex* species collected in natural and artificial caves is not common despite the epidemiological importance of these taxa. Countries where morphological identification and the use of molecular tools were combined (Austria, Germany) recovered higher diversity, but data on numbers of molecularly identified specimens are provided here (Supplementary Tables S1 and S2) and the Austrian study [33] only. Resolving the *Cx. pipiens* complex with molecular methods, we found no indication for a greater proportion of *Cx. pipiens* f. *molestus* in underground habitats. This is in line with previous reports on these taxa, that appear to be present at relatively constant frequencies in above- and underground habitats as well as over long time periods in Central Europe [2,29,33]. We found *Cx. pipiens* f. *pipiens* to be the dominating mosquito in the investigated natural and artificial subterranean sites (Supplementary Table S1). Thus, data at hand rather indicate a sympatric occurrence of the two forms in above- and underground sites. While it may be possible that in cities with significant underground constructions like London or Helsinki, such isolation may take place, we found no evidence for reproductively isolated *Cx. pipiens* f. *pipiens* and *Cx. pipiens* f. *molestus* populations in Central European artificial and natural subterranean shelters. Intriguingly, the proportions of *Cx. pipiens* f. *pipiens*, *Cx. pipiens* f. *molestus* and *Cx. pipiens* f. *pipiens* X f. *molestus* hybrids in both above-ground and subterranean habitats appear to mirror patterns that would be expected under Hardy–Weinberg equilibria of two alleles in a panmictic population [19,45]. Under reproductive isolation, patterns strongly deviating from such an equilibrium would have been expected in underground habitats The presumed reproductive isolation of the *Cx. pipiens* forms by niche differentiation, therefore, cannot be corroborated. However, to effectively test for reproductive isolation in underground habitats by assessing deviations from Hardy–Weinberg equilibrium-like states, far greater numbers of specimens need to be analyzed. More frequent and more specific sampling in underground habitats, including the collection of potentially present mosquito larvae, should be conducted to assess potential reproductive isolation between *Cx. pipiens* forms.

The subterranean realm seems to harbor a specific mosquito community recruited from the surrounding above-ground habitats. However, the available data are too limited to speculate about colonization pathways or how subterranean habitats are used. Data at hand indicate that several *Culex* and *Culiseta* species regularly use subterranean sites, sometimes accompanied by *Anopheles* species, *Uranotaenia unguiculata*, and exceptionally by *Aedes geniculatus*, a tree-hole breeding species (collected once, in summertime) (data presented here [20,34,46]). At the same time, the new records of *Cs. annulata* and *Cs. glaphyroptera* from Luxembourg demonstrates that cave habitats can be important sites for mosquito monitoring, especially when combined with molecular tools. Amongst the taxa using caves, *Cx. pipiens* f. *pipiens* reaches highest abundances [29,33]. Another *Culex* species, *Cx. torrentium*, occurs regularly and in quite high abundances in underground habitats, despite its apparent rarity in epigean habitats. In this study, we could confirm this pattern in Luxemburg and Germany in congruence with previous findings in Germany [29] and Austria [33]. However, restricted access to molecular tools seems to impede assessments of mosquito communities in caves: While there is ample information about the occasional or regular occurrence of a wide range of taxa in subterranean habitats, there is little data on the *Cx. pipiens* complex or morphologically similar taxa. Apparent absences of *Cx. torrentium* from Hungarian caves in the Bakony Balaton region, parts of Germany, or the Czech Republic could result from such limitations [18,28,36].

In addition to the lack of taxonomic resolution in the available data, the cave habitats and their potentially relevant characteristics (entrance size, temperature regimes, humidity, presence and permanence of aquatic habitats, etc.) are poorly described. Such data are necessary to evaluate which parameters drive hypogean mosquito community composition and abundance patterns. Land cover was previously shown to control mosquito community assembly at larger scales [5], but which processes lead to cave-use in mosquitoes is not clear. Data at hand points to the more frequent use of the transition zone of caves instead of the dark zone [29,33] [this study]. From all 4170 culicid specimens collected by D. Weber in a period from 2007 to 2015, 69% were collected in the transition zone, 29% in the dark zone, and only a minority of 2% in the entrance zone (Supplementary Table S1 and S2) [16]. Cave tourism may affect if and how mosquitoes are able to use subterranean habitats. Recent data indicate that higher hibernation mortality may result in this observation. Lipid reserves of overwintering *Culex* females suggest a hibernation temperature optimum ranging from 0 to 8 °C; disturbance (i.e., warming, predator attacks or human activity) interrupts hibernation and lead to increased energy demand, either by increased metabolic rates or flight activity [47], and in turn increases mortality of overwintering females. Anecdotal evidence suggests that mortality in hibernating mosquito females is generally high, and an increase may have deleterious effects on cave mosquito communities [33]. In this context, artificial urban hibernation shelters may be of greater importance for mosquito populations—and the associated vector-borne pathogens—if tourism to natural caves continues to grow.

4. Materials and Methods

In Germany, mosquitoes were retrieved and selected from a larger ongoing metabarcoding project investigating the effect of tourism on subterranean invertebrate communities. A total of 12 subterranean sites in Franconian Switzerland (2), Harz (2), Süntel (2), Swabian Jura (4), Westerwald (2) were visited, comprising of paired sets of natural and caves equipped for touristic activities. Over a period of 13.5 months between autumn 2017 and autumn 2018, all sampling sites were visited twice in each season (autumn, spring, summer). At the first visit, specimens were actively collected, and ethanol-filled Barber traps were installed to preferably capture ground-dwelling fauna. This was done separately at an outside reference, in the transition zone and in the dark zone of the cave. At each second seasonal visit, Barber traps were re-collected and long-term Barber traps until the next season were installed. Mosquitoes were sampled by hand using an ethanol-wetted brush, while only a single specimen was collected in a Barber trap (sample ID SubCul006, Supplementary Table S2).

For Luxembourg, previously undetermined material from the collection of D. Weber was investigated [16] and further ongoing collection material was included. The material was collected by hand (i.e., using a wetted brush or a vial) between 2007 and 2015. Due to non-optimal storage conditions, DNA quality was often very low, and it was thus only possible to successfully isolate enough DNA for the analyzed specimens.

DNA was extracted from one leg of each mosquito using a modified CTAB-protocol (for Austrian specimens) or one to several legs and a modified salt extraction protocol (for German and Luxembourg specimens). For all three regions, the molecular identification of *Cx. pipiens* forms and *Cx. torrentium* was performed as described in Zittra et al. [2,33], following the protocol by Bahnck and Fonseca [12]: First, *Cx. pipiens* f. *pipiens*, *Cx. pipiens* f. *molestus* and their hybrids were distinguished from *Cx. torrentium* by partial amplification (GoTaq G2 Hot Start Polymerase, Promega GmbH, Walldorf, Germany) of the ACE2 gene using the primers (synthesized at Sigma-Aldrich/Merck KGaA, Darmstadt, Germany) ACEpip, ACEpall, ACEtorr, and B1246s [13]. PCR products were separated using gel electrophoresis targeting 634 bp (*Cx. pipiens* forms) and 512 bp (*Cx. torrentium*) DNA fragments (peqGOLD agarose, VWR International LLC, Vienna, Austria). In the following step, mosquitoes were identified as taxa belonging to the *Cx. pipiens* complex were further identified to form using primers (synthesized at Sigma-Aldrich/Merck KGaA, Darmstadt, Germany) CQ11F2, pip CQ11R, and mol CQ11R. PCR products were visualized using gel electrophoresis targeting 185 bp (*Cx. pipiens* f. *pipiens*) and 241 bp (*Cx. pipiens* f. *molestus*) DNA fragments [12].

Available literature was obtained through specialized searches on GoogleScholar, using combinations of the following keywords: Mosquitoes, Diptera, Culicidae, *Aedes*, *Anopheles*, *Culex*, *Culex pipiens* complex, *Culex pipiens* assemblage (including taxon-specific queries), *Coquillettidia*, *Culiseta*, *Mansonia*, *Ochlerotatus*, *Orthopodomyia*, *Uranotaenia*, caves, subterranean, underground and Europe, and supplemented by grey literature.

5. Conclusions

Investigations on the forms of the common house mosquito *Cx. pipiens* in the UK underground channels in 1998 [10] were interpreted as indicating behavioral and ecological differentiation resulting in reproductive isolation of the forms. Our data demonstrates a comparable distribution of *Cx. pipiens* forms in above- and underground habitats. Additionally, caves harbor specific mosquito communities, even though some mosquito species are to be considered subtroglophilous. Among those are primary vectors of mosquito-borne pathogens such as the West Nile virus. Consequently, the significance of subterranean habitats in vector-pathogen dynamics should be fully explored and species unambiguously identified—it is possible that caves are primary reservoirs of vector-borne pathogens, and their epidemiological significance as well as population dynamics of individual mosquito species need to be assessed.

Supplementary Materials: The following are available online at https://www.mdpi.com/article/10.3390/pathogens10091090/s1, Table S1: Mosquitoes sampled in subterranean sites in Luxembourg (T = transition zone, d = dark zone, E = entrance); Table S2: Mosquitoes sampled in subterranean sites in Germany (T = transition zone, d = dark zone.

Author Contributions: Conceptualization, A.M.W. and C.Z.; methodology, A.M.W., S.V., J.T. and B.S.; formal analysis, C.Z.; investigation, C.Z., S.V. and A.M.W.; resources, A.M.W., D.W. and S.V.; writing—original draft preparation, C.Z. and S.V.; writing—review and editing, A.M.W., J.T., D.W.

and F.S.; supervision, A.M.W.; funding acquisition, A.M.W. and S.V. All authors have read and agreed to the published version of the manuscript.

Funding: This research was funded by the German Research Foundation (DFG), under grant number WE 6055/1-1 awarded to A.W. C.Z. acknowledges support from the FWF (P 31258-B29).

Data Availability Statement: All data are available in the supplements to this publication.

Conflicts of Interest: Authors declare no conflict of interest.

References

1. Harbach, R.E. *Culex pipiens*: Species versus species complex-taxonomic history and perspective. *J. Am. Mosq. Control. Assoc.* **2012**, *28*, 10–23. [CrossRef] [PubMed]
2. Zittra, C.; Flechl, E.; Kothmayer, M.; Vitecek, S.; Rossiter, H.; Zechmeister, T.; Fuehrer, H.P. Ecological characterization and molecular differentiation of *Culex pipiens* complex taxa and *Culex torrentium* in Eastern Austria. *Parasite. Vectors.* **2016**, *9*, 197. [CrossRef]
3. Vinogradova, E.B. *Culex pipiens pipiens Mosquitoes: Taxonomy, Distribution, Ecology, Physiology, Genetics, Applied Importance, and Control*; Pensoft Publishers: Sofia, Bulgaria, 2020; p. 250.
4. Leggewie, M.; Badusche, M.; Rudolf, M.; Jansen, S.; Börstler, J.; Krumkamp, R.; Huber, K.; Krüger, A.; Schmidt-Chanasit, J.; Tannich, E.; et al. *Culex pipiens* and *Culex torrentium* populations from Central Europe are susceptible to West Nile virus infection. *One Health* **2016**, *2*, 88–94. [CrossRef] [PubMed]
5. Zittra, C.; Vitecek, S.; Obwaller, H.; Rossiter, H.; Eigner, B.; Zechmeister, T.; Waringer, J.; Fuehrer, H.P. Landscape structure affects distribution of potential disease vectors (Diptera: Culicidae). *Parasit. Vectors.* **2017**, *10*, 205. [CrossRef]
6. Fonseca, D.M.; Keyghobadi, N.; Malcolm, C.A.; Mehmet, C.; Schaffner, F.; Mogi, M.; Fleischer, R.C.; Wilkerson, R.C. Emerging vectors in the *Culex pipiens* complex. *Science* **2004**, *303*, 1535–1538. [CrossRef]
7. Farajollahi, A.; Fonseca, D.M.; Kramer, L.D.; Marm Kilpatrick, A. "Bird biting" mosquitoes and human disease: A review of the role of *Culex pipiens* complex mosquitoes in epidemiology. *Infect. Genet. Evol.* **2011**, *11*, 1577–1585. [CrossRef]
8. Becker, N.; Petric, D.; Zgomba, M.; Boase, C.; Madon, M.; Dahl, C.; Kaiser, A. *Mosquitoes and Their Control*, 2nd ed.; Springer: Heidelberg, Germany, 2010; p. 577.
9. Legendre, J. The cave-mosquito; or the adaption of *Culex pipiens* to urban life. *Bull. Acar. Med.* **1931**, *106*, 86–89.
10. Byrne, K.; Nichols, R. *Culex pipiens* in London Underground tunnels: Differentiation between surface and subterranean populations. *Heredity* **1999**, *82*, 7–15. [CrossRef]
11. Chevillon, C.; Eritja, R.; Pasteur, N.; Raymond, M. Commensalism, adaptation and gene flow: Mosquitoes of the *Culex pipiens* complex in different habitats. *Genet. Res.* **1995**, *66*, 147–157. [CrossRef]
12. Bahnck, C.M.; Fonseca, D.M. Rapid assay to identify the two genetic forms of *Culex (Culex) pipiens* L. (Diptera: Culicidae) and hybrid populations. *Am. J. Trop. Med. Hyg.* **2006**, *75*, 251–255. [CrossRef]
13. Smith, J.L.; Fonseca, D.M. Rapid assays for identification of members of the *Culex (Culex) pipiens* complex, their hybrids, and other sibling species (Diptera: Culicidae). *Am. J. Trop. Med. Hyg.* **2004**, *70*, 339–345. [CrossRef] [PubMed]
14. Becker, N.; Jöst, A.; Weitzel, T. The *Culex pipiens* complex in Europe. *J. Am. Mosq. Control. Assoc.* **2012**, *24*, 53–67. [CrossRef] [PubMed]
15. Zaenker, S.; Weber, D.; Weigand, A. Liste der Cavernicolen Tierarten Deutschlands mit Einschluss der Grundwasserfauna (Version 1.9). 2020. Available online: https://www.hoehlentier.de/taxa.pdf (accessed on 27 April 2020).
16. Weber, D. Die Höhlenfauna Luxemburgs. *Ferrantia* **2013**, *69*, 1–408.
17. Kjaerandsen, J. Diptera in mines and other cave systems in southern Norway. *Entomol. Fenn.* **1993**, *4*, 151–160. [CrossRef]
18. Trájer, A.; Schoffhauzer, J.; Padisák, J. Diversity, Seasonal abundance and potential vector status of the cave-dwelling mosquitoes (Diptera: Culicidae) in the Bakony-Balaton region. *Acta Zool. Bulg.* **2018**, *70*, 247–258.
19. Rudolf, M.; Czajka, C.; Börstler, J.; Melaun, C.; Jöst, H.; von Thien, H.; Badusche, M.; Becker, N.; Schmidt-Chanasit, J.; Krüger, A.; et al. First nationwide surveillance of *Culex pipiens* complex and *Culex torrentium* mosquitoes demonstrated the presence of *Culex pipiens* biotype *pipiens/molestus* hybrids in Germany. *PLoS ONE* **2013**, *8*, e71832. [CrossRef]
20. Zaenker, S.; Bogon, K.; Weigand, A. *Die Höhlentiere Deutschlands: Finden-Erkennen–Bestimmen*; Quelle & Meyer Verlag: Wiebelsheim, Germany, 2020; p. 448.
21. Dobat, K. Die Höhlenfauna der Fränkischen Alb. Abhandlungen zur Karst- und Höhlenkunde. *Reihe D. Paläontologie Zool.* **1978**, *3*, 1–238.
22. Dobat, K. Die Höhlenfauna der Schwäbischen Alb. Abhandlungen zur Karst- und Höhlenkunde. *Reihe D. Paläontologie Zool.* **1975**, *2*, 260–381.
23. Weber, D. *Die Höhlenfauna und–flora des Höhlenkatastergebietes Rheinland-Pfalz/Saarland, 5. Teil*; Abhandlungen zur Karst- und Höhlenkunde: München, Germany, 2010; Volume 36.
24. Weber, D. *Die Höhlenfauna und–flora des Höhlenkatastergebietes Rheinland-Pfalz/Saarland, 4. Teil*; Abhandlungen zur Karst- und Höhlenkunde: München, Germany, 2001; Volume 33.

25. Weber, D. *Die Höhlenfauna und–flora des Höhlenkatastergebietes Rheinland-Pfalz/Saarland, 3. Teil*; Abhandlungen zur Karst- und Höhlenkunde: München, Germany, 1995; Volume 29.

26. Weber, D. *Die Höhlenfauna und–flora des Höhlenkatastergebietes Rheinland-Pfalz/Saarland, 2. Teil*; Abhandlungen zur Karst- und Höhlenkunde: München, Germany, 1989; Volume 23.

27. Weber, D. *Die Höhlenfauna und–flora des Höhlenkatastergebietes Rheinland-Pfalz/Saarland, 1. Teil*; Abhandlungen zur Karst- und Höhlenkunde: München, Germany, 1988; Volume 22.

28. Dvořák, L.; Weber, D. First record of *Culiseta glaphyroptera* (Schiffner, 1864) (Diptera: Culicidae) from Rhineland-Palatinate, Germany. *Mainzer naturwiss. Archiv.* **2019**, *56*, 303–306.

29. Dörge, D.D.; Cunze, S.; Schleifenbaum, H.; Zaenker, S.; Klimpel, S. An investigation of hibernating members from the *Culex pipiens* complex (Diptera, Culicidae) in subterranean habitats of central Germany. *Sci. Rep.* **2020**, *10*, 10276. [CrossRef]

30. Strouhal, H.; Vornatscher, J. Katalog der rezenten Höhlentiere Österreichs. *Ann. Naturhist. Mus. Wien.* **1975**, *79*, 401–542.

31. Brugman, V.A.; Hernández-Triana, L.M.; Medlock, J.M.; Fooks, A.R.; Carpenter, S.; Johnson, N. The Role of *Culex pipiens* L (Diptera: Culicidae) in Virus Transmission in Europe. *Int. J. Environ. Res. Public Health* **2018**, *15*, 389. [CrossRef] [PubMed]

32. Wu, T.-P.; Hu, Q.; Zhao, T.-Y.; Tian, J.-H.; Xue, R.-D. Morphological studies in *Culex molestus* of the *Culex pipiens* complex (Diptera: Culicidae) in underground parking lots in Wuhan, Central China. *Fla. Entomol.* **2014**, *97*, 1191–1198. [CrossRef]

33. Zittra, C.; Moog, O.; Christian, E.; Fuehrer, H.P. DNA-aided identification of *Culex* mosquitoes (Diptera: Culicidae) reveals unexpected diversity in underground cavities in Austria. *Parasitol. Res.* **2019**, *118*, 1385–1391. [CrossRef]

34. Rudolf, I.; Šebesta, O.; Straková, P.; Betášová, L.; Blažejová, H.; Venclíková, K.; Seidel, B.; Tóth, S.; Hubálek, Z.; Schaffner, F Overwintering of *Uranotaenia unguiculata* adult females in Central Europe: A possible way of persistence of the putative new lineage of West Nile virus? *J. Am. Mosq. Control. Assoc.* **2015**, *3*, 364–365. [CrossRef]

35. Polak, S.; Bedek, J.; Ozimec, R.; Zaksek, V. Subterranean Fauna of twelve Istrian caves/Fauna Sotterranea di dodici grotte istriane. *Annales: Ser. Hist. Nat.* **2012**, *22*, 7–24.

36. Dvořák, L. *Culiseta glaphyroptera* (Schiner, 1864): A common species in the southwestern Czech Republic. *Eur. Mosq. Bull.* **2012**, *30*, 66–71.

37. Dvořák, L. Invertebrates found in underground shelters of western Bohemia. I. Mosquitoes (Diptera: Culicidae). *J. Am. Mosq. Control. Assoc.* **2014**, *30*, 66–71.

38. Gazave, E.; Chevillon, C.; Lenormand, T.; Marquine, M.; Raymond, M. Dissecting the cost of insecticide resistance genes during the overwintering period of the mosquito *Culex pipiens*. *Heredity* **2001**, *87*, 441–448. [CrossRef]

39. Di Russo, C.; Carchini, G.; Rampini, M.; Lucarelli, M.; Sbordoni, V. Long term stability of a terrestrial cave community. *Int. J. Speleol.* **1999**, *26*, 75–88. [CrossRef]

40. Kowalski, K. Fauna jaskiń Tatr Polskich. *Ochr. Przyr* **1955**, *23*, 283–333.

41. Dvořák, L. New faunistic records of mosquito *Culiseta glaphyroptera* (Schiner, 1864) from subterranean habitats in Eastern Slovakia. *Biodivers. Environ.* **2020**, *12*, 4–10.

42. Košel, V. Parietal Diptera in caves of the Belianske Tatry Mts (Slovakia, the Western Carpathians) I. Introduction and species spectrum. *Acta Facultatis Ecologiae.* **2004**, *9*, 97–101.

43. Košel, V. Fauna in Medveda cave in the Slovensky Raj Mts. (Western Carpathians). *Slovenský Kras* **1976**, *14*, 105–113.

44. Jaenson, T.G. Overwintering of *Culex* mosquitoes in Sweden and their potential as reservoirs of human pathogens. *Med. Vet. Entomol.* **1987**, *1*, 151–156. [CrossRef]

45. Weinberg, H. Mendelian proportions in a mixed population. *Science* **1908**, *28*, 49–50. [CrossRef]

46. Strouhal, H. Die in den Höhlen von Warmbad Villach, Kärnten, festgestellten Tiere. *Folia Zool. Hydrobiol.* **1939**, *9*, 47–290.

47. Rozsypal, J.; Moos, M.; Rudolf, I.; Košťál, V. Do energy reserves and cold hardiness limit winter survival of *Culex pipiens*? *Comp. Biochem. Physiol. Part A Mol. Integr. Physiol.* **2021**, *255*, 110912. [CrossRef]

Article

Barcoding of the Genus *Culicoides* (Diptera: Ceratopogonidae) in Austria—An Update of the Species Inventory Including the First Records of Three Species in Austria

Carina Zittra [1,2] , Günther Wöss [3], Lara Van der Vloet [1], Karin Bakran-Lebl [1],
Bita Shahi Barogh [1], Peter Sehnal [3] and Hans-Peter Fuehrer [1,*]

[1] Institute of Parasitology, Department of Pathobiology, University of Veterinary Medicine,
    1210 Vienna, Austria; carina.zittra@univie.ac.at (C.Z.); laruschka@gmx.at (L.V.d.V.);
    Karin.Bakran-Lebl@vetmeduni.ac.at (K.B.-L.); Bita.ShahiBarogh@vetmeduni.ac.at (B.S.B.)
[2] Unit Limnology, Department of Functional and Evolutionary Ecology, University of Vienna,
    1090 Vienna, Austria
[3] Zoological Department 2, Natural History Museum Vienna, 1010 Vienna, Austria;
    g.woess@gmail.com (G.W.); peter.sehnal@nhm-wien.ac.at (P.S.)
*   Correspondence: hans-peter.fuehrer@vetmeduni.ac.at; Tel.: +43-(1)-25077-2205

Received: 7 April 2020; Accepted: 22 May 2020; Published: 23 May 2020

Abstract: Ceratopogonidae are small nematoceran Diptera with a worldwide distribution, consisting of more than 5400 described species, divided into 125 genera. The genus *Culicoides* is known to comprise hematophagous vectors of medical and veterinary importance. Diseases transmitted by *Culicoides* spp. Such as African horse sickness virus, Bluetongue virus, equine encephalitis virus (Reoviridae) and Schmallenberg virus (Bunyaviridae) affect large parts of Europe and are strongly linked to the spread and abundance of its vectors. However, *Culicoides* surveillance measures are not implemented regularly nor in the whole of Austria. In this study, 142 morphologically identified individuals were chosen for molecular analyses (barcoding) of the mitochondrial cytochrome c oxidase subunit I gene (mt COI). Molecular analyses mostly supported previous morphologic identification. Mismatches between results of molecular and morphologic analysis revealed three new *Culicoides* species in Austria, *Culicoides gornostaevae* Mirzaeva, 1984, which is a member of the Obsoletus group, *C. griseidorsum* Kieffer, 1918 and *C. pallidicornis* Kieffer, 1919 as well as possible cryptic species. We present here the first Austrian barcodes of the mt COI region of 26 *Culicoides* species and conclude that barcoding is a reliable tool with which to support morphologic analysis, especially with regard to the difficult to identify females of the medically and economically important genus *Culicoides*.

Keywords: biting midges; vector; mitochondrial cytochrome oxidase subunit I; *C. gornostaevae*; *C. griseidorsum*; *C. pallidicornis*

1. Introduction

Ceratopogonidae (Insecta: Diptera) are 0.5- to 3.0-mm long midges [1] with more than 5400 described species out of 125 genera and 38 groups [2,3]. In Europe, 567 ceratopogonid species are known [1]. Females of four genera, *Austroconops*, *Lasiohelea*, *Leptoconops* and *Culicoides*, are obligate hematophagous biting midges and vectors of pathogens of medical and veterinary importance or are at least known as a biting nuisance [4]. Especially members of the genus *Culicoides* are important vectors of economically important viruses such as African horse sickness (AHS) virus, Bluetongue (BT) virus, equine encephalitis virus (EEV) (Reoviridae) and Schmallenberg (SB) virus (Bunyaviridae) [5,6].

In particular, BT and SB viruses not only affect ruminants, but also new world camelids, whose abundance is steadily increasing in Austria, are susceptible to viral pathogens and should be considered as carriers and reservoirs [7,8]. Recently, SB virus antibodies were found in horses in Iran [9], revealing a possible unrecognized reservoir for this virus. Furthermore, *Culicoides* spp. cause common allergic dermatitis in Icelandic horses and insect bite hypersensitivity (IBH), also known as sweet itch [10]. The hibernation of the BT virus in *Culicoides* is still a frequently discussed topic. Male as well as gravid, parous and nulliparous females were found beyond the usual activity period between spring and autumn, but also during the winter season when optimal conditions prevail [11]. Moreover, successful hibernation of the BT virus with a subsequent spread in the next spring has already been observed [12].

Diseases transmitted by *Culicoides* spp. affect large parts of Europe and are strongly linked with the spread and abundance of their vectors [13]. Nevertheless, *Culicoides* surveillance in Austria (apart from a single Bluetongue surveillance program summarized in Anderle et al. [11]) is mainly implemented on a small scale, where opportunities for cost-effective continuous sampling exists [14].

The genus *Culicoides* Latreille, 1809, is distributed worldwide and includes 1365 species [2], of which 129 are confined to Europe [15], and of which species of the Obsoletus group seem to be most abundant [13]. Approximately 30 species are capable of BT virus transmission, at least under laboratory conditions [16]. Proven vectors of BT virus are *C. imicola*, *C. brevitarsis*, *C. bolitinos* (subgenus *Avaritia* Fox, 1955), *C. pulicaris* (subgenus *Culicoides*), *C. sonorensis* (subgenus *Monoculicoides* Khalaf, 1954) and *C. insignis* (subgenus *Hoffmania* Fox 1948). In the Mediterranean region only *C. obsoletus*, *C. scoticus*, *C. pulicaris* and *C. imicola* are present, with the latter accounting for approximately 90% of BT virus transmission in this region [16] In the temperate climate, *C. obsoletus* and *C. scoticus* are the most widely distributed livestock-associated *Culicoides* species [17]. *Culicoides obsoletus*, *C. scoticus* and *C. pulicaris* are also widely distributed in Austria [18,19]. At present, the Austrian species inventory consists of 32 *Culicoides* species [18], of which 19 were recorded for the first time in Austria between 2007 and 2010 within the framework of large-scale Bluetongue and *Culicoides* surveillance [11].

Initial monitoring approaches for *Culicoides* necessitate highly skilled entomologists, because species identification is impeded by a high number of cryptic species and females. Barcoding is an adequate molecular tool to supplement morphologic identification of these cryptic species or species groups and seems to be essential for further monitoring approaches. Furthermore, molecular analysis can give a first hint at revealing previously unrecognized cryptic species [20].

2. Results

The *Culicoides* monitoring during the Bluetongue surveillance yielded 30 species [11]. In our re-assessment of this sampling, a total of 77 sequences of the mitochondrial COI barcode region were obtained from 108 female and 34 male specimens of the genus *Culicoides* that initially were identified as belonging to 32 species, species complexes or hardly distinguishable species pairs (Table 1).

Table 1. *Culicoides* taxa identified by morphology and mt mitochondrial cytochrome c oxidase subunit I gene (mt COI) including sampling date, location and storage conditions.

Taxon, ID, Bold ID	Sex	Collection Date	Province	Site	Max.% Identity to GenBank Entries	Max.% Identity to Bold Entries
Culicoides chiopterus (D155) GNIA001-20	M	21.09.2009	S	Zell/See	100% (JQ898006.1)	100% (GBDP11539-12.COI-5P)
C. chiopterus (D157) GNIA002-20	F	14.06.2010	ST	Knittelfeld	100% (JQ898010.1)	100% (GBMIN29061-13.COI-5P)
C. comosioculatus (D84) GNIA003-20	F	07.06.2010	UA	Gmunden	100% (HQ824466)	100% (GBDP11574-12.COI-5P)
C. deltus (D139) GNIA004-20	F	28.06.2009	LA	Hollabrunn	99.52% (JF766303)	100% (GBMIN29039-13.COI-5P)

Table 1. *Cont.*

Taxon, ID, Bold ID	Sex	Collection Date	Province	Site	Max.% Identity to GenBank Entries	Max.% Identity to Bold Entries
C. deltus (D197) GNIA005-20	M	27.07.2009	S	Zell/See	100% (HQ824455)	100% (GBMIN29039-13.COI-5P)
C. dewulfi (D147) GNIA006-20	M	21.09.2009	S	Zell/See	100% (KF802203)	100% (GBDP11548-12)
C. dewulfi (D148) GNIA007-20	M	21.09.2009	S	Zell/See	100% (KF802203)	100% (GBDP11548-12)
C. dewulfi (D180) GNIA008-20	M	27.07.2009	S	Zell/See	100% (KT186808)	100% (GBMIN29059-13)
C. dewulfi (D188) GNIA009-20	F	07.06.2009	UA	Braunau	100% (JQ897994)	100% (GBMIN29059-13)
C. fascipennis (D101) GNIA010-20	F	18.05.2009	V	Favoriten	100% (JQ620075)	100% (GBDP11581-12)
C. fascipennis (D103) GNIA011-20	F	03.07.2009	V	Favoriten	100% (KJ767936	100% (GBDP11582-12)
C. fascipennis (D202) GNIA012-20	F	27.07.2009	S	Zell am See	100% (KJ767936)	100% (GBDP11582-12)
C. festivipennis (D113) GNIA013-20	F	13.07.2009	C	Spittal/Drau	99.77% (HM241866)	99.54% (GBDP11586-12)
C. festivipennis (D114) GNIA014-20	F	29.06.2009	UA	Urfahr-Umgebung	99.77% (HM241866)	99.54% (GBDP11586-12)
C. festivipennis (D207) GNIA015-20	F	01.06.2009	S	Spittal/Drau	99.1% (HM241866)	99.54% (GBDP11586-12)
C. festivipennis (D208) GNIA016-20	F	01.06.2009	C	Spittal/Drau	100% (JQ620084)	99.55% (GBDP11586-12)
C. festivipennis (D209) GNIA017-20	F	01.06.2009	C	Spittal/Drau	100% (HM241866)	99.77% (GBDP11586-12)
C. furcillatus (D100) GNIA018-20	F	03.07.2009	C	Spittal/Drau	99.79% (KJ624083)	99.77% (GBDP11592-12)
C. furcillatus (D128) GNIA019-20	F	29.06.2009	UA	Urfahr-Umgebung	nd	100% (GBDP11592-12)
C. furcillatus (D102) GNIA020-20	F	03.07.2009	C	Spittal/Drau	100% (KJ624083)	100% (GBDP11591-12)
C. furcillatus (D133) GNIA021-20	F	17.08.2009	UA	Urfahr-Umgebung	100% (KJ624083)	100% (GBDP11592-12)
C. furcillatus (D201) GNIA022-20	F	27.07.2009	S	Zell am See	100% (KJ624083)	100% (GBDP11592-12)
C. gornostaevae (D28)* GNIA023-20	F	07.06.2010	LA	Zwettl	100% (JQ620138)	100% (GMGRD2770-13)
C. griseidorsum (D104)* GNIA024-20	F	18.05.2009	V	Favoriten	100% (MN274523)	nd
C. griseidorsum (D106)* GNIA025-20	F	18.05.2009	V	Favoriten	100% (MN274523)	nd
C. griseidorsum (D107)* GNIA026-20	F	18.05.2009	B	Neusiedl/See	99.8% (MN274523)	nd
C. griseidorsum (D111)* GNIA028-20	F	11.05.2009	LA	Gänserndorf	99.83% (MN274523)	nd
C. griseidorsum (D138)* GNIA029-20	F	18.05.2009	V	Favoriten	100% (MN274523)	nd
C. grisescens (D145) 1 GNIA044-20	M	21.09.2009	S	Zell/See	98.52% (KJ767938)	98.9% (early release)
C. grisescens (D146) GNIA045-20	M	21.09.2009	S	Zell/See	99.85% (KJ767938)	100% (FICER050-12)
C. grisescens (D178) [1] GNIA046-20	F	27.09.2009	S	Zell am See	99.32% (HQ824453)	99.32% (GBMIN29040-13)
C. grisescens (D179) [1] GNIA047-20	F	27.09.2009	S	Zell/See	99.77% (HQ824452)	99.77% (GBDP32600-19)
C. grisescens (D184) [1] GNIA048-20	F	21.09.2009	ST	Mürzzuschlag	99.77% (HQ824452)	99.77% (GBDP32600-19)

Table 1. *Cont.*

Taxon, ID, Bold ID	Sex	Collection Date	Province	Site	Max.% Identity to GenBank Entries	Max.% Identity to Bold Entries
C. grisescens (D24) [1] GNIA049-20	F	21.09.2009	S	Zell/See	99.77% (HQ824452)	99.77% (GBDP32600-19)
C. kibunensis (D183) GNIA050-20	F	02.06.2008	V	Favoriten	99.58% (KJ624094)	99.85% (GMGRD1621-13)
C. kibunensis (D34) GNIA051-20	F	07.06.2009	LA	Zwettl	100% (JQ683272)	99.08 (private)
C. lupicaris (D74)[1] GNIA052-20	F	07.06.2010	S	Tamsweg	100% (HQ824440)	100% (GBDP11558-12)
C. lupicaris (D75) [1] GNIA05320	F	07.06.2010	S	Tamsweg	98.64% (HQ824442)	98.85% (GBDP11562-12)
C. lupicaris (D57) GNIA063-20	M	03.05.2010	UA	Kirchdorf/ Krems	100% (KF591605)	99.84 (GBDP33078-19)
C. lupicaris (D58) GNIA064-20	M	03.05.2010	UA	Kirchdorf/ Krems	100% (KF591605)	99.84 (GBDP33078-19)
C. minutissimus (D191) GNIA055-20	M	24.08.2009	LA	Tulln	no entry	no entry
C. nubeculosus (D124) GNIA065-20	F	10.08.2009	LA	Gänserndorf	nd	nd
C. nubeculosus (D125) GNIA066-20	F	10.08.2009	LA	Gänserndorf	nd	nd
C. obsoletus (D45) GNIA074-20	M	25.05.2009	C	Villach	100% (HM022818)	100% (GMGRE3865-13)
C. obsoletus (D46) GNIA075-20	M	25.05.2009	C	Villach	100% (HQ824383)	100% (GBMIN29075-13)
C. obsoletus (D48) GNIA076-20	M	28.06.2010	ST	Liezen	100% (KT186816)	100% (GMGRE3865-13)
C. pallidicornis (D35)* GNIA071-20	F	07.06.2010	LA	Zwettl	99.36% (KJ624111)	100% (early release)
C. pallidicornis (D193)* GNIA072-20	M	01.06.2009	C	Spittal/Drau	100% (JQ620154)	100% (GBDP11602-12)
C. pallidicornis (D198)* GNIA073-20	F	01.06.2009	C	Spittal/Drau	100% (JQ620154)	100% (GBDP11602-12)
C. pictipennis (D149) GNIA069-20	F	15.06.2009	T	Reutte	99.78% (JQ620162)	100% (private)
C. pictipennis (D151) GNIA070-20	F	11.05.2009	B	Güssing	99.78% (JQ620162)	100% (private)
C. poperinghensis (D181) GNIA067-20	F	14.05.2010	V	Favoriten	100% (JQ620166)	99.6% (private)
C. poperinghensis (D182) GNIA068-20	F	02.06.2008	V	Favoriten	100% (JQ620166)	99.6% (private)
C. pulicaris (D49) GNIA060-20	F	10.05.2010	LA	Zwettl	100% (JQ620183)	99.76% (GBDP11551-12)
C. pulicaris (D51) GNIA061-20	F	21.09.2010	UA	Perg	100% (AM236711)	100% (GBMIN17621-13)
C. pulicaris (D54) GNIA062-20	M	14.06.2010	ST	Bruck/Mur	100% (JF766336)	99.76% (GBDP11551-12)
C. punctatus (D129) GNIA055-20	F	18.05.2009	V	Favoriten	100% (MN274527)	98.77% (private)
C. punctatus (D130) GNIA056-20	F	18.05.2009	V	Favoriten	99.68% (MN274527)	99.69% (private)
C. punctatus (D159) GNIA057-20	F	08.06.2009	LA	Gänserndorf	99.36% (MN274527)	99.39% (private)
C. punctatus (D160) GNIA058-20	F	15.06.2009	T	Reutte	100% (KY707780)	100% (GBDP11610-12)
C. punctatus (D162) GNIA059-20	M	17.08.2009	UA	Urfahr-Umgebung	99.78% (KY707779)	98.62% (private)

Table 1. *Cont.*

Taxon, ID, Bold ID	Sex	Collection Date	Province	Site	Max.% Identity to GenBank Entries	Max.% Identity to Bold Entries
C. reconditus (D186) GNIA039-20	F	10.08.2009	UA	Linz-Land	98.48% (KJ767956)	98.4% (FICER086-12)
C. reconditus (D192) GNIA040-20	F	27.07.2009	S	Tamsweg	98.48% (KJ767956)	98.4% (FICER086-12)
C. reconditus (D194) GNIA041-20	F	17.08.2009	S	Tamsweg	100% (JQ620193)	98.78% (early release)
C. reconditus (D195) GNIA042-20	F	17.08.2009	S	Tamsweg	99.77% (HQ8245089	99.77% (GBDP11615-12)
C. reconditus (D199) GNIA043-20	F	24.08.2009	C	Spittal/Drau	99.77% (HQ824508)	99.76% (GBDP11615-12)
C. riethi (D108) GNIA027-20	M	14.09.2009	B	Neusiedl/See	99.82% (MN274531)	nd
C. riethi (D121) GNIA030-20	M	13.07.2009	LA	Mistelbach	100% (JQ620196)	nd
C. riethi (D126) GNIA031-20	F	10.08.2009	LA	Mistelbach	100% (JQ620196)	nd
C. riethi (D127) GNIA032-20	F	10.08.2009	LA	Mistelbach	100% (JQ620195)	nd
C. riouxi (D187) GNIA033-20	F	10.08.2009	LA	Melk	100% (KJ624124)	98.7% (private)
C. salinarius (D154) GNIA034-20	F	29.06.2009	LA	Hollabrunn	100% (JQ620198)	98.5% Culicoides sp. (CNGSD4978-15)
C. salinarius (D190) GNIA035-20	F	06.07.2009	St	Feldbach	99.57% (KJ624125)	99.64% (early release)
C. scoticus (D44) GNIA036-20	M	03.08.2010	ST	Liezen	100% (KT186879)	100% (GBMIN29072-13)
C. scoticus (D144) GNIA037-20	M	29.06.2009	UA	Urfahr Umgebung	100% (KT186879)	100% (GBMIN29072-13)
C. subfasciipennis (D177) GNIA038-20	F	23.06.2008	LA	Scheibbs	99.14% (JQ620238)	98.92% (private)

Abbreviations: F = female; M = male; S= Salzburg; St = Styria; UA = Upper Austria; V = Vienna; C = Carinthia; B = Burgenland; LA = Lower Austria; nd = no data for specification; * first record for Austria; [1] cryptic species. Wenk et al. [21].

Morphologic and molecular identification was congruent in 59 specimens, i.e., suggesting the same taxonomic entities for 17 of 19 taxa (Table 2), recorded in Austria during the Bluetongue surveillance [11].

Table 2. Updated list of the Austrian *Culicoides* species inventory representing 36 species.

Taxon (Author, Year)
C. alazanicus (syn. of *C. musilator*) (Dzhafarov 1961)
C. albicans (Winnertz 1852) +
C. chiopterus (Meigen 1830)
C. circumscriptus (Kieffer 1918) +
C. clastrieri (Callot, Kremer and Deduit 1962) +
C. comosioculatus (syn. of *C. chetophthalmus*) (Tokunaga, 1956) +
C. deltus (Edwards 1939) +
C. dewulfi (Goetghebuer 1933) +
C. duddingstoni (Kettle and Lawson, 1955) +
C. fascipennis (Staeger 1839)
C. festivipennis (Kieffer 1914)
C. furcillatus (Callot, Kremer and Paradis 1962) +
C. griseidorsum (Kieffer 1918) *
C. grisescens (Goetghebuer 1935) +

Table 2. *Cont.*

Taxon (Author, Year)
C. gornostaevae (Mirzaeva, 1984) *
C. kibunensis (Tokunaga, 1937)
C. lupicaris (Downes and Kettle, 1952)
C. minutissimus (Zetterstedt 1855)
C. newsteadi (Austen 1921) +
C. nubeculosus (Tokunaga 1941)
C. obsoletus (Meigen 1818)
C. pallidicornis (Kieffer, 1919) *
C. pictipennis (Staeger 1839)
C. poperinghensis (Goetghebuer 1953) +
C. pulicaris (Linnaeus 1758)
C. punctatus (Meigen 1804) +
C. reconditus (Campbell and Pelham-Clinton 1960) +
C. riethi (Kieffer, 1914) +
C. riouxi (Huttel and Huttel 1951) +
C. saevus (Kieffer, 1922)
C. salinarius (Kieffer, 1914) +
C. scoticus (Downes and Kettle, 1952) +
C. segnis (Campbell and Pelham-Clinton 1960) +
C. stigma (Meigen 1818)
C. subfasciipennis (Kieffer 1919)
C. vexans (Staeger 1839)

* First record for Austria in the framework of this study; + first record for Austria in Anderle et al. [11].

Furthermore, molecular identification revealed the presence of three species of *Culicoides* new to Austria, namely *C.* (*Avaritia*) *gornostaevae*, *C.* (*Silvaticulicoides*) *griseidorsum* and *C.* (*Wirthomyia*) *pallidicornis* (Tables 1 and 2). Molecular identification was corroborated by morphologic analysis using the key of Mathieu et al. [22]. A total of five specimens collected in May 2009, mistakenly identified by morphology as *C. clastieri* (four females) and as *C. duddingstoni* (one female), were identified by means of molecular tools as *C. griseidorsum*. This species is known from four countries bordering Austria (Germany, Italy, Slovakia and Switzerland). A retrospective reexamination of the morphologic characters using the redescription of this taxon by Szadziewski et al. [15] also identified those specimens as *C. griseidorsum*. *Culicoides pallidicornis,* generally present in bordering countries of Austria (Italy, Switzerland, Germany, Slovakia, Czech Republic and Hungary), was represented by a single female, morphologically misidentified as *C. subfasciipennis,* which had been collected in Carinthia in June 2009. Subsequent morphologic identification using the key of Mathieu et al. [22] confirmed the result of the molecular analysis. The presence of *C. gornostaevae* was confirmed only by molecular tools. This member of the Obsoletus group, a possible vector for Schmallenberg and Bluetongue virus, is so far only known from Norway, Poland, Siberia and Sweden [23].

Molecular analyzes recovered some putatively cryptic species. Four of five specimens of *C. grisescens* (D24, D178, D179 and D184) were identical to a specific type, *C. grisescens* G2, in which the specimens are morphologically similar, but differ substantially in their barcode sequences [23]. The same was observed within the Pulicaris and the Obsoletus group, where several types of mt COI sequences were observed (Table 1). Furthermore, we provide here the first barcode of *C. minutissimus* (= syn. *C. pumilus*) in the BOLD sequence database, obtained from a male specimen.

Two specimens—originally identified morphologically as *C. lupicaris*—were identical as in their barcode region to the as yet undescribed lineage *Culicoides* sp. CW2011, which was initially reported in 2011 in Switzerland [21]).

3. Discussion

Information on seasonal and spatial distribution patterns as well as autecology of different *Culicoides* species is, with exception of some single studies [11], limited in Austria. This is mostly due to identification of *Culicoides* species to complex or group-level, but not to species-level [14], as identification of *Culicoides* spp. below group-level by morphology can be difficult or impossible if reliable morphologic characters are absent–especially in those species with weakly developed wing pattern. In addition, morphology, ecology and microhabitats of most of the immature stages of different *Culicoides* species are unknown [1].

This is the first study in Austria to investigate cryptic *Culicoides* at the species-level, which we achieved using a DNA barcoding approach, in addition to exhaustive morphologic studies. DNA barcodes are already known to be able to uncover unrecognized cryptic species, especially in the suborder Nematocera (Insecta: Diptera), as well as in Chironomidae [20], Cecidomyiidae [24] and Simuliidae [25]. To derive a valid description of these new putative species and to assign a formal scientific name based on mt COI sequence divergence, is still challenging, and complicates the interpretation of results. With respect to *C. grisescens*, only *C. grisescens* G2, a lineage initially reported in 2012 [21], was examined in our study, not the nominal species. Furthermore, two types of *C. obsoletus* and three types of *C. pulicaris* were found (Table 1). The Obsoletus group consists of the morphologically similar *C. obsoletus*, *C. scoticus*, *C. dewulfi* and *C. chiopterus*. Controversially, the term *Obsoletus* complex is mainly used for *C. obsoletus*, *C. scoticus* and *C. montanus* (a species not belonging to the Obsoletus group), in which the females cannot be distinguished by morphology. However, males of *C. obsoletus*, *C. dewulfi*, *C. chiopterus* and *C. scoticus* can be differentiated reliably by morphology and both sexes by the mt COI barcode region (Table 1). In addition to the known *Obsoletus* complex diversity, several cryptic species related to *C. obsoletus* were found in Central Europe [26]; their potential presence in Austria remains to be investigated as these are not formally described or related to particular mt COI sequences. The absence of *C. montanus* from Austria can, however, be safely assumed, as this taxon is known to be rare and restricted to the southern Mediterranean region [26]. Moreover, the first finding of *C. gornostaevae*, a boreal species belonging to the Obsoletus group, was facilitated by the analysis of the barcode region. *Culicoides gornostaevae* was previously known only from Norway, Poland, Siberia and Sweden [22]. Males of this taxon can be separated from the other members of the Obsoletus group by the genitalia armature [22]. The overall lack of reliable reference data for the Obsoletus group allows us to confirm the presence of two taxa, that may either be a member of the nominal group or represent yet unknown species.

The Pulicaris group is considered to consist of 14 distinct taxa: *C. pulicaris*, *C. lupicaris*, *C. impunctatus*, *C. punctatus*, *C. grisescens*, *C. newsteadi*, *C. flavipulicaris*, *C. fagineus*, *C. subfagineus*, *C. bysta* n. sp., *C. paradoxalis* sp. nov., *C. boyi* sp. nov., *C. selandicus* sp. nov. and *C. kalix* sp. nov. [27]. In a Turkish study, three different haplotypes of *C. lupicaris* were recovered [27–30] and results obtained here indicate a similar situation in *C. pulicaris*.

Interestingly, specimens originally identified as *C. lupicaris* (Table 1) were identical to a yet unnamed lineage first reported in 2011 from Switzerland using barcoding [21]. This is particularly intriguing as these specimens displayed the morphologic characters typical of *C. lupicaris*. These results should clearly be able to spark a taxonomic reevaluation of the Pulicaris group.

In our case morphologic identification was principally congruent with molecular analysis. Mismatched results underline the difficulties of identifying female *Culicoides*, especially in species groups or complexes, and that identification of male *Culicoides* species is less susceptible to errors than females. In-depth research needs to be done within *Culicoides* species groups to correctly differentiate the group members, at least with molecular tools. Investigations on possible differences in ecology, behavior and vector competence of those newly found cryptic species could then follow. This will also be of relevance to epidemiological research as BT viruses are still to be considered in Austria. The first BT virus case was reported in the district of Schaerding, Upper Austria in November 2008, where in total 28 BT virus serotype 8 positive animals were found; extensive vaccination measures followed [31]. The Austrian BT monitoring and vaccination programs after this outbreak led to total costs of €23.6 million [32] up until 2016. In 2015, BT serotype 4 was found for the first time in the federal states Burgenland and Styria and in 2016 a BT virus serotype 4 outbreak was reported from Carinthia [32].

Correct species identification of important vectors is crucial to assess pathogen transmission potential. Cryptic species may have different ecological niches and may differ in their epidemiological relevance, thus resolving taxonomic conundrums is necessary. In addition, molecular identification through DNA barcoding may support exploration of life cycles and life history traits in the future to better characterize breeding habitats of *Culicoides* in Austria. Misidentification of these vectors can lead to substantial epidemiological implications, such as neglect of vaccination measures and loss of livestock. We conclude that morphological identification supported by molecular analysis such as DNA barcoding bases on the mt COI gene is an adequate tool for species identification in *Culicoides*. We strongly suggest using both identification approaches to deliver high quality data –necessary for both ecological research and vector-transmitted disease risk assessment. However, use of molecular tools alone is not recommended due to shortcomings of available reference libraries, which can hamper identification. Additionally, employing sets of different primers is strongly recommended within the order Diptera. General primers often fail to amplify all taxa within a family, e.g., mosquitoes, in which sets of more specific primers should be used [33]. Moreover, barcoding fails at separating the species within certain complexes (e.g., *Anopheles maculipennis* complex). In the present study, we used three different primer pairs to increase the success rate of amplification and subsequent identification. Species identification of larvae is virtually impossible as larvae lack reliable morphologic characteristics [34]. At present, larvae of most Ceratopogoninae and Palpomyiinae species are not described at species-level [1]. Only 13% of the larvae and 17% of the pupae of the genus *Culicoides* are known, but often poorly described [3]. Currently, successful hatching of larvae collected to identify the adults is a common identification strategy [35]. Molecular tools may open the way to a more comprehensive investigation of this important taxon. Accelerated identification by means of molecular barcodes will therefore support future studies on larval micro-habitats—knowledge which will be crucial for the development of preventive measures of *Culicoides*-borne diseases or the prevention of creating larval micro-habitats around farms where emerging disease outbreaks may happen [36].

This study provides the first barcodes for *Culicoides* biting midges and the first record of *Culicoides gornostaevae*, *C. griseidorsum* and *C. pallidicornis* in Austria. We demonstrate that morphologic identification is primarily congruent with molecular analyzes. However, intense taxonomic, epidemiological and ecological research efforts, ideally supported by molecular tools, are still necessary to differentiate species based on their ecology, behavior and vector competence.

4. Materials and Methods

Culicoides were sampled within a Bluetongue virus monitoring program carried out from 2007 to 2010 and realized in all Austrian provinces [11] using Onderstepoort-type black light traps [11,37] which were set weekly at a total of 54 sampling sites. Individuals were separated by sex and identified to species- or at least to group-level by morphology using the key of Delécolle [38] and were stored in 75% ethanol (EtoH). One to eight well preserved specimens from each species were sampled from different provinces and used for barcoding. After photographic documentation, DNA was extracted from one to three legs of each individual. Three 1.4 mm ceramic beads (Precellys Ceramic Kit 2.8 mm/1.4 mm, Peqlab, Erlangen, Germany) were added to each tissue sample. After homogenization with TissueLyser II (Qiagen, Hilden, Germany), DNA was extracted using the "DNeasy® Blood and Tissue" DNA isolation kit according to the manufacturer's protocol (Qiagen, Hilden, Germany). Conventional polymerase chain reaction (PCR), targeting an approximately 667 bp fragment of the mitochondrial cytochrome c oxidase subunit I gene (COI) within the BOLD-barcode region using primers H15CuliCOIFw and H15CuliCOIRv, LCO1490 and HCO2198 as well as LepF1 and LepR1 was performed as reported previously [30,39–41]. After molecular specification, *Culicoides* specimens with mismatching results in morphologic and molecular features were rechecked using an up to date online identification key for female *Culicoides* from the West Palearctic region (IIKC) by Mathieu et al. [21], which was not available at the sampling time.

Finally, morphologically and molecularly identified voucher specimens were deposited in the Diptera collection of the Natural History Museum Vienna.

Author Contributions: Conceptualization, C.Z., P.S. and H.-P.F.; methodology, C.Z., G.W., B.S.B., K.B.-L., L.V.d.V., H.-P.F.; validation, G.W., P.S. and C.Z.; resources, P.S., H.-P.F.; data curation, C.Z., K.B.L., H.-P.F.; writing—original draft preparation, C.Z.; supervision, H.-P.F.; funding acquisition, H.-P.F. All authors have read and agreed to the published version of the manuscript.

Funding: Financial support for barcoding was provided by the Austrian Federal Ministry of Education, Science and Research via an ABOL (Austrian barcode of Life; www.abol.ac.at) associated project within the framework of the "Hochschulraum–Strukturmittel" Funds. CZ acknowledges additional support from the FWF (P 31258-B29). Open Access Funding by the University of Veterinary Medicine Vienna.

Acknowledgments: The authors are grateful to Ellen R. Schoener for her indispensable help in the laboratory. We further thank Art Borkent and Elisabeth Stur for their scientific advice.

Conflicts of Interest: The authors declare no conflict of interest.

References

1. Mauch, E. Aquatische Diptera-Larven in Mittel—Nordwest-und Nordeuropa. Übersicht über die Formen und ihre Identifikation. *Lauterbornia* **2017**, *83*, 1–404.

2. Borkent, A.; Wirth, W.W. World species of biting midges (Diptera: Ceratopogonidae). *Bull Am. Mus. Nat. Hist.* **1997**, *233*, 1–257.

3. Borkent, A. *The Subgeneric Classification of Species of Culicoides-Thoughts and a Warning*; Elsevier Academic Press: Burlington, MA, USA, 2016; Available online: https://www.inhs.illinois.edu/files/5014/6532/8290/CulicoidesSubgenera.pdf (accessed on 15 February 2020).

4. Werner, D.; Kampen, H. Gnitzen (Diptera, Ceratopogonidae) und ihre medizinische Bedeutung. *Denisia* **2010**, *30*, 245–260.

5. Rasmussen, L.D.; Kirkeby, C.; Bødker, R.; Kristensen, B.; Rasmussen, T.B.; Belsham, G.; Bøtner, A. Rapid spread of Schmallenberg virus-infected biting midges (*Culicoides* spp.) across Denmark in 2012. *Transbound. Emerg. Dis.* **2013**, *61*, 12–16. [CrossRef]

6. Doceul, V.; Lara, E.; Sailleau, C.; Belbis, G.; Richardson, J.; Bréard, E.; Viarouge, C.; Dominguez, M.; Hendrikx, P.; Calavas, D.; et al. Epidemiology, molecular virology and diagnostics of Schmallenberg virus, an emerging orthobunyavirus in Europe. *Vet. Res.* **2013**, *44*, 31. [CrossRef]

7. Kriegl, C.; Klein, D.; Kofler, J.; Fuchs, K.; Baumgartner, W. Haltungs-und Gesundheitsaspekte bei Neuweltkameliden. *Wien. Tierärtl Mschr.* **2015**, *92*, 119–125.

8. Stanitznig, A.; Lambacher, B.; Eichinger, M.; Franz, S.; Wittek, T. Prevalence of important viral infections in new world camelids in Austria. *Wien. Tierärztl. Mon.* **2016**, *103*, 92–100.

9. Rasekh, M.; Sarani, A.; Hashemi, S.H. Detection of Schmallenberg virus antibody in equine population of Northern and Northeast Iran. *Vet. World* **2018**, *11*, 30–33. [CrossRef]

10. Schurink, A.; Van Der Meide, N.M.A.; Savelkoul, H.F.J.; Ducro, B.J.; Tijhaar, E. Factors associated with Culicoides obsoletus complex spp.-specific IgE reactivity in Icelandic horses and Shetland ponies. *Vet. J.* **2014**, *201*, 395–400. [CrossRef]

11. Anderle, F.; Schneemann, Y.; Sehnal, P. Culcioides spp. (Diptera, Nematocera, Ceratopogonidae) in Österreich—Resümee nach 3 Jahren Monitoring im Rahmen der Bluetongue-Überwachung. *Entomol. Austriaca* **2011**, *18*, 9–17.

12. Carpenter, S.; Szmaragd, C.; Barber, J.; Labuschagne, K.; Gubbins, S.; Mellor, P. An assessment of Culicoides surveillance techniques in northern Europe: Have we underestimated a potential bluetongue virus vector. *J. Appl. Ecol.* **2008**, *45*, 1237–1245.

13. Cuéllar, A.C.; Kjær, L.J.; Kirkeby, C.; Skovgard, H.; Nielsen, S.A.; Stockmarr, A.; Andersson, G.; Lindstrom, A.; Chirico, J.; Lühken, R.; et al. Spatial and temporal variation in the abundance of Culicoides biting midges (Diptera: Ceratopogonidae) in nine European countries. *Parasit Vectors* **2018**, *11*, 112. [CrossRef] [PubMed]

14. Brugger, K.; Rubel, F. Bluetongue disease risk assessment based on observed and projected Culicoides *obsoletus* spp. vector densities. *PLoS ONE* **2013**, *8*, e60330. [CrossRef] [PubMed]

15. Szadziewski, R.; Filatov, S.; Dominiak, P. A redescription of Culicoides griseidorsum Kieffer, 1918, with comments on subgeneric position of some European taxa (Diptera: Ceratopogonidae). *Zootaxa* **2016**, *4107*, 413–422. [CrossRef] [PubMed]

16. Meiswinkel, R.; Gomulski, L.M.; Delécolle, J.C.; Gasperi, G. The taxonomy of *Culicoides* vector complexes—Unfinished business. *Vet. Ital.* **2004**, *40*, 151–159. [PubMed]

17. Versteirt, V.; Balenghien, T.; Tack, W.; Wint, W. A first estimation of Culicoides imicola and Culicoides obsoletus/Culicoides scoticus seasonality and abundance in Europe. *EFSA Supporting Publ.* **2017**, *14*, 1182. [CrossRef]

18. Anderle, F.; Sehnal, P.; Schneemann, Y.; Schindler, M.; Wöss, G.; Marschler, M. Culicoides surveillance in Austria (Diptera: Ceratopogonidae)—A snap-shot. *Beiträge Zur Entomofaunist.* **2008**, *9*, 67–79.

19. Sehnal, P.; Schweiger, S.; Schindler, M.; Anderle, F.; Schneemann, Y. Bluetongue: Vector surveillance in Austria in 2007. *Wien Klin Wochenschr* **2008**, *120*, 34–39. [CrossRef]

20. Xiao-Long, L.; Stur, E.; Ekrem, T. DNA-barcodes and morphology reveal unrecognized species in Chironomidae (Diptera). *Insect Syst. Evol.* **2018**, *49*, 329–398.

21. Wenk, C.E.; Kaufmann, C.; Schaffner, F.; Mathis, A. Molecular characterization of Swiss Ceratopogonidae (Diptera) and evaluation of real-time PCR assays for the identification of Culicoides biting midges. *Vet. Parasitol* **2012**, *184*, 258–266. [CrossRef]

22. Mathieu, B.; Cêtre-Sossah, C.; Garros, C.; Chavernac, D.; Balenghien, T.; Carpenter, S.; Setier-Rio, M.L.; Vignes Lebbe, R.; Ung, V.; Candolfi, E.; et al. Development and validation of IIKC: An interactive identification key for Culicoides (Diptera: Ceratopogonidae) females from the Western Palaearctic region. *Parasit. Vectors* **2012**, *5*, 137. [CrossRef] [PubMed]

23. Kirkeby, C.; Dominiak, P. Culicoides (Avaritia) gornostaevae Mirzaeva, 1984 (Diptera: Ceratopogonidae)—A possible vector species of the Obsoletus group new to the European fauna. *Parasit. Vectors* **2014**, *7*, 445. [CrossRef] [PubMed]

24. Mathur, S.; Cook, M.A.; Sinclair, B.J.; Fitzpatrick, S.M. DNA barcodes suggest cryptic speciation in Dasineura oxyoccana (Diptera: Cecidomyiidae) on cranberry, Vaccinium macrocarpon, and blueberry, V. corymbosum. *Fla. Entomol.* **2019**, *95*, 387–394. [CrossRef]

25. Hernández-Triana, L.M.; Chaverri, L.G.; Rodríguez-Pérez, M.A.; Prosser, S.W.J.; Hebert, P.D.N.; Gregory, T.R.; Johnson, N. DNA barcodes of neotropical black flies (Diptera: Simuliidae): Species identification and discovery of cryptic diversity in Msoaerica. *Zootaxa* **2015**, *3936*, 093–114. [CrossRef]

26. Meiswinkel, R.; De Bree, F.; Bossers-De Vries, R.; Elbers, A.R.W. An unrecognized species of the Culicoides obsoletus complex feeding on livestock in The Netherlands. *Vet. Parasitol.* **2015**, *207*, 324–328. [CrossRef]

27. Yildirim, A.; Dik, B.; Duzlu, O.; Onder, Z.; Ciloglu, A.; Yetismis, G.; Inci, A. Genetic diversity of Culicoides species within the Pulicaris complex (Diptera: Ceratopogonidae) in Turkey inferred from mitochondrial COI gene sequences. *Acta Trop.* **2018**, *190*, 380–388. [CrossRef]

28. Sarvašová, A.; Kočišová, A.; Candolfi, E.; Mathieu, B. Description of Culicoides (Culicoides) bysta n. sp., a new member of the Pulicaris group (Diptera: Ceratopogonidae) from Slovakia. *Parasites Vectors* **2017**, *10*, 279. [CrossRef]

29. Nielsen, S.A.; Kristensen, M.; Pape, T. Three new Scandinavian species of Culicoides (Culicoides): C. boyi sp. nov., C. selandicus sp. nov. and C. kalix sp. nov. (Diptera: Ceratopogonidae). *Biodivers. Data J.* **2015**, *3*, e5823. [CrossRef]

30. Ramilo, D.; Garros, C.; Mathieu, B.; Benedet, C.; Allène, X.; Silva, E.; Alexandre-Pires, G.; Pereira Da Fonseca, I.; Carpenter, S.; Rádrová, J.; et al. Description of Culicoides paradoxalis sp. nov. from France and Portugal (Diptera: Ceratopogonidae). *Zootaxa* **2013**, *3745*, 243–256. [CrossRef]

31. Loitsch, A.; Sehnal, P.; Winkler, M.; Revilla-Fernandez, S.; Schwarz, M.; Kloud, T.; Stockreiter, S.; Winkler, U. BTV-Abschlussbericht 2009: Durchführung der Bluetongue Überwachung in Österreich (Project report, in German). Available online: https://www.verbrauchergesundheit.gv.at/tiere/krankheiten/BTV-Abschlussbericht_20093229.pdf?63xzm0 (accessed on 10 March 2020).

32. Pinior, B.; Firth, C.L.; Loitsch, A.; Stockreiter, S.; Hutter, S.; Richter, V.; Lebl, K.; Schwermer, H.; Käsbohrer, A. Cost distribution of bluetongue surveillance and vaccination programmes in Austria and Switzerland (2007–2016). *Vet. Rec.* **2018**, *182*, 257. [CrossRef]

33. Zittra, C.; Flechl, E.; Kothmayer, M.; Vitecek, S.; Rossiter, H.; Zechmeister, T.; Fuehrer, H.P. Ecological characterization and molecular differentiation of Culex pipiens complex taxa and Culex torrentium in eastern Austria. *Parasit. Vectors* **2016**, *9*, 197. [CrossRef] [PubMed]

34. Schwenkenbecher, J.M.; Mordue, A.J.; Piertney, S.B. Phylogenetic analysis indicates that Culicoides dewulfi should not be considered part of the Culicoides obsoletus complex. *Bull. Entomol. Res.* **2009**, *99*, 371–375. [CrossRef] [PubMed]

35. Zimmer, J.-Y.; Haubruge, E.; Francis, F.; Bortels, J.; Simonon, G.; Losson, B.; Mignon, B.; Paternostre, J.; De Deken, R.; De Deken, G.; et al. Breeding sites of bluetongue vectors in northern Europe. *Vet. Rec.* **2008**, *162*, 131. [CrossRef] [PubMed]

36. Zimmer, J.-Y.; Brostaux, Y.; Haubruge, E.; Francis, F. Larval development sites of the main Culicoides species (Diptera: Ceratopogonidae) in northern Europe and distribution of coprophilic species larvae in Belgian pastures. *Vet. Parasitol.* **2014**, *205*, 676–686. [CrossRef]

37. Venter, G.J.; Meiswinkel, R. The virtual absence of Culicoides imicola (Diptera: Ceratopogonidae) in a light trap survey of the colder, high-lying area of the Eastern Orange Free State, South Africa, and implications for the transmission of arboviruses. *Onderstepoort J. Vet. Res.* **1994**, *61*, 327–340.

38. Delécolle, J.C. Nouvelle contribution à l'étude systématique et iconographique des espèces du genre Culicoides du nord-est de la France. *Dissertation* **1985**, *1*, 238.

39. Folmer, O.; Black, W.; Hoeh, W.; Lutz, R.; Vrijenhoeck, R. DNA primers for amplifications of mitochondrial cytochrome c oxidase subunit I from diverse metazoan invertebrates. *Mol. Mar. Biol. Biotechnol.* **1994**, *3*, 294–299.

40. Hebert, P.D.; Penton, E.H.; Burns, J.M.; Janzen, D.H.; Hallwachs, W. Ten Species in One: DNA Barcoding Reveals Cryptic Species in the Neotropical Skipper Butterfly Astraptes Fulgerator. *Proc. Natl. Acad. Sci. USA* **2004**, *101*, 14812–14817. [CrossRef]

41. Werblow, A.; Flechl, E.; Klimpel, S.; Zittra, C.; Lebl, K.; Kieser, K.; Laciny, A.; Silbermayr, K.; Melaun, C.; Fuehrer, H.-P. Direct PCR of indigenous and invasive mosquito species: A time and cost-effective technique of mosquito barcoding. *Med. Vet. Entomol.* **2016**, *30*, 8–13. [CrossRef]

 *pathogens*

Article

Integrative Approach to *Phlebotomus mascittii* Grassi, 1908: First Record in Vienna with New Morphological and Molecular Insights

Edwin Kniha [1], Vít Dvořák [2], Petr Halada [3], Markus Milchram [4], Adelheid G. Obwaller [5], Katrin Kuhls [6,7,†], Susanne Schlegel [6,8,†], Martina Köhsler [1], Wolfgang Poeppl [9], Karin Bakran-Lebl [10], Hans-Peter Fuehrer [10], Věra Volfová [2], Gerhard Mooseder [9], Vladimir Ivovic [11], Petr Volf [2] and Julia Walochnik [1,*]

[1] Institute of Specific Prophylaxis and Tropical Medicine, Center for Pathophysiology, Infectiology and Immunology, Medical University of Vienna, 1090 Vienna, Austria; edwin.kniha@meduniwien.ac.at (E.K.); martina.koehsler@meduniwien.ac.at (M.K.)

[2] Department of Parasitology, Faculty of Science, Charles University Prague, 128 43 Prague, Czech Republic; vidvorak@natur.cuni.cz (V.D.); veravolf@seznam.cz (V.V.); volf@cesnet.cz (P.V.)

[3] BioCeV, Institute of Microbiology of the Czech Academy of Sciences, 252 50 Vestec, Czech Republic; halada@biomed.cas.cz

[4] Department of Integrative Biology and Biodiversity Research, Institute of Zoology, University of Natural Resources and Life Sciences Vienna, 1180 Vienna, Austria; markusmilchram@gmx.net

[5] Federal Ministry of Defence, Division of Science, Research and Development, 1090 Vienna, Austria; adelheid.obwaller@bmlv.gv.at

[6] Division Molecular Biotechnology and Functional Genomics, Technical University of Applied Sciences Wildau, 15745 Wildau, Germany; katrin.kuhls@th-wildau.de (K.K.); susanne.schlegel@th-wildau.de (S.S.)

[7] Research Platform "Models & Simulation", Leibniz Centre for Agricultural Landscape Research (ZALF), 15374 Müncheberg, Germany

[8] Division Microsystems Engineering, Technical University of Applied Sciences Wildau, 15745 Wildau, Germany

[9] Department of Dermatology and Tropical Medicine, Military Medical Cluster East, Austrian Armed Forces, 1210 Vienna, Austria; wolfgang.poeppl@bmlv.gv.at (W.P.); gerhard.mooseder@bmlv.gv.at (G.M.)

[10] Department of Pathobiology, Institute of Parasitology, University of Veterinary Medicine Vienna, 1210 Vienna, Austria; karin.bakran-lebl@vetmeduni.ac.at (K.B.-L.); hans-peter.fuehrer@vetmeduni.ac.at (H.-P.F.)

[11] Department of Biodiversity, FAMNIT, University of Primorska, 6000 Koper-Capodistria, Slovenia; vladimir.ivovic@famnit.upr.si

* Correspondence: julia.walochnik@meduniwien.ac.at

† Current affiliation.

Received: 17 November 2020; Accepted: 4 December 2020; Published: 9 December 2020

Abstract: Sand flies (Diptera: Psychodidae: Phlebotominae) are blood-feeding insects that transmit the protozoan parasites *Leishmania* spp. and various arthropod-borne (arbo) viruses. While in Mediterranean parts of Europe the sand fly fauna is diverse, in Central European countries including Austria mainly *Phlebotomus mascittii* is found, an assumed but unproven vector of *Leishmania infantum*. To update the currently understudied sand fly distribution in Austria, a sand fly survey was performed and other entomological catches were screened for sand flies. Seven new trapping locations of *Ph. mascittii* are reported including the first record in Vienna, representing also one of the first findings of this species in a city. Morphological identification, supported by fluorescence microscopy, was confirmed by two molecular approaches, including sequencing and matrix-assisted laser desorption/ionization-time of flight mass spectrometry (MALDI-TOF MS) protein profiling. Sand fly occurrence and activity were evaluated based on surveyed locations, habitat requirements and climatic parameters. Moreover, a first comparison of European *Ph. mascittii* populations was made by two marker genes, cytochrome c oxidase subunit 1 (*COI*), and cytochrome b (*cytb*), as well as

MALDI-TOF mass spectra. Our study provides new important records of *Ph. mascittii* in Austria and valuable data for prospective entomological surveys. MALDI-TOF MS protein profiling was shown to be a reliable tool for differentiation between sand fly species. Rising temperatures and globalization demand for regular entomological surveys to monitor changes in species distribution and composition. This is also important with respect to the possible vector competence of *Ph. mascittii*.

Keywords: *Transphlebotomus*; Central Europe; autoimmunofluorescence; MALDI-TOF mass spectrometry; genotyping; leishmaniasis

1. Introduction

Phlebotomine sand flies (Diptera: Psychodidae: Phlebotominae) occur in tropical, subtropical, as well as temperate regions. They are of significant medical importance as vectors of *Leishmania* spp., bacteria and several arthropod-borne (arbo) viruses in various regions of both Old and New World. Leishmaniasis is among the most prominent and yet neglected infectious diseases [1,2].

In Europe, sand flies are endemic throughout most of the Mediterranean countries while their occurrence north of the Alps and in Central Europe was overlooked for a long time. *Phlebotomus mascittii* Grassi, 1908, and *Phlebotomus perniciosus* Newstead, 1911, were recorded in Germany for the first time in 1999 and 2001, respectively [3,4]. A decade later, *Ph. mascittii* was recorded in Austria and its presence further confirmed by several entomological surveys [5–7], while a single *Ph. mascittii* specimen was also collected in Slovakia close to the Austrian border [8]. Very recently, *Phlebotomus simici* Nitzulescu, 1931, was recorded for the first time in Austria [9]. Apart from these findings, knowledge on sand fly distribution, species diversity and ecological factors determining their occurrence in Central Europe is scarce.

Ph. mascittii is the most widely distributed species in Europe and known to occur in Spain, France including Corsica, Italy, Switzerland, Germany, Belgium, Austria, Slovakia, Slovenia, Croatia, Hungary, and Serbia [3,6,8,10–19], as well as Algeria in North Africa [20]. It is also the sand fly species with the northernmost distribution in the Palearctic, occurring in Germany as far as 50° North [11,15,21].

Ph. mascittii is a suspected vector for *Leishmania infantum* [22,23], however, its vector competence has not yet been experimentally proven, albeit repeated reports of autochthonous leishmaniasis cases in Germany and Austria may indicate possible involvement of *Ph. mascittii* in transmission of *Leishmania infantum* [24–27].

To update the hitherto underreported sand fly distribution in Austria, dedicated entomological field studies were performed and bycatches from other entomological surveys were investigated. Species identification of obtained specimens was achieved by both, morphology and two different molecular approaches, including DNA sequencing of the two mitochondrial genes *COI* and *cytb*, and matrix-assisted laser desorption/ionization-time of flight mass spectrometry (MALDI-TOF MS) protein profiling. Ecological and climatic parameters of *Ph. mascittii* records were evaluated to promote prospective trapping success.

2. Results

2.1. Sand Fly Trapping and Identification

Altogether, 65 sampling sites in six federal states of Austria, namely Vienna, Lower Austria, Burgenland, Styria, Upper Austria, and Vorarlberg were surveyed (Table S1). At seven (10.8%) of these locations altogether 28 sand flies were trapped (Figure 1). Two (7.1%) were male and 26 (92.9%) were female, of which one specimen was engorged and three were gravid. The earliest capture was observed on 5 July and the latest capture was observed on 24 August.

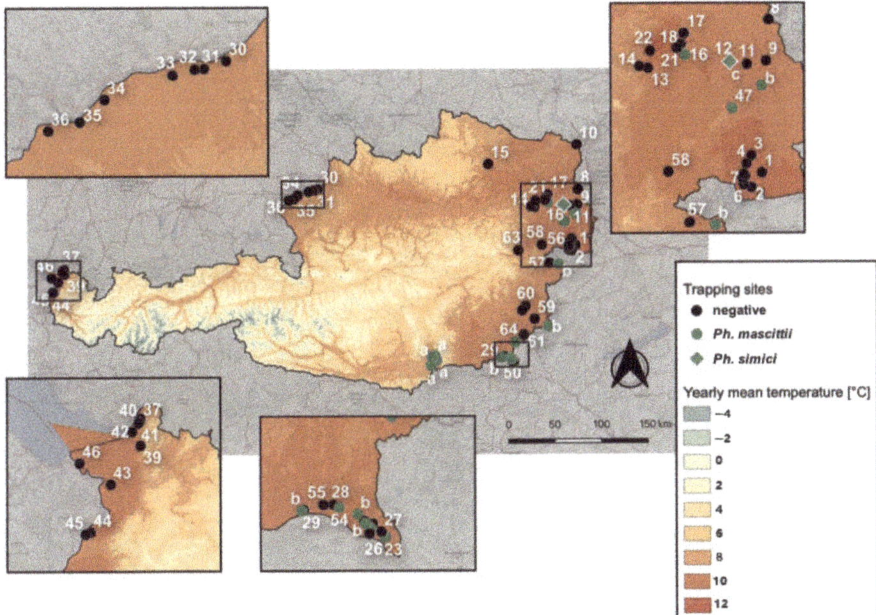

Figure 1. Distribution map of all published *Ph. mascittii* and *Ph. simici* records in Austria including all surveyed locations presented in this study. The climate map of Austria shows mean annual temperatures (1971–2000). Locations sampled within this study are marked with digits (Table S1), prior published records are marked with letters, (**a**) Naucke et al. 2011 [5], (**b**) Poeppl et al. 2013 [6], (c) Kniha et al. 2020 [9].

In survey 1, two of the 46 (4.3%) sampled locations were positive. A single specimen was trapped in the 11th district of Vienna, the capital of Austria, on 17 July 2019 at a horse farm outside a barn used for hay storage. A second location was found positive in Laafeld, Styria, where one and two specimens were trapped in front of a chicken shed on 6 August 2018 and 5 August 2019, respectively. (Table 1). All locations surveyed in Upper Austria and Vorarlberg were negative (Figure 1).

In survey 2, only one location, namely Kaisersteinbruch in Lower Austria was sampled, where two specimens were caught on 11 and 26 July 2013 outside of a dog kennel (Table 1).

In survey 3, three of nine (33.3%) sampled locations were positive, which were all located in the federal district of Styria. In Hummersdorf, six and eight specimens were caught on 5 and 6 July 2015, respectively, inside and outside a chicken barn. In Bad Radkersburg, two specimens were caught on 6 July 2015 inside an old chicken shed and in Unterpurkla, four specimens were caught on 6 July 2015 inside a barn next to a chicken shed (Table 1).

In survey 4, screening bycatches of a mosquito monitoring, one of nine (11.1%) sampled locations were positive. Two sand fly specimens were trapped on 29 July and 24 August 2019 with a BG sentinel trap baited with CO_2 in Neuhaus in the federal state of Burgenland (Table 1). The trap was set in the garden of a private property without animal barns or sheds. No animals were reported to live on the property and only rodents were observed to be present under a pile of wood close to the trap by the owners.

The majority of trapping sites, including those without sand fly catches, were animal farms with various domestic and farm animals as well as rodents present, being typical and suitable sand fly trapping sites.

Table 1. Sand fly positive locations surveyed in Austria.

Id	Location (Federal State)	Latitude, Longitude	Altitude (m.a.s.l.)	Trapping Site (Potential Host) [a]	Specimen (Male/Female)	Species	Trap Date [b]	Traps Set (Type)
47	Kaiser-steinbruch (Burgenland)	47.9911, 16.7103	179 m	outside, dog pound (dog)	1 (0m/1f)	*Ph. mascittii*	11.07.13	1 (CDC light)
					1 (0m/1f)	*Ph. mascittii*	26.07.13	1 (CDC light)
64	Neuhaus/Klausenbach (Burgenland)	46.8683, 16.0232	274 m	outside, human dwelling (rodents possibly)	1 (0m/1f)	*Ph. mascittii*	30.07.19	1 (BG sentinel & CO_2)
					1 (0m/1f)	*Ph. mascittii*	25.08.19	1 (BG sentinel & CO_2)
16	Neu Albern (Vienna)	48.1676, 16.4801	157 m	outside, roofed shed with hay (horse, dog)	1 (1m/0f)	*Ph. mascittii*	17.07.19	1 (CDC light & dry ice)
25	Laafeld (Styria)	46.6867, 16.0069	207 m	outside, chicken shed (chicken, geese, deer)	1 (0m/1f)	*Ph. mascittii*	06.08.18	1 (CDC light & dry ice)
					2 (0m/2f)	*Ph. mascittii*	05.08.19	1 (CDC light & dry ice)
48	Hummers-dorf (Styria)	46.7076, 15.9812	209 m	inside and outside, barn (chicken, dog)	6 (0m/6f)	*Ph. mascittii*	05.07.15	1 (CDC light)
					8 (1m/7f)	*Ph. mascittii*	06.07.15	2 (CDC light)
49	Bad Radkersburg (Styria)	46.7012, 15.9756	209 m	inside, old chicken shed (no)	2 (0m/2f)	*Ph. mascittii*	06.07.15	1 (CDC light)
54	Unterpurkla (Styria)	46.7319, 15.9062	223 m	inside, barn (chicken, cat)	4(0m/4f)	*Ph. mascittii*	06.07.15	1 (CDC light)

[a] potential host present within a 50 m radius. [b] date of the evening when trap was set.

Of all 28 caught specimens, 22 (78.6%) were caught with standard CDC light traps, four (14.3%) with standard CDC light traps with additional dry ice and two (7.1%) were caught as bycatch with BG sentinel traps using CO_2 as bait. However, due to the low number of caught specimens and the fact that not all sampling techniques were applied at all locations, statistical analysis on trapping success by type of trap was not possible.

All specimens were morphologically identified as *Phlebotomus mascittii*. Further confirmation was achieved by fluorescence microscopy, all three applied light spectra providing detailed images of the spermathecae (Figure 2). Correct morphological identification as *Ph. mascittii* was confirmed by DNA sequencing of all specimens. Obtained sequences were compared to reference sequences stored in GenBank. Sequence similarities with sequences in GenBank ranged from 99.84% to 100% (KY848831.1, KX981913.1) for cytochrome c oxidase subunit 1 (*COI*) and from 98.44% to 100% (MG800324.1, KR336656.1) for cytochrome b (*cytb*).

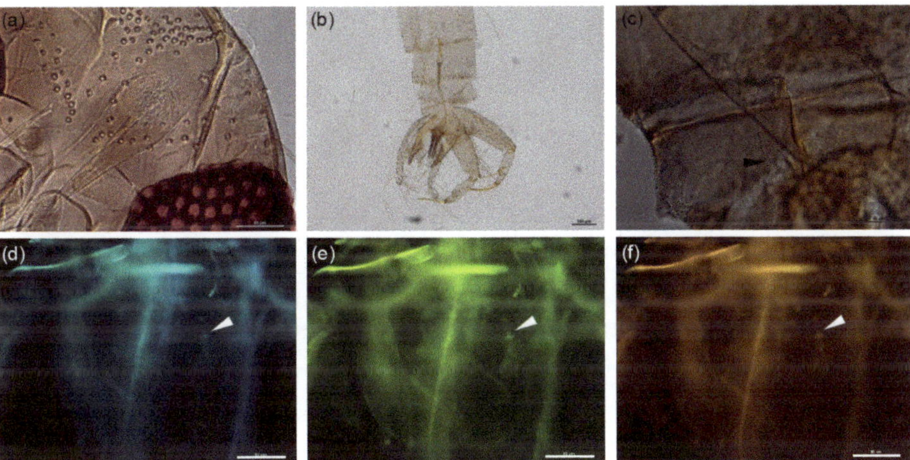

Figure 2. *Phlebotomus mascittii* characters for morphological identification. Pharynx with large irregular teeth (**a**), aedeagus with distal cup-like expansion (**b**), spermatheca not ringed, without neck indicated by black arrow (**c**), spermatheca under 330–380 nm light (**d**), spermatheca under 420–490 nm light (**e**), spermatheca under 585 nm light (**f**), tip of spermatheca with missing head is indicated by white arrow (**d–f**).

MALDI-TOF MS protein profiling provided reproducible mass spectra with a high number of intense signals for all analyzed specimens (Figure 3), males and females producing similar protein profiles. All 5 spectra of specimens from Austria confirmed the species identification of *Ph. mascittii*. When these protein profiles were compared with those of *Ph. mascittii* originating from three European countries (France, Slovenia, and Serbia), the dendrogram generated by cluster analysis provided clear clustering according to the geographical origin with distinct clusters for all four countries. The protein profile of *Ph. killicki*, a closely related species of the subgenus *Transphlebotomus*, formed a sister cluster to all analyzed *Ph. mascittii* specimens (Figure 4).

Leishmania spp. DNA was not detected by PCR in any of the female specimens.

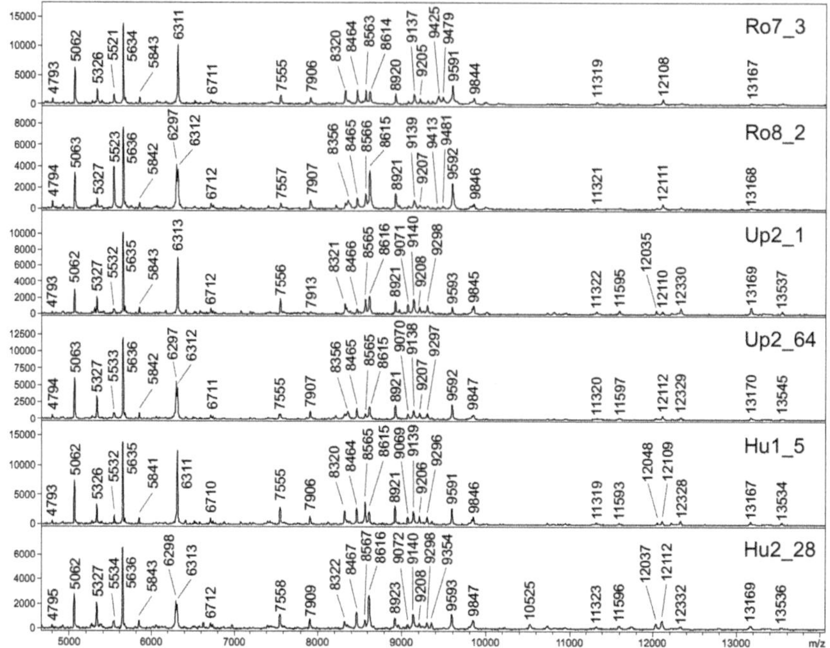

Figure 3. Matrix-assisted laser desorption/ionization-time of flight (MALDI-TOF) mass spectra of *Ph. mascittii* specimens originating from three different Austrian localities (Ro = Rohrau, Up = Unterpurkla, Hu = Hummersdorf).

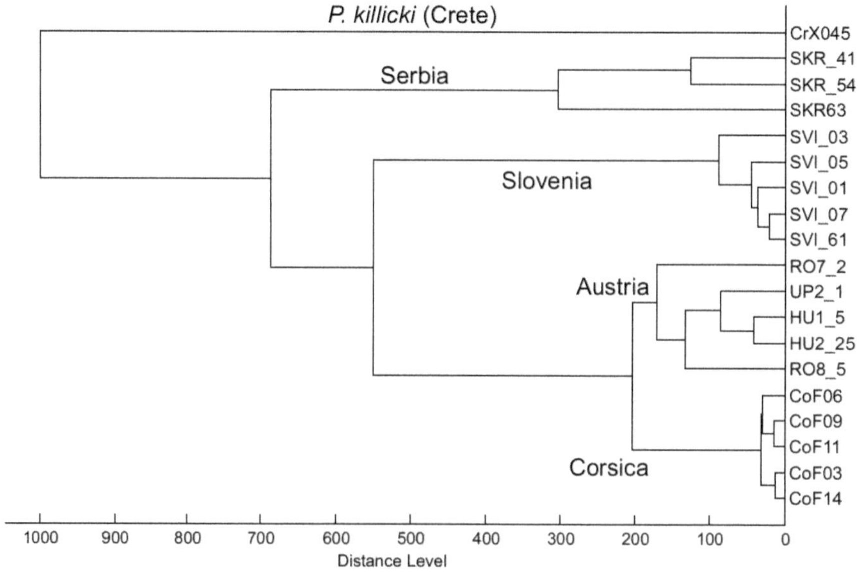

Figure 4. Dendrogram generated by cluster analysis of MALDI-TOF MS protein profiles of four geographically distant European populations of *Ph. mascittii*. As an outgroup, a mass spectrum of *Phlebotomus killicki* collected in Crete was used.

2.2. Pairwise Distances of Transphlebotomus Species

Overall, 27 *COI* sequences of all five *Transphlebotomus* species with a final length of 566 bp were included in the comparative analyses (Table S3). Mean intraspecific distances ranged from 0.09% to 0.6%, the lowest being calculated for *Ph. mascittii* (Table 2). In general, mean interspecific distances ranged from 6.6% to 15.4% and from 10.7% to 15.4% between *Ph. mascittii* and other *Transphlebotomus* species (Table 2).

Table 2. Interspecific mean genetic distances (%) of the cytochrome c oxidase subunit 1 (*COI*) and cytochrome *b* (*cytb*) between the five *Transphlebotomus* species based on the Kimura-2-parameter model (*COI/cytb*). Diagonal bold values indicate intraspecific mean distances.

	Species	1	2	3	4	5
1	*Phlebotomus mascittii*	**0.09/0.3**				
2	*Phlebotomus canaaniticus*	10.7/14.7	**–/–** [a]			
3	*Phlebotomus anatolicus*	10.9/13.9	6.6/8.7	**0.2/0.2**		
4	*Phlebotomus killicki*	12.2/12.9	11.4/13.7	12.1/12.7	**0.6/1.4**	
5	*Phlebotomus economidesi*	15.4/13.8	13.2/13.3	14.5/12.8	13.4/9.9	**– [a]/1.9**

[a] only one sequence available, included sequences are shown in Tables S2 and S3.

Altogether, 31 *cytb* sequences of all five *Transphlebotomus* species with a final length of 475 bp were included in the comparative analysis (Table S4). Mean intraspecific distances ranged from 0.2% to 1.9% (Table 2). Overall, mean interspecific distances ranged from 8.7% to 13.9% and from 12.9% to 14.7% between *Ph. mascittii* and other *Transphlebotomus* species (Table 2).

2.3. Haplotype Analysis of Ph. Mascittii from Austria and Other Countries

Overall, 12 *COI* sequences with a final length of 613 bp were included in the analysis (Table S2) with an overall haplotype diversity (Hd) of 0.53 and an overall nucleotide diversity (π) of 0.53 (Table 3). The sequences of *Ph. mascittii*, originating from four different countries, namely Austria, Slovakia, Slovenia and Serbia, revealed two haplotypes (COI_1, COI_2) defined by one polymorphic site (Pos: 106, A/G) (Figure 5a). Interestingly, both haplotypes were found among the Austrian specimens. The specimens from the Austrian federal states Vienna and Styria shared one haplotype (COI_1) with specimens from Pernek, Slovakia, and Vojdovina, Serbia. The other Austrian specimens from Lower Austria, Burgenland, and Styria shared the other haplotype (COI_2) with specimens from Velike Zabjle and Cetore, Slovenia, as well as Krasava, Serbia (Table 3).

Table 3. Included *Ph. mascittii* specimens to the haplotype network analysis based on cytochrome c oxidase subunit 1 (*COI*) and cytochrome b (*cytb*) DNA sequences.

Country, Location	COI GenBank	COI Haplotype	cytb GenBank	cytb Haplotype	Reference
France, Cévennes	-	-	KR336654.1	Cytb_2	Kasap et al. (2015)
Corsica, Porto Vecchio 1	-	-	KR336655.1	Cytb_3	Kasap et al. (2015)
Corsica, Porto Vecchio 2	-	-	KR336656.1	Cytb_1	Kasap et al. (2015)
Belgium, Saint-Cécile	-	-	KR336656.1	Cytb_1	Kasap et al. (2015)
Germany, Neuenburg	-	-	KR336656.1	Cytb_1	Kasap et al. (2015)
France, Haute-Pyrénées	-	-	HQ023281.1	Cytb_1	Mahamdallie et al. (2011)
Serbia, Vojdovina	KY848831.1	COI_1	-	-	Vaselek et al. (2017)
Serbia, Krasava	MN003381.1	COI_2	MK991774.1	Cytb_1	Vaselek et al. (2019)
Slovakia, Pernek	KX963380.1	COI_1	-	-	Dvořák et al. (2016)
Slovenia, Truske	-	-	MG800323.1	Cytb_4	Praprotnik et al. (2019)
Slovenia, Cetore	KX981916.1	COI_2	MG800324.1	Cytb_5	Hlavackova (GenBank)/Praprotnik et al. (2019)
Slovenia, Velike Zablje	KX869078.1	COI_2	-	-	Hlavackova (GenBank)
Austria, Burgenland, Kaisersteinbruch	MN812827.1	COI_2	MN812832.1	Cytb_1	present study

Table 3. *Cont.*

Country, Location	COI		cytb		Reference
	GenBank	Haplotype	GenBank	Haplotype	
Austria, Burgenland, Neuhaus/Klausen	MN812828.1	COI_2	MN812833.1	Cytb_1	present study
Austria, Vienna	MN812829.1	COI_1	MN812834.1	Cytb_1	present study
Austria, Styria, Laafeld	MN812830.1	COI_2	MN812835.1	Cytb_1	present study
Austria, Styria, Hummersdorf	MT332686.1	COI_1	MT332689.1	Cytb_1	present study
Austria, Styria, Bad Radkersburg	MT332687.1	COI_1	MT332690.1	Cytb_1	present study
Austria, Styria, Unterpurkla	MT332688.1	COI_2	MT332691.1	Cytb_1	present study

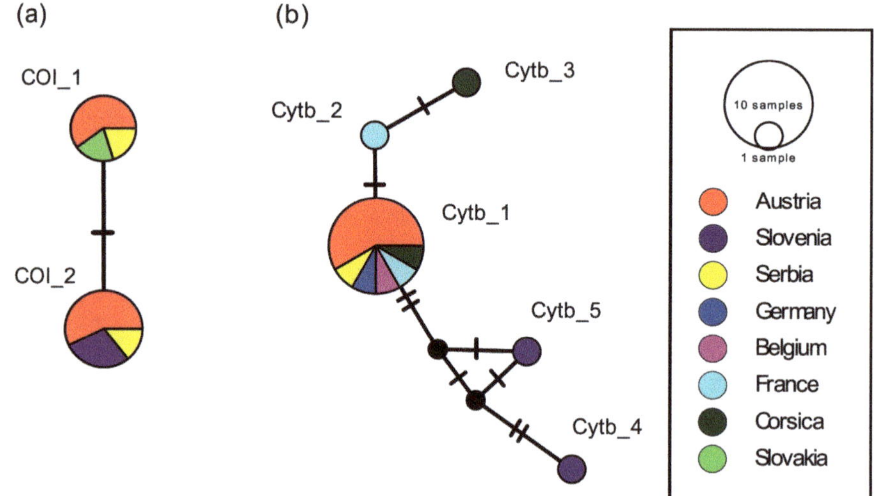

Figure 5. Haplotype networks of *Ph. mascittii* based on 12 *COI* sequences originating from 4 countries (**a**) and based on 16 *cytb* sequences originating from 6 countries (**b**).

Altogether, 16 *cytb* sequences with a final length of 475 bp were used for the analysis (Table S3) with an overall haplotype diversity (Hd) of 0.45 and an overall nucleotide diversity (π) of 0.19 (Table 3). Sequences of *Ph. mascittii* originating from six different countries, namely France including Corsica, Belgium, Germany, Austria, Slovenia, and Serbia, revealed five haplotypes (Cytb_1–Cytb_5) defined by seven variable sites, of which three were parsimony informative (Figure 5b). In the *cytb* analysis, all Austrian specimens belonged to a single haplotype (Cytb_1), shared with specimens from Belgium, France including Corsica, Germany, and Serbia (Table 3).

2.4. Climatic Parameters

No significant differences were observed between annual mean temperature (Kruskal–Wallis test, $p = 0.941$), mean June temperature (Kruskal–Wallis test, $p = 0.548$), mean July temperature (Kruskal–Wallis test, $p = 0.889$) and mean August temperature (Kruskal–Wallis test, $p = 0.883$) at successful and unsuccessful trapping sites (Figure 6a–d).

Further, no significant differences were observed between mean day temperatures (Kruskal–Wallis test, $p = 0.344$) and mean night temperatures (Kruskal–Wallis test, $p = 0.435$) at successful and unsuccessful trapping sites (Figure 6e,f). Sand flies were found to be active between 14.1 and 22.9 °C (mean = 19.8, SD = 2.9) mean night temperature (Figure 6f). Minimum night temperatures during successful trap nights ranged from 13.5 to 21.4 °C.

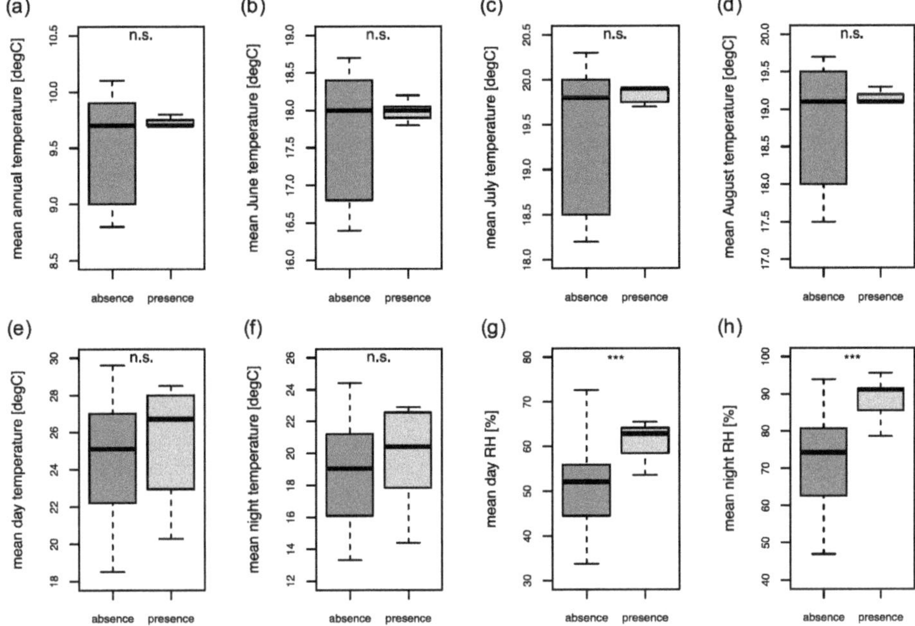

Figure 6. Comparison of climatic parameters at positive and negative surveyed locations analyzed by Kruskal–Wallis test. Mean annual temperature (**a**), mean June temperature (**b**), mean July temperature (**c**) and mean August temperature (**d**), parameters on temperature (**e,f**) and on relative humidity (RH) (**g,h**). Statistical differences between absence and presence are shown as not significant (n.s.), or as significant by asterisks (*** $p < 0.001$).

However, significant differences based on successful and unsuccessful trap nights were observed between mean day relative humidity (RH) (Kruskal–Wallis test, $p < 0.001$) and mean night RH (Kruskal–Wallis test, $p < 0.001$) (Figure 6g,h). Sand flies were found to be active between 78.6% and 95.7% (mean = 88.0, SD = 7.8) mean night RH (Figure 6h).

3. Discussion

This study reports new findings of *Phlebotomus mascittii* in Austria, including the first record of this species in the capital Vienna, representing the northernmost record in Austria. In total, *Ph. mascittii* was found at seven new locations within four surveys, suggesting that its distribution in Austria may be wider than previously thought. We provide an integrative approach for its identification, using morphological as well as molecular methods.

Ph. mascittii is the type species of the subgenus *Transphlebotomus* Artemiev, 1984, which comprises four other species, namely *Phlebotomus canaaniticus* Adler and Theodor, 1931, *Phlebotomus economidesi* Léger, Depaquit, and Ferté, 2000, *Phlebotomus killicki* Dvořák, Votýpka and Volf, 2015, and *Phlebotomus anatolicus* Kasap, Depaquit, and Alten, 2015. A recent description of the latter two species from localities in the eastern Mediterranean region where composition of sand fly fauna had been repeatedly studied in the past underlines the challenges of species identification based on morphology within this subgenus [28]. Therefore, molecular approaches are useful to confirm species identification and the natrium dehydrogenase subunit 4 (*NADH4*) gene was presented as an informative marker for interspecific and intraspecific analysis within the *Transphlebotomus* subgenus [11].

Here, we applied three different techniques, including morphology, sequencing of the two commonly used mitochondrial marker genes *COI* and *cytb* and MALDI-TOF MS protein profiling.

All specimens were morphologically identified to the species level, however, some individuals had hardly visible spermathecae, making morphological discrimination challenging. We further evaluated the usage of fluorescence microscopy to detect the autofluorescent spermathecae of *Ph. mascittii*, which we had already used for the identification of *Phlebotomus simici* [9]. By using different wavelengths of light the typical microstructures of the spermatheca of *Ph. mascittii* were visualized in high detail. Verification of morphological results was achieved with both mitochondrial marker genes used, *COI* and *cytb*. In general, interspecific distances exceeded 10% in both marker genes, with one exception, interspecific distances between *Ph. canaaniticus* and *Ph. anatolicus* were 6.6 and 8.7% for *COI* and *cytb*, respectively. This is in concordance with Kasap et al. [28], who observed similar interspecific distances with both marker genes and identified these two species as sister species.

MALDI-TOF MS protein profiling also confirmed species identification of Austrian specimens, providing distinct, species-specific protein profiles. This mass spectrometry method represents a currently emerging tool for species identification of various organisms including medically important arthropods that becomes favorable as a cost- and labor-effective alternative approach especially in field studies. However, in sand flies, it has been mostly used in studies focusing on the species diversity in rather small areas, often in foci of human leishmaniasis [29,30]. We attempted for the first time to compare protein profiles of distant populations of one species on a much larger geographical scale, analyzing specimens of *Ph. mascittii* from Austria, the French island of Corsica, Slovenia and Serbia. Our results clearly show that this method differentiates between these geographically distinct populations while successfully identifying all as *Ph. mascittii* when compared to the closely related species *Ph. killicki*. Specimens from a single locality in Corsica form a very compact cluster while specimens from Austria and Slovenia that originate from several localities show minor differences and form longer branches within their clusters. Specimens from Serbia were all collected at one locality, yet they also form longer branches, presumably due to the lower quality of obtained spectra because of compromised storage conditions.

Altogether, only seven out of the 64 surveyed locations were found positive for *Ph. mascittii*, even though most of the sampled locations were considered ideal for sand fly occurrence regarding annual mean temperature and potential breeding sites. Typically, only small trapping numbers were observed, as has been commonly reported from previous entomological surveys on sand flies in Austria [5,6] and bordering countries [8,31]. The attractiveness of light traps to *Ph. mascittii* is still a matter of discussion. However, eight individuals were trapped by a standard CDC light trap during a single night in Hummersdorf, indicating that light trapping is an appropriate technique for *Ph. mascittii*. In addition to this trapping method, trappings were also successful using CDC light traps with additional dry ice and BG sentinel traps baited with CO_2 as an attractant. Successful trappings have also been reported with other methods, e.g., sticky traps [12,18,32], malaise traps [23] and emergence traps in a cellar in Switzerland [15], all nevertheless resulting in rather low trapping numbers. High numbers were reported from trappings with light traps in Sessa, southern Switzerland, where over 280 *Ph. mascittii* specimens were caught during a survey from 1987 to 1989 [15]. Comparative field studies are necessary to optimize trapping methods and attractants for *Ph. mascittii*. Based on our findings of *Ph. mascittii* in the frame of mosquito monitoring and the variety of suitable trapping methods for this species, the inspection of bycatch of entomological studies presents an opportunity to detect small overlooked *Ph. mascittii* populations.

The generally low population densities found might not only depend on the trapping method, but might be due to *Ph. mascittii* having specific ecological niches within typical habitats or might even depend on other yet unknown factors. As also observed in this study, *Ph. mascittii* in Central Europe is frequently found in old barns and sheds close to human dwellings and animals [3,5,6,8], even though *Ph. mascittii* has been proposed to be cavernicolous [11]. *Ph. mascittii* has been found to be active during winter time in a cave-like tunnel on the island of Corsica, where constant climatic conditions might increase population density and the time of activity [13]. A small number of individuals of this species have recently also been found in a small cave in Germany [31]. Despite these findings, records of such trapping sites are rather scarce. On the other hand, *Ph. mascittii* has been found to be present at other trapping sites such as wall crevices in Italy [12,32], dry stone walls at illegal waste sites in Slovenia [33] and in the basement of a house in Switzerland [15].

Despite the fact that global warming might promote the expansion of sand fly populations in Central Europe, the constant renovation and demolition of typical trapping sites such as old barns and sheds in Austria might be a potential limitation to the further dispersal, if other niches are not inhabited. This was observed while trapping in Upper Austria and Vorarlberg, where locating traditional old barns remained challenging, possibly one of the main reasons for not detecting sand flies in these areas so far, rather than climatic conditions.

In Vienna *Ph. mascittii* was found on a horse farm, of which there are many on the outskirts of Vienna but only very few towards the center. Even though climatic conditions are optimal in Vienna, typical breeding sites are scarce. Thus, alternative breeding sites should be evaluated and sampled within Vienna, to assess the potential spread of *Ph. mascittii* and the potential presence of other sand fly species within the city. The presence of sand flies in Vienna, with currently 1.9 million inhabitants, might be of public health relevance [22].

Temperature is one of the driving factors for sand fly occurrence and the 10 °C-annual-isotherm was proposed to be the boundary for northward dispersal of European sand fly species [34]. However, sand fly larvae overwinter in the ground and therefore we expect that temperatures during the summer season are more relevant for sand fly distribution. In this study, mean June, July, and August temperatures were slightly higher at successful trapping sites, but no significant differences compared to unsuccessful trapping sites were observed. Of all climatic parameters included in the analyses, only the relative humidity was significantly higher at successful sampling sites than at unsuccessful ones. Nonetheless, not all negative trapping sites can be explained by climatic conditions. *Phlebotomus mascitti* was observed to be active at night temperatures as low as 14.2 °C mean, which probably marks the lower boundary of sand fly activity and might affect population densities. The observed large activity spectrum between 14.2 °C and 22.9 °C is not surprising, as *Ph. mascittii* occurs in a variety of geographic regions including Central Europe, Southern Europe, and Northern Africa, where climatic conditions differ considerably. In addition, other factors such as relative humidity, precipitation and air pressure have an impact on sand fly activity during their active season and should be considered for field work [35,36].

Even though we did not detect *Leishmania* DNA in any of the sand fly females, such a monitoring is important. An anthropophilic behavior of *Ph. mascittii* was shown in Southern Switzerland [15] and the detection of *Leishmania infantum* DNA in an unfed female specimen in Austria [22] and on the Italian island of Montecristo [23] urge for further clarification of its suspected vector competence. Moreover, the frequent import of often asymptomatic *Leishmania*-infected dogs [37] adds another significant risk factor for the further spread of leishmaniasis.

The failure to trap *Ph. mascittii* in the federal states Vorarlberg and Upper Austria, of course, does not exclude that populations were overlooked. Further trappings over longer periods of time are necessary for confirmation. The absence of *Ph. mascittii* and other sand fly species in certain parts of Central Europe is most certainly a result of biogeographic events during and after the last glacial period [38]. During the last glacial period, Central Europe was most certainly free of sand flies and their return from Mediterranean refugial areas possibly started around 10,000 to 8000 years ago [39]. A maximum sand fly distribution in Central Europe was possibly reached during the Holocene optima around 6500 and 4500 years ago, respectively, and sand flies remained restricted to small favorable microclimatic habitats until today [34]. This hypothesis is supported by our haplotype networks which demonstrate close genetic relationships between European *Ph. mascittii* populations, indicating a rather recent dispersal. Although both marker genes, *COI* and *cytb*, are frequently used in phylogenetic studies [40], we could not detect substantial differences at the population level to further elucidate the origin of *Ph. mascittii* in Austria. Due to such very recent dispersal of *Ph. mascittii* within Europe,

COI and *cytb* might not be ideal as marker genes and genes with higher mutation rates, such as the internal transcribed spacer 2, might be more suitable [41].

Overall, the Danube valley might constitute a corridor for sand fly dispersal, thereby connecting eastern and western populations. Whether eastern and western parts of Europe were post-glacially colonized from the same refugial area should be addressed in a detailed phylogeographic study, which would be crucial to understand the current sand fly distribution in Central Europe and to predict further sand fly dispersal in the future.

4. Material and Methods

4.1. Sand Fly Trapping and Available Material

The present study combines data of four different trapping surveys to update the current knowledge of sand fly distribution in Austria. Firstly, extensive field surveys to monitor sand flies were performed from 8 July to 21 August 2018 and 1 July to 25 August 2019, in six different federal states of Austria (survey 1). Trapping locations were chosen based on prevailing mean annual temperatures above 8 °C. In a second step, within these areas, suitable trapping sites were selected including farms, old barns, animal stables, private properties with and without animals present as well as two zoos. Every location was sampled for 2 to 3 consecutive nights based on weather conditions. Battery-operated CDC miniature light traps with fine gossamer collection bags (model #512, John W. Hock Company, Gainesville, FL, USA) were used. At outside locations, an additional source of CO_2 was provided by a cooling bottle filled with ~700 g of dry ice/trap.

Secondly, material from prior sand fly trapping surveys in the years 2013 (survey 2) and 2015 (survey 3) with CDC light traps was included in this study.

Thirdly, bycatch of mosquito monitoring in the federal district Burgenland in July and August 2019 performed with BG sentinel traps baited with CO_2 (Biogents AG, Regensburg, Germany) was screened for sand flies and included in this study (survey 4). Ethics Approval and Consent to Participate: Not applicable.

4.2. Climate Data

Mean annual, mean June, mean July and mean August temperatures of the trapping sites were accessed online (https://de.climate-data.org). Hourly temperature and relative humidity data of sampled locations were retrospectively obtained from the Central Institute for Meteorology and Geodynamics (ZAMG). Mean and minimum temperature and relative humidity were calculated for days (6 a.m. to 21 p.m.) and nights (22 p.m. to 5 a.m.).

4.3. Mapping of Sand Fly Distribution

Positive trapping sites as well as all previously published trapping locations in Austria were incorporated into a distribution map using Quantum GIS 3.4.11 [42].

4.4. Morphological Identification

Head and terminal segments of the abdomen of all caught sand fly specimens were dissected and slide-mounted in CMCP-10 mountant (Polysciences, Inc., Warrington, PA, USA). Identification was based on published morphological keys and descriptions of male genitalia, female spermatheca and pharyngeal armature [11,43]. Additionally, fluorescence microscopy was performed with a NIKON Eclipse E800 to identify hardly visible female spermatheca, as proposed in [9].

4.5. Molecular Identification by PCR and Sequencing

Molecular identification was performed by sequencing of the mitochondrial cytochrome c oxidase subunit 1 (*COI*) and cytochrome b (*cytb*) gene regions.

DNA was isolated from the remaining bodies with a QIAamp® DNA Mini Kit 250 (QIAGEN, Hilden, Germany). PCR amplifications of the *COI* and *cytb* gene regions were performed with the primer pairs LCO-1490/CoxUniEr and CytbEf1/CytbEr2, respectively, as published in [9].

All PCR amplifications were performed with an Eppendorf Mastercycler (Eppendorf AG, Hamburg, Germany). The PCR products were subjected to electrophoresis in 2% agarose gels stained with GelRed® Nucleic Acid Gel Stain (Biotium, Inc., Hayward, CA, USA). For further sequencing bands were analyzed with a Gel Doc™ XR+ Imager (Bio-Rad Laboratories, Inc., Hercules, CA, USA), cut out from the gel and purified with the Illustra™ GFX™ PCR DNA and Gel Purification Kit (GE Healthcare, Buckinghamshire, UK). Sanger sequencing was performed with a Thermo Fisher Scientific SeqStudio (Thermo Fisher Scientific, Waltham, MA, USA). Sequences were obtained from both DNA strands and a consensus sequence was generated in GenDoc 2.7.0. Obtained sequences were compared to available sequences in the GenBank using the Basic Local Alignment Search Tool (BLAST) (https://blast.ncbi.nlm.nih.gov/Blast.cgi). One sequence of each location was submitted to GenBank: *COI* (MN812827.1–MN812830.1, MT332686.1–MT332688.1) and *cytb* (MN812832.1–MN812835.1, MT332689.1–MT332691.1).

To confirm the suitability of *COI* and *cytb* as marker genes to distinguish between *Transphlebotomus* species, mean interspecific and intraspecific Kimura-2-parameter (K2P) distances were calculated with MEGAX [44].

4.6. Identification by MALDI-TOF MS Protein Profiling

To confirm species identification of chosen specimens at surveyed localities (survey 3), MALDI-TOF MS protein profiling was applied according to previously optimized protocols for trapping and sample preparation [45,46]. Thoraxes of specimens trapped by CDC light traps were manually grinded by disposable pestles in 1.5 mL microtubes with 10 µL of 25% formic acid and briefly centrifuged at 10,000× g. Two µL of the homogenate were mixed with 2 µL of freshly prepared MALDI matrix, an aqueous 60% acetonitrile/0.3% TFA solution of sinapinic acid (30 mg/mL; Bruker Daltonics, Billerica, MA, USA). One µL of this mixture was applied on a steel MALDI plate in duplicates, air-dried and measured by Autoflex Speed MALDI-TOF spectrometer (Bruker Daltonics, Billerica, MA, USA) in a mass range of 4–25 kDa (20 × 300 laser shots from different positions of the sample spot). Obtained spectra were visualized by FlexAnalysis 3.4 software (Bruker Daltonics, Billerica, MA, USA), processed by MALDI Biotyper 3.1 and an in-house database that comprises reference spectra of 25 different sand fly species was searched to identify the species. Log score value (LSV) > 2.0 was decided as a threshold for an unambiguous identification. For MSP dendrogram creation, an individual main spectrum (MSP) was generated from each analyzed spectrum.

In total, 5 specimens from two Austrian localities (Hummersdorf and Unterpurkla) were subjected to MALDI-TOF MS protein profiling. For subsequent comparison of geographically distant European populations, we also included protein profiles of 4 specimens from another Austrian location (Rohrau) collected in 2018 [7] and three other European countries: 5 specimens (all males) collected in 2019 from a railway tunnel close to the road from Solenzara to Ste. Lucie de Porto Vecchio, Corsica, France [13], 3 specimens (1 male, 2 females) collected in 2016 from Krasava, Serbia [17] and 5 specimens (all females) collected in 2020 from Godovic, Ilirska Bistrica, Vrhnika, Zovnek and Velike Zablje, Slovenia. As an outgroup, a protein profile of *Phlebotomus killicki*, a closely related species of the subgenus *Transphlebotomus*, collected in 2019 in Xerokampos, Crete, Greece, was used [47] (Table S4).

4.7. Leishmania spp. Screening

Female specimens were screened by PCR for the presence of *Leishmania* parasites using the primers LITSR/L5.8S targeting the internal transcribed spacer 1 (ITS1) of the ribosomal DNA following the PCR protocol of protocol of El Tai et al. [48]. Five µL of extracted DNA from female sand flies was used in all PCR reactions. Microbial DNA free water (QIAGEN, Hilden, Germany) and 5 µL DNA of a male *Ph. mascittii* specimen were used as negative controls, 2 µL DNA of *Leishmania infantum* MHOM/TR/2000/OG-VL was used as a positive control.

4.8. Sequence Analyses

Obtained sequences were aligned with ClustalX 2.1 and edited with GeneDoc 2.7.0. for further analysis. DnaSP v.5 [49] was used to identify unique haplotypes. Median Joining Networks [50] were calculated and visualized with Popart v.1.7 [51].

4.9. Statistical Analyses

Data was analyzed with the R environment for Mac [52]. Descriptive statistics were performed with a one sample proportions test. A Shapiro–Wilk test was used to check for normality of the data. A nonparametric Kruskal–Wallis test was used to compare climate data. A two-sided p-value < 0.05 was considered statistically significant.

Supplementary Materials: The following are available online at http://www.mdpi.com/2076-0817/9/12/1032/s1, Table S1. Surveyed locations in Austria. Table S2. Included *COI* sequences of *Transphlebotomus* for inter- and intraspecific distance calculations. Table S3. Included *cyt b* sequences of *Transphlebotomus* for inter- and intraspecific distance calculations. Table S4. Included sand fly specimens to MALDI-TOF mass spectrometry.

Author Contributions: E.K., A.G.O., W.P., G.M., and J.W. designed the study. E.K., A.G.O., K.K., S.S., K.B.-L., H.-P.F., V.V., V.I., and P.V. conducted field work. E.K., V.D., and P.H. performed laboratory work. E.K., V.D., P.H., M.M., and M.K. analyzed the data. E.K., V.D., P.V., and J.W. wrote the manuscript. All authors have read and agreed to the published version of the manuscript.

Funding: The study was supported by the Austrian Federal Ministry of Defence and the Medical University of Vienna, Austria. The additional support from the European Regional Development Fund (projects BIOCEV CZ.1.05/1.1.00/02.0109 and CePaViP CZ.02.1.01/0.0/0.0/16_019/0000759) and from the Czech Infrastructure for Integrative Structural Biology (LM2015043 funded by MEYS CR) is gratefully acknowledged. Cooperative field work was funded by an Austrian–Czech Republic joint project (grant nr. CZ02/2020) granted by the Austrian Agency for International Cooperation in Education and Research (OEAD) and by the German Federal Ministry of Education and Research (BMBF) (grant nr. FKZ01DK14021). The funders had no role in the study design, data collection and analysis, decision to publish, or preparation of the manuscript. Edwin Kniha is a recipient of a DOC fellowship and funded by the Austrian Academy of Sciences (ÖAW). His stay at the Department of Parasitology, Charles University, was funded by the European Union's Horizon 2020 research and innovation programme Infravec2, research infrastructures for the control of insect vector-borne diseases under grant agreement No 731060.

Acknowledgments: We would like to thank the Tiergarten Schönbrunn (Vienna), the Blumengarten Hirschstetten (Vienna), the Nationalpark Donauauen (Lower Austria/Vienna), and the Nationalpark Neusiedler See—Seewinkel (Burgenland) for their collaboration and their great support during fieldwork. We thank the owners of the properties were trappings have been performed. Special thanks to the ZAMG (Central Institute for Meteorology and Geodynamics) for collaboration and providing climate data. The authors are also grateful to Iveta Haefeli for technical support.

Conflicts of Interest: The authors declare that they have no competing interest.

Abbreviations: *COI*: cytochrome c oxidase subunit I, *cytb*: cytochrome b, Hd: haplotype diversity, MALDI-TOF MS: matrix-assisted laser desorption/ionization time of flight mass spectrometry, RH: relative humidity, SD: standard deviation.

References

1. Akhoundi, M.; Kuhls, K.; Cannet, A.; Votýpka, J.; Marty, P.; Delaunay, P.; Sereno, D. A Historical Overview of the Classification, Evolution, and Dispersion of Leishmania Parasites and Sandflies. *PLoS Negl. Trop. Dis.* **2016**, *10*, e0004349. [CrossRef] [PubMed]

2. Ready, P.D. Biology of Phlebotomine Sand Flies as Vectors of Disease Agents. *Annu. Rev. Entomol.* **2013**, *58*, 227–250. [CrossRef] [PubMed]

3. Naucke, T.J.; Pesson, B. Presence of *Phlebotomus* (*Transphlebotomus*) *mascittii* Grassi, 1908 (Diptera: Psychodidae) in Germany. *Parasitol. Res.* **2000**, *86*, 335–336. [CrossRef] [PubMed]

4. Naucke, T.J.; Schmitt, C. Is leishmaniasis becoming endemic in Germany? *Int. J. Med. Microbiol.* **2004**, *293*, 179–181. [CrossRef]

5.   Naucke, T.J.; Lorentz, S.; Rauchenwald, F.; Aspöck, H. *Phlebotomus (Transphlebotomus) mascittii* Grassi, 1908, in Carinthia: First record of the occurrence of sandflies in Austria (Diptera: Psychodidae: Phlebotominae). *Parasitol. Res.* **2011**, *109*, 1161–1164. [CrossRef] [PubMed]

6.   Poeppl, W.; Obwaller, A.G.; Weiler, M.; Burgmann, H.; Mooseder, G.; Lorentz, S.; Rauchenwald, F.; Aspöck, H.; Walochnik, J.; Naucke, T.J. Emergence of sandflies (Phlebotominae) in Austria, a Central European country. *Parasitol. Res.* **2013**, *112*, 4231–4237. [CrossRef]

7.   Obwaller, A.G.; Poeppl, W.; Naucke, T.J.; Luksch, U.; Mooseder, G.; Aspöck, H.; Walochnik, J. Stable populations of sandflies (Phlebotominae) in Eastern Austria: A comparison of the trapping seasons 2012 and 2013. *Trends Entomol.* **2014**, *2*, 1–5.

8.   Dvořák, V.; Hlavackova, K.; Kocisova, A.; Volf, P. First record of *Phlebotomus (Transphlebotomus) mascittii* in Slovakia. *Parasite* **2016**, *23*, 48. [CrossRef]

9.   Kniha, E.; Dvořák, V.; Milchram, M.; Obwaller, A.G.; Koehsler, M.; Poeppl, W.; Antoniou, M.; Chaskopoulou, A.; Paronyan, L.; Stefanovska, J.; et al. *Phlebotomus (Adlerius) simici* Nitzulescu, 1931: First record in Austria and phylogenetic relationship with other *Adlerius* species. *Parasit. Vectors* **2020**, (in press).

10.   Aransay, A.M.; Testa, J.M.; Morillas-Márquez, F.; Lucientes, J.; Ready, P.D. Distribution of sandfly species in relation to canine leishmaniasis from the Ebro Valley to Valencia, northeastern Spain. *Parasitol. Res.* **2004**, *94*, 416–420. [CrossRef]

11.   Depaquit, J.; Naucke, T.J.; Schmitt, C.; Ferté, H.; Léger, N. A molecular analysis of the subgenus Transphlebotomus Artemiev, 1984 (Phlebotomus, Diptera, Psychodidae) inferred from ND4 mtDNA with new northern records of *Phlebotomus mascittii* Grassi, 1908. *Parasitol. Res.* **2005**, *95*, 113–116. [CrossRef]

12.   Veronesi, E.; Pilani, R.; Carrieri, M.; Bellini, R. Trapping sand flies (Diptera: Psychodidae) in the Emilia-Romagna region of northern Italy. *J. Vector Ecol.* **2007**, *32*, 313–318. [CrossRef]

13.   Naucke, T.J.; Menn, B.; Massberg, D.; Lorentz, S. Winter activity of *Phlebotomus (Transphlebotomus) mascittii*, Grassi 1908 (Diptera: Psychodidae) on the island of Corsica. *Parasitol. Res.* **2008**, *103*, 477–479. [CrossRef]

14.   Bosnić, S.; Gradoni, L.; Khoury, C.; Maroli, M. A review of leishmaniasis in Dalmatia (Croatia) and results from recent surveys on phlebotomine sandflies in three southern counties. *Acta Trop.* **2006**, *99*, 42–49. [CrossRef] [PubMed]

15.   Grimm, F.; Gessler, M.; Jenni, L. Aspects of sandfly biology in southern Switzerland. *Med. Vet. Entomol.* **1993**, *7*, 170–176. [CrossRef] [PubMed]

16.   Praprotnik, E.; Zupan, S.; Ivović, V. Morphological and Molecular Identification of *Phlebotomus mascittii* Grassi, 1908 Populations From Slovenia. *J. Med. Entomol.* **2019**, *56*, 565–568. [CrossRef] [PubMed]

17.   Vaselek, S.; Dvořák, V.; Hlavackova, K.; Ayhan, N.; Halada, P.; Oguz, G.; Ivović, V.; Ozbel, Y.; Charrel, R.N.; Alten, B.; et al. A survey of sand flies (Diptera, Phlebotominae) along recurrent transit routes in Serbia. *Acta Trop.* **2019**, *197*, 105063. [CrossRef] [PubMed]

18.   Farkas, R.; Tánczos, B.; Bongiorno, G.; Maroli, M.; Dereure, J.; Ready, P.D. First Surveys to Investigate the Presence of Canine Leishmaniasis and Its Phlebotomine Vectors in Hungary. *Vector-Borne Zoonotic Dis.* **2011**, *11*, 823–834. [CrossRef]

19.   Dantas-Torres, F.; Tarallo, V.D.; Latrofa, M.S.; Falchi, A.; Lia, R.P.; Otranto, D. Ecology of phlebotomine sand flies and *Leishmania infantum* infection in a rural area of southern Italy. *Acta Trop.* **2014**, *137*, 67–73. [CrossRef]

20.   Berdjane-Brouk, Z.; Charrel, R.N.; Bitam, I.; Hamrioui, B.; Izri, A. Record of *Phlebotomus (Transphlebotomus) mascittii* Grassi, 1908 and *Phlebotomus (Laroussius) chadli* Rioux, Juminer & Gibily, 1966 female in Algeria. *Parasite* **2011**, *18*, 337–339.

21.   Melaun, C.; Krüger, A.; Werblow, A.; Klimpel, S. New record of the suspected leishmaniasis vector *Phlebotomus (Transphlebotomus) mascittii* Grassi, 1908 (Diptera: Psychodidae: Phlebotominae)—The northernmost phlebotomine sandfly occurrence in the Palearctic region. *Parasitol. Res.* **2014**, *113*, 2295–2301. [CrossRef]

22.   Obwaller, A.G.; Karakus, M.; Poeppl, W.; Töz, S.; Özbel, Y.; Aspöck, H.; Walochnik, J. Could *Phlebotomus mascittii* play a role as a natural vector for *Leishmania infantum*? New data. *Parasit. Vectors* **2016**, *9*, 458. [CrossRef] [PubMed]

23.   Zanet, S.; Sposimo, P.; Trisciuoglio, A.; Giannini, F.; Strumia, F.; Ferroglio, E. Epidemiology of *Leishmania infantum*, *Toxoplasma gondii*, and *Neospora caninum* in *Rattus rattus* in absence of domestic reservoir and definitive hosts. *Vet. Parasitol.* **2014**, *199*, 247–249. [CrossRef] [PubMed]

24. Bogdan, C.; Schönian, G.; Bañuls, A.L.; Hide, M.; Pratlong, F.; Lorenz, E.; Röllinghoff, M.; Mertens, R. Visceral leishmaniasis in a German child who had never entered a known endemic area: Case report and review of the literature. *Clin. Infect. Dis.* **2001**, *32*, 302–306. [CrossRef] [PubMed]

25. Koehler, K.; Stechele, M.; Hetzel, U.; Domingo, M.; Schönian, G.; Zahner, H.; Burkhardt, E. Cutaneous leishmaniosis in a horse in southern Germany caused by *Leishmania infantum*. *Vet. Parasitol.* **2002**, *109*, 9–17. [CrossRef]

26. Kollaritsch, H.; Emminger, W.; Zaunschirm, A.; Aspöck, H. Suspected Autochthonous Kala-azar in Austria. *Lancet* **1989**, *1*, 901–902. [CrossRef]

27. Beyreder, J. Ein Fall von Leishmaniose in Niederösterreich. *Wien. Med. Wochenschr.* **1962**, *115*, 900–901.

28. Kasap, O.E.; Dvořák, V.; Depaquit, J.; Alten, B.; Votypka, J.; Volf, P. Phylogeography of the subgenus *Transphlebotomus* Artemiev with description of two new species, *Phlebotomus anatolicus* n. sp. and *Phlebotomus killicki* n. sp. *Infect. Genet. Evol.* **2015**, *34*, 467–479. [CrossRef]

29. Lafri, I.; Almeras, L.; Bitam, I.; Caputo, A.; Yssouf, A.; Forestier, C.L.; Izri, A.; Raoult, D.; Parola, P. Identification of Algerian Field-Caught Phlebotomine Sand Fly Vectors by MALDI-TOF MS. *PLoS Negl. Trop. Dis.* **2016**, *10*, e0004351. [CrossRef]

30. Pareyn, M.; Dvořák, V.; Halada, P.; Van Houtte, N.; Medhin, G.; de Kesel, W.; Merdekios, B.; Massebo, F.; Leirs, H.; Volf, P. An integrative approach to identify sand fly vectors of leishmaniases in Ethiopia by morphological and molecular techniques. *Parasit. Vectors* **2020**, *13*, 580. [CrossRef]

31. Oerther, S.; Jöst, H.; Heitmann, A.; Lühken, R.; Krüger, A.; Steinhausen, I.; Brinker, C.; Lorentz, S.; Marx, M.; Schmidt-Chanasit, J.; et al. Phlebotomine sand flies in Southwest Germany: An update with records in new locations. *Parasit. Vectors* **2020**, *13*, 173. [CrossRef] [PubMed]

32. Rossi, E.; Bongiorno, G.; Ciolli, E.; Di Muccio, T.; Scalone, A.; Gramiccia, M.; Gradoni, L.; Maroli, M. Seasonal phenology, host-blood feeding preferences and natural *Leishmania* infection of *Phlebotomus perniciosus* (Diptera, Psychodidae) in a high-endemic focus of canine leishmaniasis in Rome province, Italy. *Acta Trop.* **2008**, *105*, 158–165. [CrossRef]

33. Ivović, V.; Kalan, K.; Zupan, S.; Bužan, E. Illegal waste sites as a potential micro foci of Mediterranean Leishmaniasis: First records of phlebotomine sand flies (Diptera: Psychodidae) from Slovenia. *Acta Vet. Brno* **2015**, *65*, 348–357.

34. Trájer, A.J.; Sebestyén, V. The changing distribution of *Leishmania infantum* Nicolle, 1908 and its Mediterranean sandfly vectors in the last 140 kys. *Sci. Rep.* **2019**, *9*, 11820. [CrossRef] [PubMed]

35. Simsek, F.M.; Alten, B.; Caglar, S.S.; Ozbep, Y.; Aytekin, A.M.; Kaynas, S.; Belen, A.; Kasap, O.E.; Yaman, M.; Rastgeldi, S. Distribution and altitudinal structuring of phlebotomine sand flies (Diptera: Psychodidae) in southern Anatolia, Turkey: Their relation to human cutaneous leishmaniasis. *J. Vector Ecol.* **2007**, *32*, 269–279. [CrossRef]

36. Tichy, H.; Kallina, W. Insect hygroreceptor responses to continuous changes in humidity and air pressure. *J. Neurophysiol.* **2010**, *103*, 3274–3286. [CrossRef] [PubMed]

37. Leschnik, M.; Löwenstein, M.; Edelhofer, R.; Kirtz, G. Imported non-endemic, arthropod-borne and parasitic infectious diseases in Austrian dogs. *Wien. Klin. Wochenschr.* **2008**, *120*, 59–62. [CrossRef] [PubMed]

38. Aspöck, H.; Walochnik, J. When sandflies move north. *Public Health* **2009**, *20*, 24–31.

39. Aspöck, H. Postglacial formation and fluctuations of the biodiversity of Central Europe in the light of climate change. *Parasitol. Res.* **2008**, *103*, 10–13. [CrossRef] [PubMed]

40. Depaquit, J. Molecular systematics applied to Phlebotomine sandflies: Review and perspectives. *Infect. Genet. Evol.* **2014**, *28*, 744–756. [CrossRef] [PubMed]

41. Depaquit, J.; Ferté, H.; Léger, N.; Lefranc, F.; Alves-Pires, C.; Hanafi, H.; Maroli, M.; Morillas-Márquez, F.; Rioux, J.A.; Svobodova, M.; et al. ITS 2 sequences heterogeneity in *Phlebotomus sergenti* and *Phlebotomus similis* (Diptera, Psychodidae): Possible consequences in their ability to transmit *Leishmania tropica*. *Int. J. Parasitol.* **2002**, *32*, 1123–1131. [CrossRef]

42. QGIS Development Team QGIS Geographic Information System. Open Source Geospatial Foundation Project. Available online: http://qgis.osgeo.org (accessed on 29 October 2019).

43. Lewis, D.J. A taxonomic review of the genus *Phlebotomus* (Diptera: Psychodidae). *Bull. Br. Mus. Nat. Hist.* **1982**, *45*, 121–209.

44. Kumar, S.; Stecher, G.; Li, M.; Knyaz, C.; Tamura, K. MEGA X: Molecular evolutionary genetics analysis across computing platforms. *Mol. Biol. Evol.* **2018**, *35*, 1547–1549. [CrossRef] [PubMed]

45. Dvořák, V.; Halada, P.; Hlacvackiva, K.; Dokianakis, E.; Antoniou, M.; Volf, P. Identification of phlebotomine sand flies (Diptera: Psychodidae) by matrix-assisted laser desorption/ionization time of flight mass spectrometry. *Parasit. Vectors* **2014**, *7*, 21. [CrossRef]
46. Halada, P.; Hlavackova, K.; Risueño, J.; Berriatua, E.; Volf, P.; Dvořák, V. Effect of trapping method on species identification of phlebotomine sandflies by MALDI-TOF MS protein profiling. *Med. Vet. Entomol.* **2018**, *32*, 388–392. [CrossRef]
47. Dvořák, V.; Tsirigotakis, N.; Pavlou, C.; Dokianakis, E.; Akhoundi, M.; Halada, P.; Volf, P.; Depaquit, J.; Antoniou, M. Sand fly fauna of Crete and the description of *Phlebotomus (Adlerius) creticus* n. sp. *Parasit. Vectors* **2020**, *13*, 547. [CrossRef]
48. El Tai, N.O.; Osman, F.O.; El Far, M.; Presber, W.; Schönian, G. Genetic heterogeneity of ribosomal internal transcribed spacer in clinical samples of *Leishmania donovani* spotted on filter paper as revealed by single-strand conformation polymorphisms and sequencing. *Trans. R. Soc. Trop. Med. Hyg.* **2000**, *94*, 575–579. [CrossRef]
49. Librado, P.; Rozas, J. DnaSP v5: A software for comprehensive analysis of DNA polymorphism data. *Bioinformatics* **2009**, *25*, 1451–1452. [CrossRef]
50. Bandelt, H.-J.; Forster, P.; Röhl, A. Median-joining networks for inferring intraspecific phylogenies. *Mol. Biol. Evol.* **1999**, *16*, 37–48. [CrossRef]
51. Leigh, J.W.; Bryant, D. Popart: Full-feature software for haplotype network construction. *Methods Ecol. Evol.* **2015**, *6*, 1110–1116. [CrossRef]
52. R Core Team R: A Language and Environment for Statistical Computing. R Foundation for Statistical Computing: Vienna, Austria. Available online: https://www.R-project.org/ (accessed on 5 May 2020).

Publisher's Note: MDPI stays neutral with regard to jurisdictional claims in published maps and institutional affiliations.

 *pathogens*

Review

One Health Approach to Leishmaniases: Understanding the Disease Dynamics through Diagnostic Tools

Ahyun Hong [1], Ricardo Andrade Zampieri [1], Jeffrey Jon Shaw [2], Lucile Maria Floeter-Winter [1] and Maria Fernanda Laranjeira-Silva [1,*]

[1]   Department of Physiology, Institute of Biosciences, University of São Paulo, São Paulo 05508-090, Brazil; avery.ahyun.hong@ib.usp.br (A.H.); ricardoz@ib.usp.br (R.A.Z.); lucile@ib.usp.br (L.M.F.-W.)

[2]   Department of Parasitology, Institute of Biomedical Sciences, University of São Paulo, São Paulo 05508-000, Brazil; jeffreyj@usp.br

*   Correspondence: mfernandals@usp.br

Received: 28 July 2020; Accepted: 21 September 2020; Published: 1 October 2020

Abstract: Leishmaniases are zoonotic vector-borne diseases caused by protozoan parasites of the genus *Leishmania* that affect millions of people around the globe. There are various clinical manifestations, ranging from self-healing cutaneous lesions to potentially fatal visceral leishmaniasis, all of which are associated with different *Leishmania* species. Transmission of these parasites is complex due to the varying ecological relationships between human and/or animal reservoir hosts, parasites, and sand fly vectors. Moreover, vector-borne diseases like leishmaniases are intricately linked to environmental changes and socioeconomic risk factors, advocating the importance of the One Health approach to control these diseases. The development of an accurate, fast, and cost-effective diagnostic tool for leishmaniases is a priority, and the implementation of various control measures such as animal sentinel surveillance systems is needed to better detect, prevent, and respond to the (re-)emergence of leishmaniases.

Keywords: *Leishmania*; protozoan parasite; epidemiology; environment; diagnosis

1. Introduction

Leishmaniases are vector-borne diseases caused by protozoan parasites of the genus *Leishmania* and are transmitted amongst mammalian hosts by phlebotomine sandflies. They are endemic in 98 countries and are estimated to affect over 350 million people around the globe [1,2]. The diseases can be categorized into two types according to the primary reservoir hosts of the human infection: zoonosis and anthroponosis. Zoonosis refers to an infectious disease of animals that can be transmitted to humans, and anthroponosis refers to a naturally occurring infectious disease among humans [3,4]. The majority of the *Leishmania* species are involved in zoonotic transmission. Infected animal reservoir hosts are often introduced into the human population and spillover events result in zoonotic diseases (Figure 1). Twenty-two *Leishmania* species belonging to the subgenera *L. (Leishmania)*, *L. (Mundinia)*, and *L. (Viannia)* [5] are found in humans. Among these, just two *L. (L.) donovani* and *L. (L.) tropica* are associated with an anthroponotic cycle (Table 1). However, infections of both species have been reported in livestock animals in Nepal [6,7], and in hyraxes and dogs in the Mediterranean Basin and several countries in Africa [8].

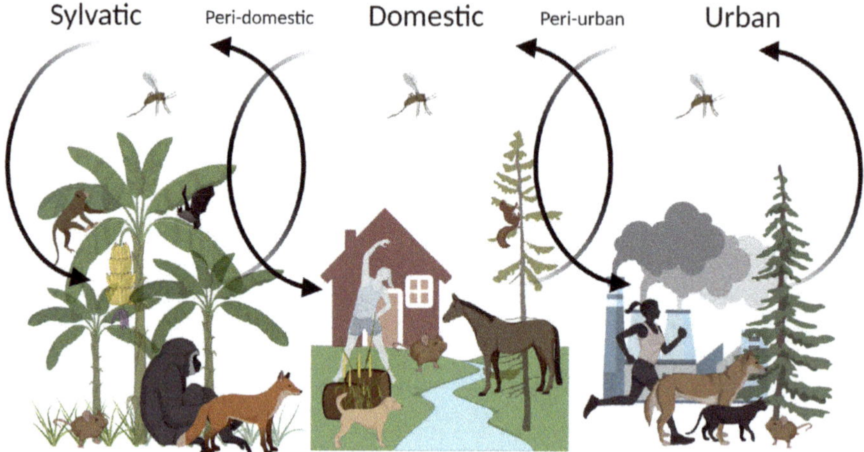

Figure 1. The transmission cycles of zoonotic leishmaniases. Sylvatic leishmaniases can spill over into humans living in proximity to forest foci of transmission, mainly due to deforestation or other factors affecting the ecological balance. As depicted by arrows, sand fly vectors, whose primary forests are their natural breeding sites, adapt to peri-domestic and domestic environments and eventually invade densely populated urban environments (modified from [9]).

Table 1. Clinical manifestations, reservoir host, and geographical distribution of *Leishmania* species (modified from [1]).

	Species	Clinical Manifestation	Reservoir Host	Country/Region
Eurasia (Old World)	*L. (L.) donovani*	AVL, PKDL, CL	Human	VL: West and Central Asia, China, The Indian subcontinent, The Mediterranean Basin, East Africa; CL: The Mediterranean Basin; ML: North Africa; PKDL: The Indian subcontinent, East and North Africa
	L. (L.) infantum	AVL, ZVL, CL	Human, Dog, Fox, Jackal, Badger, Rodent, Cat, Opossum	VL: Central and West Asia, China, The Mediterranean Basin, Africa; CL: The Mediterranean Basin, West Asia, China, West Africa
	L. (L.) major	ZCL	Rodent	West and Central Asia, The Indian subcontinent, The Mediterranean Basin, Africa
	L. (L.) tropica	ACL, ZCLAVL	Human, HyraxHuman	Central, South and West Asia, The Mediterranean Basin, East Africa West Asia
	L. (L.) killicki	CL	Unknown	The Mediterranean Basin
	L. (L.) aethiopica	ZCL, DCL, ML	Hyrax, Rodent	CL: East Africa (Ethiopia and Kenya); ML: Ethiopia
	L. (M.) orientalis	CL, DL, VL	Unknown	Thailand
Americas (New World)	*L. (L.) infantum chagasi*	ZVL, CL	Dog, Cat, Fox, Opossum	South and Central America, Mexico
	L. (L.) mexicana	ZCL, MCL, DCL	Rodent, Opossum	Americas
	L. (L.) pifanol	DCL	Unknown	Venezuela
	L. (L.) venezuelensis	CL	Unknown	Venezuela
	L. (L.) garnhami	ZCL	Unknown	Central America, Venezuela
	L. (L.) amazonensis	ZCL, DCL, CL	Rodent	South America
	L. (V.) braziliensis	ZCL, MCL, DL	Dog, Horse, Donkey, Mule, Rodent, Opossum	South and Central America, Mexico
	L. (L.) waltoni	DCL	Unknown	Dominican Republic
	L. (V.) guyanesis	ZCL, MCL	Sloth, Anteater, Opossum	South America
	L. (V.) panamensis	ZCL, MCL	Dog, Sloth, Opossum, Tamandua	South and Central America
	L. (V.) shawi	ZCL	Sloth, Primate	Brazil
	L. (V.) naiffi	ZCL	Armadillo	Brazil, French Guiana
	L. (V.) lainsoni	ZCL	Rodent	South America
	L. (V.) lindenbergi	ZCL	Unknown	Brazil
	L. (V.) peruviana	ZCL, MCL	Dog, Opossum, Rodent	Peru
	L. (M.) martiniquensis	CL	Unknown	French Guiana
	Endotrypanum colombiensis	ZCL	Sloth	Colombia, Venezuela, Panama

ACL, anthroponotic cutaneous leishmaniasis; AVL, anthroponotic visceral leishmaniasis; CL, cutaneous leishmaniasis; DL, disseminated cutaneous leishmaniasis; DCL, diffuse (anergic) cutaneous leishmaniasis; MCL, mucocutaneous leishmaniasis; ML, mucosal leishmaniasis; PKDL, post-kala-azar dermal leishmaniasis; VL, visceral leishmaniasis; ZCL, zoonotic cutaneous leishmaniasis; ZVL, zoonotic visceral leishmaniasis.

Leishmaniases present a broad spectrum of clinical manifestations, ranging from self-healing localized or multiple cutaneous lesions to mucosal lesions and potentially fatal visceral forms. These different forms are often associated with a particular species or subgenus, nonetheless, they are not unique to a species [10]. In most cases, cutaneous leishmaniasis (CL) skin lesions are self-healing and leave permanent scars. However, some species can lead to more severe pathologies such as mucocutaneous (MCL), diffuse (DCL), or disseminated (DL) cutaneous leishmaniases. Visceral leishmaniasis (VL), also known as *kala-azar*, is the most severe form of leishmaniases and can be fatal unless treated. Common clinical signs include non-tender splenomegaly, with or without hepatomegaly, and individuals with pre-existing health conditions may develop post-kala-azar dermal leishmaniasis (PKDL) consequent to the treatment [1].

Due to the complex relationship between human, animal hosts, parasites, and sand fly vectors, the transmission of *Leishmania* spp. is intricate. Moreover, vector-borne diseases are influenced by environmental changes and socioeconomic factors such as poor housing and sanitary conditions, malnutrition, or population movement. Anthropogenic factors tend to reorient the composition and behavior of sand fly vectors. To date, there are at least 50 different sand fly species known to transmit leishmaniases. In general, each sand fly species has its preferred ecological niche and transmits a certain *Leishmania* species (reviewed in [11]). Furthermore, zoonotic leishmaniases have a broad mammalian reservoir diversity in different parts of the world [12]. The sylvatic transmission is affected by the wildlife population in and around human settlements. Divergent species of sylvatic, domestic, and synanthropic animals have been reported as reservoir hosts for various *Leishmania* species around the globe—rodents, foxes, dogs, cats, primates, hyraxes, and bats are among those maintaining the transmission of *Leishmania* [1] (Figure 1). *Leishmania* species may infect a distinct mammalian host, yet, in the northeast region of Brazil, a mosaic of different sylvatic and synanthropic rodents appear to be reservoirs of *L. (V.) braziliensis* [12]. Events such as deforestation due to urbanization can create new breeding habitats for vectors, which can lead to spillovers across ecosystem boundaries [13,14]. With over 60% of human infectious diseases being zoonotic [15], recognizing the interdependence and connections between humans, animals, and the environment that the hosts and vectors inhabit is indisputably essential. Hence, adopting a 'One Health' approach becomes imperative to control leishmaniases.

The One Health approach is a global strategy for advocating multi-sectoral and trans-disciplinary collaborations in all aspects of human, animal, and environmental health while recognizing their interconnectedness [16,17] (Figure 2). The phenomenon of emerging and re-emerging infectious diseases is driven by various, often inadvertently, anthropogenic constituents, such as environmental factors (i.e., climate change, deforestation), population movement (i.e., migration, increased international travel), socioeconomic and political-driven factors (i.e., poverty, lack of political will, war/conflict), and genetic factors including host adaptation and susceptibility to infection [18]. This multiplicity of components driving disease emergence was first described by The National Academy of Sciences as a "convergence model" [19] and later defined as follows: "ecological instabilities arise from the way we alter the physical and biological environment, the microbial and animal tenants (humans included) of these environments, and our interactions (including hygienic and therapeutic interventions) with the parasites" [20].

Thus, this review aims to explore the complexities of *Leishmania* transmission and to provide an overview of various diagnostic methods and their uses in epidemiological studies to support leishmaniases control.

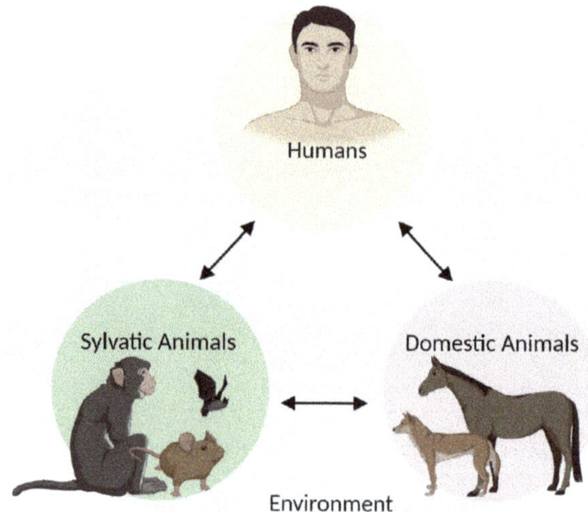

Figure 2. The One Health diagram illustrating the interactions between humans, sylvatic, and domestic animals within a shared environment.

2. Clinical Manifestations

2.1. Cutaneous/Mucocutaneous Leishmaniases

Cutaneous leishmaniasis (CL) is endemic in over 90 countries worldwide, with approximately 0.7 to 1.2 million cases occurring every year. Yet 70–75% of cases emerge in just nine countries: Afghanistan, Algeria, Brazil, Colombia, Iran, Pakistan, Peru, Saudi Arabia, and Syria [2]. CL is caused by multiple and phylogenetically distinct *Leishmania* species, such as *L. (L.) infantum, L. (L.) tropica, L. (L.) major,* and *L. (L.) aethiopica*, which are endemic in Eurasia, and species, such as *L. (L.) amazonensis, L. (L.) mexicana, L. (V.) braziliensis*, or *L. (V.) guyanensis*, which are endemic in the Americas [1]. CL caused by *L. (L.) donovani* is unusual but some cases do occasionally arise in endemic countries such as in the southwest region of India and the northwest region of Yemen, where the majority of CL cases are caused by *L. (L.) tropica* [21–23]. Moreover, several atypical cases caused by *L. (L.) infantum* have been reported in the Mediterranean Basin and by *L. (L.) infantum chagasi* in Central and South American countries [24–27]. The majority of species predominantly associated with CL in humans are zoonotic, with the exception of *L. (L.) tropica* [1].

The clinical manifestations of leishmaniases depend not only on the parasite species but also on the host's immune response, directing macrophages polarization towards the M1 classically activated or M2 alternatively activated phenotypes [28–32]. CL, particularly, can be characterized by diverse forms including localized, mucocutaneous, diffused, and disseminated forms. A large proportion of these infections are asymptomatic and/or generate minimal or no pain. However, recovered patients often experience substantial trauma and social stigmatization [33]. Localized CL lesions vary in the severity and timing of the clinical presentation and prognosis. Small erythema appears on the site of inoculation, which then develops into a papule and ulcerates over several weeks to months. Lesions caused by *L. (L.) major, L. (L.) mexicana, L. (V.) braziliensis, L. (V.) panamensis, L. (V.) guyanensis,* and *L. (V.) peruviana* often self-heal within two to eight months, whereas lesions caused by *L. (L.) infantum* and *L. (L.) tropica* tend to begin healing spontaneously after one year of disease onset, and up to five years for *L. (L.) aethiopica* [1,34].

Diffuse (anergic) cutaneous leishmaniasis (DCL) is characterized by multiple non-tender, non-ulcerated skin lesions, predominantly nodular, that contain large numbers of parasites, which may resemble lepromatous leprosy. The skin lesions typically manifest at the primary infection site, and mucosal involvement is rare, as less than five percent of these patients develop mucosal lesions [35]. *L. (L.) aethiopica*, *L. (L.) mexicana*, and *L. (L.) amazonensis* are responsible for the majority of DCL although other *Leishmania* species can cause this diffused form of the disease in Human Immunodeficiency Virus (HIV)-coinfected patients and present atypical manifestations such as ulceration [36,37]. Multiple lesion CL generally emerges after initially successful treatment but relapses and becomes resistant to further treatment [1,35,38]. Disseminated cutaneous leishmaniasis (DL) is largely associated with *L. (V.) braziliensis*, *L. (V.) panamenensis*, *L. (V.) guyanensis*, and *L. (L.) amazonensis*, and occurs mainly in Central and South America (Table 1). The disease is distinguished by multiple ulcerated papules and acneiform lesions which appear in a different site from the primary foci, and it usually transpires when the initial lesions begin to develop [35]. Patients who have recovered from *L. (L.) tropica* infections may develop a chronic form of anthroponotic CL called leishmaniasis recidivans, also known as tuberculoid leishmaniasis due to its clinical resemblance to cutaneous tuberculosis [39].

Mucocutaneous leishmaniasis (MCL) is a secondary stage of CL, in which 1–20% of cutaneous lesions may develop into mucosal lesions [40–43]. Nearly 90% of MCL occur in Bolivia, Brazil, and Peru [44]. Risk factors include male, older than 22 years, duration of CL, site of primary skin lesion above the waistline, absence, or delayed treatment of the primary lesions [1,45]. MCL is associated principally with *L. (V.) braziliensis* and to a lesser degree with *L. (V.) guyanensis* and *L. (V.) panamensis*. The disease may manifests itself when the parasites metastasize to the mucosal tissues of the aerodigestive tract via the lymphatic or bloodstream [1]. The clinical signs range from mild edema of the nose, upper lip, palate, and frequent local lymphadenopathy to severe mutilation with obstruction and/or destruction of the nose, pharynx, and larynx. MCL does not heal spontaneously and is potentially life-threatening with secondary bacterial infections including pneumonia and tuberculosis [34,38].

2.2. Visceral Leishmaniasis

Visceral leishmaniasis (VL) is caused by *L. (L.) donovani*, *L. (L.) infantum*, and *L. (L.) infantum chagasi* [1] (Table 1). It is noteworthy that the nomenclature of the etiological agent of VL in the Americas as a result of a decade-long discussion between experts regarding its origin [46,47]. Some suggest that *L. (L.) chagasi* is synonymous with *L. (L.) infantum*, and was introduced in the Americas during Spanish and Portuguese colonization [48,49], while some argue that *L. (L.) chagasi* existed in the Americas before colonization [50]. Regardless, *L. (L.) infantum chagasi* is the most widely accepted name for the causal species of VL in the Americas today [46].

Despite nearly 70 countries around the world being endemic for VL, over 90% of the new cases occur in seven countries: Brazil, Ethiopia, India, Kenya, Somalia, South Sudan, and Sudan [44]. It has been estimated that VL results in 0.2 to 0.4 million cases each year with a fatality rate of 10–20% when the affected individuals have access to treatment [2]. In endemic areas of East Africa and the Indian subcontinent, *L. (L.) donovani* infections are predominantly anthroponoses, whereas, across Europe, North Africa, and the Americas, *L. (L.) infantum* and *L. (L.) infantum chagasi* [1,51] are zoonoses. Moreover, there are a few records [52–55] of atypical visceral infections caused by *L. (L.) tropica* that are generally associated with CL. The clinical manifestations of VL range from asymptomatic infection to severe systemic cases associated with fever, fatigue, weight loss, pancytopenia, and non-tender splenomegaly, with or without hepatomegaly [13,56]. Hyperpigmentation of the patient's skin can also be observed in the Indian subcontinent; hence, the disease is also called 'kala-azar', meaning 'black-fever' in Hindi [57].

In southern Europe, up to 70% of the cases in adults are linked to HIV infection. It has been shown that HIV infection increases the risk of developing VL by 100 to 2320 times [58]. Moreover, HIV-coinfected patients may have an impact on disease transmission as several studies have demonstrated that these patients are highly infectious to sandflies [1]. Furthermore, coinfection with VL can lead to faster progression to Acquired Immune Deficiency Syndrome (AIDS) [58]. HIV coinfection can also often induce atypical leishmaniases, for instance, infections with *L. (L.) amazonensis* [59] and *L. (V.) braziliensis* [58] can cause VL. Several reported cases in endemic countries show VL can be transmitted through a non-vector transmission route, such as blood transfusion [60,61] or via the sharing of needles between intravenous drug users [62]. Congenital VL cases, although extremely rare in humans, were found to be the predominant route of transmission among canines in the U.S. [63], suggesting *Leishmania*'s potential in invading the placenta [64] (reviewed in [65]).

2.3. Post-Kala-Azar Dermal Leishmaniasis

Post-kala-azar dermal leishmaniasis (PKDL) is a complication that arises from 6 to 12 months after the recovery of VL. It occurs in all areas endemic for *L. (L.) donovani*, particularly in East Africa and the Indian subcontinent, which are responsible for up to 50% and 10–20% of the estimated VL cases, respectively [66]. However, a few cases caused by *L. (L.) infantum* in immunosuppressed individuals were reported, with one case implying that highly active antiretroviral treatment (HAART) may lead to the development of PKDL [67,68]. Clinical manifestations of PKDL vary due to one's immune responses but generally can be described as one of the three main manifestations: hypopigmented macules, malar rash on the face, and nodular lesions [69,70]. Yet, rare signs of the disease have been reported from endemic regions, including oral and genital mucosa lesions [69]. Macules and lesions from these patients often contain a few parasites that seem to persist in the skin after treatment [71]. Having said that, those individuals are important reservoirs for anthroponotic VL, although PKDL is not fatal [72]. A recent xenodiagnoses study from the Indian subcontinent emphasized the significance of PKDL and its contribution to the ongoing transmission of *L. (L.) donovani*. Three patients with PKDL, maculopapular and/or nodular forms, were infectious to sand fly vectors [73].

2.4. Asymptomatic Infection

Asymptomatic infections of both VL and CL (reviewed in [74]) are common in endemic countries. Although it has not yet been proven that human asymptomatic carriers can transmit *Leishmania*, asymptomatic canines have been shown to infect sandflies at lower rates—18.3% versus 51.9% for symptomatic canines [75]. Thus, human asymptomatic cases may have the potential to alter the transmission dynamics, making it harder to estimate the global burden of the disease [76]. Several studies have recorded the ratio of asymptomatic to symptomatic infections of VL in endemic areas: 4:1 in Kenya [77], 7.9:1 to 8.9:1 in India and Nepal [78,79], 6.5:1 to 89:1 in Brazil [13,80,81], and 13:1 in Iran [82]. A recent active mass survey from a highly endemic district for VL in India reported that, among individuals who tested positive, 42% of them were asymptomatic cases, and nearly 10% of the asymptomatic cases developed symptomatic VL during 36 months of follow-up [78]. Moreover, within the same district, researchers found that two percent of the patients who developed PKDL had no known history of VL, suggesting that asymptomatic cases of VL may develop into PKDL [72]. Additionally, a cohort study conducted after the leishmaniasis outbreak in Madrid (Spain) in 2009 observed the prevalence rate of asymptomatic infection to be nearly 20%. Moreover, 38.3% of the participants reported that they had close contact with dogs, and 24.6% of the 38.3% of participants were asymptomatic carriers. As domestic dogs are the most common reservoir host of *L. (L.) infantum* in Spain, these findings suggest that having frequent contact with dogs may increase the likelihood of becoming an asymptomatic carrier of VL [75].

Asymptomatic carriers can also transmit *Leishmania* through blood transfusions. Nearly 10% and 5% of the blood donors from Granada (Spain) and Salvador (Brazil), respectively, tested positive for asymptomatic *Leishmania* infection, raising concerns about the safety of the blood supplies in endemic regions [60,83].

3. Risk Factors for Leishmaniases

3.1. Socioeconomic Factors and Malnutrition

Although leishmaniases are widespread across continents, the risk of these diseases is much greater for those living in poverty. Treatment for leishmaniases is comparatively more expensive than other poverty-related diseases. Malaria treatment, for instance, costs between USD 0.10 and USD 2.40 per course [84], whereas, for leishmaniasis, the drug alone typically costs between USD 30 and USD 1500 [85]. In Nepal, the average cost of anti-leishmanial drugs is greater than the median annual household per capita income, and many households opt to sell their livestock to cover the treatment cost, leading to a devastating financial impact for affected families [86]. Moreover, many of those in poverty cannot afford the cost of prevention. A survey report from Afghanistan revealed that 78% of respondents said they cannot afford a bed net [87].

Undeniably, poor living conditions are often associated with a higher risk of *Leishmania* infection [88]. A recent study in Nepal where the researchers examined housing structures and land lots in endemic districts found that houses with natural floors increased the risk of infection by eightfold, walls made from straw, leaves, and/or bamboos increased by threefold, walls with cracks, especially in the bedroom, increased by threefold and proximity to a livestock shed, particularly located in front or at the side of a dwelling, was shown to increase the risk by fourfold [89]. Furthermore, living in overcrowded homes may also increase the risk of infection since anthropophilic sand fly vectors are attracted by human kairomones and carbon dioxide, which results in increased density of sand fly vectors in these homes [90,91]. Likewise, a study from rural Bangladesh reported that living with a recent anthroponotic VL patient increases the risk of infection by 26-fold [92]. While overcrowded households may have a direct influence on their increased risk, there are other shared risk factors such as malnourished households that contribute to the likelihood of becoming infected [92].

Malnutrition increases the host susceptibility and is also a determinant for the severity and clinical manifestations of the disease [1]. An in vivo study with mice fed on altered protein, iron, zinc, and calorie intakes indicated that malnutrition can cause a failure of the lymph node barrier function after infection with *L. (L.) donovani*, and promote early visceralization [88]. These findings are supported by field observations in Ethiopia that showed malnourished individuals were three times more susceptible to developing VL [93]. Similar cases were also seen in animals; dogs with poor nutritional status were about 13 times more susceptible to *L. (L.) donovani* infection than those in good health [94]. Furthermore, malnutrition has also been associated with an increased risk of various complications of leishmaniases including MCL and PKDL [1].

3.2. Migration

Anthroponotic leishmaniases occur predominantly in rural and urban areas, and the diseases are usually characterized by large outbreaks in densely populated cities, especially in war and conflict zones, refugee camps, and in settings where there is large-scale population migration [1]. Several studies revealed a strong relationship between civil unrest and VL epidemics. For instance, one of the most devastating VL outbreaks occurred during 22 years of civil war in South Sudan, forcing the exodus of more than four million people (reviewed in [95]). A community-based longitudinal study in South Sudan divulged a significantly higher leishmaniases mortality rate in areas of civil war, unrest, and human displacement [2]. Furthermore, due to the sudden influx of displaced refugees, the number of VL cases reported in neighboring Eastern African countries doubled. More than 400,000 South Sudanese refugees fled to the woodland region of the state of White Nile in Sudan, and another 300,000 refugees migrated along the Omo Valley and Gambela regions in Ethiopia. These migrations resulted in the re-emergence of the VL epidemic in those two regions [95]. A similar trend was also observed in Lebanon between 2001 and 2014 due to the Syrian conflict and the subsequent influx of refugees. In previous years, the average number of reported leishmaniases cases was zero to six per year, whereas in 2013, more than 1000 new cases were reported [96].

The rapid development of road transportation systems encourages the migration of people and their *Leishmania*-infected dogs to non-endemic regions. In Brazil, 95.4% of zoonotic VL cases were restricted to the north and northeast regions of the country until 2003. The major road infrastructure development across the country resulted in a mass migration of people seeking better job opportunities in the central, south, and southeast regions of Brazil. The cases reported in north and northeast regions decreased to 78.1% while in the central and southeast regions of the country, observed cases increased to 21.8% [97,98]. In the state of São Paulo, located in the southeast region of Brazil, there was a gradual increase in human VL along the Marechal Rondon highway, Novoeste railway, and the Bolivia–Brazil gas pipeline [98]. It is noteworthy that replacing the animal agriculture with sugar cane plantations in São Paulo may have contributed to the spread of the disease, as harvesting the cane involves migrant labor, often from the northeast region [99].

3.3. Environmental Changes

Changes in the environment also have a strong impact on leishmaniases, either due to climate change caused by global warming or man-made ecological alterations. An abrupt climate change exerts influence on the migration of sand fly vectors and reservoir animals. Numerous studies have shown that leishmaniases disease distribution has been shifting and will shift as a response to climate change [100]. Moreover, population growth appears to be the significant driver of clearing forests due to the increased demand for agricultural products and the expansion of transmigration settlements [101]. It is indisputable that widespread deforestation has led to an expeditious increase in leishmaniases cases, and peri-domestic, peri-urban, and urban transmission [1]. As such, the number of annual cases of CL in Costa Rica has increased by 46% from 2002 to 2007 [2], and the incidence rate of CL has increased by 30% from 1998 to 2002 in Brazil [102], with both cases assumed to be associated with the result of deforestation. The transmission rate for zoonotic leishmaniases is the highest among the marginalized population and small frontier farmers living at the edge of natural foci nearby human dwellings and wild habitats [1,103]. For instance, the city of Manaus in the state of Amazonas (Brazil) is the largest urban settlement within the Amazon rainforest, with a population of over 1.7 million. In Amazonas, where more than 40% of the leishmaniases cases occur in Brazil, *L. (V.) guyanensis* is the prevalent species [104]. Due to the rapid urbanization and expansion of the city and suburbs on the edge of the primary forest, researchers suspect that the sylvatic transmission of *L. (V.) guyanensis* had spilled over into a peri-domestic transmission cycle, and eventually reached the urban transmission cycle [100].

The relationship between agricultural expansion and leishmaniases seems evident, as the abundance of sand fly vectors and increased exposure to infection with *Leishmania* have been recorded in several coffee plantations in Brazil, Colombia, and Mexico [105–107]. Moreover, the development of new irrigation technologies appears to promote environmental changes, resulting in the attraction of certain sand fly species from neighboring regions/countries. In Central Tunisia, where *Phlebotomus perniciousus* and *P. papatasi* used to be the principal vectors of *L. (L.) infantum*, *P. perfiliewi* became the most abundant sand fly species in irrigated areas. Reportedly, the overall abundance of *P. perfiliewi* in Central Tunisia was at 5% prior to irrigation development [108,109].

4. Diagnostic Tools for Leishmaniases

Despite the recent advances in diagnostic tools, diagnosing leishmaniases still imposes substantial challenges in the remote areas of endemic countries around the globe. Moreover, due to its complex transmission cycle, involving the various biological entities, identifying the responsible *Leishmania* species is crucial in disease control and interventions [1]. Early and accurate detection as well as improving patients' outcomes [110] also provides imperative data for eco-epidemiological studies to monitor and assess the outbreak and evaluate current control measures that are in place in endemic regions. A diagnosis of leishmaniases is often made by evaluating the clinical manifestations of the disease in the patient, followed by one or more laboratory diagnostic tests (Figure 3).

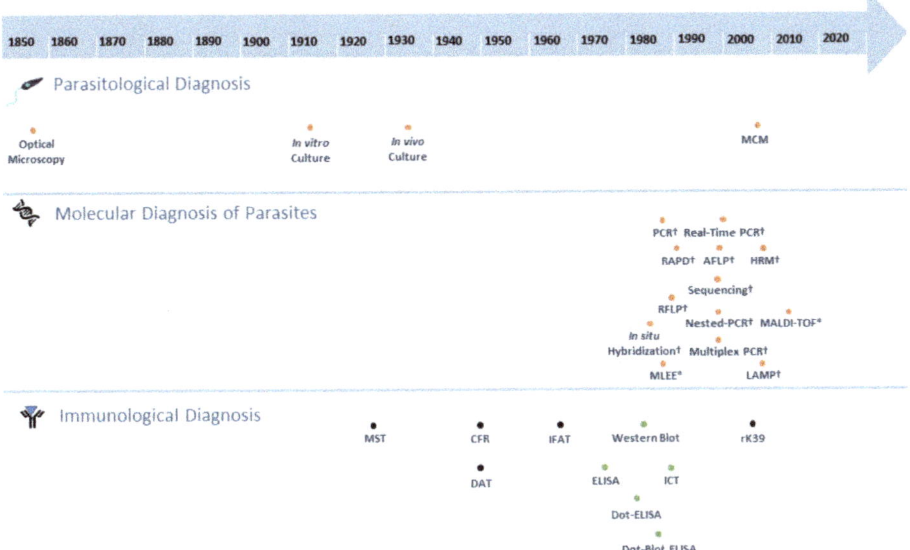

Figure 3. Timeline of the development of diagnostic tests for leishmaniases. Orange dot, direct method; black dot, indirect method; green dot, direct and/or indirect method; *, protein-based molecular method; †, nucleic acid-based molecular method; Parasitological Diagnosis: 1885, Optical Microscopy [111–113]; 1912, in vitro culture [114]; 1935, in vivo culture [115]; 2004, MCM (Microcapillary Culture Method) [116]. Molecular Diagnosis of Parasites: 1987, in situ Hybridization [117]; 1990, PCR (Polymerase Chain Reaction) [118]; 1990, MLEE (Multilocus Enzyme Electrophoresis) [49]; 1991, RFLP [119]; 1993, RAPD (Random Amplification of Polymorphic DNA) [120]; 1997, AFLP (Amplified Fragment Length Polymorphism) [121]; 1998, Nested-PCR [122]; 1998, Multiplex PCR [123,124]; 1999, Sequencing [125]; 2001, Real-Time PCR [126]; 2009, LAMP (Loop-mediated Isothermal Amplification) [127]; 2010, HRM (High Resolution Melting) [8,128]; 2014, MALDI-TOF (Matrix-Assisted Laser Desorption Ionization-Time-of-Flight Mass Spectrometry) [129]. Immunological Diagnosis: 1926, MST (Montenegro Skin Test) [130]; 1947, CFR (Complement Fixation Reaction) [131]; 1947, DAT (Direct Agglutination Test) [132]; 1964, IFAT (Indirect Immunofluorescence Antibody Test) [133,134]; 1978, ELISA (Enzyme-Linked Immunosorbent Assay) [135]; 1983, Dot-ELISA [136]; 1984, Western Blot [137]; 1987, Dot-Blot-ELISA [138]; 1990, ICT (Immunochromatographic Test) [139]; 2000, rK39 (rapid anti-K29 antibody Immunochromatographic strip-test) [140].

4.1. Parasitological Diagnoses

Parasitological diagnoses employ tissue aspirates from the spleen, bone marrow, lymph nodes, or peripheral blood of suspected individuals with VL, or skin biopsies/smears from ulcers/lesions of suspected individuals with CL or PKDL [1]. The presence of the parasites in samples can either be directly visualized by optical microscopy or cultured in appropriate culture media for later microscopic visualization (in vitro culture) [1]. The viability of the parasites present in the tissue samples can also be assessed by inoculation into susceptible animals followed by infectivity analysis (in vivo culture) [115]. Microscopic examination provides high genus specificity, although its sensitivity can vary depending on the different tissue aspirates. It also presents a relatively high risk of contamination and requires experienced technicians when performing the examination [66,141]. Parasite culturing is not as routinely used in a clinical setting as microscopic examination, yet culturing can enhance the detection sensitivity and isolated parasites can then be identified to the genotype and species level [34]. Among the various culture media to cultivate desired parasites, Solid Novy–MacNeal–Nicolle (NNN) medium is the most widely used for isolating and culturing the parasites although improvements were found using a more nutritious media [142]. However, in recent decades, a Microcapillary Culture Method (MCM) surfaced to increase the sensitivity of parasite culturing techniques by using peripheral blood mononuclear cells and buffy coats. Noteworthily, MCM only requires a small volume of culture medium, and it is cheaper than other media [143].

4.2. Molecular Diagnosis of Parasites

The identification of *Leishmania* species by various methods exploring specific characteristics of nucleic acids has been described in recent decades. One of the earliest applications of molecular diagnosis for discriminating *Leishmania* species employed Restriction Fragment Length Polymorphism (RFLP). Researchers used P32-labelled radioactive kDNA probes to hybridize the digested fragments of DNA by Southern blotting [119]. Shortly after, the in situ hybridization technique was described for the detection of *Leishmania* in blood and smear samples from in vivo studies using non-radioactive DNA probes [117]. However, molecular diagnoses then were not widely implemented in clinical settings due to the complexities, cost, and time requirements, until the Polymerase Chain Reaction (PCR) was developed [144]. The detection of parasites' nucleic acids by PCR has not only accelerated the confirmation of the diagnosis but also constituted a cogent tool to enhance control and surveillance of leishmaniases by identifying causative species [145], sylvatic origins [12,146], and their geographical distribution. The advancement of PCR diversified diagnostic assays targeting regions of the parasites' DNA in the mitochondrial genome (e.g., kinetoplast DNA or kDNA) [147,148], nuclear genome (e.g., SSU rDNA [149,150], *glucose-6-phosphate dehydrogenase* (*g6pd*) [151], *70kDa heat-shock protein* (*hsp70*) [152,153], *amino acid permease 3* (*aap3*) [154], *cysteine proteinase B* (*cpb*) [155], or intergenic genomic regions (ITSs) [156]). The main advantage of the PCR method is its high sensitivity and specificity. On the other hand, the limitations of this method are the relatively higher risk of contamination and its restricted detection range, which may result in variations in the method's sensitivity [157,158].

Subsequently, numerous PCR-based diagnostic methods were developed to further improve the detection sensitivity [159]. These methods include DNA hybridization coupled with conventional PCR, Randomly Amplified Polymorphic DNA (RAPD), and Amplified Fragment Length Polymorphism (AFLP) [160,161]. However, these methods also present some limitations. For instance, RAPD has poor reproducibility and requires the laborious standardization of PCR conditions [162–164].

Standard PCR protocols were also modified to achieve improved outcomes, such as the nested PCR approach. This approach was first used in the late 1990s to distinguish the Eurasia species *L. (L.) tropica*, *L. (L.) infantum*, and *L. (L.) major* based on their amplicon sizes using kDNA as a target [122]. In another study, researchers designed specific primer sets targeting G6PD to distinguish between the two subgenera of *L. (Leishmania)* spp. and *L. (Viannia)* spp., then *L. (V.) braziliensis* from other *L. (Viannia)* spp. [151]. It is noteworthy that the discrimination of *L. (V.) braziliensis* from the subgenus *L. (Viannia)* is important for disease prognosis because infections by *L. (V.) braziliensis* often lead to MCL [151]. The main advantage of nested PCR is its higher accuracy, sensitivity, and specificity compared to conventional PCR. Nevertheless, a higher risk of contamination can be a major disadvantage of this method [143]. Additionally, the multiplex PCR approach/technique was developed. The application of multiplex PCR is particularly useful for rapid in-country identification of *Leishmania* parasites in endemic areas where more than one species resides. For instance, researchers developed a one-step differentiation for the American species of *L. (V.) braziliensis*, *L. (L.) mexicana*, and *L. (L.) donovani*, while minimizing the time and resources. As the multiplex PCR method allows for the detection of multiple targets simultaneously, the main advantages of the method are cost and time effectiveness. However, the standardization of PCR conditions can be laborious and this method is relatively less sensitive that other PCR-based methods [123].

On the other hand, a loop-mediated isothermal amplification (LAMP) flourished as an alternative technique for other PCR-based diagnostic tools. It is a species-specific DNA amplification method comparable to other PCR-based methods. Unlike PCR, in which the amplification is carried out with repeated thermal cycling, isothermal amplification is carried out at a constant temperature without needing a thermal cycler, which can be advantageous for field use [165]. In addition, compared to other PCR-based methods, LAMP has shorter reaction times [166]. However, the biggest drawback of this technique is that the amplified products cannot be used for downstream applications. Moreover, depending on the sample used, this technique can be less sensitive than conventional PCR (reviewed in [165]).

Sequencing became indispensable not only for the precise identification of *Leishmania* species [167] but also to discriminate *Leishmania* from other trypanosomatids. In the state of Mato Grosso do Sul (Brazil), naturally occurring co-infections of *L. (L.) infantum chagasi* and *Trypanosoma evansi* in dogs were identified by PCR associated with sequencing of target SSU rDNA [168]. Sequencing is also efficacious in distinguishing *Leishmania*-like parasites presenting leishmaniases-like disease [169]. Detecting *Leishmania*-like parasites is extremely challenging as these organisms morphologically resemble *Leishmania* under a microscope and show a positive reaction to the rK39 test. Moreover, these *Leishmania*-like parasites' DNA can also be amplified by several commonly used primers targeting different genomic regions of *Leishmania*. In a recent study, a group from Brazil identified a *Crithidia*-related, non-*Leishmania* parasite from a patient with the VL-like disease by whole-genome sequencing, underlining the need for the accurate identification of *Leishmania*-like parasites [169]. Further difficulties were also encountered when detecting species of the subgenus *L. (Mundinia)* in patients. It was initially thought that isolates of *L. (M.) martiniquensis* were monoxenous trypanosomatids [170], until it was later discovered that they too belonged to the genus *Leishmania*.

The use of Real-Time PCR emerged as an advanced PCR-based method in the early 2000s and was first described for the quantification of parasite load in the liver of *L. (L.) infantum*-infected mice [126]. Real-Time PCR allows the continuous monitoring of the synthesized PCR products [143,171]. In a study, researchers undertook Real-Time PCR tests targeting *g6pd* to identify and quantify the responsible species for the American CL [172]. Similarly, another study was able to distinguish nine different species from both *L. (Leishmania)* and *L. (Viannia)* subgenera and to quantify parasite load for in vivo studies with Real-Time PCR targeting *aap3* coupled with a multiplex strategy [154]. The main advantages of Real-Time PCR are its increased sensitivity and reduced contamination risk, as the method does not require post-amplification handling, besides the possibility of adapting it to a high throughput format (reviewed in [171]). However, some limitations are the requirement of expensive and specialized equipment, and experienced personnel to analyze the results [158].

High-resolution melting (HRM) analysis is one of the most recent developments in molecular diagnosis as a variation on Real-Time PCR. This is a robust technique for the detection of polymorphisms, mutations, and epigenetic differences in DNA samples by monitoring dissociation kinetics of PCR products (reviewed in [173]). While the use of HRM analysis for *Leishmania* spp. is still infrequently described in the literature, this method has been applied in identification protocols with results that vary in terms of approach and its discriminatory capacity. The use of an HRM-like approach to discriminate the Eurasian species *L. (L.) infantum*, *L. (L.) donovani*, *L. (L.) tropica*, and *L. (L.) major* by targeting their kDNA polymorphisms was first described in the early 2000s [174]. Subsequently, the use of kDNA enabled the discrimination of the subgenera *L. (Leishmania)*, *L. (Viannia)*, and the species *L. (L.) amazonensis* from *L. (L.) infantum* in isolated strains and clinical samples [175,176]. Some recent studies targeting ITS-1 rRNA and *hsp70* allowed the discrimination of six different American *Leishmania* species, *L. (L.) infantum chagasi*, *L. (L.) amazonensis*, *L. (L.) mexicana*, *L. (V.) braziliensis*, *L. (V.) guyanensis*, and *L. (V.) panamensis* [177]. Moreover, the melting and dissociation profiles of three different regions in *hsp70* or *aap3* enabled the differentiation of the main causative species of leishmaniases in Africa, Eurasia, and the Americas [178,179]. In addition to the advantages of Real-Time PCR, HRM provides even higher specificity than using Real-Time PCR alone, as the method allows better discrimination of genetic variants [173].

The analysis of electrophoretic profiles of isoenzymes, or Multilocus Enzyme Electrophoresis (MLEE), is considered as a gold standard technique by the WHO [1]. This method is based on the differences in the electrophoretic profiles of several enzymes obtained from promastigotes grown in culture, which can generate highly informative profiles regarding the identity of the tested organisms. This technique made significant contributions to the taxonomy of the genus *Leishmania* and discriminated between Eurasian and American species [143], yet it is not suitable for routine diagnosis, since it depends on the establishment of a culture of parasites and the requirement of specialized equipment [156,180].

Lastly, Matrix-Assisted Laser Desorption Ionization-Time-of-Flight (MALDI-TOF) mass spectrometry (MS) has emerged in recent years as a robust tool for the identification of microorganisms. The protein spectra, or "fingerprint", of an isolate are compared to a reference spectral database, definitively identifying the isolates within an hour [129]. This technique, also known as MALDI BioTyper Systems, has been successfully applied to various pathogens in clinical laboratory settings [181]. Several studies have demonstrated the use of MALDI-TOF MS to accurately discriminate *Leishmania* spp. [129,182]. Nevertheless, like MLEE, this technique requires the isolation and cultivation of parasites and, due to the high cost of equipment and maintenance, it is currently only available in larger research laboratories or reference centers [129].

4.3. Immunological Diagnoses

Immunological diagnoses are based on either the presence of host-specific antibodies (indirect) and/or the detection of the parasite's antigens (direct) (reviewed in [172]) (Figure 3). Methods include the Montenegro Skin Test (MST), also called the *Leishmanin* Skin Test (LST), Complement Fixation Reaction (CFR), Direct Agglutination Test (DAT), Indirect Immunofluorescence Antibody Test (IFAT), various forms of Enzyme-Linked Immunosorbent Assay (ELISA), Western blotting, Immunochromatographic Test (ICT), and the rK39 antigen-based immunochromatographic test. Immunological diagnoses show comparatively good diagnostic accuracy, especially during the acute stage of VL. On the contrary, they are not widely used for CL due to their low sensitivity and variable specificity, as cutaneous lesions often show lower levels of antibodies [1,183]. This could be since antibodies are partly species specific, and if a heterologous antigen is used, its detection may be low or undetectable [184]. Nonetheless, in the diagnosis of CL, Dot-Blot [138], and IFAT [185] have demonstrated high sensitivity. The Montenegro skin test is the most frequently used method for CL as it can detect the occurrence of a delayed-type hypersensitivity response [1,143]. In the diagnosis of VL, ELISA is the most widely used and sensitive immunological diagnostic method. In a study comparing sensitivities of standard ELISA, Dot-ELISA, and CFR to detect *L. (L.) donovani*, Dot-ELISA showed the highest sensitivity of 89%, followed by the standard ELISA and CFR, 80% and 72%, respectively [136]. Irrespective of the method, the sensitivity of serological tests depends on the antigens used [186]. For instance, when a study evaluated the sensitivity and specificity of the recombinant proteins rK9, rK26, rK39, and Crude Soluble Antigen (CSA), researchers found sensitivities of 78%, 38%, 100%, and 80% and specificities of 84%, 80%, 96%, and 72%, respectively [183,187]. Furthermore, rK39 was developed for field use, and it is recognized for its low-cost, quick and easy procedures, and relatively high reproductivity [1,71]. Nevertheless, the biggest drawback of utilizing any of these immunological tests is that they are unreliable for the diagnosis of relapses as specific antibodies can remain detectable up to several years post-recovery [1], and they are not species specific.

5. Use of Diagnostic Tools in Epidemiological Studies

Diagnostic tools have many important applications in population-based studies and disease surveillance for assessing the endemicity in specific areas, estimating the prevalence, and other various epidemiologic studies such as quantitative risk assessment [188].

Disease surveillance is critical in monitoring the disease burden, providing early warnings of an outbreak, determining risk factors, and evaluating the effectiveness of implemented control measures [189]. Active sampling is required for surveillance, followed by laboratory tests to improve the accuracy of the system. From 2008 to 2014 in northern Spain, researchers collected spleen samples from deceased wolves and small carnivores to test for *L. (L.) infantum* DNA using conventional

PCR. The study, which was a part of the wildlife sanitary surveillance programs, not only revealed a widespread presence of the parasite in previously evaluated as non-endemic areas but also promoted the use of the wolf as a sentinel for leishmaniases [190]. Moreover, geographical information system (GIS) techniques can be applied to enhance surveillance systems and identify their causes. GIS integrates a broad range of information from different sources including diagnostic results and environmental data, to analyze and display the geographically referenced or spatial distribution of the parasite, host, or vector, hence enabling the identification of disease occurrence patterns and probable risk factors [191]. In the Central Western regions of São Paulo (Brazil), road-killed wild animals were subjected as sentinels for detecting *L. (L.) infantum chagasi* DNA using PCR. It is noteworthy that monitoring road-killed wild animals not only provides insights about their health and the environment but also about their behaviors as, often, their natural habitat is disrupted by road construction. The geographic coordinates of the infected animals were recorded by using GIS, which gave rise to a better understanding of pathogen distribution [192]. In another study from Nepal, researchers combined DAT results with GIS to evaluate the exposure of human and domestic animals to *L. (L.) donovani* and showed the occurrence of spatial clustering of seropositive humans and domestic animals [6]. Further to this, several studies in endemic countries observed a correlation between meteorological factors (e.g., temperature, precipitation, humidity) and the incidence and distribution of serological titers of the infected canines [193].

The disease prevalence estimate is the proportion of positive results obtained from diagnostic tests to the entire tested population, and the seroprevalence estimate of the disease serves as a proxy for the outcome of interest [188]. In a longitudinal study in the state of Minas Gerais (Brazil), researchers employed ELISA to screen and IFAT to confirm the infection, and a Dual-Path Platform (DPP) immunoassay was performed with the stored sera samples after the end of the collection period. A high prevalence of seropositive dogs was found in areas with a high density of free-roaming dogs and significantly contributed to the continuous circulation of leishmaniases among the human population [194].

To estimate the exposure effect of a response variable based on the expected risk prevalence factor and diseases, a risk prediction study can be performed with cohort, case–control, or cross-sectional study designs [188]. In west Ethiopia, a cross-sectional study was conducted to determine the seroprevalence of human VL and associated risk factors. Researchers used ITleish, a rapid immunochromatographic test, and the Montenegro skin test for the survey, along with semi-structured questionnaires to identify associated risk factors [195]. In another study, researchers performed a cohort study among U.S. soldiers who were deployed to Iraq between 2002 and 2011 to determine the prevalence and asymptomatic VL risk factors among these soldiers. Several tests including ELISA, rK39, and Real-Time PCR were used to test blood samples, and a detailed risk factor survey was conducted for the study. Nearly 20% of the soldiers tested positive for asymptomatic VL, and those who were deployed to the Ninewa governorate showed a strong association with infection [196]. Furthermore, to assess the disease risk associated with the movement of host reservoirs and/or vectors, a quantitative risk assessment can be performed [188]. In South Africa, researchers performed IFAT and Real-Time PCR for an initial screening, followed by PCR and sequencing analysis to diagnose leishmaniasis in imported dogs and evaluate the risk of introducing canine leishmaniasis in the country [197].

6. Conclusions

The control of leishmaniases is an ongoing global challenge complicated by many different biological and environmental factors involved in the circulation of the diseases. Furthermore, the leishmaniases are re-emerging in endemic areas and emerging in non-endemic areas due to the increasing human influence on the environment. Taking the complexity of the disease into account, integrating the One Health approach is the essential key to controlling the disease. Moreover, diagnosing the leishmaniases is arduous due to its broad spectrum of clinical manifestations that often overlap with other co-endemic infectious diseases. The rapid advancement of various diagnostic tests in recent decades has improved early detection and accurate species identification. Diagnostic tests are also pivotal in performing various epidemiological studies, including active surveillance, and identifying risk factors. Considering the relationship between zoonotic diseases and spillover events (Figure 1), the use of synanthropic animals as sentinels is vital for surveilling zoonotic leishmaniases in the human population. However, detecting zoonotic spillover events requires rapid and large-scale testing, which require special equipment and trained personnel that are often not available in the resource-poor countries most affected by the disease. Future efforts must be directed towards the development of cost-effective and more accurate, accessible tests designed for field use in remote areas of endemic regions.

Author Contributions: Conceptualization: A.H., R.A.Z., L.M.F.-W. and M.F.L.-S.; writing—original draft preparation: A.H. and R.A.Z.; writing—review and editing: A.H. and M.F.L.-S.; formal analysis: J.J.S. and L.M.F.-W.; supervision: L.M.F.-W. and M.F.L.-S. All authors have read and agreed to the published version of the manuscript.

Funding: This work was supported by funding from the Brazilian Research Foundation "Fundação de Amparo à Pesquisa do Estado de São Paulo" (FAPESP) (2017/23933-3, 2018/23512-0) and the Brazilian Research Council (CNPq). The funders had no role in either the study design, data collection, and analysis, the decision to publish, or the preparation of the manuscript.

Conflicts of Interest: The authors declare no conflict of interest.

References

1. WHO. *Control of the Leishmaniases: Report of a Meeting of the WHO Expert Committee on the Control of Leishmaniases*, 949th ed.; WHO Press: Geneva, Switzerland, 2010.
2. Alvar, J.; Velez, I.D.; Bern, C.; Herrero, M.; Desjeux, P.; Cano, J.; Jannin, J.; den Boer, M. Leishmaniasis worldwide and global estimates of its incidence. *PLoS ONE* **2012**, *7*, e35671. [CrossRef]
3. Palatnik-de-Sousa, C.B.; Santos, W.R.; França-Silva, J.C.; da Costa, R.T.; Barbosa Reis, A.; Palatnik, M.; Mayrink, W.; Genaro, O. Impact of canine control on the epidemiology of canine and human visceral leishmaniasis in Brazil. *Am. Soc. Trop. Med. Hyg.* **2001**, *65*, 510–517. [CrossRef] [PubMed]
4. Hubálek, Z. Emerging Human Infectious Diseases: Anthroponoses, Zoonoses, and Sapronoses. *Emerg. Infect. Dis.* **2003**, *9*, 403–404. [CrossRef] [PubMed]
5. Espinosa, O.A.; Serrano, M.G.; Camargo, E.P.; Teixeira, M.M.G.; Shaw, J.J. An appraisal of the taxonomy and nomenclature of trypanosomatids presently classified as Leishmania and Endotrypanum. *Parasitology* **2018**, *145*, 430–442. [CrossRef] [PubMed]
6. Khanal, B.; Picado, A.; Bhattarai, N.R.; Van Der Auwera, G.; Das, M.L.; Ostyn, B.; Davies, C.R.; Boelaert, M.; Dujardin, J.C.; Rijal, S. Spatial analysis of Leishmania donovani exposure in humans and domestic animals in a recent kala azar focus in Nepal. *Parasitology* **2010**, *137*, 1597–1603. [CrossRef]
7. Bhattarai, N.R.; Van der Auwera, G.; Rijal, S.; Picado, A.; Speybroeck, N.; Khanal, B.; De Doncker, S.; Das, M.L.; Ostyn, B.; Davies, C.; et al. Domestic animals and epidemiology of visceral leishmaniasis, Nepal. *Emerg. Infect. Dis.* **2010**, *16*, 231–237. [CrossRef]
8. Talmi-Frank, D.; Kedem-Vaanunu, N.; King, R.; Bar-Gal, G.K.; Edery, N.; Jaffe, C.L.; Baneth, G. Leishmania tropica infection in golden jackals and red foxes, Israel. *Emerg. Infect. Dis.* **2010**, *16*, 1973–1975. [CrossRef]
9. Lainson, R. Ecological interactions in the transmission of the leishmaniases. *Philos. Trans. R. Soc. Lond. B Biol. Sci.* **1988**, *321*, 389–404.
10. Colmenares, M.; Kar, S.; Goldsmith-Pestana, K.; McMahon-Pratt, D. Mechanisms of pathogenesis: Differences amongst Leishmania species. *Trans. R. Soc. Trop. Med. Hyg.* **2002**, *96* (Suppl. 1), S3–S7. [CrossRef]
11. Rangel, E.F.; Shaw, J.J. (Eds.) *Brazilian Sand Flies: Biology, Taxonomy, Medical Importance and Control*, 1st ed.; Springer International Publishing: Cham, Switzerland, 2018; Volume 1, p. 494.
12. Andrade, M.S.; Courtenay, O.; Brito, M.E.; Carvalho, F.G.; Carvalho, A.W.; Soares, F.; Carvalho, S.M.; Costa, P.L.; Zampieri, R.; Floeter-Winter, L.M.; et al. Infectiousness of Sylvatic and Synanthropic Small Rodents Implicates a Multi-host Reservoir of Leishmania (Viannia) braziliensis. *PLoS. Negl. Trop. Dis.* **2015**, *9*, e0004137. [CrossRef]

13. Badaró, R.; Jones, T.C.; Lorenço, R.; Cerf, B.J.; Sampaio, D.; Carvalho, E.M.; Rocha, H.; Teixeira, R.; Johnson, W.D., Jr. A prospective study of visceral leishmaniasis in an endemic area of Brazil. *J. Infect. Dis.* **1986**, *154*, 639–649. [CrossRef] [PubMed]

14. Sang, R.; Arum, S.; Chepkorir, E.; Mosomtai, G.; Tigoi, C.; Sigei, F.; Lwande, O.W.; Landmann, T.; Affognon, H.; Ahlm, C.; et al. Distribution and abundance of key vectors of Rift Valley fever and other arboviruses in two ecologically distinct counties in Kenya. *PLoS. Negl. Trop. Dis.* **2017**, *11*, e0005341. [CrossRef] [PubMed]

15. Taylor, L.H.; Latham, S.M.; Woolhouse, M.E. Risk factors for human disease emergence. *Philos. Trans. R. Soc. Lond. B Biol. Sci.* **2001**, *356*, 983–989. [CrossRef] [PubMed]

16. CDC. One Health Basics. Available online: https://www.cdc.gov/onehealth/basics/index.html (accessed on 18 April 2020).

17. Kahn, L.H. The need for one health degree programs. *Infect. Ecol. Epidemiol.* **2011**, 1. [CrossRef] [PubMed]

18. Gebreyes, W.A.; Dupouy-Camet, J.; Newport, M.J.; Oliveira, C.J.; Schlesinger, L.S.; Saif, Y.M.; Kariuki, S.; Saif, L.J.; Saville, W.; Wittum, T.; et al. The global one health paradigm: Challenges and opportunities for tackling infectious diseases at the human, animal, and environment interface in low-resource settings. *PLoS Negl. Trop. Dis.* **2014**, *8*, e3257. [CrossRef]

19. Institute of Medicine Committee on Emerging Microbial Threats to H. In *Emerging Infections: Microbial Threats to Health in the United States*; Lederberg, J., Shope, R.E., Oaks, S.C., Jr., Eds.; National Academies Press (US): Washington, DC, USA, 1992.

20. *Institute of Medicine (US) Committee on Emerging Microbial Threats to Health in the 21st Century*; National Academies Press (US): Washington, DC, USA, 2003; p. 367.

21. Kumar, N.P.; Srinivasan, R.; Anish, T.S.; Nandakumar, G.; Jambulingam, P. Cutaneous leishmaniasis caused by Leishmania donovani in the tribal population of the Agasthyamala Biosphere Reserve forest, Western Ghats, Kerala, India. *J. Med. Microbiol.* **2015**, *64 Pt 2*, 157–163. [CrossRef]

22. Sharma, N.L.; Mahajan, V.K.; Kanga, A.; Sood, A.; Katoch, V.M.; Mauricio, I.; Singh, C.D.; Parwan, U.C.; Sharma, V.K.; Sharma, R.C. Localized cutaneous leishmaniasis due to Leishmania donovani and Leishmania tropica: Preliminary findings of the study of 161 new cases from a new endemic focus in himachal pradesh, India. *Am. J. Trop. Med. Hyg.* **2005**, *72*, 819–824. [CrossRef]

23. Khatri, M.L.; Di Muccio, T.; Fiorentino, E.; Gramiccia, M. Ongoing outbreak of cutaneous leishmaniasis in northwestern Yemen: Clinicoepidemiologic, geographic, and taxonomic study. *Int. J. Dermatol.* **2016**, *55*, 1210–1218. [CrossRef]

24. Del Giudice, P.; Marty, P.; Lacour, J.P.; Perrin, C.; Pratlong, F.; Haas, H.; Dellamonica, P.; Le Fichoux, Y. Cutaneous leishmaniasis due to Leishmania infantum. Case reports and literature review. *Arch. Dermatol.* **1998**, *134*, 193–198. [CrossRef]

25. Crowe, A.; Slavin, J.; Stark, D.; Aboltins, C. A case of imported Leishmania infantum cutaneous leishmaniasis; an unusual presentation occurring 19 years after travel. *BMC Infect. Dis.* **2014**, *14*, 597. [CrossRef]

26. Convit, J.; Ulrich, M.; Pérez, M.; Hung, J.; Castillo, J.; Rojas, H.; Viquez, A.; Araya, L.N.; Lima, H.D. Atypical cutaneous leishmaniasis in Central America: Possible interaction between infectious and environmental elements. *Trans. R. Soc. Trop. Med. Hyg.* **2005**, *99*, 13–17. [CrossRef] [PubMed]

27. Gitari, J.W.; Nzou, S.M.; Wamunyokoli, F.; Kinyeru, E.; Fujii, Y.; Kaneko, S.; Mwau, M. Leishmaniasis recidivans by Leishmania tropica in Central Rift Valley Region in Kenya. *Int. J. Infect. Dis.* **2018**, *74*, 109–116. [CrossRef] [PubMed]

28. Afonso, L.C.; Scott, P. Immune responses associated with susceptibility of C57BL/10 mice to Leishmania amazonensis. *Infect. Immun.* **1993**, *61*, 2952–2959. [CrossRef] [PubMed]

29. Aoki, J.I.; Laranjeira-Silva, M.F.; Muxel, S.M.; Floeter-Winter, L.M. The impact of arginase activity on virulence factors of Leishmania amazonensis. *Curr. Opin. Microbiol.* **2019**, *52*, 110–115. [CrossRef]

30. Scorza, B.M.; Carvalho, E.M.; Wilson, M.E. Cutaneous Manifestations of Human and Murine Leishmaniasis. *Int. J. Mol. Sci.* **2017**, *18*, 1296. [CrossRef]

31. Scott, P.; Novais, F.O. Cutaneous leishmaniasis: Immune responses in protection and pathogenesis. *Nat. Rev. Immunol.* **2016**, *16*, 581–592. [CrossRef]

32. Tomiotto-Pellissier, F.; Bortoleti, B.; Assolini, J.P.; Gonçalves, M.D.; Carloto, A.C.M.; Miranda-Sapla, M.M.; Conchon-Costa, I.; Bordignon, J.; Pavanelli, W.R. Macrophage Polarization in Leishmaniasis: Broadening Horizons. *Front. Immunol.* **2018**, *9*, 2529. [CrossRef]

33. Yanik, M.; Gurel, M.S.; Simsek, Z.; Kati, M. The psychological impact of cutaneous leishmaniasis. *Clin. Exp. Dermatol.* **2004**, *29*, 464–467. [CrossRef]
34. Reithinger, R.; Dujardin, J.C.; Louzir, H.; Pirmez, C.; Alexander, B.; Brooker, S. Cutaneous leishmaniasis. *Lancet Infect. Dis.* **2007**, *7*, 581–596. [CrossRef]
35. Carvalho, E.M.; Barral, A.; Costa, J.M.; Bittencourt, A.; Marsden, P. Clinical and immunopathological aspects of disseminated cutaneous leishmaniasis. *Acta Trop.* **1994**, *56*, 315–325. [CrossRef]
36. Develoux, M.; Diallo, S.; Dieng, Y.; Mane, I.; Huerre, M.; Pratlong, F.; Dedet, J.P.; Ndiaye, B. Diffuse cutaneous leishmaniasis due to Leishmania major in Senegal. *Trans. R. Soc. Trop. Med. Hyg.* **1996**, *90*, 396–397. [CrossRef]
37. Alcover, M.M.; Rocamora, V.; Guillén, M.C.; Berenguer, D.; Cuadrado, M.; Riera, C.; Fisa, R. Case Report: Diffuse Cutaneous Leishmaniasis by Leishmania infantum in a Patient Undergoing Immunosuppressive Therapy: Risk Status in an Endemic Mediterranean Area. *Am. J. Trop. Med. Hyg.* **2018**, *98*, 1313–1316. [CrossRef] [PubMed]
38. Sangueza, O.P.; Sangueza, J.M.; Stiller, M.J.; Sangueza, P. Mucocutaneous leishmaniasis: A clinicopathologic classification. *J. Am. Acad. Dermatol.* **1993**, *28*, 927–932. [CrossRef]
39. Dassoni, F.; Daba, F.; Naafs, B.; Morrone, A. Leishmaniasis recidivans in Ethiopia: Cutaneous and mucocutaneous features. *J. Infect. Dev. Ctries.* **2017**, *11*, 106–110. [CrossRef] [PubMed]
40. Davies, C.R.; Reithinger, R.; Campbell-Lendrum, D.; Feliciangeli, D.; Borges, R.; Rodriguez, N. The epidemiology and control of leishmaniasis in Andean countries. *Cadernos de Saude Publica* **2000**, *16*, 925–950. [CrossRef]
41. Marsden, P.D. Mucosal leishmaniasis ("espundia" Escomel, 1911). *Trans. R. Soc. Trop. Med. Hyg.* **1986**, *80*, 859–876. [CrossRef]
42. Osorio, L.E.; Castillo, C.M.; Ochoa, M.T. Mucosal leishmaniasis due to Leishmania (Viannia) panamensis in Colombia: Clinical characteristics. *Am. J. Trop. Med. Hyg.* **1998**, *59*, 49–52. [CrossRef]
43. David, C.; Dimier-David, L.; Vargas, F.; Torrez, M.; Dedet, J.P. Fifteen years of cutaneous and mucocutaneous leishmaniasis in Bolivia: A retrospective study. *Trans. R. Soc. Trop. Med. Hyg.* **1993**, *87*, 7–9. [CrossRef]
44. WHO. Leishmaniasis: Epidemiological Situation. Available online: https://www.who.int/leishmaniasis/burden/en/ (accessed on 1 May 2020).
45. Machado-Coelho, G.L.; Caiaffa, W.T.; Genaro, O.; Magalhães, P.A.; Mayrink, W. Risk factors for mucosal manifestation of American cutaneous leishmaniasis. *Trans. R. Soc. Trop. Med. Hyg.* **2005**, *99*, 55–61. [CrossRef]
46. Shaw, J.J. Further thoughts on the use of the name Leishmania (Leishmania) infantum chagasi for the aetiological agent of American visceral leishmaniasis. *Memorias do Instituto Oswaldo Cruz* **2006**, *101*, 577–579. [CrossRef]
47. Dantas-Torres, F. Leishmania infantum versus Leishmania chagasi: Do not forget the law of priority. *Memorias do Instituto Oswaldo Cruz* **2006**, *101*, 117–118, discussion 118. [CrossRef] [PubMed]
48. Killick-Kendrick, R. Some epidemiological consequences of the evolutionary fit between Leishmaniae and their phlebotomine vectors. *Bull. Soc. Pathol. Exot. Filiales.* **1985**, *78 Pt 2*, 747–755.
49. Rioux, J.A.; Lanotte, G.; Serres, E.; Pratlong, F.; Bastien, P.; Perieres, J. Taxonomy of Leishmania. Use of isoenzymes. Suggestions for a new classification. *Ann. Parasitol. Hum. Comp.* **1990**, *65*, 111–125. [CrossRef]
50. Lainson, R.; Rangel, E.F. Lutzomyia longipalpis and the eco-epidemiology of American visceral leishmaniasis, with particular reference to Brazil: A review. *Memorias do Instituto Oswaldo Cruz* **2005**, *100*, 811–827. [CrossRef] [PubMed]
51. Lukes, J.; Mauricio, I.L.; Schonian, G.; Dujardin, J.C.; Soteriadou, K.; Dedet, J.P.; Kuhls, K.; Tintaya, K.W.; Jirku, M.; Chocholova, E.; et al. Evolutionary and geographical history of the Leishmania donovani complex with a revision of current taxonomy. *Proc. Natl. Acad. Sci. USA* **2007**, *104*, 9375–9380. [CrossRef] [PubMed]
52. Alborzi, A.; Rasouli, M.; Shamsizadeh, A. Leishmania tropica-isolated patient with visceral leishmaniasis in southern Iran. *Am. J. Trop. Med. Hyg.* **2006**, *74*, 306–307. [CrossRef]
53. Alborzi, A.; Pouladfar, G.R.; Fakhar, M.; Motazedian, M.H.; Hatam, G.R.; Kadivar, M.R. Isolation of Leishmania tropica from a patient with visceral leishmaniasis and disseminated cutaneous leishmaniasis, southern Iran. *Am. J. Trop. Med. Hyg.* **2008**, *79*, 435–437. [CrossRef]

54. Sarkari, B.; Bavarsad Ahmadpour, N.; Moshfe, A.; Hajjaran, H. Molecular Evaluation of a Case of Visceral Leishmaniasis Due to Leishmania tropica in Southwestern Iran. *Iran. J. Parasitol.* **2016**, *11*, 126–130.

55. Magill, A.J.; Grögl, M.; Gasser, R.A., Jr.; Sun, W.; Oster, C.N. Visceral infection caused by Leishmania tropica in veterans of Operation Desert Storm. *N. Engl. J. Med.* **1993**, *328*, 1383–1387. [CrossRef]

56. Singh, O.P.; Hasker, E.; Sacks, D.; Boelaert, M.; Sundar, S. Asymptomatic Leishmania infection: A new challenge for Leishmania control. *Clin. Infect. Dis.* **2014**, *58*, 1424–1429. [CrossRef]

57. Stauch, A.; Sarkar, R.R.; Picado, A.; Ostyn, B.; Sundar, S.; Rijal, S.; Boelaert, M.; Dujardin, J.C.; Duerr, H.P. Visceral leishmaniasis in the Indian subcontinent: Modelling epidemiology and control. *PLoS. Negl. Trop. Dis.* **2011**, *5*, e1405. [CrossRef] [PubMed]

58. Alvar, J.; Aparicio, P.; Aseffa, A.; Den Boer, M.; Cañavate, C.; Dedet, J.P.; Gradoni, L.; Ter Horst, R.; López-Vélez, R.; Moreno, J. The relationship between leishmaniasis and AIDS: The second 10 years. *Clin. Microbiol. Rev.* **2008**, *21*, 334–359. [CrossRef] [PubMed]

59. Barral, A.; Pedral-Sampaio, D.; Grimaldi, D., Jr.; Momen, H.; McMahon-Pratt, D.; de Jesus, A.R.; Almeida, R.; Badaro, R.; Barral-Netto, M.; Carvalho, E.M.; et al. Leishmaniasis in Bahia, Brazil: Evidence that Leishmania amazonensis Produces a Wide Spectrum of Clinical Disease. *Am. J. Trop. Med. Hyg.* **1991**, *44*, 536–546. [CrossRef] [PubMed]

60. Aliaga, L.; Ceballos, J.; Sampedro, A.; Cobo, F.; López-Nevot, M.; Merino-Espinosa, G.; Morillas-Márquez, F.; Martín-Sánchez, J. Asymptomatic Leishmania infection in blood donors from the Southern of Spain. *Infection* **2019**, *47*, 739–747. [CrossRef]

61. França, A.O.; Pompilio, M.A.; Pontes, E.; de Oliveira, M.P.; Pereira, L.O.R.; Lima, R.B.; Goto, H.; Sanchez, M.C.A.; Fujimori, M.; Lima-Júnior, M.; et al. Leishmania infection in blood donors: A new challenge in leishmaniasis transmission? *PLoS ONE* **2018**, *13*, e0198199. [CrossRef]

62. Pineda, J.A.; Macías, J.; Morillas, F.; Fernandez-Ochoa, J.; Cara, J.; de La Rosa, R.; Mira, J.A.; Martín-Sánchez, J.; González, M.; Delgado, J.; et al. Evidence of increased risk for leishmania infantum infection among HIV-seronegative intravenous drug users from southern Spain. *Eur. J. Clin. Microbiol. Infect. Dis.* **2001**, *20*, 354–357. [CrossRef]

63. Vida, B.; Toepp, A.; Schaut, R.G.; Esch, K.J.; Juelsgaard, R.; Shimak, R.M.; Petersen, C.A. Immunologic progression of canine leishmaniosis following vertical transmission in United States dogs. *Vet. Immunol. Immunopathol.* **2016**, *169*, 34–38. [CrossRef]

64. Eltoum, I.A.; Zijlstra, E.E.; Ali, M.S.; Ghalib, H.W.; Satti, M.M.; Eltoum, B.; el-Hassan, A.M. Congenital kala-azar and leishmaniasis in the placenta. *Am. J. Trop. Med. Hyg.* **1992**, *46*, 57–62. [CrossRef]

65. Berger, B.A.; Bartlett, A.H.; Saravia, N.G.; Galindo Sevilla, N. Pathophysiology of Leishmania Infection during Pregnancy. *Trends Parasitol.* **2017**, *33*, 935–946. [CrossRef]

66. Barrett, M.P.; Croft, S.L. Management of trypanosomiasis and leishmaniasis. *Br. Med. Bull.* **2012**, *104*, 175–196. [CrossRef]

67. Zijlstra, E.E.; Musa, A.M.; Khalil, E.A.; el-Hassan, I.M.; el-Hassan, A.M. Post-kala-azar dermal leishmaniasis. *Lancet Infect. Dis.* **2003**, *3*, 87–98. [CrossRef]

68. Stark, D.; Pett, S.; Marriott, D.; Harkness, J. Post-kala-azar dermal leishmaniasis due to Leishmania infantum in a human immunodeficiency virus type 1-infected patient. *J. Clin. Microbiol.* **2006**, *44*, 1178–1180. [CrossRef] [PubMed]

69. Rathi, S.K.; Pandhi, R.K.; Khanna, N.; Chopra, P. Mucosal and peri-orificial involvement in post-kala-azar dermal leishmaniasis. *Indian J. Dermatol. Venereol. Leprol.* **2004**, *70*, 280–282. [PubMed]

70. Zijlstra, E.E.; Alves, F.; Rijal, S.; Arana, B.; Alvar, J. Post-kala-azar dermal leishmaniasis in the Indian subcontinent: A threat to the South-East Asia Region Kala-azar Elimination Programme. *PLoS Negl. Trop. Dis.* **2017**, *11*, e0005877. [CrossRef]

71. Burza, S.; Croft, S.L.; Boelaert, M. Leishmaniasis. *Lancet* **2018**, *392*, 951–970. [CrossRef]

72. Ganguly, S.; Saha, P.; Chatterjee, M.; Roy, S.; Ghosh, T.K.; Guha, S.K.; Kundu, P.K.; Bera, D.K.; Basu, N.; Maji, A.K. PKDL–A Silent Parasite Pool for Transmission of Leishmaniasis in Kala-azar Endemic Areas of Malda District, West Bengal, India. *PLoS Negl. Trop. Dis.* **2015**, *9*, e0004138. [CrossRef]

73. Molina, R.; Ghosh, D.; Carrillo, E.; Monnerat, S.; Bern, C.; Mondal, D.; Alvar, J. Infectivity of Post-Kala-azar Dermal Leishmaniasis Patients to Sand Flies: Revisiting a Proof of Concept in the Context of the Kala-azar Elimination Program in the Indian Subcontinent. *Clin. Infect. Dis.* **2017**, *65*, 150–153. [CrossRef]

74. Andrade-Narvaez, F.J.; Loría-Cervera, E.N.; Sosa-Bibiano, E.I.; Van Wynsberghe, N.R. Asymptomatic infection with American cutaneous leishmaniasis: Epidemiological and immunological studies. *Memorias do Instituto Oswaldo Cruz* **2016**, *111*, 599–604. [CrossRef]

75. Da Costa-Val, A.P.; Cavalcanti, R.R.; de Figueiredo Gontijo, N.; Michalick, M.S.; Alexander, B.; Williams, P.; Melo, M.N. Canine visceral leishmaniasis: Relationships between clinical status, humoral immune response, haematology and Lutzomyia (Lutzomyia) longipalpis infectivity. *Vet. J.* **2007**, *174*, 636–643. [CrossRef]

76. Topno, R.K.; Das, V.N.; Ranjan, A.; Pandey, K.; Singh, D.; Kumar, N.; Siddiqui, N.A.; Singh, V.P.; Kesari, S.; Bimal, S.; et al. Asymptomatic infection with visceral leishmaniasis in a disease-endemic area in bihar, India. *Am. J. Trop. Med. Hyg.* **2010**, *83*, 502–506. [CrossRef]

77. Schaefer, K.U.; Kurtzhals, J.A.; Gachihi, G.S.; Muller, A.S.; Kager, P.A. A prospective sero-epidemiological study of visceral leishmaniasis in Baringo District, Rift Valley Province, Kenya. *Trans. R. Soc. Trop. Med. Hyg.* **1995**, *89*, 471–475. [CrossRef]

78. Saha, P.; Ganguly, S.; Chatterjee, M.; Das, S.B.; Kundu, P.K.; Guha, S.K.; Ghosh, T.K.; Bera, D.K.; Basu, N.; Maji, A.K. Asymptomatic leishmaniasis in kala-azar endemic areas of Malda district, West Bengal, India. *PLoS Negl. Trop. Dis.* **2017**, *11*, e0005391. [CrossRef] [PubMed]

79. Ostyn, B.; Gidwani, K.; Khanal, B.; Picado, A.; Chappuis, F.; Singh, S.P.; Rijal, S.; Sundar, S.; Boelaert, M. Incidence of symptomatic and asymptomatic Leishmania donovani infections in high-endemic foci in India and Nepal: A prospective study. *PLoS Negl. Trop. Dis.* **2011**, *5*, e1284. [CrossRef]

80. Evans, T.G.; Teixeira, M.J.; McAuliffe, I.T.; Vasconcelos, I.; Vasconcelos, A.W.; Sousa Ade, A.; Lima, J.W.; Pearson, R.D. Epidemiology of visceral leishmaniasis in northeast Brazil. *J. Infect. Dis.* **1992**, *166*, 1124–1132. [CrossRef] [PubMed]

81. Maia, Z.; Viana, V.; Muniz, E.; Gonçalves, L.O.; Mendes, C.M.; Mehta, S.R.; Badaro, R. Risk Factors Associated with Human Visceral Leishmaniasis in an Urban Area of Bahia, Brazil. *Vector Borne Zoonotic Dis.* **2016**, *16*, 368–376. [CrossRef]

82. Davies, C.R.; Mazloumi Gavgani, A.S. Age, acquired immunity and the risk of visceral leishmaniasis: A prospective study in Iran. *Parasitology* **1999**, *119 Pt 3*, 247–257. [CrossRef]

83. Fukutani, K.F.; Figueiredo, V.; Celes, F.S.; Cristal, J.R.; Barral, A.; Barral-Netto, M.; de Oliveira, C.I. Serological survey of Leishmania infection in blood donors in Salvador, Northeastern Brazil. *BMC Infect. Dis.* **2014**, *14*, 422. [CrossRef]

84. Arrow, K.J.; Panosian, C.B.; Gelband, H. (Eds.) *Saving Lives, Buying Time: Economics of Malaria Drugs in an Age of Resistance*; National Academies Press (US): Washington, DC, USA, 2004; Volume 2.

85. Alvar, J.; Yactayo, S.; Bern, C. Leishmaniasis and poverty. *Trends Parasitol.* **2006**, *22*, 552–557. [CrossRef]

86. Rijal, S.; Koirala, S.; Van der Stuyft, P.; Boelaert, M. The economic burden of visceral leishmaniasis for households in Nepal. *Trans. R. Soc. Trop. Med. Hyg.* **2006**, *100*, 838–841. [CrossRef]

87. Reithinger, R.; Aadil, K.; Kolaczinski, J.; Mohsen, M.; Hami, S. Social impact of leishmaniasis, Afghanistan. *Emerg. Infect. Dis.* **2005**, *11*, 634–636. [CrossRef]

88. Anstead, G.M.; Chandrasekar, B.; Zhao, W.; Yang, J.; Perez, L.E.; Melby, P.C. Malnutrition alters the innate immune response and increases early visceralization following Leishmania donovani infection. *Infect. Immun.* **2001**, *69*, 4709–4718. [CrossRef]

89. Saha, S.; Ramachandran, R.; Hutin, Y.J.; Gupte, M.D. Visceral leishmaniasis is preventable in a highly endemic village in West Bengal, India. *Trans. R. Soc. Trop. Med. Hyg.* **2009**, *103*, 737–742. [CrossRef] [PubMed]

90. Pinto, M.C.; Campbell-Lendrum, D.H.; Lozovei, A.L.; Teodoro, U.; Davies, C.R. Phlebotomine sandfly responses to carbon dioxide and human odour in the field. *Med. Vet. Entomol.* **2001**, *15*, 132–139. [CrossRef]

91. Tavares, D.D.S.; Salgado, V.R.; Miranda, J.C.; Mesquita, P.R.R.; Rodrigues, F.M.; Barral-Netto, M.; de Andrade, J.B.; Barral, A. Attraction of phlebotomine sandflies to volatiles from skin odors of individuals residing in an endemic area of tegumentary leishmaniasis. *PLoS ONE* **2018**, *13*, e0203989. [CrossRef] [PubMed]

92. Bern, C.; Hightower, A.W.; Chowdhury, R.; Ali, M.; Amann, J.; Wagatsuma, Y.; Haque, R.; Kurkjian, K.; Vaz, L.E.; Begum, M.; et al. Risk factors for kala-azar in Bangladesh. *Emerg. Infect. Dis.* **2005**, *11*, 655–662. [CrossRef] [PubMed]

93. Bantie, K.; Tessema, F.; Tafere, Y. Factors Associated with Visceral Leishmaniasis Infection in North Gondar Zone, Amhara Region, North West Ethiopia, Case Control Study. *Sci. J. Public Health* **2014**, *2*, 560.

94. Islam, A.; Rahman, M.H.; Islam, S.; Debnath, P.; Alam, M.; Hassan, M. Sero-prevalence of visceral leishmaniasis (VL) among dogs in VL endemic areas of Mymensingh district, Bangladesh. *J. Adv. Vet. Anim. Res.* **2018**, *4*, 241–248. [CrossRef]

95. Al-Salem, W.; Herricks, J.R.; Hotez, P.J. A review of visceral leishmaniasis during the conflict in South Sudan and the consequences for East African countries. *Parasit. Vectors* **2016**, *9*, 460. [CrossRef]

96. Alawieh, A.; Musharrafieh, U.; Jaber, A.; Berry, A.; Ghosn, N.; Bizri, A.R. Revisiting leishmaniasis in the time of war: The Syrian conflict and the Lebanese outbreak. *Int. J. Infect. Dis.* **2014**, *29*, 115–119. [CrossRef]

97. Saúde, M.D. *Manual de Vigilância e Controle da Leishmaniose Visceral*, 1st ed.; Editora do Ministério da Saúde: Brasília-DF, Brasil, 2006; p. 120.

98. Cardim, M.F.; Rodas, L.A.; Dibo, M.R.; Guirado, M.M.; Oliveira, A.M.; Chiaravalloti-Neto, F. Introduction and expansion of human American visceral leishmaniasis in the state of Sao Paulo, Brazil, 1999–2011. *Rev. Saude Publica* **2013**, *47*, 691–700. [CrossRef]

99. Barata, R.B. Cem anos de endemias e epidemias. *Ciênc Saúde Coletiva* **2000**, *5*, 333–345. [CrossRef]

100. Barrett, T.V.; Senra, M.S. Leishmaniasis in Manaus, Brazil. *Parasitol. Today* **1989**, *5*, 255–257. [CrossRef]

101. Defries, R.; Rudel, T.; Uriarte, M.; Hansen, M. Deforestation driven by urban population growth and agricultural trade in the twenty-first century. *Nat. Geosci.* **2010**, *3*, 178–181. [CrossRef]

102. Desjeux, P. Leishmaniasis: Current situation and new perspectives. *Comp. Immunol. Microbiol. Infect. Dis.* **2004**, *27*, 305–318. [CrossRef] [PubMed]

103. Desjeux, P. The increase in risk factors for leishmaniasis worldwide. *Trans. R. Soc. Trop. Med. Hyg.* **2001**, *95*, 239–243. [CrossRef]

104. Penna, G.; Pinto, L.F.; Soranz, D.; Glatt, R. High incidence of diseases endemic to the Amazon region of Brazil, 2001-2006. *Emerg. Infect. Dis.* **2009**, *15*, 626–632. [CrossRef]

105. Warburg, A.; Montoya-Lerma, J.; Jaramillo, C.; Cruz-Ruiz, A.L.; Ostrovska, K. Leishmaniasis vector potential of Lutzomyia spp. in Colombian coffee plantations. *Med. Vet. Entomol.* **1991**, *5*, 9–16. [CrossRef]

106. Alexander, B.; Oliveria, E.B.; Haigh, E.; Almeida, L.L. Transmission of Leishmania in coffee plantations of Minas Gerais, Brazil. *Memórias do Instituto Oswaldo Cruz* **2002**, *97*, 627–630. [CrossRef]

107. Alexander, B.; Agudelo, L.A.; Navarro, J.F.; Ruiz, J.F.; Molina, J.; Aguilera, G.; Klein, A.; Quiñones, M.L. Relationship between coffee cultivation practices in Colombia and exposure to infection with Leishmania. *Trans. R. Soc. Trop. Med. Hyg.* **2009**, *103*, 1263–1268. [CrossRef]

108. Barhoumi, W.; Chelbi, I.; Zhioua, E. Effects of the development of irrigation systems in the arid areas on the establishment of Phlebotomus (Larroussius) perfiliewi Parrot, 1939. *Bull. Soc. Pathol. Exot.* **2012**, *105*, 403–405. [CrossRef]

109. Zhioua, E.; Kaabi, B.; Chelbi, I. Entomological investigations following the spread of visceral leishmaniasis in Tunisia. *J. Vector. Ecol.* **2007**, *32*, 371–374. [CrossRef]

110. Vink, M.M.T.; Nahzat, S.M.; Rahimi, H.; Buhler, C.; Ahmadi, B.A.; Nader, M.; Zazai, F.R.; Yousufzai, A.S.; van Loenen, M.; Schallig, H.; et al. Evaluation of point-of-care tests for cutaneous leishmaniasis diagnosis in Kabul, Afghanistan. *EBioMedicine* **2018**, *37*, 453–460. [CrossRef] [PubMed]

111. Cunningham, D.D. (Ed.) *On the Presence of Peculiar Parasitic Organisms in the Tissue of a Specimen of Delhi*; Superintendent of Government: Calcutta, India, 1885; Volume 1, pp. 21–31.

112. Ross, R. Further notes on leishman's bodies. *Br. Med. J.* **1903**, *2*, 1401. [CrossRef] [PubMed]

113. Ross, R. Note on the bodies recently described by leishman and donovan. *Br. Med. J.* **1903**, *2*, 1261–1262. [CrossRef] [PubMed]

114. Row, R. A simple haemoglobinized saline culture medium: For the growth of leishmania and allied protozoa. *Br. Med. J.* **1912**, *1*, 1119–1120. [CrossRef]

115. Forkner, C.E.; Zia, L.S. Further studies on kala-azar: Leishmania in nasal and oral secretions of patients and the bearing of this finding on the transmission of the disease. *J. Exp. Med.* **1935**, *61*, 183–203. [CrossRef]

116. Allahverdiyev, A.M.; Uzun, S.; Bagirova, M.; Durdu, M.; Memisoglu, H.R. A sensitive new microculture method for diagnosis of cutaneous leishmaniasis. *Am. J. Trop. Med. Hyg.* **2004**, *70*, 294–297. [CrossRef]

117. Van Eys, G.J.; Schoone, G.J.; Ligthart, G.S.; Laarman, J.J.; Terpstra, W.J. Detection of Leishmania parasites by DNA in situ hybridization with non-radioactive probes. *Parasitol. Res.* **1987**, *73*, 199–202. [CrossRef]

118. Rodgers, M.R.; Popper, S.J.; Wirth, D.F. Amplification of kinetoplast DNA as a tool in the detection and diagnosis of Leishmania. *Exp. Parasitol.* **1990**, *71*, 267–275. [CrossRef]

119. Arnot, D.E.; Barker, D.C. Biochemical identification of cutaneous leishmanias by analysis of kinetoplast DNA. II. Sequence homologies in Leishmania kDNA. *Mol. Biochem. Parasitol.* **1981**, *3*, 47–56. [CrossRef]

120. Tibayrenc, M.; Neubauer, K.; Barnabé, C.; Guerrini, F.; Skarecky, D.; Ayala, F.J. Genetic characterization of six parasitic protozoa: Parity between random-primer DNA typing and multilocus enzyme electrophoresis. *Proc. Natl. Acad. Sci. USA* **1993**, *90*, 1335–1339. [CrossRef] [PubMed]

121. Qubain, H.I.; Saliba, E.K.; Oskam, L. Visceral leishmaniasis from Bal'a, Palestine, caused by Leishmania donovani s.1. identified through polymerase chain reaction and restriction fragment length polymorphism analysis. *Acta Trop.* **1997**, *68*, 121–128. [CrossRef]

122. Noyes, H.A.; Reyburn, H.; Bailey, J.W.; Smith, D. A nested-PCR-based schizodeme method for identifying Leishmania kinetoplast minicircle classes directly from clinical samples and its application to the study of the epidemiology of Leishmania tropica in Pakistan. *J. Clin. Microbiol.* **1998**, *36*, 2877–2881. [CrossRef] [PubMed]

123. Harris, E.; Kropp, G.; Belli, A.; Rodriguez, B.; Agabian, N. Single-step multiplex PCR assay for characterization of New World Leishmania complexes. *J. Clin. Microbiol.* **1998**, *36*, 1989–1995. [CrossRef]

124. Belli, A.; Rodriguez, B.; Aviles, H.; Harris, E. Simplified polymerase chain reaction detection of new world Leishmania in clinical specimens of cutaneous leishmaniasis. *Am. J. Trop. Med. Hyg.* **1998**, *58*, 102–109. [CrossRef]

125. Robert-Gangneux, F.; Baixench, M.T.; Piarroux, R.; Pratlong, F.; Tourte-Schaefer, C. Use of molecular tools for the diagnosis and typing of a Leishmania major strain isolated from an HIV-infected patient in Burkina Faso. *Trans. R. Soc. Trop. Med. Hyg.* **1999**, *93*, 396–397. [CrossRef]

126. Bretagne, S.; Durand, R.; Olivi, M.; Garin, J.F.; Sulahian, A.; Rivollet, D.; Vidaud, M.; Deniau, M. Real-time PCR as a new tool for quantifying Leishmania infantum in liver in infected mice. *Clin. Diagn. Lab Immunol.* **2001**, *8*, 828–831. [CrossRef]

127. Takagi, H.; Itoh, M.; Islam, M.Z.; Razzaque, A.; Ekram, A.R.; Hashighuchi, Y.; Noiri, E.; Kimura, E. Sensitive, specific, and rapid detection of Leishmania donovani DNA by loop-mediated isothermal amplification. *Am. J. Trop. Med. Hyg.* **2009**, *81*, 578–582. [CrossRef]

128. Nasereddin, A.; Jaffe, C.L. Rapid diagnosis of Old World Leishmaniasis by high-resolution melting analysis of the 7SL RNA gene. *J. Clin. Microbiol.* **2010**, *48*, 2240–2242. [CrossRef]

129. Mouri, O.; Morizot, G.; Van der Auwera, G.; Ravel, C.; Passet, M.; Chartrel, N.; Joly, I.; Thellier, M.; Jauréguiberry, S.; Caumes, E.; et al. Easy identification of leishmania species by mass spectrometry. *PLoS Negl. Trop. Dis.* **2014**, *8*, e2841. [CrossRef]

130. Montenegro, J. A cutis-reação na Leishmaniose. *Anais da Faculdade de Medicina de São Paulo* **1926**, *1*, 9.

131. Ghosh, H.; Ghosh, N.N. Complement-fixation reaction in sera of rabbits actively immunized with living culture of Leishmania donovani. *Ann. Biochem. Exp. Med.* **1947**, *7*, 1. [PubMed]

132. Ghosh, H.; Ghosh, N.N. Agglutination reaction in sera of rabbits immunized with different strains of Leishmania donovani. *Ann. Biochem. Exp. Med.* **1947**, *7*, 3–6. [PubMed]

133. Duxbury, R.E.; Sadun, E.H. Fluorescent antibody test for the serodiagnosis of visceral leishmaniasis. *Am. J. Trop. Med. Hyg.* **1964**, *13*, 525–529. [CrossRef]

134. Shaw, J.J.; Voller, A. The detection of circulating antibody to kala-azar by means of immunofluorescent techniques. *Trans. R. Soc. Trop. Med. Hyg.* **1964**, *58*, 349–352. [CrossRef]

135. Baldelli, B.; Orfei, A.B.; Fioretti, D.P.; Polidori, G.A.; Ambrosi, M. Serological diagnosis of human leishmaniasis by ELISA (enzyme-linked immunosorbent assay). *Parassitologia* **1978**, *20*, 91–99.

136. Pappas, M.G.; Hajkowski, R.; Hockmeyer, W.T. Dot enzyme-linked immunosorbent assay (Dot-ELISA): A micro technique for the rapid diagnosis of visceral leishmaniasis. *J. Immunol. Methods* **1983**, *64*, 205–214. [CrossRef]

137. Jaffe, C.L.; Bennett, E.; Grimaldi, G., Jr.; McMahon-Pratt, D. Production and characterization of species-specific monoclonal antibodies against Leishmania donovani for immunodiagnosis. *J. Immunol.* **1984**, *133*, 440–447.

138. Handman, E.; Mitchell, G.F.; Goding, J.W. Leishmania major: A very sensitive dot-blot ELISA for detection of parasites in cutaneous lesions. *Mol. Biol. Med.* **1987**, *4*, 377–383.

139. Reed, S.G.; Shreffler, W.G.; Burns, J.M., Jr.; Scott, J.M.; Orge Mda, G.; Ghalib, H.W.; Siddig, M.; Badaro, R. An improved serodiagnostic procedure for visceral leishmaniasis. *Am. J. Trop. Med. Hyg.* **1990**, *43*, 632–639. [CrossRef]

140. Sundar, S.; Pai, K.; Sahu, M.; Kumar, V.; Murray, H.W. Immunochromatographic strip-test detection of anti-K39 antibody in Indian visceral leishmaniasis. *Ann. Trop. Med. Parasitol.* **2002**, *96*, 19–23. [CrossRef] [PubMed]

141. Boelaert, M.; Bhattacharya, S.; Chappuis, F.; El Safi, S.H.; Hailu, A.; Mondal, D.; Rijal, S.; Sundar, S.; Wasunna, M.; Peeling, R.W. Evaluation of rapid diagnostic tests: Visceral leishmaniasis. *Nat. Rev. Microbiol.* **2007**, *5*, S30–S39. [CrossRef]

142. Walton, B.C.; Shaw, J.J.; Lainson, R. Observations on the in vitro cultivation of Leishmania braziliensis. *J. Parasitol.* **1977**, *63*, 1118–1119. [CrossRef] [PubMed]

143. Thakur, S.; Joshi, J.; Kaur, S. Leishmaniasis diagnosis: An update on the use of parasitological, immunological and molecular methods. *J. Parasit. Dis.* **2020**, *44*, 1–20. [CrossRef]

144. Patrinos, G.P.; Ansorge, W.J. *Molecular Diagnostic*, 2nd ed.; Academic Press: Cambridge, MA, USA, 2010.

145. Uliana, S.R.; Ishikawa, E.; Stempliuk, V.A.; de Souza, A.; Shaw, J.J.; Floeter-Winter, L.M. Geographical distribution of neotropical Leishmania of the subgenus Leishmania analysed by ribosomal oligonucleotide probes. *Trans. R. Soc. Trop. Med. Hyg.* **2000**, *94*, 261–264. [CrossRef]

146. Brandão-Filho, S.P.; Brito, M.E.; Carvalho, F.G.; Ishikawa, E.A.; Cupolillo, E.; Floeter-Winter, L.; Shaw, J.J. Wild and synanthropic hosts of Leishmania (Viannia) braziliensis in the endemic cutaneous leishmaniasis locality of Amaraji, Pernambuco State, Brazil. *Trans. R. Soc. Trop. Med. Hyg.* **2003**, *97*, 291–296. [CrossRef]

147. Lopez, M.; Inga, R.; Cangalaya, M.; Echevarria, J.; Llanos-Cuentas, A.; Orrego, C.; Arevalo, J. Diagnosis of Leishmania using the polymerase chain reaction: A simplified procedure for field work. *Am. J. Trop. Med. Hyg.* **1993**, *49*, 348–356. [CrossRef]

148. Disch, J.; Pedras, M.J.; Orsini, M.; Pirmez, C.; de Oliveira, M.C.; Castro, M.; Rabello, A. Leishmania (Viannia) subgenus kDNA amplification for the diagnosis of mucosal leishmaniasis. *Diagn. Microbiol. Infect. Dis.* **2005**, *51*, 185–190. [CrossRef]

149. Uliana, S.R.; Affonso, M.H.; Camargo, E.P.; Floeter-Winter, L.M. Leishmania: Genus identification based on a specific sequence of the 18S ribosomal RNA sequence. *Exp. Parasitol.* **1991**, *72*, 157–163. [CrossRef]

150. Uliana, S.R.; Nelson, K.; Beverley, S.M.; Camargo, E.P.; Floeter-Winter, L.M. Discrimination amongst Leishmania by polymerase chain reaction and hybridization with small subunit ribosomal DNA derived oligonucleotides. *J. Eukaryot. Microbiol.* **1994**, *41*, 324–330. [CrossRef]

151. Castilho, T.M.; Shaw, J.J.; Floeter-Winter, L.M. New PCR assay using glucose-6-phosphate dehydrogenase for identification of Leishmania species. *J. Clin. Microbiol.* **2003**, *41*, 540–546. [CrossRef] [PubMed]

152. Garcia, L.; Kindt, A.; Bermudez, H.; Llanos-Cuentas, A.; De Doncker, S.; Arevalo, J.; Wilber Quispe Tintaya, K.; Dujardin, J.C. Culture-independent species typing of neotropical Leishmania for clinical validation of a PCR-based assay targeting heat shock protein 70 genes. *J. Clin. Microbiol.* **2004**, *42*, 2294–2297. [CrossRef] [PubMed]

153. Montalvo, A.M.; Fraga, J.; Monzote, L.; Montano, I.; De Doncker, S.; Dujardin, J.C.; Van der Auwera, G. Heat-shock protein 70 PCR-RFLP: A universal simple tool for Leishmania species discrimination in the New and Old World. *Parasitology* **2010**, *137*, 1159–1168. [CrossRef] [PubMed]

154. Tellevik, M.G.; Muller, K.E.; Løkken, K.R.; Nerland, A.H. Detection of a broad range of Leishmania species and determination of parasite load of infected mouse by real-time PCR targeting the arginine permease gene AAP3. *Acta Trop.* **2014**, *137*, 99–104. [CrossRef] [PubMed]

155. Quispe Tintaya, K.W.; Ying, X.; Dedet, J.P.; Rijal, S.; De Bolle, X.; Dujardin, J.C. Antigen genes for molecular epidemiology of leishmaniasis: Polymorphism of cysteine proteinase B and surface metalloprotease glycoprotein 63 in the Leishmania donovani complex. *J. Infect. Dis.* **2004**, *189*, 1035–1043. [CrossRef]

156. Cupolillo, E.; Grimaldi Júnior, G.; Momen, H.; Beverley, S.M. Intergenic region typing (IRT): A rapid molecular approach to the characterization and evolution of Leishmania. *Mol. Biochem. Parasitol.* **1995**, *73*, 145–155. [CrossRef]

157. Moreira, M.A.; Luvizotto, M.C.; Garcia, J.F.; Corbett, C.E.; Laurenti, M.D. Comparison of parasitological, immunological and molecular methods for the diagnosis of leishmaniasis in dogs with different clinical signs. *Vet. Parasitol.* **2007**, *145*, 245–252. [CrossRef]

158. De Paiva-Cavalcanti, M.; de Morais, R.C.; Pessoa, E.S.R.; Trajano-Silva, L.A.; Gonçalves-de-Albuquerque Sda, C.; Tavares Dde, H.; Brelaz-de-Castro, M.C.; Silva Rde, F.; Pereira, V.R. Leishmaniases diagnosis: An update on the use of immunological and molecular tools. *Cell Biosci.* **2015**, *5*, 31. [CrossRef]

159. Rodríguez, N.; Guzman, B.; Rodas, A.; Takiff, H.; Bloom, B.R.; Convit, J. Diagnosis of cutaneous leishmaniasis and species discrimination of parasites by PCR and hybridization. *J. Clin. Microbiol.* **1994**, *32*, 2246–2252. [CrossRef]

160. Kumar, A.; Boggula, V.R.; Misra, P.; Sundar, S.; Shasany, A.K.; Dube, A. Amplified fragment length polymorphism (AFLP) analysis is useful for distinguishing Leishmania species of visceral and cutaneous forms. *Acta Trop.* **2010**, *113*, 202–206. [CrossRef]

161. Mueller, U.G.; Wolfenbarger, L.L. AFLP genotyping and fingerprinting. *Trends Ecol. Evol.* **1999**, *14*, 389–394. [CrossRef]

162. Schönian, G.; Schweynoch, C.; Zlateva, K.; Oskam, L.; Kroon, N.; Gräser, Y.; Presber, W. Identification and determination of the relationships of species and strains within the genus Leishmania using single primers in the polymerase chain reaction. *Mol. Biochem. Parasitol.* **1996**, *77*, 19–29. [CrossRef]

163. Macedo, A.M.; Melo, M.N.; Gomes, R.F.; Pena, S.D. DNA fingerprints: A tool for identification and determination of the relationships between species and strains of Leishmania. *Mol. Biochem. Parasitol.* **1992**, *53*, 63–70. [CrossRef]

164. Wang, J.Y.; Ha, Y.; Gao, C.H.; Wang, Y.; Yang, Y.T.; Chen, H.T. The prevalence of canine Leishmania infantum infection in western China detected by PCR and serological tests. *Parasit. Vectors* **2011**, *4*, 69. [CrossRef] [PubMed]

165. Sahoo, P.R.; Sethy, K.; Mohapatra, S.; Panda, D. Loop mediated isothermal amplification: An innovative gene amplification technique for animal diseases. *Vet. World* **2016**, *9*, 465–469. [CrossRef] [PubMed]

166. Verma, S.; Avishek, K.; Sharma, V.; Negi, N.S.; Ramesh, V.; Salotra, P. Application of loop-mediated isothermal amplification assay for the sensitive and rapid diagnosis of visceral leishmaniasis and post-kala-azar dermal leishmaniasis. *Diagn. Microbiol. Infect. Dis.* **2013**, *75*, 390–395. [CrossRef]

167. Savani, E.S.; de Oliveira Camargo, M.C.; de Carvalho, M.R.; Zampieri, R.A.; dos Santos, M.G.; D'Auria, S.R.; Shaw, J.J.; Floeter-Winter, L.M. The first record in the Americas of an autochthonous case of Leishmania (Leishmania) infantum chagasi in a domestic cat (Felix catus) from Cotia County, São Paulo State, Brazil. *Vet. Parasitol.* **2004**, *120*, 229–233. [CrossRef]

168. Savani, E.S.; Nunes, V.L.; Galati, E.A.; Castilho, T.M.; Araujo, F.S.; Ilha, I.M.; Camargo, M.C.; D'Auria, S.R.; Floeter-Winter, L.M. Occurrence of co-infection by Leishmania (Leishmania) chagasi and Trypanosoma (Trypanozoon) evansi in a dog in the state of Mato Grosso do Sul, Brazil. *Memorias do Instituto Oswaldo Cruz* **2005**, *100*, 739–741. [CrossRef]

169. Maruyama, S.R.; de Santana, A.K.M.; Takamiya, N.T.; Takahashi, T.Y.; Rogerio, L.A.; Oliveira, C.A.B.; Milanezi, C.M.; Trombela, V.A.; Cruz, A.K.; Jesus, A.R.; et al. Non-Leishmania Parasite in Fatal Visceral Leishmaniasis-Like Disease, Brazil. *Emerg. Infect. Dis.* **2019**, *25*, 2088–2092. [CrossRef] [PubMed]

170. Boisseau-Garsaud, A.M.; Cales-Quist, D.; Desbois, N.; Jouannelle, J.; Jouannelle, A.; Pratlong, F.; Dedet, J.P. A new case of cutaneous infection by a presumed monoxenous trypanosomatid in the island of Martinique (French West Indies). *Trans. R. Soc. Trop. Med. Hyg.* **2000**, *94*, 51–52. [CrossRef]

171. Mortarino, M.; Franceschi, A.; Mancianti, F.; Bazzocchi, C.; Genchi, C.; Bandi, C. Quantitative PCR in the diagnosis of Leishmania. *Parassitologia* **2004**, *46*, 163–167. [PubMed]

172. Castilho, T.M.; Camargo, L.M.; McMahon-Pratt, D.; Shaw, J.J.; Floeter-Winter, L.M. A real-time polymerase chain reaction assay for the identification and quantification of American Leishmania species on the basis of glucose-6-phosphate dehydrogenase. *Am. J. Trop. Med. Hyg.* **2008**, *78*, 122–132. [CrossRef] [PubMed]

173. Reed, G.H.; Kent, J.O.; Wittwer, C.T. High-resolution DNA melting analysis for simple and efficient molecular diagnostics. *Pharmacogenomics* **2007**, *8*, 597–608. [CrossRef] [PubMed]

174. Nicolas, L.; Milon, G.; Prina, E. Rapid differentiation of Old World Leishmania species by LightCycler polymerase chain reaction and melting curve analysis. *J. Microbiol. Methods* **2002**, *51*, 295–299. [CrossRef]

175. Pita-Pereira, D.; Lins, R.; Oliveira, M.P.; Lima, R.B.; Pereira, B.A.; Moreira, O.C.; Brazil, R.P.; Britto, C. SYBR Green-based real-time PCR targeting kinetoplast DNA can be used to discriminate between the main etiologic agents of Brazilian cutaneous and visceral leishmaniases. *Parasit. Vectors* **2012**, *5*, 15. [CrossRef]

176. Ceccarelli, M.; Galluzzi, L.; Migliazzo, A.; Magnani, M. Detection and characterization of Leishmania (Leishmania) and Leishmania (Viannia) by SYBR green-based real-time PCR and high resolution melt analysis targeting kinetoplast minicircle DNA. *PLoS ONE* **2014**, *9*, e88845. [CrossRef]

177. Hernández, C.; Alvarez, C.; González, C.; Ayala, M.S.; León, C.M.; Ramírez, J.D. Identification of six New World Leishmania species through the implementation of a High-Resolution Melting (HRM) genotyping assay. *Parasit. Vectors* **2014**, *7*, 501. [CrossRef]

178. Zampieri, R.A.; Laranjeira-Silva, M.F.; Muxel, S.M.; Stocco de Lima, A.C.; Shaw, J.J.; Floeter-Winter, L.M. High Resolution Melting Analysis Targeting hsp70 as a Fast and Efficient Method for the Discrimination of Leishmania Species. *PLoS Negl. Trop. Dis.* **2016**, *10*, e0004485. [CrossRef]

179. Müller, K.E.; Zampieri, R.A.; Aoki, J.I.; Muxel, S.M.; Nerland, A.H.; Floeter-Winter, L.M. Amino acid permease 3 (aap3) coding sequence as a target for Leishmania identification and diagnosis of leishmaniases using high resolution melting analysis. *Parasit. Vectors* **2018**, *11*, 421. [CrossRef]

180. Grimaldi, G.; McMahon-Pratt, D. Monoclonal antibodies for the identification of New World Leishmania species. *Memorias do Instituto Oswaldo Cruz* **1996**, *91*, 37–42. [CrossRef]

181. Sogawa, K.; Watanabe, M.; Sato, K.; Segawa, S.; Ishii, C.; Miyabe, A.; Murata, S.; Saito, T.; Nomura, F. Use of the MALDI BioTyper system with MALDI-TOF mass spectrometry for rapid identification of microorganisms. *Anal. Bioanal. Chem.* **2011**, *400*, 1905–1911. [CrossRef] [PubMed]

182. Cassagne, C.; Pratlong, F.; Jeddi, F.; Benikhlef, R.; Aoun, K.; Normand, A.C.; Faraut, F.; Bastien, P.; Piarroux, R. Identification of Leishmania at the species level with matrix-assisted laser desorption ionization time-of-flight mass spectrometry. *Clin. Microbiol. Infect.* **2014**, *20*, 551–557. [CrossRef] [PubMed]

183. Singh, S.; Dey, A.; Sivakumar, R. Applications of molecular methods for Leishmania control. *Expert Rev. Mol. Diagn.* **2005**, *5*, 251–265. [CrossRef] [PubMed]

184. Shaw, J.J. A partnership that worked: The Wellcome Trust and the Instituto Evandro Chagas and beyond. *Rev. Pan-Amazônica de Saúde* **2016**, *7*, 23–42. [CrossRef]

185. Walton, B.C. Evaluation of chemotherapy of American leishmaniasis by the indirect fluorescent antibody test. *Am. J. Trop. Med. Hyg.* **1980**, *29*, 747–752. [CrossRef]

186. Singh, O.P.; Sundar, S. Developments in Diagnosis of Visceral Leishmaniasis in the Elimination Era. *J. Parasitol. Res.* **2015**, *2015*, 239469. [CrossRef]

187. Mohapatra, T.M.; Singh, D.P.; Sen, M.R.; Bharti, K.; Sundar, S. Compararative evaluation of rK9, rK26 and rK39 antigens in the serodiagnosis of Indian visceral leishmaniasis. *J. Infect. Dev. Ctries.* **2010**, *4*, 114–117. [CrossRef]

188. Greiner, M.; Gardner, I.A. Application of diagnostic tests in veterinary epidemiologic studies. *Prev. Vet. Med.* **2000**, *45*, 43–59. [CrossRef]

189. Pan-American Health Organization (PAHO). Plan of Action to Strengthen the Surveillance and Control of Leishmaniasis in the Americas 2017–2022. Available online: https://iris.paho.org/bitstream/handle/10665.2/34147/PlanactionLeish20172022-eng.pdf?sequence=5&isAllowed=y (accessed on 1 May 2020).

190. Oleaga, A.; Zanet, S.; Espí, A.; Pegoraro de Macedo, M.R.; Gortázar, C.; Ferroglio, E. Leishmania in wolves in northern Spain: A spreading zoonosis evidenced by wildlife sanitary surveillance. *Vet. Parasitol.* **2018**, *255*, 26–31. [CrossRef]

191. Ostad, M.; Shirian, S.; Pishro, F.; Abbasi, T.; Ai, A.; Azimi, F. Control of Cutaneous Leishmaniasis Using Geographic Information Systems from 2010 to 2014 in Khuzestan Province, Iran. *PLoS ONE* **2016**, *11*, e0159546. [CrossRef]

192. Richini-Pereira, V.B.; Marson, P.M.; Hayasaka, E.Y.; Victoria, C.; da Silva, R.C.; Langoni, H. Molecular detection of Leishmania spp. in road-killed wild mammals in the Central Western area of the State of São Paulo, Brazil. *J. Venom. Anim. Toxins. Incl. Trop. Dis.* **2014**, *20*, 27. [CrossRef]

193. Salahi-Moghaddam, A.; Mohebali, M.; Moshfae, A.; Habibi, M.; Zarei, Z. Ecological study and risk mapping of visceral leishmaniasis in an endemic area of Iran based on a geographical information systems approach. *Geospat. Health* **2010**, *5*, 71–77. [CrossRef] [PubMed]

194. Melo, S.N.; Teixeira-Neto, R.G.; Werneck, G.L.; Struchiner, C.J.; Ribeiro, R.A.N.; Sousa, L.R.; de Melo, M.O.G.; Carvalho Júnior, C.G.; Penaforte, K.M.; Manhani, M.N.; et al. Prevalence of visceral leishmaniasis in A population of free-roaming dogs as determined by multiple sampling efforts: A longitudinal study analyzing the effectiveness of euthanasia. *Prev. Vet. Med.* **2018**, *161*, 19–24. [CrossRef] [PubMed]

195. Bsrat, A.; Berhe, M.; Gadissa, E.; Taddele, H.; Tekle, Y.; Hagos, Y.; Abera, A.; G/micael, M.; Alemayhu, T.; Gugsa, G.; et al. Serological investigation of visceral Leishmania infection in human and its associated risk factors in Welkait District, Western Tigray, Ethiopia. *Parasite Epidemiol. Control.* **2018**, *3*, 13–20. [CrossRef] [PubMed]

196. Mody, R.M.; Lakhal-Naouar, I.; Sherwood, J.E.; Koles, N.L.; Shaw, D.; Bigley, D.P.; Co, E.A.; Copeland, N.K.; Jagodzinski, L.L.; Mukbel, R.M.; et al. Asymptomatic Visceral Leishmania infantum Infection in US Soldiers Deployed to Iraq. *Clin. Infect. Dis.* **2019**, *68*, 2036–2044. [CrossRef]
197. Latif, A.A.; Nkabinde, B.; Peba, B.; Matthee, O.; Pienaar, R.; Josemans, A.; Marumo, D.; Labuschagne, K.; Abdelatief, N.A.; Krüger, A.; et al. Risk of establishment of canine leishmaniasis infection through the import of dogs into South Africa. *Onderstepoort. J. Vet. Res.* **2019**, *86*, e1–e11. [CrossRef]

Article

Optimization of DNA Extraction from Field-Collected Mammalian Whole Blood on Filter Paper for *Trypanosoma cruzi* (Chagas Disease) Detection

Bonnie E. Gulas-Wroblewski [1,2], Rebecca B. Kairis [1,3,4], Rodion Gorchakov [1,5], Anna Wheless [1,6] and Kristy O. Murray [1,3,*]

1 Department of Pediatrics, Section of Pediatric Tropical Medicine, National School of Tropical Medicine, Baylor College of Medicine and Texas Children's Hospital, Houston, TX 77030, USA; bonnie.gulas@ag.tamu.edu (B.E.G.-W.); Rebecca.b.kairis@uth.tmc.edu (R.B.K.); rodion.gorchakov@kaust.edu.sa (R.G.); awheless@unc.edu (A.W.)
2 Texas A&M Natural Resources Institute, College Station, TX 77843, USA
3 The William T. Shearer Center for Human Immunobiology, Texas Children's Hospital, Houston, TX 77030, USA
4 The University of Texas Health Science Center at Houston, Houston, TX 77030, USA
5 Health, Safety and Environment Department, King Abdullah University of Science and Technology, Thuwal 23955, Saudi Arabia
6 Department of Biochemistry and Biophysics, University of North Carolina at Chapel Hill, Chapel Hill, NC 27599, USA
* Correspondence: kmurray@bcm.edu

check for **updates**

Citation: Gulas-Wroblewski, B.E.; Kairis, R.B.; Gorchakov, R.; Wheless, A.; Murray, K.O. Optimization of DNA Extraction from Field-Collected Mammalian Whole Blood on Filter Paper for *Trypanosoma cruzi* (Chagas Disease) Detection. *Pathogens* **2021**, *10*, 1040. https://doi.org/10.3390/pathogens10081040

Academic Editor: Vito Colella

Received: 28 April 2021
Accepted: 7 August 2021
Published: 17 August 2021

Publisher's Note: MDPI stays neutral with regard to jurisdictional claims in published maps and institutional affiliations.

Abstract: Blood filter paper strips are cost-effective materials used to store body fluid specimens under challenging field conditions, extending the reach of zoonotic pathogen surveillance and research. We describe an optimized procedure for the extraction of parasite DNA from whole blood (WB) stored on Type I Advantec Nobuto strips from both experimentally spiked and field-collected specimens from canine and skunks, respectively. When comparing two commercial kits for extraction, Qiagen's DNeasy Blood & Tissue Kit performed best for the detection of parasite DNA by PCR from *Trypanosoma cruzi*-spiked canine WB samples on Nobuto strips. To further optimize recovery of β-actin from field-collected skunk WB archived on Nobuto strips, we modified the extraction procedures for the Qiagen kit with a 90 °C incubation step and extended incubation post-addition of proteinase K, a method subsequently employed to identify a *T. cruzi* infection in one of the skunks. Using this optimized extraction method can efficaciously increase the accuracy and precision of future molecular epidemiologic investigations targeting neglected tropical diseases in field-collected WB specimens on filter strips.

Keywords: blood filter paper; Chagas disease; DNA extraction; Nobuto strip; *Trypanosoma cruzi*; mammalian surveillance; neglected tropical diseases; PCR

1. Introduction

The parasitic protozoan *Trypanosoma cruzi* is the etiologic agent of Chagas disease, which is maintained in domestic, peridomestic, and sylvatic transmission cycles by a diversity of triatomine vectors and mammalian hosts [1]. This neglected tropical parasite infects an estimated 6–7 million people across the Americas, making the zoonosis one of the most significant in terms of disease burden and public health importance in the western hemisphere [2]. Chagas disease and other neglected tropical diseases (NTDs) are particularly devastating for impoverished populations in remote regions with limited public health infrastructure [3–6]. In addition, the zoonotic nature of NTDs necessitates their investigation and control within a One Health framework, which holistically integrates domestic animal, wildlife, environmental, ecological, and public health [7,8]. Consequently, public health workers and epidemiologists face costly and logistical challenges in not only adequately sampling human and animal populations, but also preserving biological specimens of high enough quality for genomic applications under adverse field conditions and for prolonged periods of time [9,10]. However, NTD research and elimination programs are relatively underfunded, especially in comparison to those targeting pandemic-producing pathogens, exacerbating the challenges of NTD field-to-laboratory workflows [10–12]. In order to facilitate an inexpensive, reliable, and sensitive method option for polymerase chain reaction (PCR)-based NTD epidemiologic surveillance, we evaluated the performance of three commercial extraction kits with adjusted DNA extraction protocols for the recovery of *T. cruzi* DNA from canine whole blood (WB) preserved on filter paper. We then validated the optimized procedure via PCR recovery of β-actin from WB specimens collected from skunks (Mammalia: Mephitidae) and archived on filter papers. Developing optimized protocols for *T. cruzi* DNA extraction from blood filter papers greatly expands the efficiency and effectiveness of field investigations into the molecular epidemiology and surveillance of Chagas disease among mammalian reservoirs.

2. Results

2.1. Optimization with T. cruzi-Spiked Canine WB Samples

T. cruzi DNA was successfully extracted and recovered via quantitative (q) PCR for the spiked WB samples processed with each extraction method from both the Zymo Research (Zymo Research, Irvine, CA, USA) and Qiagen (Qiagen, Germantown, MD, USA) kits, though each procedure varied in the recovery of target DNA as measured by quantification cycle (Cq) values of the qPCR output (Figure 1, Supplementary Table S1). DNA extraction optimization methods employing the Qiagen DNeasy Blood & Tissue Kit outperformed those relying on Zymo Research kits for both WB and WB stored on Nobuto strips (Supplementary Table S1). At medium spiking loads, the difference between the mean Cq value recovered for WB versus WB-saturated Nobuto strips was substantially lower for extraction optimization method A (0.25) compared to the variances evidenced with extraction optimization method B (1.77), the Quick-DNA/RNA Pathogen Miniprep kit (1.1), and the ZR-Duet DNA/RNA Miniprep Plus kit (3.95). Similarly, differences between the mean Cq values generated by high spiking loads for WB and WB stored on Nobuto blood filter paper were lower for the Qiagen DNeasy Blood & Tissue Kit methods (0.65 for extraction optimization method A and 1.74 for extraction optimization method B) than for either of the Zymo Research kits (2.31 for ZR-Duet DNA/RNA Miniprep Plus kit and 5.22 for the Quick-DNA/RNA Pathogen Miniprep kit) (Supplementary Table S1).

2.2. Optimization with Skunk WB Samples

Once we established the enhanced capacity of the Qiagen DNeasy Blood & Tissue Kit to recover *T. cruzi* DNA from spiked WB specimens stored on Nobuto strips, we repeated extraction optimization methods A and B with an additional alternate extraction procedure on samples of skunk WB archived on Nobuto blood filter paper (Supplementary Table S2). Since the *T. cruzi* status of these animals was unknown, we employed a qPCR assay developed to detect β-actin-encoding DNA from mammals (Table 1) [13].

Each of the extraction optimization methods for the Qiagen kit recovered β-actin DNA as indicated by qPCR analysis (Figure 2, Supplementary Table S3). Extraction optimization A outperformed the other methods for all three samples, recovering an average of 60% more DNA than the other two protocols. In the case of relatively higher concentrations of DNA (samples from the western spotted (*Spilogale gracilis*) and striped skunks (*Mephitis mephitis*)), extraction optimization B extracted more DNA than did method C, whereas the latter method excelled with the comparatively lower amount of β-actin DNA present in the American hog-nosed skunk (*Conepatus leuconotus*) sample (Figure 2). Subsequent evaluation of these WB samples for the presence of *T. cruzi* DNA was performed using extraction optimization A, identifying parasitic infection in the western spotted skunk

(Cq value of 27.8) and not detecting *T. cruzi* DNA from either the American hog-nosed or striped skunk samples.

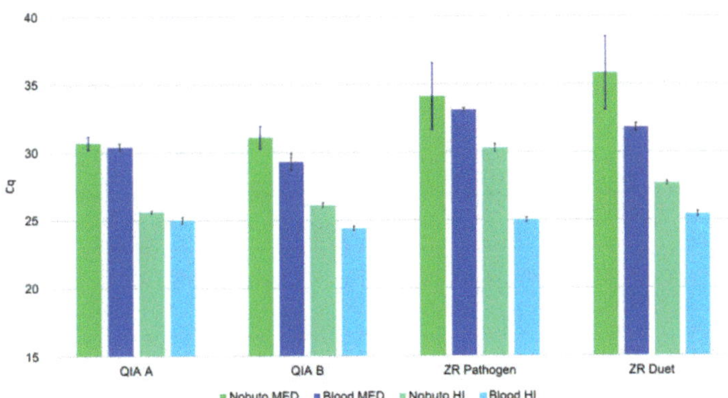

Figure 1. *Trypanosoma cruzi* DNA recovery from spiked canine whole blood specimens. Bars indicate one standard deviation. As detailed in the text for use with Qiagen DNeasy Blood & Tissue Kit: QIA A = extraction optimization method A; QIA B = extraction optimization method B. ZR Pathogen = Zymo Research Quick-DNA/RNA Pathogen Miniprep; ZR Duet = Zymo Research ZR-Duet DNA/RNA Miniprep Plus kit. Cq = quantification cycle; Nobuto = whole blood samples processed from Nobuto blood filter papers; Blood = whole blood samples processed directly; MED = medium spiking load; HI = high spiking load.

Table 1. Primer and probe sets used in qPCR analysis for detection of *Trypanosoma cruzi* DNA and β-actin. Cruzi TaqMan assay developed by Piron et al. (2007) [14]. β-actin TaqMan assay developed by Piorkowski et al. (2014) [13].

TaqMan Assay	Primer	Sequence (5′–3′)	Probe	Sequence (5′–3′)
T. cruzi	Cruzi 1	ASTCGGCTGATCGTTTTCGA	Cruzi 3	CACACACTGGACACCAA
	Cruzi 2	AATTCCTCCAGCAGCGGATA		
β-actin	Act. f	GTSTGGATYGGHGGHTCBATC	Act. p	ACCTTCCAGCAGATGTGGATC
	Act. r	GAYTCRTCRTAYTCCTSCTTG		

Figure 2. β-actin DNA recovered from skunk whole blood archived on Nobuto blood filter paper strips. DNA extracted according to protocols outlined in the text for Qiagen DNeasy Blood & Tissue Kit: Method A = extraction optimization method A; Method B = extraction optimization method B; Method C = extraction optimization method C. Cq = quantification cycle.

3. Discussion

Our objective was to evaluate the most effective method for extracting DNA from WB preserved on blood filter paper, comparing three commonly used extraction kits with three modifications on samples from four mammalian species analyzed for two types of DNA. Our comparison of extraction methodologies using spiked and archived specimens demonstrates that using the Qiagen DNeasy Blood & Tissue Kit with an initial 90 °C incubation step and extended 56 °C incubation step after addition of proteinase K provided the optimal recovery of both β-actin and *T. cruzi* DNA from WB stored on Nobuto strips. Incubation for an extended period of time and/or at elevated temperatures during the blood cells lysis and removal steps of DNA extraction from WB can increase the purity of the DNA recovered [15]. Stowell et al. (2018) recovered the most DNA from WB stored on various filter paper types by combining the QIAamp kit, Nobuto strips, and an incubation modification step of post-ATL overnight shaking at 37 °C and a 2-h incubation at 56 °C following application of proteinase K. The authors also report consistently high recovery of DNA with all high incubation modifications they employed on Classic FTA and Nobuto filter papers [16]. By adjusting these conditions within the optimized extraction protocols, we successfully increased the quantity of recovered β-actin and *T. cruzi* DNA detectable by qPCR analysis.

We focused our DNA extraction optimization analyses on Nobuto blood sampling filter paper due to this product's relatively low cost and wide availability through online markets. For instance, Nobuto blood filter paper is currently available through online marketplaces for as low as USD 0.34 per filter strip. The cost-effectiveness and ease with which filter papers can be used to store blood products and other body fluid specimens without pre-preparation or temperature control constraints makes them a convenient collection method for diagnostic sampling in the field, particularly in cases of remote work in warm climates and with geographically-isolated populations [9,10,17]. The ability of filter papers to preserve extractable DNA in WB specimens for multiple years at room temperature further facilitates their application in biobanking and retrospective analyses [10,17]. In practice, DNA stored on filter paper has been extensively employed in population genetics analyses [16,18–21] and wildlife disease detection [22,23], while its application in anthropocentric epidemiology has been largely limited to serological diagnostic screening, pharmaceutical development, and drug monitoring [9,10,24]. Previous field studies reported the limitation of sample volume available for extractions and the reduced integrity, stability, and purity of extracted DNA as potential factors contributing to the loss of sensitivity of PCR-based investigations using blood filter papers [9,10,25]. By increasing the yield of PCR-detectable parasite DNA from WB archived on Nobuto strips, our optimized extraction protocol advances the ease and reliability of using this relatively inexpensive method in field-to-laboratory epidemiological surveillance programs.

Stringent validation guidelines would mandate the use of blood samples from a greater range of mammalian species than our study encompasses to assess the potential influence of matrix heterogeneity on the DNA extraction yield. However, we addressed the putative influence of matrix heterogeneity on DNA recovery indirectly via testing WB collected from a domestic dog and three skunk species. Furthermore, a DNA extraction optimization study using samples of blood from domestic dogs, elk (*Cervus elaphus*), bighorn sheep (*Ovis canadensis*), and mule deer (*Odocoileus hemionus*) did not detect any variations in DNA yield between the taxa when performing extractions from multiple types of filter paper [16]. Blood matrix properties that affect DNA recovery most likely differ at higher phylogenetic levels [16]. Future studies, such as those requiring stringent validation for clinical research purposes or those surveying non-human animals, would benefit from repeating our optimization experiments on samples from several individuals of the same species and/or on a wider variety of species, particularly those from other Classes (e.g., avians, reptiles, amphibians).

Future research would also benefit from evaluating the performance of our optimized extraction protocol for the recovery of DNA from other body fluids and biological samples stored on blood filter paper. In addition to elevating the ease and inclusivity of sampling via non-invasive means of collection, the use of non-WB specimens holds great potential for expansion of NTD surveillance across different stages of disease progression [26–28]. The utility of filter paper for the storage of a diversity of biological samples is illustrated by its use to preserve DNA from manta ray (*Manta birostris*) mucus [19] and human buccal cells [29]. We expect our extended incubation steps to especially elevate DNA yield from urine, semen, breast milk, and other body fluids with complex specimen matrices [28,30].

4. Materials and Methods

4.1. Samples

All WB samples were collected on Type I Advantec® (Tokyo, Japan) Nobuto blood sampling filter paper. Canine WB from a female German Shepherd domestic dog was scavenged from excess WB collected for clinical purposes and donated by a local veterinary hospital (Houston, TX, USA). Skunk WB samples were collected from an American hog-nosed skunk, a striped skunk, and a western spotted skunk as part of a biobanking project conducted by Angelo State Natural History Collections in the Department of Biology, Angelo State University, San Angelo, TX (Supplementary Table S2). *T. cruzi* parasites used for spiking were the Vero cell culture supernatants of the parasite strain TD25 isolated from a triatomine collected in Texas [31].

4.2. Canine WB Samples Spiked with T. cruzi

Fresh WB from a *T. cruzi*-negative domestic dog was spiked with *T. cruzi* parasites cultured on Vero cells. WB was prepared in 1 mL aliquots accordingly: one unspiked control and duplicates for each of the *T. cruzi* spiking loads, medium (MED; expected cycle threshold (Cq = 29–30) and high (HI; expected Cq = 24–25) spiked with 20 μL of the respective dilution of parasite. The medium spiking load (MED) used in our experiments corresponds to Cq values at the lower end of the assay's dynamic range [8]. Medium and high spiking loads were confirmed through controls combining 20 μL of the respective dilution of parasite with 1 mL of phosphate-buffered saline solution. Spiked and control samples were processed in duplicate either directly as WB (50 μL) or post-application to Nobuto blood filter strips. In the latter case, 50 μL of each treatment of WB was applied to Nobuto strips and thoroughly dried in a biosafety cabinet for at least 30 min to replicate field preparation of filter paper specimens. Each Nobuto strip was cut into four, equally-sized pieces that were combined with ATL lysis buffer in a 1.5 mL tube for further processing. One aliquot was processed directly without spiking in order to confirm the negative *T. cruzi* status of the canine patient.

For DNA extractions, we tested three extraction kits to determine optimized methods for extracting *T. cruzi* DNA from the spiked canine WB samples. We followed manufacturer instructions for DNA extraction from WB for the Quick-DNA/RNA Pathogen Miniprep and ZR-Duet DNA/RNA Miniprep Plus kits. Two alternate DNA extraction optimizations were assessed for the DNeasy Blood & Tissue Kit. In extraction optimization method A for the Qiagen kit, samples were mixed with 150 μL or 180 μL of ATL for direct WB or Nobuto strips, respectively, and incubated at 90 °C for 15 min. Following this incubation period, 20 μL of proteinase K was added and the sample was incubated at 56 °C for 60 min. Next, we added 200 μL of AL buffer, incubated the sample for 10 min at 56 °C, vortexed and spun the sample at maximum speed for 1 min, collected the supernatant, and proceeded to follow manufacturer instructions. Extraction optimization Method B resembled extraction optimization method A with the exclusion of the initial 15 min 90 °C incubation period and the reduction of the 56 °C incubation period post-addition of proteinase K from 60 min to 15 min.

For each evaluation of sample DNA content, five microliters of extracted DNA were run in duplicate in a 20-μL reaction with TaqMan Fast Advanced Master Mix (Thermo Fisher Scientific, Waltham, MA, USA) on a ViiA 7 Real Time PCR System (Thermo Fisher Scientific). Duplicate extraction negative controls, extraction positive controls, and no template controls were included in each qPCR analysis. Mean DNA recovery estimates were calculated from the average Cq values from the duplicates of each sample tested.

Detection of *T. cruzi* was performed using primers Cruzi 1 and Cruzi 2 and the probe Cruzi 3 described by Piron et al. (2007) [14], which amplifies a fragment of 166 base pairs (bp) in the satellite DNA of all *T. cruzi* lineages (Table 1). The assay exhibits a specificity of 100% and a four-log dynamic range with its lower end positioned at 10 parasites/mL of blood corresponding to Cq values about 30 [14].

4.3. DNA Extraction from Skunk WB Samples

In preparation for DNA extraction, the biobanked skunk Nobuto strips were processed to ensure equivalent quantities of WB were sampled per extraction replicate from each strip. One section approximately 5 × 5 mm in length was cut from each WB-saturated Nobuto strip and further divided into four, equally-sized pieces that were transferred into a 1.5 mL tube with ATL lysis buffer for continued processing. DNA was extracted from each strip in accordance with extraction optimization A and B as described above. In addition, we evaluated a third extraction optimization with the Qiagen DNeasy Blood & Tissue Kit for each of these archived WB specimens. Extraction optimization C replicates the steps of extraction optimization B with extension of the 56 °C incubation period following the addition of proteinase K from 15 min to 16 h. The DNA extracted from each treatment was then run in duplicate on qPCR with negative and positive controls.

5. Conclusions

Optimization of DNA extraction from WB preserved on blood filter papers can extend the reach, efficacy, and reliability of infectious disease research and surveillance, particularly for investigations in which field constraints typically limit the capacity for body fluid collection and preservation. Developing these procedures for use with commercially-available and user-friendly kits and Nobuto strips further enhances their cost-efficiency and simplicity of application to molecular epidemiological studies. We identify one such optimized DNA extraction method: the Qiagen DNeasy Blood & Tissue Kit with the addition of an initial 90 °C incubation step and extended, post-proteinase K 56 °C incubation step, which provided the most accurate and precise recovery of *T. cruzi* DNA. Future work should adopt and continuously adapt this protocol to maximize the sensitivity of these techniques across a range of storage and extraction products, body fluid specimens, target pathogens, and host species sampled.

Supplementary Materials: The following are available online at https://www.mdpi.com/article/10.3390/pathogens10081040/s1, Table S1: Quantitative PCR results for *Trypanosoma cruzi* assays performed on spiked canine whole blood specimens; Table S2: Collection information pertaining to skunk whole blood samples used in DNA extraction optimization testing; Table S3: Quantitative PCR results for β-actin assays performed on skunk whole blood samples archived on Nobuto blood filter strips and processed using three optimized Qiagen DNeasy Blood & Tissue Kit DNA extraction methods as detailed in text.

Author Contributions: Conceptualization, R.G. and B.E.G.-W.; methodology, R.G.; software; validation, formal analysis, investigation, and data curation, B.E.G.-W., R.B.K., R.G. and A.W.; writing—original draft preparation, B.E.G.-W.; writing—review and editing, B.E.G.-W., R.B.K., R.G., A.W. and K.O.M.; visualization, B.E.G.-W., R.B.K. and R.G.; supervision, project administration, funding acquisition, and resources: K.O.M. All authors have read and agreed to the published version of the manuscript.

Funding: This research was funded in part from NIH/NIAID, grant number R03AI123650.

Institutional Review Board Statement: This study was classified as exempt by the Institutional Animal Care and Use Committee (IACUC) at Texas A&M University since our methodology did not constitute "use of animals" as defined by IACUC.

Informed Consent Statement: Not applicable.

Acknowledgments: We thank Melinda D. Luper from Fur & Feather Veterinary Hospital for the donation of canine WB used in our spiking experiment. We are also grateful to R.C. Dowler and the Angelo State Natural History Collections for access to archived skunk WB samples. Finally, we would like to thank the William T. Shearer for Human Immunobiology for providing the facilities and resources needed to carry out this study. This study was integrated into B.E.G.-W.'s PhD dissertation, which was approved by the Department of Wildlife and Fisheries Sciences, Texas A&M University. Finally, we are grateful for suggestions from four anonymous reviewers that greatly improved the quality of this manuscript.

Conflicts of Interest: The authors declare no conflict of interest. The funders had no role in the design of the study; in the collection, analyses, or interpretation of data; in the writing of the manuscript, or in the decision to publish the results.

References

1. Bern, C.; Messenger, L.A.; Whitman, J.D.; Maguire, J.H. Chagas Disease in the United States: A Public Health Approach. *Clin. Microbiol. Rev.* **2019**, *33*, e00023-19. [CrossRef]
2. World Health Organization. Chagas Disease (American Trypanosomiasis), Fact Sheet No. 340. 2020. Available online: https://www.who.int/en/news-room/fact-sheets/detail/chagas-disease-(american-trypanosomiasis) (accessed on 19 November 2020).
3. Hanford, E.J.; Zhan, F.B.; Lu, Y.; Giordano, A. Chagas disease in Texas: Recognizing the significance and implications of evidence in the literature. *Soc. Sci. Med.* **2007**, *65*, 60–79. [CrossRef]
4. Sarkar, S.; Strutz, S.E.; Frank, D.M.; Rivaldi, C.L.; Sissel, B.; Sánchez-Cordero, V. Chagas disease risk in Texas. *PLoS Negl. Trop. Dis.* **2010**, *4*, e836. [CrossRef] [PubMed]
5. Hotez, P.J.; Bottazzi, M.E.; Dumonteil, E.; Valenzuela, J.G.; Kamhawi, S.; Ortega, J.; Rosales, S.P.D.L.; Cravioto, M.B.; Tapia-Conyer, R. Texas and Mexico: Sharing a Legacy of Poverty and Neglected Tropical Diseases. *PLoS Negl. Trop. Dis.* **2012**, *6*, e1497. [CrossRef] [PubMed]
6. Hotez, P.J.; Dumonteil, E.; Woc-Colburn, L.; Serpa, J.A.; Bezek, S.; Edwards, M.S.; Hallmark, C.J.; Musselwhite, L.W.; Flink, B.J.; Bottazzi, M.E. Chagas disease: "The new HIV/AIDS of the Americas". *PLoS Negl. Trop. Dis.* **2012**, *6*, e1498. [CrossRef] [PubMed]
7. Daszak, P.; Cunningham, A.A.; Hyatt, A.D. Emerging infectious diseases of wildlife threats to endangerment. *Conserv. Biol.* **2000**, *20*, 1349–1357.
8. Cunningham, A.A.; Daszak, P.; Wood, J.L. One Health, emerging infectious diseases and wildlife: Two decades of progress? *Philos. Trans. R. Soc. B Biol. Sci.* **2017**, *372*, 20160167. [CrossRef]
9. Bereczky, S.; Färnert, A.; Mårtensson, A.; Gil, J.P. Short report: Rapid dna extraction from archive blood spots on filter paper for genotyping of plasmodium falciparum. *Am. J. Trop. Med. Hyg.* **2005**, *72*, 249–251. [CrossRef]
10. Lim, M.D. Dried Blood Spots for Global Health Diagnostics and Surveillance: Opportunities and Challenges. *Am. J. Trop. Med. Hyg.* **2018**, *99*, 256–265. [CrossRef]
11. Gyapong, J.; Gyapong, M.; Yellu, N.; Anakwah, K.; Amofah, G.; Bockarie, M.; Adjei, S. Integration of control of neglected tropical diseases into health-care systems: Challenges and opportunities. *Lancet* **2010**, *375*, 160–165. [CrossRef]
12. Ehrenberg, J.P.; Zhou, X.-N.; Fontes, G.; Rocha, E.M.M.; Tanner, M.; Utzinger, J. Strategies supporting the prevention and control of neglected tropical diseases during and beyond the COVID-19 pandemic. *Infect. Dis. Poverty* **2020**, *9*, 1–7. [CrossRef] [PubMed]
13. Piorkowski, G.; Baronti, C.; de Lamballerie, X.; de Fabritus, L.; Bichaud, L.; Pastorino, B.; Bessaud, M. Development of generic Taqman PCR and RT-PCR assays for the detection of DNA and mRNA of β-actin-encoding sequences in a wide range of animal species. *J. Virol. Methods* **2014**, *202*, 101–105. [CrossRef]
14. Piron, M.; Fisa, R.; Casamitjana, N.; López-Chejade, P.; Puig, L.; Vergés, M.; Gascon, J.; Prat, J.G.I.; Portús, M.; Sauleda, S. Development of a real-time PCR assay for Trypanosoma cruzi detection in blood samples. *Acta Trop.* **2007**, *103*, 195–200. [CrossRef]
15. Qamar, W.; Khan, M.R.; Arafah, A. Optimization of conditions to extract high quality DNA for PCR analysis from whole blood using SDS-proteinase K method. *Saudi J. Biol. Sci.* **2017**, *24*, 1465–1469. [CrossRef] [PubMed]
16. Stowell, S.M.L.; Bentley, E.G.; Gagne, R.B.; Gustafson, K.D.; Rutledge, L.Y.; Ernest, H.B. Optimal DNA extractions from blood on preservation paper limits conservation genomic but not conservation genetic applications. *J. Nat. Conserv.* **2018**, *46*, 89–96. [CrossRef]
17. Michaud, V.; Gil, P.; Kwiatek, O.; Prome, S.; Dixon, L.; Romero, L.; Le Potier, M.-F.; Arias, M.; Couacy-Hymann, E.; Roger, F.; et al. Long-term storage at tropical temperature of dried-blood filter papers for detection and genotyping of RNA and DNA viruses by direct PCR. *J. Virol. Methods* **2007**, *146*, 257–265. [CrossRef]
18. Sacks, B.N.; Brown, S.K.; Ernest, H.B. Population structure of California coyotes corresponds to habitat-specific breaks and illuminates species history. *Mol. Ecol.* **2004**, *13*, 1265–1275. [CrossRef]
19. Kashiwagi, T.; Maxwell, E.A.; Marshall, A.; Christensen, A.B. Evaluating manta ray mucus as an alternative DNA source for population genetics study: Underwater-sampling, dry-storage and PCR success. *PeerJ* **2015**, *3*, e1188. [CrossRef]
20. Nunziata, S.O.; Wallenhorst, P.; Barrett, M.A.; Junge, R.E.; Yoder, A.D.; Weisrock, D.W. Population and conserva-tion genetics in an Endangered lemur, Indri indri, across three forest reserves in Madagascar. *Int. J. Primatol.* **2016**, *37*, 688–702. [CrossRef]

21. Stowell, S.M.L.; Gagne, R.B.; McWhirter, D.; Edwards, W.; Ernest, H.B. Bighorn sheep genetic structure in Wyoming reflects geography and management. *J. Wildl. Manag.* **2020**, *84*, 1072–1090. [CrossRef]
22. Forzán, M.; Wood, J. Low detection of ranavirus dna in wild postmetamorphic green frogs, rana (lithobates) clamitans, despite previous or concurrent tadpole mortality. *J. Wildl. Dis.* **2013**, *49*, 879–886. [CrossRef]
23. LeClaire, S.; Menard, S.; Berry, A. Molecular characterization of Babesia and Cytauxzoon species in wild South-African meerkats. *Parasitology* **2014**, *142*, 543–548. [CrossRef] [PubMed]
24. Lindstrom, B.; Ericsson, O.; Alvan, G.; Rombo, L.; Ekman, L.; Rais, M.; Sjoqvist, F. Determination of chloroquine and its de-sethyl metabolite in whole blood: An application for samples collected in capillary tubes and dried on filter paper. *Ther. Drug Monit.* **1985**, *7*, 207–210. [CrossRef] [PubMed]
25. Färnert, A.; Arez, A.P.; Correia, A.T.; Björkman, A.; Snounou, G.; Rosário, V.D. Sampling and storage of blood and the detection of malaria parasites by polymerase chain reaction. *Trans. R. Soc. Trop. Med. Hyg.* **1999**, *93*, 50–53. [CrossRef] [PubMed]
26. Mfuh, K.O.; Yunga, S.T.; Esemu, L.F.; Bekindaka, O.N.; Yonga, J.; Djontu, J.C.; Mbakop, C.D.; Taylor, D.W.; Nerurkar, V.R.; Leke, R.G.F. Detection of Plasmodium falciparum DNA in saliva samples stored at room temperature: Potential for a non-invasive saliva-based diagnostic test for malaria. *Malar. J.* **2017**, *16*, 434. [CrossRef]
27. Bezerra, G.S.N.; Barbosa, W.L.; da Silva, E.D.; Leal, N.C.; Medeiros, Z. Urine as a promising sample for Leishmania DNA extraction in the diagnosis of visceral leishmaniasis—A review. *Braz. J. Infect. Dis.* **2019**, *23*, 111–120. [CrossRef]
28. Ronca, S.E.; Gulas-Wroblewski, B.E.; Kairis, R.B.; Murray, K.O. RNA extraction techniques of different body fluids for Zika virus: Blood, genitourinary specimens, saliva, and other relevant fluids. In *Zika Virus Impact, Diagnosis, Control, and Models*; Academic Press: Cambridge, MA, USA, 2021; pp. 243–253.
29. He, H.; Argiro, L.; Dessein, H.; Chevillard, C. Improved technique that allows the performance of large-scale SNP genotyping on DNA immobilized by FTA® technology. *Infect. Genet. Evol.* **2007**, *7*, 128–132. [CrossRef]
30. Gorchakov, R.; Gulas-Wroblewski, B.E.; Ronca, S.E.; Ruff, J.C.; Nolan, M.S.; Berry, R.; Alvarado, R.E.; Gunter, S.M.; Murray, K.O. Optimizing PCR Detection of West Nile Virus from Body Fluid Specimens to Delineate Natural History in an Infected Human Cohort. *Int. J. Mol. Sci.* **2019**, *20*, 1934. [CrossRef]
31. Talavera-López, C.; Messenger, L.A.; Lewis, M.D.; Yeo, M.; Reis-Cunha, J.L.; Matos, G.M.; Bartholomeu, D.C.; Calzada, J.E.; Saldaña, A.; Ramírez, J.D.; et al. Repeat-Driven Generation of Antigenic Diversity in a Major Human Pathogen, Trypanosoma cruzi. *Front. Cell. Infect. Microbiol.* **2021**, *11*. [CrossRef]

pathogens

Article

Antitrypanosomal and Antileishmanial Activity of Chalcones and Flavanones from *Polygonum salicifolium*

Ahmed M. Zheoat [1,2], Samya Alenezi [1], Ehab Kotb Elmahallawy [3,4], Marzuq A. Ungogo [3,5], Ali H. Alghamdi [3,6], David G. Watson [1], John O. Igoli [1,7], Alexander I. Gray [1], Harry P. de Koning [3,*] and Valerie A. Ferro [1]

1   Strathclyde Institute of Pharmacy and Biomedical Sciences, University of Strathclyde, Glasgow G4 0RE, UK; ahmedpharm.2013@gmail.com (A.M.Z.); s-alenezi@strath.ac.uk (S.A.); D.G.Watson@strath.ac.uk (D.G.W.); igolij@gmail.com (J.O.I.); a.i.gray345@gmail.com (A.I.G.); v.a.ferro@strath.ac.uk (V.A.F.)
2   Al-Manara College for Medical Sciences, Misan 10028, Iraq
3   Institute of Infection, Immunity and Inflammation, College of Medical, Veterinary and Life Sciences, University of Glasgow, Glasgow G12 8TA, UK; eehaa@unileon.es (E.K.E.); 2226184U@student.gla.ac.uk (M.A.U.); Ayfan24@hotmail.com (A.H.A.)
4   Department of Zoonoses, Faculty of Veterinary Medicine, Sohag University, Sohag 82524, Egypt
5   Department of Veterinary Pharmacology and Toxicology, Faculty of Veterinary Medicine, Ahmadu Bello University, Zaria 810107, Nigeria
6   Biology Department, Faculty of Science, Albaha University, Albaha 7738-65799, Saudi Arabia
7   Phytochemistry Research Group, Department of Chemistry, University of Agriculture, Makurdi 2373, Nigeria
*   Correspondence: Harry.de-Koning@glasgow.ac.uk

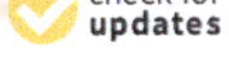

check for updates

Citation: Zheoat, A.M.; Alenezi, S.; Elmahallawy, E.K.; Ungogo, M.A.; Alghamdi, A.H.; Watson, D.G.; Igoli, J.O.; Gray, A.I.; de Koning, H.P.; Ferro, V.A. Antitrypanosomal and Antileishmanial Activity of Chalcones and Flavanones from *Polygonum salicifolium*. *Pathogens* **2021**, *10*, 175. https://doi.org/10.3390/pathogens10020175

Academic Editors: Vyacheslav Yurchenko and Vito Colella

Received: 9 December 2020
Accepted: 2 February 2021
Published: 5 February 2021

Publisher's Note: MDPI stays neutral with regard to jurisdictional claims in published maps and institutional affiliations.

Abstract: Trypanosomiasis and leishmaniasis are a group of neglected parasitic diseases caused by several species of parasites belonging to the family Trypansomatida. The present study investigated the antitrypanosomal and antileishmanial activity of chalcones and flavanones from *Polygonum salicifolium*, which grows in the wetlands of Iraq. The phytochemical evaluation of the plant yielded two chalcones, 2′,4′-dimethoxy-6′-hydroxychalcone and 2′,5′-dimethoxy-4′,6′-dihydroxychalcone, and two flavanones, 5,7-dimethoxyflavanone and 5,8-dimethoxy-7-hydroxyflavanone. The chalcones showed a good antitrypanosomal and antileishmanial activity while the flavanones were inactive. The EC_{50} values for 2′,4′-dimethoxy-6′-hydroxychalcone against *Trypanosoma brucei brucei* (0.5 µg/mL), *T. congolense* (2.5 µg/mL), and *Leishmania mexicana* (5.2 µg/mL) indicated it was the most active of the compounds. None of the compounds displayed any toxicity against a human cell line, even at 100 µg/mL, or cross-resistance with first line clinical trypanocides, such as diamidines and melaminophenyl arsenicals. Taken together, our study provides significant data in relation to the activity of chalcones and flavanones from *P. salicifolium* against both parasites in vitro. Further future research is suggested in order to investigate the mode of action of the extracted chalcones against the parasites.

Keywords: *Polygonum salicifolium*; chalcone; flavanone; *Leishmania mexicana*; *Trypanosoma brucei brucei*; *Trypanosoma congolense*

1. Introduction

The marshes of Iraq are considered to be the largest ecosystem in the Middle East and Western Eurasia [1]. More than one hundred species of aquatic and amphibious plants have been recorded in this area. Ancient Mesopotamians used some of the plants for a range of medicinal and culinary purposes. In modern Mesopotamia, Marsh Arabs also use plants from the marshes for medicinal and healing purposes [1]. *Polygonum salicifolium* is a common species found in the wetlands, and it is an important source of food in the localities [2]. There is little information on the phytochemical constituents and the potential biological activities of this plant. Previous studies on *P. salicifolium* indicated flavonoid glycosides [3] and flavonol glycosides [4] as predominant constituents in the aerial parts of the plant. These classes of compounds are well known for their antioxidant effects [4]. Midiwo et al. [5] described the wide use of Polygonaceae as ethnobotanical treatments for a variety of wounds and ailments, including as anthelminthics. Specifically, four *Polygonum* spp. were mentioned, with applications against ectoparasites and syphilis among other uses [5,6]. *Polygonum acuminatum* Kunth has also been reported to possess antimalarial properties [7]. Very recently, other antiparasite properties of *Polygonum* species have been reported, including anthelminthic [8,9] and antiprotozoal activities, the latter against an important parasite of fresh water fish, *Ichthyophthirius multifiliis* [10]. However, *Polygonum*-derived antiparasitics have yet to be isolated and positively identified.

This report focusses on the isolation of phytochemicals from the aerial parts of *P. salicifolium*, and their antiparasitic activities. Specifically, we tested compounds against *Trypanosoma brucei brucei*, *T. congolense*, and *Leishmania mexicana*. The *T. brucei* subspecies *T. b. gambiense* and *T. b. rhodesiense* are the etiological agents of human African trypanosomiasis [11], commonly known as sleeping sickness, whereas *T. congolense* is the most important pathogen causing the important livestock disease, nagana, in Sub-Saharan Africa [12]. *L. mexicana* is one of the parasites causing cutaneous leishmaniasis, a condition that is highly prevalent throughout the Middle East [13], whereas visceral leishmaniasis is also increasing around the Mediterranean Sea [14] and in Black Sea countries [15,16]. Drugs for these conditions are old and inadequate [17–19], and the current report is part of our investigations into whether new medicines can be developed from local medicinal plants [20–23] or propolis [24–26] as a sustainable solution for developing countries [27], and to validate ethnopharmacological practice [21].

2. Results

2.1. Isolation and Identification of Compounds

The compound 2′,4′-dimethoxy-6′-hydroxychalcone (**1**) (Figure 1) was obtained as a yellow solid (50 mg) from the combined column fractions 9–11 of the hexane extract. It was purified by preparative thin layer chromatography (PTLC), using 30% EtOAc in hexane as the mobile phase. On TLC (Rf = 0.48), the compound appeared as a yellow spot, but under short UV, it appeared as a dark spot. However, the compound appeared as a brown spot after spraying with p-anisaldehyde-sulfuric acid and heating. Its mass ion $[M+H]^+$ observed at m/z 285.0, suggested a molecular formula of $C_{17}H_{16}O_4$ (Supplementary Materials Figure S1). The proton spectrum indicated the presence of seven aromatic protons, which needed to be from two phenyl rings (Supplementary Materials Figure S2). Based on integration and ^1H-^1H couplings in the COSY spectrum, one of the rings was tetra and the other mono substituted (Supplementary Materials Figure S3). Protons H-3, H-4, and H-5 on the mono-substituted ring appeared as a multiplets between δ_H 7.37 and 7.43 ppm (3H, m), while protons H-2 and H-6 appeared as a doublet at 7.62 (2H, dd, J = 7.6, 1.8). The signals at 3.86 (3H, s) and 3.95 (3H, s) were assigned to 4′- and 2′-OCH$_3$, respectively, while the one at δ_H 14.30 ppm was attributed to the H-bonded or chelated 6′-OH. The meta-coupled aromatic protons at 6.14 (1H, d, J = 2.4) and 5.99 (1H, d, J = 2.4) were assigned to H-5′ and H-3′, respectively. Two trans-coupled olefinic protons were observed at δ_H 7.91 (1H, d, J = 15.6, α-H) and 7.83 (1H, d, J = 15.6, β-H). The carbon spectrum indicated the presence of 12 aromatic and two olefinic carbon signals (Supplementary Materials Figure S4). The signal at δ_C 192.5 was attributed to the carbonyl carbon of the chalcone, while the signals at 127.4 and 142.2 ppm were assigned to the α- and β-olefinic carbons, respectively. The rest of the signals were for the aromatic ring carbons. These assignments were further supported by the HMBC and HSQC spectra for the compound (Supplementary Materials Figure S5 and S6). The hydroxyl proton showed long range correlations (^3J) to C-5′ and C-1′, and (^2J) to C-6′. Hence, it must be attached at C-6′. The methoxy group protons at δ_H 3.95 and 3.86 showed long range correlations to the carbons at δ_C 162.4 (C-2′) and 166.1 (C-4′), respectively; hence, they must also be attached to these carbons. The full chemical shift assignments are given in Supplementary Table S1. It has been previously reported from Kava Plant (*Piper methysticum*) [28].

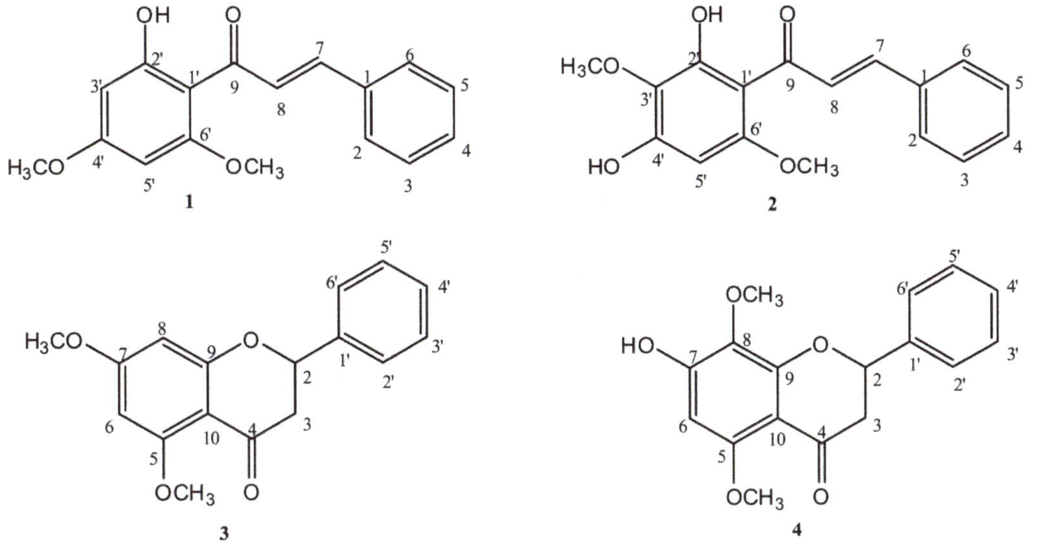

Figure 1. Structures of the test compounds isolated from *Polygonum salicifolium*.

Compound **2**, identified as 2′,5′-dimethoxy-4′,6′-dihydroxychalcone (Figure 1), was also obtained as a yellow solid (40 mg) from the combined column fractions 33–37 of the hexane extract. It was also purified by PTLC using 40% EtOAc in hexane as the mobile phase. Compound **2** also appeared as a yellow spot on TLC (Rf = 0.53). Under short UV, it appeared as a dark spot, which turned into a brown spot after spraying with p-anisaldehyde-sulfuric acid, followed by heating. The positive mode HRLC-MS spectrum showed a molecular ion [M+H]$^+$ at m/z 301.1100 (Calc 301.1076, $C_{17}H_{17}O_5$), suggesting a molecular formula of $C_{17}H_{16}O_5$ (Supplementary Materials Figure S7). The proton spectrum was similar to that of compound **1**, except for the presence of six aromatic protons (Supplementary Materials Figure S8). The difference was due to an extra substitution in the A ring, as the monosubstituted ring protons were still identical. Based on the integration and ^{1}H-^{1}H couplings in the COSY spectrum, this was confirmed by the disappearance of the meta-coupled protons in compound **1**, now replaced by a proton singlet in compound **2** (Supplementary Materials Figure S9). The two methoxy group signals appeared at δ_H 3.83 ppm (3H, s) and 3.86 ppm (3H, s), while the chelated 6′-OH was at 14.28 ppm. The aromatic singlet at δ 5.99 ppm (s) was assigned to H-3′. The trans-olefinic protons were also observed at δ_H 7.82 ppm (1H, d, J = 15.6, α-H) and 7.72 (1H, d, J = 15.6, β-H). The ^{13}C spectrum of this compound showed 17 signals identical to compound **1**, but with one aromatic CH less and replaced by a quaternary carbon signal at 128.4 ppm (Supplementary material Figure S10). Long range correlations in the HMBC (Supplementary Materials Figure S11), indicated the hydroxyl proton at δ_H 14.20 ppm showed correlations with C1′, C-6′, and C-5′, while the methoxy protons at 3.83 and 3.86 showed correlations to C-2′ (δ_C 158.8) and C-5′ (δ_C 128.4), respectively. Using these long-range correlations and the HSQC (Supplementary Materials Figure S12) spectrum, the complete chemical shift assignments (Table S2) were made. Compound **2** was previously isolated from the leaves of *P. limbatum* [29,30].

Compound **3**, identified as 5,7-dimethoxyflavanone (Figure 1), was obtained as a yellow solid (15 mg) from the combined column fractions 70–76 of the hexane extract. It was similarly purified by PTLC using 70% EtOAc in hexane as the mobile phase. On TLC (Rf = 0.37), it appeared as a dark spot when visualized under short UV and as a light blue under long UV. The spot of the compound turned yellow after spraying with p-anisaldehyde-sulfuric acid reagent followed by heating. Its mass ion [M+H]$^+$ was observed at m/z 285.1, suggesting a molecular formula of $C_{17}H_{16}O_4$ (Supplementary Materials Spectrum S13). The proton spectrum showed the presence of seven aromatic protons (Supplementary Materials Spectrum S14), which was suggested to be from a flavanone nucleus. Based on the integration and ^{1}H-^{1}H couplings (Supplementary Materials Figure S15) in the COSY spectrum, the proton signal at δ_H 5.44 ppm (1H, dd, J = 13.1, 2.8) showed an ABX spin system with the axial proton at δ_H 3.05 ppm (1H, dd, J = 16.5, 13.2) and the equatorial one at δ_H 2.83 ppm (1H- dd, J =16.5, 2.8). Protons H-2′, H-3′, H-4′, H-5′, and H-6′ on the mono-substituted B ring appeared as a multiplet, between δ_H 7.37 and 7.45 ppm (5H, m). The meta-coupled aromatic protons at δ_H 6.13 ppm (1H, d, J = 2.3) and 6.19 ppm (1H, d, J = 2.3) were assigned to H-6 and H-8, respectively. The signals at δ 3.85 ppm (3H, s) and 3.98 ppm (3H, s) were assigned to C-7 and C-5-OCH$_3$, respectively. The ^{13}C spectrum of this compound (Supplementary Materials Figure S16) showed the presence of 17 carbon atoms, corresponding to carbon atoms of the flavanone moiety and two methoxy groups. The methoxy group protons at δ_H 3.85 and 3.98 ppm showed long range correlations to the carbons at δc at 166.0 (C-7) and 162.3 (C-5), respectively (Supplementary Materials Figure S17 and S18); hence, they must also be attached to these carbons. These assignments were further supported by the HMBC spectrum for the compound, and the full chemical shift assignments are given in Table S3. Compound **3** was also isolated from Kava *(Piper methysticum)* roots [31].

Compound **4**, identified as 5,8-dimethoxy-7-hydroxyflavanone (Figure 1), was obtained as a white solid (10.0 mg) from the PTLC purification of fractions 90-97 of the hexane extract of *P. salicifolium*, using 10% *(v/v)* MeOH in EtOAc as the mobile phase. On the TLC (Rf = 0.32 in EtOAc: hexane 7:3), it appeared as a dark spot under short UV and a white spot under long UV, which turned yellow upon spraying with p-anisaldehyde-sulfuric acid reagent followed by heating. Its mass ion [M+H]$^+$ at m/z 301.2 supported a molecular formula of $C_{17}H_{16}O_5$ (Supplementary Materials Figure S19). The proton spectrum was similar to that of compound **3**, but showed six aromatic protons (Supplementary Materials Figure S20). The difference was due to an extra substitution in the A ring, whereas the mono-substituted ring protons were still identical. This was confirmed by the disappearance of the meta-coupled protons in compound **3**, now replaced by a proton singlet in compound **4**. The proton signal at δ_H 5.45 ppm (1H, dd, J = 12.9, 3.0) also showed an ABX spin system with the axial proton at δ_H 3.0 ppm (1H, dd, J = 16.6, 13.0) and the equatorial one at δ_H 2.83 ppm (1H, dd, J = 16.6, 3.1; Supplementary Materials Figure S21). The two methoxy groups appeared at δ_H 3.86 and 3.85 ppm. The appearance of the proton at δ_H 6.19 ppm as a singlet (1H, s), and the long-range coupling with a hydroxyl bearing carbon (C-7) and with the carbon bearing a methoxy group (C-5) confirmed the penta substitution of the A ring. The signals between δ_H 7.35 and 7.47 ppm (5H, m) were attributed to the five protons of the unsubstituted B ring. The signals at δ_H 3.85 ppm (3H, s) and 3.86 ppm (3H s,) were assigned to the C-5 and C-8-OCH$_3$ groups, respectively. The ^{13}C spectrum of the compound (Supplementary Materials Spectrum S22) showed the presence of 17 signals, corresponding to carbon atoms of the flavanone structure and two methoxy groups. Using the HMBC and HSQC spectra for the compound (Supplementary Materials Figures S23 and S24), complete chemical shift assignments (Table S4) were made. Compound **4** was previously isolated from the aerial parts of *Polygonum senegalensis* [5].

2.2. Anti-Kinetoplastid and Cytotoxic Activity of the Isolated Compounds

The four compounds were tested for activity against the following three kinetoplastid pathogens: *Trypanosoma brucei brucei*, *Trypanosoma congolense*, and *Leishmania mexicana*. For the two *Trypanosoma* species, the compounds were tested in parallel against strains that were resistant to the most common trypanocides. *T. b. brucei* B48 is highly resistant to the entire classes of diamidines and melaminophenyl arsenicals [28], and *T. congolense* DA-Res was rendered resistant to diminazene aceturate by means of in vitro exposure to the drug. The highest activity was observed against *T. b. brucei*, followed by *T. congolense*, but only a moderate activity was observed against *L. mexicana* (Table 1). For the drug-resistant trypanosome strains, highly significant resistance to diamidines was confirmed, but there was no cross-resistance with the chalcones and flavanones. None of the four compounds displayed measurable toxicity against the human foreskin fibroblast (HFF) cell line at the highest concentration tested (100 µg/mL; Table 1). Compound **1** was the most active against all kinetoplastid species and strains, displaying a promising EC_{50} of 0.58 µg/mL (2.04 µM) against *T. b. brucei* in our standard resazurin-based assay. Chalcone **2** was the second-most active, showing about 8-fold, 4-fold, and 5-fold less activity against *T. b. brucei*, *T. congolense*, and *L. mexicana*, respectively (Table 1). While this dataset of two chalcones is self-evidently insufficient for a structure–activity relationship (SAR) analysis, it is clear that the position of the hydroxy and methoxy groups on ring A (Figure 1) influenced the trypanocidal activity without increasing the toxicity.

Table 1. Anti-kinetoplastid effects of the tested chalcones and flavanones.

		T. b. brucei EC_{50} (µM)		*T. congolense* EC_{50} (µM)		*L. mexicana* EC_{50} (µM)	HFF	
Compound	MW	427-WT	B48	IL3000-WT	DA-Res	WT	EC_{50} (µM)	SI (*Tbb*)
1	284.3	2.04 ± 0.07	1.80 ± 011	8.8 ± 0.39	8.8 ± 0.28	18.2 ± 1.0	>350	>172
2	300.3	14.6 ± 0.80	13.9 ± 0.60	34.0 ± 1.2	28.6 ± 1.4 *	83.6 ± 0.7	>330	>22.8
3	284.3	30.1 ± 0.95	30.3 ± 1.1	137 ± 10	106 ± 15	271 ± 31	>350	>11.7
4	300.3	55.3 ± 1.6	51.9 ± 1.6	63.3 ± 7.9	48.0 ± 6.2	338 ± 9	>330	>6.0
Diminazene		0.15 ± 0.01	2.4 ± 0.36 **	0.15 ± 0.01	1.43 ± 0.03 ***	ND	ND	–
Pentamidine		0.0034 ± 0.0004	0.72 ± 0.03 ***	0.72 ± 0.07	ND	0.56 ± 0.07	ND	–
PAO		ND	ND	ND	ND	ND	1.31 + 0.08	–

All of the data listed are the average and standard error of the mean (SEM) of at least three independent determinations. PAO—phenylarsine oxide; ND—not done. Statistical significance was determined using an unpaired Student's *t*-test; * $p < 0.05$; ** $p < 0.01$; ***, $p < 0.001$. Selectivity index (SI) = EC_{50}(HFF)/EC_{50}(Tbb).

3. Discussion

Leishmaniasis and trypanosomiasis are a heterogeneous group of neglected parasitic diseases of public health concern [29]. These diseases remain endemic in several countries worldwide [29–31]. The search for novel drugs remains one of the major control strategies for combating these diseases [32], as safe and effective treatment of the various forms of leishmaniasis and trypanosomiasis remains a major challenge [19,32]. The present study provides data related to the investigation into the effect of *P. salicifolium* chalcones and flavanones against *Trypanosoma* species and *Leishmania* promastigotes. This work also investigated the effect of compounds against a human cell line and cross-resistance with first line clinical trypanocides such as diamidines and melaminophenyl arsenicals. The phytochemical investigation of the hexane extract of *P. salicifolium* led to the isolation of two chalcones—compound (**1**), and compound (**2**)—and two flavanones—compound (**3**) and compound (**4**). This is an initial report of their isolation from *Polygonum salicifolium*.

The anti-kinetoplastid activities of the compounds isolated from *P. salicifolium* indicate that flavanones **3** and **4** displayed a moderate activity against *Trypanosoma* species, and their activity against *Leishmania* promastigotes was poor. However, chalcone **1** displayed a promising activity against *T. b. brucei* at ~2 µM, as well as a moderate activity against *T. congolense* and a reasonable activity against *L. mexicana*. No cross-resistance with the current trypanocides was observed, which is very important, as resistance to the old anti-kinetoplastid drugs is a key driver of the need for new treatments [11,12,18]. The consistently lower anti-protozoal activity of chalcone **2** suggests that a systematic investigation of the structure–activity relationship of chalcones, including substitutions on the chalcone rings, could yield compounds with substantially improved efficacy against parasites. Importantly, neither of the chalcones showed toxicity against the human cell line HFF at 100 µg/mL, and the in vitro selectivity index of 1 was >172. An important advantage of the chalcones over the flavanones is that their synthesis, and thus SAR, should be more accessible than that of most natural compounds because of the lack of chiral centers [33].

4. Materials and Methods

4.1. General Experimental Procedures

The ^{1}H and ^{13}C NMR spectra were run on a Bruker AVANCE III 500 MHz spectrophotometer, operating at 500 MHz (^{1}H) and 125 MHz (^{13}C), respectively, using $CDCl_3$ as the solvent and TMS as the internal standard. The mass spectra were recorded using a Thermo LTQ Orbitrap, while the exact masses were determined using a Thermo Exactive Orbitrap mass spectrometer. Column chromatographic separations were performed in glass columns using silica gel MN-60 (Macherey-Nagel GmbH, KG, Düren, Germany). TLC and PTLC were carried out using pre-coated silica gel 60 Aluminium sheets (Merck Chemicals, Bedfont Lakes Business Park Feltham, U.K.). The spots on the TLC were visualized using an anisaldehyde-H_2SO_4 reagent. Solvents, hexane, ethyl acetate, and methanol were purchased from Sigma-Aldrich, U.K.

4.2. Collection of Plant Material

The plant *P. salicifolium* Brouss ex Wild was collected from the banks of the River Tigris in Southern Iraq in April 2015, and was identified and deposited at the College of Science, University of Diyala by Assist. Prof. Dr Khazal Dh. Wadi Al-Jibouri.

4.3. Preparation of Extracts

The aerial parts of the plant were dried and finely powdered with an IKA grinder (IKA Werke GmbH and Co. KG, Staufen im Breisgau, Germany). The ground material (50 g) was extracted (500 mL, 72 h each) with hexane, ethyl acetate (EtOAc), and then methanol (MeOH) using a Soxhlet apparatus. The extracts were then filtered and dried at 40 °C using a rotary evaporator (Büchi, Flawil, Switzerland).

4.4. Isolation and Identification of Compounds

The hexane extract (1 g; 2% yield) was subjected to silica gel column chromatography eluting gradient wise with hexane, followed by increasing amounts (10–90% *v/v*) of EtOAc in hexane and then EtOAc (100%). A total of 150 fractions (5 mL each) were collected and, based on TLC results, similar fractions were combined. Further purification of the compounds was carried out using preparative thin layer chromatography (PTLC) with 30–70% (*v/v*) EtOAc in hexane and 10% (*v/v*) MeOH in EtOAc. The characterization of the compounds was carried out using NMR (1D and 2D) on a Bruker AVIII HD 500 spectrophotometer using 5–20 mg samples dissolved in chloroform-*d*. The ethyl acetate and methanol extracts did not yield any significant results in the preliminary assays and spectroscopic analysis, and were thus not purified any further.

4.5. Parasites and Cultures

T. b. brucei strain Lister 427 (427-WT) was used as the standard drug-sensitive (wild-type; WT) laboratory strain [34]. This strain was previously adapted for multi-drug resistance by deletion of the TbAT1/P2 drug transporter [35], subsequently followed by in vitro exposure to increasing concentrations of pentamidine [28]. For *T. congolense*, the culture-adapted Savannah strain IL3000 was used [22], as well as a clonal line, 6C3, adapted from IL3000 in vitro to diminazene aceturate (Sigma), leading to a ~10-fold resistance. All *Trypanosoma* strains were cultured and used as bloodstream forms; *T. b. brucei* in a full HMI-9 medium supplemented with 10% Foetal Bovine Serum (FBS) [27] and *T. congolense* in TC-BSF1 medium with 20% goat serum at 34 °C, as described by Coustou et al. [36]. The promastigotes of *L. mexicana* strain MNY/BZ/62/M379 were cultured at 25 °C in a minimal essential medium, HO-MEMO-MEME, supplemented with 10% FBS and 1% of a penicillin/streptomycin solution (Gibco), as described previously [37,38].

4.6. Anti-Protozoal Drug Testing

The anti-trypanosomal activity of the compounds was tested using the Alamar Blue (resazurin) assay in plastic 96-well plates, as described [39]. The assay was based on the reduction of blue, non-fluorescent resazurin sodium salt (Sigma) by live, but not by dead cells, to the red fluorescent metabolite resorufin [40]. Briefly, dilutions of the test compounds and control drugs were distributed in the first wells of the respective plate rows, and doubling dilution was carried out over two rows in the appropriate medium for *T. b. brucei* or *T. congolense*, leaving the last rows as the drug-free negative control (i.e., 23 doubling dilutions). Then, 10^5 trypanosomes were added to each well, followed by incubation of the plates at 37 °C/5% CO_2 (*T. b. brucei*) or 34.5 °C/5% CO_2 (*T. congolense*) for 48 h before the addition of the resazurin dye (20 µL of 125 mg/L), and a further incubation under the same conditions for 24 h. The *T. congolense* IL3000 WT and the diminazene resistant *T. congolense* IL3000 DA-Res (previously adapted to diminazene; clone 6C3) were utilized. Fluorescence was measured using a FLUOstar Optima plate reader at excitation and emission wavelengths of 544 nm and 590 nm, respectively, and the EC_{50} of the compounds was then calculated using GraphPad Prism 5, using an equation for a sigmoid curve with a variable slope. The assay for *L. mexicana* promastigotes was performed as for *T. b. brucei* [41], except that incubation times of 72 h and 48 h were used before and after the addition of resazurin, respectively, because of the slower metabolism of the dye by *Leishmania* promastigotes [40].

4.7. Drug Toxicity Assay

Toxicity of drugs to mammalian cells was carried out in the human cell line HFF, using a previously described method [41], with slight modifications. Briefly, HFF cells were grown in a medium consisting of 500 mL Dulbecco's Modified Eagle's Medium (DMEM; Sigma), 50 mL new-born calf serum (NBCS; Gibco, Cleveland, TN, USA), 5 mL penicillin/streptomycin (Gibco), and 5 mL L-Glutamax (200 mM, Gibco), at 37 °C/5% CO_2 up to ~80% confluence in vented flasks. For the assay, 100 µL of the cell suspension (3×10^5 cells/mL) was added to each well of a 96-well plate. The plate was incubated at 37 °C/5% CO_2 for 24 h to allow for cell adhesion, after which 100 µL of a serial drug dilution was added (prepared in a separate sterile plate). Phenylarsine oxide (PAO; Sigma) was used as the positive control. The cells were then incubated for a further 30 h before the addition of 10 µL of 125 mg/L resazurin solution, and underwent a final incubation for 24 h. Fluorescence measurements and data analysis were performed, as described above. The selectivity index was calculated as EC_{50} (HFF)/EC_{50}(427-WT).

5. Conclusions

The present data showed that chalcones exhibited interesting activity against *T. b. brucei*; *T. congolense*; and, to a lesser extent, *L. mexicana*, and provide further evidence for the potential uses of natural plant extracts for combating these global parasitic diseases. Future work should concentrate on exploring the SAR of anti-protozoal chalcones and identifying their cellular targets. In addition, future research should test for the potential activity of *P. salicifolium* extracts against different protozoan species of medical and veterinary importance.

Supplementary Materials: The following are available online at https://www.mdpi.com/2076-081 7/10/2/175/s1: Figures Spectrum S1–S24 and Tables S1–S4.

Author Contributions: D.G.W., J.O.I., A.I.G., H.P.d.K. and V.A.F. were involved in the conception of the research idea and performed the methodology design, supervision, data analysis, and interpretation. A.M.Z., S.A., E.K.E., M.A.U. and A.H.A. participated in the methodology, sampling, and data analysis. H.P.d.K. drafted and prepared the manuscript for publication and revision. The funders had no role in the data collection and analysis, decision to publish, or preparation of the manuscript. All authors have read and agreed to the published version of the manuscript.

Funding: E.K.E. was funded through a Newton-Mosharafa Researcher Links Travel Grant awarded through a collaboration between the U.K. government and the Egyptian government, represented by the Science Technology Development Fund (STDF). M.A.U. is funded through a studentship from the Petroleum Technology Development Fund (PTDF), Abuja, Nigeria. We thank the University of Misan, Amarah, Iraq, and the Ministry of Higher Education and Scientific Research of Iraq, for funding the PhD studentship of A.Z. The University of Albaha, Albaha, Saudi Arabia, funded the studentship of A.H.A.

Conflicts of Interest: The authors declare that they have no conflict of interest.

References

1. Al-Mudaffar Fawzi, N.; Goodwin, K.P.; Mahdi, B.A.; Stevens, M.L. Effects of Mesopotamian Marsh (Iraq) desiccation on the cultural knowledge and livelihood of Marsh Arab women. *Ecosyst. Health Sustain.* **2016**, *2*, e01207. [CrossRef]
2. Hamdan, M.A.; Asada, T.; Hassan, F.M.; Warner, B.G.; Douabul, A.; Al-Hilli, M.R.; Alwan, A. Vegetation response to re-flooding in the Mesopotamian Wetlands, Southern Iraq. *Wetlands* **2010**, *30*, 177–188. [CrossRef]
3. Calis, I.; Kuruüzüm, A.; Demirezer, L.Ö.; Sticher, O.; Ganci, W.; Rüedi, P. Phenylvaleric Acid and Flavonoid Glycosides from *Polygonum salicifolium*. *J. Nat. Prod.* **1999**, *62*, 1101–1105. [CrossRef] [PubMed]
4. Hussein, S.R.; Mohamed, A.A. Antioxidant activity and phenolic profiling of two Egyptian medicinal herbs *Polygonum salicifolium* Brouss ex Wild and Polygonum senegalense Meisn. *An. Univ. din Oradea Fasc. Biol.* **2013**, *20*, 59–63.
5. Midiwo, J.O.; Yenesew, A.; Juma, B.; Derese, S.; Ayoo, J.; Aluoch, A.; Guchu, S. Bioactive compounds from some Kenyan ethnomedicinal plants: Myrsinaceae, Polygonaceae and *Psiadia punctulata*. *Phytochem. Rev.* **2002**, *1*, 311–323. [CrossRef]
6. Kokwaro, J.O. *Medicinal Plants of East Africa*; East African Publishing Bureau: Nairobi, Kenya, 1976.
7. Calderón, A.I.; Simithy-Williams, J.; Gupta, M.P. Antimalarial natural products drug discovery in Panama. *Pharm. Biol.* **2012**, *50*, 61–71. [CrossRef]
8. Hu, Y.; Ji, J.; Ling, F.; Chen, Y.; Lu, L.; Zhang, Q.; Wang, G. Screening medicinal plants for use against *Dactylogyrus intermedius* (Monogenea) infection in goldfish. *J. Aquat. Anim. Health* **2014**, *26*, 127–136. [CrossRef] [PubMed]
9. Gürağaç Dereli, F.T.; Ilhan, M.; Kozan, E.; Küpeli Akkol, E. Effective eradication of pinworms (*Syphacia obvelata* and *Aspiculuris tetraptera*) with *Polygonum cognatum* Meissn. *Exp. Parasitol.* **2019**, *196*, 63–67. [CrossRef] [PubMed]
10. Zhou, S.Y.; Liu, Y.M.; Zhang, Q.Z.; Fu, Y.W.; Lin, D.J. Evaluation of an antiparasitic compound extracted from *Polygonum cuspidatum* against *Ichthyophthirius multifiliis* in grass carp. *Vet. Parasitol.* **2018**, *253*, 22–25. [CrossRef]
11. De Koning, H.P. The drugs of sleeping sickness: Their mechanisms of action and resistance, and a brief history. *Trop. Med. Infect. Dis.* **2020**, *5*, 14. [CrossRef] [PubMed]
12. Giordani, F.; Morrison, L.J.; Rowan, T.G.; De Koning, H.P.; Barrett, M.P. The animal trypanosomiases and their chemotherapy: A review. *Parasitology* **2016**, *143*, 1862–1889. [CrossRef]
13. Tabbabi, A. Review of leishmaniasis in the Middle East and North Africa. *Afr. Health Sci.* **2019**, *19*, 1329–1337. [CrossRef]
14. Monge-Maillo, B.; Norman, F.F.; Cruz, I.; Alvar, J.; Lopez-Velez, R. Visceral leishmaniasis and HIV coinfection in the Mediterranean region. *PLoS Negl. Trop. Dis.* **2014**, *8*, e3021. [CrossRef] [PubMed]
15. Babuadze, G.; Alvar, J.; Argaw, D.; De Koning, H.P.; Iosava, M.; Kekelidze, M.; Tsertsvadze, N.; Tsereteli, D.; Chakhunashvili, G.; Mamatsashvili, T. Epidemiology of visceral leishmaniasis in Georgia. *PLoS Negl. Trop. Dis.* **2014**, *8*, e2725. [CrossRef]
16. Babuadze, G.; Farlow, J.; De Koning, H.P.; Carrillo, E.; Chakhunashvili, G.; Murskvaladze, M.; Kekelidze, M.; Karseladze, I.; Kokaia, N.; Kalandadze, I. Seroepidemiology and molecular diversity of *Leishmania donovani* complex in Georgia. *Parasites Vectors* **2016**, *9*, 279. [CrossRef] [PubMed]
17. Delespaux, V.; De Koning, H.P. Drugs and drug resistance in African trypanosomiasis. *Drug Resist. Updates* **2007**, *10*, 30–50. [CrossRef]
18. Burza, S.; Croft, S.L.; Boelaert, M. Leishmaniasis–Authors' reply. *Lancet* **2019**, *393*, 872–873. [CrossRef]
19. Elmahallawy, E.K.; Agil, A. Treatment of leishmaniasis: A review and assessment of recent research. *Curr. Pharm. Des.* **2015**, *21*, 2259–2275. [CrossRef]
20. Dike, V.T.; Vihiior, B.; Bosha, J.A.; Yin, T.M.; Ebiloma, G.U.; De Koning, H.P.; Igoli, J.O.; Gray, A.I. Antitrypanosomal activity of a novel taccalonolide from the tubers of *Tacca leontopetaloides*. *Phytochem. Anal.* **2016**, *27*, 217–221. [CrossRef]
21. Ebiloma, G.U.; Igoli, J.O.; Katsoulis, E.; Donachie, A.-M.; Eze, A.; Gray, A.I.; De Koning, H.P. Bioassay-guided isolation of active principles from Nigerian medicinal plants identifies new trypanocides with low toxicity and no cross-resistance to diamidines and arsenicals. *J. Ethnopharmacol.* **2017**, *202*, 256–264. [CrossRef]
22. Ebiloma, G.U.; Katsoulis, E.; Igoli, J.O.; Gray, A.I.; De Koning, H.P. Multi-target mode of action of a Clerodane-type diterpenoid from *Polyalthia longifolia* targeting African trypanosomes. *Sci. Rep.* **2018**, *8*, 4613. [CrossRef] [PubMed]
23. Ebiloma, G.U.; Ayuga, T.D.; Balogun, E.O.; Gil, L.A.; Donachie, A.; Kaiser, M.; Herraiz, T.; Inaoka, D.K.; Shiba, T.; Harada, S. Inhibition of trypanosome alternative oxidase without its N-terminal mitochondrial targeting signal (ΔMTS-TAO) by cationic and non-cationic 4-hydroxybenzoate and 4-alkoxybenzaldehyde derivatives active against *T. brucei* and *T. congolense*. *Eur. J. Med. Chem.* **2018**, *150*, 385–402. [CrossRef]
24. Omar, R.; Igoli, J.O.; Zhang, T.; Gray, A.I.; Ebiloma, G.U.; Clements, C.J.; Fearnley, J.; Ebel, R.E.; Paget, T.; De Koning, H.P. The chemical characterization of Nigerian propolis samples and their activity against *Trypanosoma brucei*. *Sci. Rep.* **2017**, *7*, 923. [CrossRef]

25. Siheri, W.; Ebiloma, G.U.; Igoli, J.O.; Gray, A.I.; Biddau, M.; Akrachalanont, P.; Alenezi, S.; Alwashih, M.A.; Edrada-Ebel, R.; Muller, S. Isolation of a novel flavanonol and an alkylresorcinol with highly potent anti-trypanosomal activity from Libyan propolis. *Molecules* **2019**, *24*, 1041. [CrossRef] [PubMed]

26. Siheri, W.; Zhang, T.; Ebiloma, G.U.; Biddau, M.; Woods, N.; Hussain, M.Y.; Clements, C.J.; Fearnley, J.; Ebel, R.E.; Paget, T. Chemical and antimicrobial profiling of propolis from different regions within Libya. *PLoS ONE* **2016**, *11*, e0155355. [CrossRef] [PubMed]

27. Cerone, M.; Uliassi, E.; Prati, F.; Ebiloma, G.U.; Lemgruber, L.; Bergamini, C.; Watson, D.G.; de AM Ferreira, T.; Roth Cardoso, G.S.H.; Soares Romeiro, L.A. Discovery of sustainable drugs for neglected tropical diseases: Cashew nut shell liquid (CNSL)-based hybrids target mitochondrial function and ATP production in *Trypanosoma brucei*. *ChemMedChem* **2019**, *14*, 621–635. [CrossRef]

28. Jhoo, J.; Freeman, J.P.; Heinze, T.M.; Moody, J.D.; Schnackenberg, R.K.; Berger, R.D.; Dragull, K.; Tang, C.; Ang, C.Y.W. In vitro cytotoxicity of nononpolar constituents from different parts of kava plant (*Piper methysticum*). *J. Agric. Food Chem.* **2006**, *54*, 3157–3162. [CrossRef] [PubMed]

29. Dzoyem, J.P.; NKuete, A.H.; Kuete, V.; Tala, M.F.; Wabo, H.K.; Guru, S.K.; Rajput, V.S.; Sharma, A.; Tane, P.; Khan, I.A. Cytotoxicity and antimicrobial activity of the methanol extract and compounds from *Polyg. Limbatum*. *Planta Med.* **2012**, *78*, 787–792. [PubMed]

30. Dzoyem, J.P.; Nkuete, A.H.; Ngameni, B.; Eloff, J.N. Anti-inflammatory and anticholinesterase activity of six flavonoids isolated from *Polygonum* and *Dorstenia* species. *Arch. Pharmacal. Res.* **2017**, *40*, 1129–1134. [CrossRef] [PubMed]

31. Xuan, T.D.; Fukuta, M.; Wei, A.C.; Elzaawely, A.A.; Khanh, T.D.; Tawata, S. Efficacy of extracting solvents to chemical components of kava (*Piper methysticum*) roots. *J. Nat. Med.* **2008**, *62*, 188. [CrossRef]

32. Mitra, A.K.; Mawson, A.R. Neglected tropical diseases: Epidemiology and global burden. *Trop. Med. Infect. Dis.* **2017**, *2*, 36. [CrossRef] [PubMed]

33. De Koning, H.P.; MacLeod, A.; Barrett, M.P.; Cover, B.; Jarvis, S.M. Further evidence for a link between melarsoprol resistance and P2 transporter function in African trypanosomes. *Mol. Biochem. Parasitol.* **2000**, *106*, 181–185. [CrossRef]

34. Matovu, E.; Stewart, M.L.; Geiser, F.; Brun, R.; Mäser, P.; Wallace, L.J.; Burchmore, R.J.; Enyaru, J.C.; Barrett, M.P.; Kaminsky, R. Mechanisms of arsenical and diamidine uptake and resistance in *Trypanosoma brucei*. *Eukaryot. Cell* **2003**, *2*, 1003–1008. [CrossRef]

35. Coustou, V.; Guegan, F.; Plazolles, N.; Baltz, T. Complete in vitro life cycle of *Trypanosoma congolense*: Development of genetic tools. *PLoS Negl. Trop. Dis.* **2010**, *4*, e618. [CrossRef] [PubMed]

36. Al-Salabi, M.I.; Wallace, L.J.; De Koning, H.P. A *Leishmania major* nucleobase transporter responsible for allopurinol uptake is a functional homolog of the *Trypanosoma brucei* H2 transporter. *Mol. Pharmacol.* **2003**, *63*, 814–820. [CrossRef]

37. Alzahrani, K.J.; Matyugina, E.S.; Khandazhinskaya, A.L.; Kochetkov, S.N.; Seley-Radtke, K.L.; De Koning, H.P. Evaluation of the antiprotozoan properties of 5′-norcarbocyclic pyrimidine nucleosides. *Bioorg. Med. Chem. Lett.* **2017**, *27*, 3081–3086. [CrossRef] [PubMed]

38. Eze, A.A.; Igoli, J.; Gray, A.I.; Skellern, G.G.; De Koning, H.P. The individual components of commercial isometamidium do not possess stronger trypanocidal activity than the mixture, nor bypass isometamidium resistance. *Int. J. Parasitol. Drugs Drug Resist.* **2019**, *9*, 54–58. [CrossRef]

39. Gould, M.K.; Vu, X.L.; Seebeck, T.; De Koning, H.P. Propidium iodide-based methods for monitoring drug action in the kinetoplastidae: Comparison with the Alamar Blue assay. *Anal. Biochem.* **2008**, *382*, 87–93. [CrossRef]

40. Alzahrani, K.J.; Ali, J.A.; Eze, A.A.; Looi, W.L.; Tagoe, D.N.; Creek, D.J.; Barrett, M.P.; De Koning, H.P. Functional and genetic evidence that nucleoside transport is highly conserved in *Leishmania* species: Implications for pyrimidine-based chemotherapy. *Int. J. Parasitol. Drugs Drug Resist.* **2017**, *7*, 206–226. [CrossRef]

41. Rodenko, B.; Wanner, M.J.; Alkhaldi, A.A.; Ebiloma, G.U.; Barnes, R.L.; Kaiser, M.; Brun, R.; McCulloch, R.; Koomen, G.-J.; De Koning, H.P. Targeting the parasite's DNA with methyltriazenyl purine analogs is a safe, selective, and efficacious antitrypanosomal strategy. *Antimicrob. Agents Chemother.* **2015**, *59*, 6708–6716. [CrossRef]

Article

Comparison of Diagnostic Tests for *Onchocerca volvulus* in the Democratic Republic of Congo

An Hotterbeekx [1,*], Jolien Perneel [1], Michel Mandro [2,3], Germain Abhafule [3], Joseph Nelson Siewe Fodjo [1], Alfred Dusabimana [1], Steven Abrams [1], Samir Kumar-Singh [4] and Robert Colebunders [1]

1   Global Health Institute, University of Antwerp, 2000 Antwerp, Belgium; jolien.perneel@student.uantwerpen.be (J.P.); JosephNelson.SieweFodjo@uantwerpen.be (J.N.S.F.); Alfred.Dusabimana@uantwerpen.be (A.D.); Steven.Abrams@uantwerpen.be (S.A.); robert.colebunders@uantwerpen.be (R.C.)
2   Provincial Health Division Ituri, Ministry of Health, Bunia 185 DRC 57, Democratic Republic of Congo; Michel.MandroNdahura@student.uantwerpen.be
3   Centre de Recherche en Maladies Tropicales, Rethy Box 143, Democratic Republic of Congo; abhafule@gmail.com
4   Molecular Pathology Group, Laboratory of Cell Biology & Histology, Faculty of Medicine and Health Sciences, University of Antwerp, 2000 Antwerp, Belgium; Samir.Kumar-Singh@uantwerpen.be
*   Correspondence: an.hotterbeekx@uantwerpen.be; Tel.: +32-3-265-27-52; Fax: +32-3-265-26-63

Received: 28 April 2020; Accepted: 1 June 2020; Published: 2 June 2020

Abstract: Onchocerciasis is diagnosed by detecting microfilariae in skin snips or by detecting OV16 IgG4 antibodies in blood by either enzyme linked immunosorbent assay (ELISA) or a rapid diagnostic test (RDT). Here, we compare the sensitivity and specificity of these three tests in persons with epilepsy living in an onchocerciasis endemic region in the Democratic Republic of Congo. Skin snips and blood samples were collected from 285 individuals for onchocerciasis diagnosis. Three tests were performed: the OV16 RDT (SD Bioline) and the OV16 ELISA both on serum samples, and microscopic detection of microfilariae in skin snips. The sensitivity and specificity of each test was calculated with the combined other tests as a reference. Microfilariae were present in 105 (36.8%) individuals, with a median of 18.5 (6.5–72.0) microfilariae/skin snip. The OV16 RDT and OV16 ELISA were positive in, respectively, 112 (39.3%) and 143 (50.2%) individuals. The OV16 ELISA had the highest sensitivity among the three tests (83%), followed by the OV16 RDT (74.8%) and the skin snip (71.4%). The OV16 RDT had a higher specificity (98.6%) compared to the OV16 ELISA (84.8%). Our study confirms the need to develop more sensitive tests to ensure the accurate detection of ongoing transmission before stopping elimination efforts.

Keywords: onchocerciasis; Onchocerca volvulus; antibodies; diagnosis; OV16 testing; microfilariae; epilepsy

---

1. Introduction

Onchocerciasis is a disabling disease caused by infection with the filarial nematode *Onchocerca volvulus* and is linked to skin disease, blindness and epilepsy in remote areas of Africa and Latin America [1,2]. To reduce the onchocerciasis disease burden, the World Health Organization (WHO) and African Program for Onchocerciasis Control (APOC), now part of the Expanded Special Program for Elimination of Neglected Tropical Diseases (ESPEN), have started rigorous elimination campaigns with the community distribution of ivermectin (CDTI) [1,3–5]. When a country achieves the required interruption of onchocerciasis transmission to discontinue CDTI, many years of post-treatment surveillance still have to follow to ensure permanent elimination [6]. Current post-treatment surveillance guidelines to screen for ongoing transmission include the PCR pool screening of the blackfly vector and serological screening of children younger than 10 years old for the presence of OV16 antibodies [6–8].

OV16 IgG4 antibodies can be detected in dried blood spots or serum by an enzyme linked immunosorbent assay (ELISA), or using a rapid diagnostic test (RDT) [7,9]. The OV16 serology only detects exposure to the *O. volvulus* parasite and is therefore not informative about the current infection status. The sensitivity of the OV16 RDT is reported to be approximately 60–80%, whereas the specificity is estimated to be 99% [7,8,10]. This sensitivity is not high enough to detect the <0.1% seroprevalence proposed to stop onchocerciasis elimination efforts [6]. Moreover, it is not clear when seroconversion occurs: before or after the maturation of the adult worm or when the first microfilariae are produced [9,10]. Currently the OV16 RDT is used to determine transmission rates of onchocerciasis in epidemiological studies and is well accepted by the community [11]. Active onchocerciasis infection is diagnosed by the detection of microfilariae in skin snips usually taken from the left and right iliac crests. Although diagnosis by skin snip is highly specific and is considered to be the gold standard for onchocerciasis, it also has major disadvantages. For example, it is labor intensive and requires a well-trained lab technician, might be painful, is logistically challenging, time consuming and has a low sensitivity in areas with low microfilariae loads, such as after multiple rounds of CDTI [11–13].

In this study, we compare serological results obtained with the OV16 RDT and the OV16 ELISA with skin snips results from persons with epilepsy in an onchocerciasis-endemic region the Democratic Republic of Congo (DRC).

2. Materials and Methods

2.1. Study Setting and Design

Samples were collected during a cross-sectional onchocerciasis assessment in persons with epilepsy (PWE), as part of a clinical trial conducted in onchocerciasis-endemic villages in the Logo health zone, Ituri province, DRC [14,15]. In these villages (Draju, Kanga, Wala, Tedheja, and Ulyeko), ivermectin mass drug administration was never implemented. Previously, a high epilepsy prevalence (4.6%, 95% confidence interval: 3.6–5.8) had been documented in the area [16]. Among the 420 persons with epilepsy examined by Lenaerts et al., 67.6% met the diagnostic criteria of onchocerciasis associated epilepsy [17].

The study sites were essentially rural communities, with several fast-flowing rivers providing suitable breeding grounds for the blackfly vectors. The main economic activity of the residents was farming. All individuals who agreed to take part in the screening for the aforementioned clinical trial were eligible, even those who did not meet the inclusion criteria for the trial.

2.2. Study Participants and Sample Collection

Persons with epilepsy were asked to participate in the study and after informed consent was obtained, participants were interviewed and clinical data collected on a standardised questionnaire. Local health centres were used as recruitment grounds, where the research team established mobile clinics. Skin snips were taken from the left and the right iliac crests with a sterile corneoscleral punch (Holt, 2 mm) [18]. Blood samples were also obtained from each participants, immediately placed in a cold flask with ice, and transferred to a refrigerator upon returning to the laboratory on the same day. All procedures were performed following rigorous aseptic conditions.

2.3. Detection of O. volvulus in Skin Snips by Microscopy

Each skin snip was transferred to a single well of a microtitre plate and a few drops of saline was added. Biopsies were incubated for 24 h to allow the microfilariae to emerge from the tissue and a count was performed under a microscope. A person was considered to be infected when *O. volvulus* microfilariae were present in his skin.

2.4. Detection of Antibodies against O. volvulus

The OV16 rapid diagnostic test (OV16 RDT) was performed by trained health care workers, according to the manufacturer protocol (SD Bioline Onchocerciasis IgG4 rapid test, Abbott Standard Diagnostics, Inc., Yongin, Republic of Korea). This test qualitatively detects IgG4 antibodies against the OV16 antigen and the results are immediately available after the recommended 20 minute incubation. Additionally, serum was screened for OV16 IgG4 antibodies by a non-commercial enzyme-linked immunosorbent assay (ELISA) based on horseradish peroxidase-based product detection, as described earlier [19]. Commercially available recombinant OV16 IgG4 (Bio-Rad AbD Serotec, Puchheim, Germany) was used to make a standard curve in every assay (Figure 1A), which was used to calculate the OV16 IgG4 antibody concentration in the serum samples [19]. Briefly, 96-well plates (Thermofisher Scientific, Merelbeke, Belgium) were coated with OV16 antigen (CUSABIO, Cambridge, UK) overnight. Serum samples were diluted 1:200 in each assay. HRP-conjugated mouse anti-human IgG4 Fc antibodies (HP6025, Abcam, Cambridge, UK) were used as a detection antibody, diluted 1:10000 and incubated for 1h. The outcome of the OV16 ELISA was compared to skin snip positivity, and to the OV16 RDT.

2.5. Data Processing and Statistical Analysis

Data processing and statistical analysis were performed in R version 3.6.3 and Microsoft Excel 2010. Only individuals with test results for all three diagnostic tests were included. Sensitivity and specificity were calculated for each test, and, to do so, true and false results were determined by comparing the result of a given test with the combined results of the other two with the following considerations: a positive skin snip is always a true positive even when serology is negative for both OV16 tests. In this case, the serology is considered false negative. A negative skin snip but two positive OV16 tests (RDT and ELISA) was considered as an active *O. volvulus* infection with low microfilarial load, in the absence of ivermectin treatment. In case of discrepancy between the RDT and ELISA OV16 tests, the skin snip result is used as a golden standard to determine which result is true and which is false. Median concentrations of OV16 antibodies were compared for different groups using the Wilcoxon Rank Sum test. P-values below 0.05 were considered significant. The inter- and intraplate coefficient of variation was calculated as a quality control for the OV16 ELISA.

2.6. Ethical Considerations and Informed Consent

The study was approved by the Ethics Committee of the School of Health of the University of Kinshasa, DRC, and the University of Antwerp, Antwerp, Belgium. All eligible candidates provided a written informed consent before enrolment into the study.

3. Results

3.1. Demographic and Clinical Information of the Study Participants

In total, 285 persons with epilepsy (PWE) were included, with a median age of 21 years old (IQR: 15–30) (Table 1). Of those, 147 (51.6%) were male. Only six (2%) individuals of the total study population reported a history of ivermectin use at the time of the study. Clinical examination revealed that the majority of people (85.3%) had normal skin, although 105 (36.8%) reported itching. Forty-two (14.7%) individuals were reported with skin abnormalities, 8 (19.0%) presented with papular skin, 10 (23.8%) had leopard skin, 12 (28.6%) had lizard skin and 18 (42.8%) did not specify the abnormality. Palpable onchocercal nodules were observed in 17 (5.8%) of all PWE, whereas 105 (36.8%) had microfilariae in their skin snips, with a median infection load of 18.5 (IQR: 6.5–72.0) microfilariae/skin snip. Notably, only 44 persons with a positive skin snip also reported itching, whereas 60 persons with a positive skin snip did not report itching and 61 persons who reported itching had a negative skin snip. There was no significant correlation between both ($p = 0.1557$). Serology revealed 112 (39.3%) positive OV16 RDT tests and 143 (50.2%) positive OV16 ELISA results.

Table 1. Demographic and clinical information of the study population.

Characteristics	Persons with Epilepsy (n = 285)
Age: median (IQR)	21 (15–30)
Male: n (%)	147 (51.6%)
Weight (kg): median (IQR)	45.8 (35–51)
Height (cm): median (IQR)	152 (143–160)
Received ivermectin: n, (%)	6 (2%)
Itching [1]: n (%)	105 (36.8%)
Abnormal skin [#]: n (%)	42 (14.7%)
Papular skin: n (%)	8 (19.0%)
Leopard skin: n (%)	10 (23.8%)
Lizard skin: n (%)	12 (28.6%)
Unspecified: n (%)	18 (42.8%)
Onchocerca nodules [2]: n (%)	17 (5.9%)
O. volvulus microfilariae present in skin snip: n (%)	105 (36.8%)
Microfilariae load: median (IQR)	18.5 (6.5–72.0)
Positive OV16 ELISA: n (%)	143 (50.2%)
Concentration OV16 antibodies (μg/mL): median (IQR)	1.66 (0.47–4.46)
Positive OV16 RDT: n (%)	112 (39.3%)
O. volvulus infection *: n (%)	147 (51.6%)

[1] 8 missing data; [2] 7 missing data; [#] Some individuals reported multiple skin abnormalities; IQR: interquartile range; RDT: Rapid diagnostic test; * *O. volvulus* infection was determined by a positive skin snip and/or a positive result for both OV16 tests.

3.2. Comparison of the Three Diagnostic Tests

The OV16 ELISA had the highest sensitivity among the three tests (83%), followed by the OV16 RDT (74.8%) and the skin snip (71.4%) (Table 2). The skin snip had the highest specificity (100%) followed closely by the OV16 RDT (98.6%) and ELISA (84.8%). Twenty-three (8.1%) participants had not (yet) developed an antibody response despite the presence of microfilariae in their skin snips, whereas 42 (14.7%) did not have microfilariae in their skin snips but developed an antibody response detected by both RDT and ELISA. There was no significant difference in the median skin mf load between the persons with a positive OV16 serology (15.0 mf/skin snip, IQR: 6.5–57.4) or a negative OV16 serology (52.5 mf/skin snip, IQR: 9.0–115.3; $p = 0.054$). Similarly, there was also no difference in median age between the persons with positive OV16 serology but negative skin snip (24.5 years, IQR: 18.5–30.75) and the persons with a positive skin snip and negative OV16 serology (21 years, IQR: 15.5–31.0; $p = 0.438$). Overall, 147 (51.6%) PWE were infected with *O. volvulus* (true positive + false negative results).

Table 2. Comparison of the three diagnostic tests. A positive skin snip is always considered a true positive, and a negative skin snip with two positive OV16 tests is considered a false negative. The skin snip test result is followed in case of discrepancy between the OV16 tests.

	Skin Snip	OV16 ELISA	OV16 RDT	Total
All negative	-	-	-	115
All OV16 false negative	+	-	-	23
ELISA false positive	-	+	-	21
RDT false negative	+	+	-	14
RDT false positive	-	-	+	2
ELISA false negative	+	-	+	2
Skin snip false negative	-	+	+	42
All positive	+	+	+	66
True Negative	138	117	136	NA
False Negative	42	25	37	NA
True Positive	105	122	110	NA
False Positive	0	21	2	NA
Sensitivity	71.4%	83%	74.8%	NA
Specificity	100%	84.8%	98.6%	NA

-: Negative test result; +: Positive test result; NA: not available.

3.3. OV16 Antibody Concentration

There was a significant difference in the median OV16 antibody concentration, depending on how many tests were positive ($p < 0.001$, Figure 1B). The median OV16 antibody concentrations were the highest in the individuals with a positive result for all three tests (3.457 µg/mL) and with a positive result for both OV16 tests (2.084 µg/mL) (Figure 1B). The lowest median OV16 antibody concentration (0.402 µg/mL) was found in the individuals with only a positive result for the OV16 ELISA test, which is considered a false positive test result. However, the median OV16 antibody concentration of the false positive ELISA test was not significantly different from those with a false negative OV16 RDT and a positive skin snip result (median = 0.911; $p = 0.828$) and those with a false negative skin snip result and a positive OV16 RDT result (median = 0.895; $p = 0.104$, Figure 1B). Figure 1B shows the concentration-dependent absorbance values of the positive control used in the OV16 ELISA assay. The lowest concentration tested was 0.78125 ng/mL. The interplate coefficient of variation was 18.5% and the intraplate coefficient of variation was 4.2%.

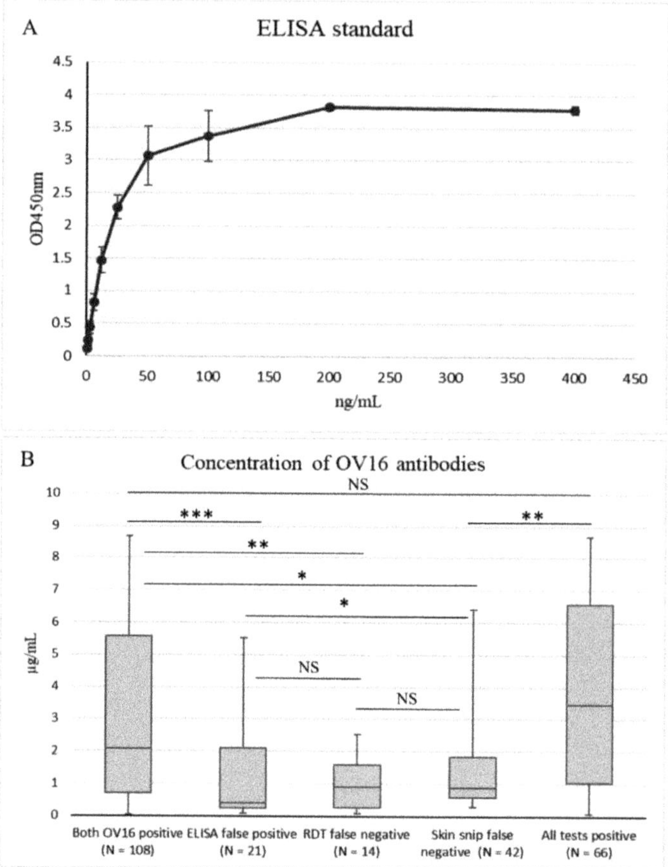

Figure 1. (**A**) Concentration-dependent absorbance values of the OV16 ELISA standards. The lowest concentration tested was 0.78125 ng/mL. (**B**) Concentration of OV16 antibodies. Both OV16 positive: all samples with 2 positive OV16 tests and positive or negative skin snip results; ELISA false positive: negative for OV16 RDT and skin snip; RDT false negative: positive for OV16 ELISA and skin snip; skin snip false negative: negative result for skin snip but positive for both OV16 tests. NS: not significant; * $p < 0.05$; ** $p < 0.01$; *** $p < 0.001$. Wilcoxon Rank Sum test.

4. Discussion

In our study, the OV16 ELISA test had a slightly higher sensitivity (83%) compared to the OV16 RDT (74.8%), and the RDT had a slightly higher specificity (98.6%) compared to the ELISA test (84%). The performance of the serology tests was in the same range as published earlier [7,8,10]. Only 17 (11.6%) of the infected individuals, who were never exposed to ivermectin, had palpable onchocercal nodules and 42 (14.7%) showed abnormal skin. A potential explanation could be that the population is relatively young and is therefore more likely to have a relatively recent infection and not yet developed prominent skin abnormalities. Indeed, both the number of nodules and skin abnormalities are increasing with age [20–22]. However, it is possible that some nodules have been missed because it is not easy to detect nodules that are small or located in areas where they are uncommon to occur, such as deep tissues.

Forty-two individuals had negative skin snip results but a positive serology. However, the OV16 antibody concentration observed in these individuals was lower compared to individuals who had positive skin snip results. These individuals either had very low microfilarial densities, below the detection limit, or had a recent infection while already having antibody production but no microfilariae production. It has been reported that antibodies become detectable 1–2 years before microfilariae appeared in the skin [23]. On the other hand, in our study, 23 individuals had a positive skin snip but had not (yet) developed a detectable antibody response against the OV16 antigen. The mechanism behind the generation of antibodies to specific parasite compounds is dependent on the parasite as well as the immune response, and, sometimes, the antibodies appear after the microfilariae [23]. We do not know whether some of these 23 individuals might have developed OV16 IgG4 antibodies later, because no follow up study was done. However, the most likely explanation of the false negative OV16 result is that the test is not sensitive enough. Individuals with a false negative OV16 RDT had low OV16 antibody concentrations, similar to those with a false negative skin snip result, because the limit of detection is lower in the ELISA test. Individuals with a false positive ELISA test had the lowest antibody concentrations, raising the question of whether these are truly false positive results or the actual sensitivity of this test might be higher than estimated based on the other tests. A test combining different onchocerciasis biomarkers will be needed to increase the sensitivity to detect ongoing transmission [24].

A limitation of this study is that skin snips were not tested for the presence of *O. volvulus* DNA by quantitative PCR, the golden standard for the detection of an *O. volvulus* infection. Indeed, the sensitivity of skin snip microscopy was found to be 62.3% compared to a quantitative *O. volvulus* PCR test on skin snips and 40.4% in skins with <2 microfilariae/skin snip [25].

In conclusion, the skin snip microscopy and OV16 RDT show equal performance to detect *O. volvulus* infection in a population where ivermectin has never been distributed. The OV16 ELISA test has a slightly higher sensitivity but lower specificity compared to the other two tests. More research is needed to determine the optimal tools for onchocerciasis diagnosis, especially in regions with ongoing elimination programs and low community microfilariae loads, to ensure the accurate detection of ongoing transmission before deciding to stop elimination efforts.

Author Contributions: Conceptualization and drafting the manuscript: A.H. and R.C. Data analysis: A.H., J.P. and A.D. Sample collection and processing: M.M. and G.A. Writing–review and editing: all authors, Funding acquisition: R.C. All authors have read and agreed to the published version of the manuscript.

Funding: The study was funded by a grant from the European Research Council (ERC 671055). The study sponsor had no role in the design, execution, interpretation, or writing of the study.

Acknowledgments: We thank the health care workers of the Logo health zone for their involvement in the study and the people of the health zone for their participation.

Conflicts of Interest: The authors declare no conflict of interest.

References

1. World Health Organisation. *Onchocerciasis and Its Control*; Technical report series no. 852; WHO: Geneva, Switzerland, 1995.
2. Colebunders, R.; Siewe, F.N.; Hotterbeekx, A. Onchocerciasis-Associated Epilepsy, an Additional Reason for Strengthening Onchocerciasis Elimination Programs. *Trends Parasitol.* **2018**, *34*, 208–216. [CrossRef] [PubMed]
3. WHO. Progress report on the elimination of human onchocerciasis, 2015–2016. *Wkly. Epidemiol. Rec.* **2016**, *91*, 505–514.
4. Tekle, A.H.; Zouré, H.G.; Mounkaila, N.; Boussinesq, M.; Coffeng, L.E.; Stolk, W.; Remme, J.H.F. Progress towards onchocerciasis elimination in the participating countries of the African Programme for Onchocerciasis Control: Epidemiological evaluation results. *Infect. Dis. Poverty* **2016**, *5*, 66. [CrossRef] [PubMed]
5. Amazigo, U. The African Programme for Onchocerciasis Control (APOC). *Ann. Trop. Med. Parasitol.* **2008**, *102*, 19–22. [CrossRef]
6. WHO. *Guidelines for Stopping Mass Drug Administration and Verifying Elimination of Human Onchocerciasis: Criteria and Procedures*; WHO Press: Geneva, Switzerland, 2016.

7.   Weil, G.J.; Steel, C.; Liftis, F.; Li, B.-W.; Mearns, G.; Lobos, E.; Nutman, T.B. A Rapid-Format Antibody Card Test for Diagnosis of Onchocerciasis. *J. Infect. Dis.* **2000**, *182*, 1796–1799. [CrossRef]

8.   Lipner, E.M.; Dembele, N.; Souleymane, S.; Alley, W.S.; Prevots, D.R.; Toe, L.; Boatin, B.; Weil, G.J.; Nutman, T.B. Field Applicability of a Rapid-Format Anti–Ov-16 Antibody Test for the Assessment of Onchocerciasis Control Measures in Regions of Endemicity. *J. Infect. Dis.* **2006**, *194*, 216–221. [CrossRef]

9.   Cama, V.A.; Feleke, S.M.; McDonald, C.; Wiegand, R.E.; Cantey, P.T.; Arcury-Quandt, A.; Eberhard, M.; Abanyie, F.; Smith, J.; Jenks, M.H.; et al. Evaluation of an OV-16 IgG4 Enzyme-Linked Immunosorbent Assay in Humans and Its Application to Determine the Dynamics of Antibody Responses in a Non-Human Primate Model of Onchocerca volvulus Infection. *Am. J. Trop. Med. Hyg.* **2018**, *99*, 1041–1048. [CrossRef]

10.   Lont, Y.L.; Coffeng, L.E.; De Vlas, S.J.; Golden, A.; Santos, T.D.L.; Domingo, G.J.; Stolk, W. Modelling Anti-Ov16 IgG4 Antibody Prevalence as an Indicator for Evaluation and Decision Making in Onchocerciasis Elimination Programmes. *PLoS Negl. Trop. Dis.* **2017**, *11*, e0005314. [CrossRef]

11.   Dieye, Y.; Storey, H.L.; Barrett, K.L.; Gerth-Guyette, E.; Di Giorgio, L.; Golden, A.; Faulx, D.; Kalnoky, M.; Ndiaye, M.K.N.; Sy, N.; et al. Feasibility of utilizing the SD BIOLINE Onchocerciasis IgG4 rapid test in onchocerciasis surveillance in Senegal. *PLoS Negl. Trop. Dis.* **2017**, *11*, e0005884. [CrossRef]

12.   Golden, A.; Faulx, D.; Kalnoky, M.; Stevens, E.; Yokobe, L.; Peck, R.; Karabou, P.; Banla, M.; Rao, R.; Adade, K.; et al. Analysis of age-dependent trends in Ov16 IgG4 seroprevalence to onchocerciasis. *Parasites Vectors* **2016**, *9*, 338. [CrossRef]

13.   Richards, F.; Aziz, N.A.; Katabarwa, M.; Rodríguez-Pérez, M.A.; Unnasch, T.R.; Silva, R.L.; Fernández-Santos, N.A.; Bekele, F.; Monroy, Z.M.; Tadesse, Z.; et al. Operational Performance of the Onchocerca volvulus "OEPA" Ov16 ELISA Serological Assay in Mapping, Guiding Decisions to Stop Mass Drug Administration, and Posttreatment Surveillance Surveys. *Am. J. Trop. Med. Hyg.* **2018**, *99*, 749–752. [CrossRef] [PubMed]

14.   Colebunders, R.; Mandro, M.; Mukendi, D.; Dolo, H.; Suykerbuyk, P.; Van Oijen, M.; Hopkins, A.; Kaiser, C.; Chaccour, C. Ivermectin Treatment in Patients With Onchocerciasis-Associated Epilepsy: Protocol of a Randomized Clinical Trial. *JMIR Res. Protoc.* **2017**, *6*, e137. [CrossRef] [PubMed]

15.   Mandro, M.; Fodjo, J.N.S.; Dusabimana, A.; Mukendi, D.; Haesendonckx, S.; Lokonda, R.; Nakato, S.; Nyisi, F.; Abhafule, G.; Wonya'Rossi, D.; et al. Single versus Multiple Dose Ivermectin Regimen in Onchocerciasis-Infected Persons with Epilepsy Treated with Phenobarbital: A Randomized Clinical Trial in the Democratic Republic of Congo. *Pathogens* **2020**, *9*, 205. [CrossRef] [PubMed]

16.   Lenaerts, E.; Mandro, M.; Mukendi, D.; Suykerbuyk, P.; Dolo, H.; Wonya'Rossi, D.; Nyisi, F.; Ensoy-Musoro, C.; Laudisoit, A.; Hotterbeekx, A.; et al. High prevalence of epilepsy in onchocerciasis endemic health areas in Democratic Republic of the Congo. *Infect. Dis. Poverty* **2018**, *7*, 68. [CrossRef]

17.   Fodjo, J.N.S.; Mandro, M.; Mukendi, D.; Tepage, F.; Menon, S.; Nakato, S.; Nyisi, F.; Abhafule, G.; Wonya'Rossi, D.; Anyolito, A.; et al. Onchocerciasis-associated epilepsy in the Democratic Republic of Congo: Clinical description and relationship with microfilarial density. *PLoS Negl. Trop. Dis.* **2019**, *13*, e0007300. [CrossRef]

18.   Prost, A.; Prod'hon, J. Parasitological diagnosis of onchocerciasis. A critical review of present methods (author's transl). *Med. Trop. Rev. Corps Sante Colonial* **1978**, *38*, 519–532.

19.   Golden, A.; Stevens, E.J.; Yokobe, L.; Faulx, D.; Kalnoky, M.; Peck, R.; Valdez, M.; Steel, C.; Karabou, P.; Banla, M.; et al. A Recombinant Positive Control for Serology Diagnostic Tests Supporting Elimination of Onchocerca volvulus. *PLoS Negl. Trop. Dis.* **2016**, *10*, e0004292. [CrossRef]

20.   Somo, R.M.; Fobi, G.; Ngosso, A.; Dinga, J.S.; LaFleur, C.; Enyong, P.A.; Ngolle, E.M.; Agnamey, P. A Study of Onchocerciasis with Severe Skin and Eye Lesions in a Hyperendemic Zone in the Forest of Southwestern Cameroon: Clinical, Parasitologic, and Entomologic Findings. *Am. J. Trop. Med. Hyg.* **1993**, *48*, 14–19. [CrossRef]

21.   Stingl, P. Onchocerciasis: Clinical presentation and host parasite interactions in patients of Southern Sudan. *Int. J. Dermatol.* **1997**, *36*, 23–28. [CrossRef]

22.   Murdoch, M.E.; Murdoch, I.E.; Evans, J.R.; Yahaya, H.; Njepuome, N.; Cousens, S.; Jones, B.R.; Abiose, A. Pre-control relationship of onchocercal skin disease with onchocercal infection in Guinea Savanna, Northern Nigeria. *PLoS Negl. Trop. Dis.* **2017**, *11*, e0005489. [CrossRef]

23.   Lobos, E.; Weiss, N.; Karam, M.; Taylor, H.R.; Ottesen, E.; Nutman, T. An immunogenic Onchocerca volvulus antigen: A specific and early marker of infection. *Science* **1991**, *251*, 1603–1605. [CrossRef] [PubMed]

24. Bennuru, S.; Oduro-Boateng, G.; Osigwe, C.; Del Valle, P.; Golden, A.; Ogawa, G.M.; Cama, V.; Lustigman, S.; Nutman, T.B. Integrating Multiple Biomarkers to Increase Sensitivity for the Detection of Onchocerca volvulus Infection. *J. Infect. Dis.* **2019**, *221*, 1805–1815. [CrossRef] [PubMed]
25. Thiele, E.A.; Cama, V.A.; Sleshi, M.; Abanyie, F.; Lakwo, T.; Mekasha, S.; Cantey, P.T.; Kebede, A. Detection of Onchocerca volvulus in Skin Snips by Microscopy and Real-Time Polymerase Chain Reaction: Implications for Monitoring and Evaluation Activities. *Am. J. Trop. Med. Hyg.* **2016**, *94*, 906–911. [CrossRef] [PubMed]

 *pathogens*

Article

Comparison of Diagnostic Tools for the Detection of *Dirofilaria immitis* Infection in Dogs

Rossella Panarese [1], Roberta Iatta [1], Jairo Alfonso Mendoza-Roldan [1], Donald Szlosek [2], Jennifer Braff [2], Joe Liu [2], Frédéric Beugnet [3], Filipe Dantas-Torres [4], Melissa J. Beall [2] and Domenico Otranto [1,5,*]

1   Department of Veterinary Medicine, University of Bari, 70010 Valenzano, Bari, Italy; rossella.panarese@uniba.it (R.P.); roberta.iatta@uniba.it (R.I.); jairo.mendozaroldan@uniba.it (J.A.M.-R.)
2   IDEXX Laboratories, Inc., Westbrook, ME 04092, USA; Donald-Szlosek@idexx.com (D.S.); Jennifer-Braff@idexx.com (J.B.); Joe-Liu@idexx.com (J.L.); Melissa-Beall@idexx.com (M.J.B.)
3   Boehringer Ingelheim Animal Health, 69007 Lyon, France; Frederic.BEUGNET@boehringer-ingelheim.com
4   Aggeu Magalhães Institute, Oswaldo Cruz Foundation, 50740-465 Recife, Brazil; filipe.dantas@cpqam.fiocruz.br
5   Faculty of Veterinary Sciences, Bu-Ali Sina University, Hamedan 6516738695, Iran
*   Correspondence: domenico.otranto@uniba.it; Tel.: +39-080-4679944/9839

Received: 28 May 2020; Accepted: 17 June 2020; Published: 22 June 2020

Abstract: In the last two decades, reports of canine heartworm (HW) infection have increased even in non-endemic areas, with a large variability in prevalence data due to the diagnostic strategy employed. This study evaluated the relative performance of two microtiter plate ELISA methods for the detection of HW antigen in determining the occurrence of *Dirofilaria immitis* in a dog population previously tested by the modified Knott's test and SNAP 4Dx Plus test. The prevalence of this infection in the sheltered dog population (n = 363) from a high-risk area for HW infection was 44.4% according to the modified Knott's test and 58.1% according to a point-of-care antigen ELISA. All serum samples were then evaluated by a microtiter plate ELISA test performed with and without immune complex dissociation (ICD). The prevalence increased from 56.5% to 79.6% following ICD, indicating a high proportion of samples with immune complexing. Comparing these results to that of the modified Knott's test, the samples negative for microfilariae (mfs) and those positive only for *D. repens* mfs demonstrated the greatest increase in the proportion of positive results for *D. immitis* by ELISA following ICD. While the ICD method is not recommended for routine screening, it may be a valuable secondary strategy for identifying HW infections in dogs.

Keywords: *Dirofilaria immitis*; modified Knott's test; ELISA; immune complex dissociation; serological assays

---

1. Introduction

Dirofilaria immitis and *Dirofilaria repens* are widespread mosquito-borne filarial worms that may infect and cause mild to severe diseases in a vast range of mammals, including dogs, cats and humans [1–3]. *Dirofilaria immitis* is the causative agent of heartworm disease (HWD), while *D. repens* is that of subcutaneous dirofilariosis (SCD) in dogs. *Dirofilaria* spp. infections in humans are oftentimes underestimated or misdiagnosed, but symptomatic cases are usually detected in areas with a high prevalence of infection in dogs [4,5]. In fact, the risk of *Dirofilaria* spp. infections in humans is strongly linked to the presence of infected dogs and competent mosquito vectors [6,7].

In the last two decades, reports of clinical cases of heartworm (HW) infection in dogs have increased even in non-endemic areas with new endemic foci detected in the Mediterranean region [8–10].

However, several studies have recorded a large variability in prevalence data [11–13], which depends on epidemiological factors (e.g., presence and abundance of competent vectors and absence of chemoprophylaxis treatments) as well as on the diagnostic methods employed. Indeed, the frequency of *Dirofilaria* spp. infections in dogs may differ according to the diagnostic method used, as demonstrated in Slovakia, where 36% and 64% of the dogs were found positive using the modified Knott's method and heartworm (HW) antigen test, respectively [12]. In fact, HW antigen tests are also able to detect occult infections characterized by amicrofilaremia.

Dirofilaria immitis infection in dogs is usually diagnosed by parasitological and serological assays, and eventually confirmed by molecular analyses. The modified Knott's test is the most popular parasitological method among concentration tests (e.g., acetone, 5% Tween 20 solution, distilled water and 1% or 0.1% SDS) [14,15], being based on the detection and identification of microfilariae (mfs) of *Dirofilaria* spp. in blood samples. However, the results of the modified Knott's test may be impaired by occult infections as has been observed in up to 67% of dogs positive for *D. immitis* adults [16].

Serological tests (i.e., enzyme-linked immunosorbent assay (ELISA) and immunochromatographic tests (ICT)) for the detection of somatic and female antigens of *D. immitis* adults are available for point-of care (POC) testing or for reference diagnostic laboratories [1,12,17,18]. The antigen tests are highly specific to *D. immitis*. Nevertheless, cross-reactions may occur with antigens of other nematodes (e.g., *D. repens*, *Angiostrongylus vasorum* and *Spirocerca lupi*). These tests are also highly sensitive, particularly when samples are subjected to immune complex dissociation (ICD) methods prior to testing [19,20]. Indeed, the preheating method has been shown to improve antigen detection in both experimental and natural *D. immitis* infections by releasing HW antigen that is bound to host antibodies [21–23]. From this perspective, the aims of this study were to evaluate the relative performance of two different microtiter plate ELISA methods for the detection of HW antigen in a sheltered dog population previously tested by a modified Knott's test in order to evaluate the infection prevalence in this population using combinations of different methods (i.e., modified Knott's test plus HW antigen tests), and to determine the effect of preheating on HW antigen detection.

2. Results

The overall prevalence of *Dirofilaria* spp. infection was 49.3% (179/363) according to the modified Knott's test, with 44.4% (161/363) positive for *D. immitis*, 7.2% (26/363) for *D. repens* and one positive for a *Dirofilaria* sp. that could not be identified morphologically. Nine cases (already included in their respective totals above) were co-infected by *D. immitis* and *D. repens* (2.5%, 9/363). No other filarial species (e.g., *Acanthocheilonema* spp.) were detected in the canine blood samples analyzed. Table 1 shows the results of the HW antigen detection by the SNAP®4Dx®Plus test and the microtiter plate ELISA with and without preheating of the samples scored negative on the modified Knott's test (n = 184).

Table 1. Modified Knott's test microfilaria negative samples (n = 184) tested for *D. immitis* antigen by SNAP 4Dx Plus and microtiter plate ELISA with and without preheating.

Knott's Test Negative Samples (n = 184)	SNAP 4Dx Plus	HW Ag ELISA without Preheating	HW Ag ELISA with Preheating
50	Positive	Positive	Positive
9	Negative	Positive	Positive
56	Negative	Negative	Positive
2	Negative	Positive	Negative
67	Negative	Negative	Negative

The proportion of *D. immitis*-positive samples increased to 58.1% (211/363) when Knott's positive samples were combined with those positive by the SNAP 4Dx Plus test. Out of 134 negative samples by SNAP 4Dx Plus and modified Knott's tests, 11 and 65 serum samples tested positive by the microtiter plate ELISA without and with preheating, respectively. Of these, nine samples were simultaneously positive by both the microtiter plate ELISA and two samples were positive only by the microtiter plate ELISA without preheating (Table 1).

Overall, the proportion of samples that tested positive for antigens was of 56.5% (205/363) and 79.6% (289/363) on the microtiter plate ELISA without and with preheating, respectively. The positive and the negative agreements between these methods were 70% (95% CI: 0.65–0.75) and 97% (95% CI: 0.91–1.00), respectively, with a statistically significant (McNemar test, $p < 0.0001$) difference between the methods.

The impact of immune complexes on antigen detection relative to each species of mfs identified by a modified Knott's test was evaluated based on the number of samples found to be positive by the microtiter plate ELISA with preheating (n = 289). Samples were classified into three groups for the analysis. The first group included all samples positive for *D. immitis* mfs, including nine samples co-infected with *D. repens* (n = 159). The second group consisted of samples positive only for *D. repens* mfs (n = 15), whereas the third group included samples found to be negative by a modified Knott's test (n = 115). The proportion of samples positive by the microtiter plate ELISA without preheating relative to those positive after heating was higher for dogs with *D. immitis* mfs (86%; 95% CI: 0.80–0.91) compared to samples positive only for *D. repens* mfs (47%; 95% CI: 0.25–0.70) or negative for mfs (51%; 95% CI 0.42–0.60). The latter two groups had an increased proportion of positive results on the microtiter plate ELISA with preheating (Table 2).

Table 2. Comparison of heartworm antigen detection by microtiter plate ELISA without and with preheating relative to Knott's test results for those samples that tested antigen positive after preheating.

Modified Knott's Method	HW Ag ELISA without Preheating	HW Ag ELISA with Preheating	Increase in Positive Samples with vs. without Preheating (%)
Dirofilaria immitis and co-infected [a]	137	159	16%
Dirofilaria repens only	7	15	114%
No microfilaria observed	59	115	95%
Total	**205**	**289**	**42%**

[a] This includes nine dogs positive for both *D. immitis* and *D. repens*.

3. Discussion

The prevalence of *Dirofilaria* spp. in the studied sheltered dog population from a high-risk geographical area for HWD varies according to the diagnostic method employed, ranging from 44.4% (modified Knott's method) to 79.6% (plate ELISA test with preheating), indicating how important it is the use of a multi-test diagnostic strategy for detecting positive dogs in a given population. All serum samples that tested positive using the SNAP 4Dx Plus test were positive by the microtiter plate ELISA reflecting the accuracy of the antigen detection of this POC test. A fairly similar proportion of infected dogs was obtained by combining the number of positive samples by the modified Knott's test and the SNAP 4Dx Plus test with those analysed only by the microtiter plate ELISA without preheating (58.1% vs 56.5%, respectively). As in previous studies, the microtiter plate ELISA, regardless of the test used, provided a higher proportion of antigen-positive results than the POC ELISA test [20,24,25]. Nonetheless, the microtiter plate ELISA test with preheating detected the most positive test results (79.6%) in the studied dog population.

The difference in the proportion of positive results between the two microtiter plate ELISA methods is likely due to antigen–antibody immune complexes, which are known to interfere with the detection of HW antigen in immunoassays [21,22,26,27]. Antigen–antibody immune complexes occur when the dog's antibodies bind to the carbohydrate epitopes of the HW antigens, making these regions unavailable for the antibodies used in the diagnostic assay [21]. As observed in this study, the negative agreement between the two microtiter plate ELISA tests was high (97%), indicating that samples negative for HW antigen after heating were also negative without heating. Contrarily, the low positive percent agreement (70%) reflects those samples that were only positive for antigen after heating, suggesting that this procedure acted by disrupting immune complexes and liberated the antigens, allowing them to be detected by the capture antibody present in the assay.

One concern that has been raised regarding the use of the ICD method is the potential for detecting similar carbohydrate antigens from other nematode parasites. Indeed, a previous study has shown that immunoassays for the detection of *D. immitis* antigen can detect excretory/secretory antigens obtained in vitro from other parasites, including *D. repens* [19]. Several different immunoassays demonstrated improved detection of *D. repens* antigen after heating [19,20,25]. However, in the current study, the modified Knott's test identified both *D. immitis* and *D. repens* mfs in the examined dog blood samples, with the former infection predominating. This may be consistent with the number of occult *D. immitis* infections detected by the POC ELISA. Nevertheless, the true frequency of co-infections is difficult to determine, considering that the microtiter plate ELISA with preheating may have dissociated either *D. immitis* or *D. repens* antigen from immune complexes.

Given the relatively large proportion of samples that converted from antigen negative to positive after heating in the studied dog population, it was of interest to evaluate the test results based on the identification of the mfs by the modified Knott's test. Based on the Knott's results, the increase in the number of antigen-positive results after heating was greater in samples negative for mfs and in samples positive only for *D. repens* mfs as compared to those containing *D. immitis* mfs. The relative amounts of antigen and antibody in circulation determine the degree of immune complex formation and the residual amount of free antigen available for detection by serological tests. Several hypotheses for future studies could be suggested from these observations. First, patent *D. immitis* infections might be expected to have higher circulating antigen concentrations given that the antigen is shed from the uterus of the mature female parasite as the mfs are released [23]. On the other hand, *D. repens* infections are typically localized to the subcutaneous tissues and may release lower concentrations of antigen directly into circulation. From the results obtained in this study, it appears that amicrofilaremic dogs that were HW antigen positive would fall somewhere in between these two scenarios.

The modified Knott's test detected fewer *D. immitis* infections than the microtiter plate ELISA antigen tests in this study. In particular, 30.3% of dogs negative for microfilaria using the Knott's test in a previous study [10] reverted to a positive result using the plate ELISA test with preheating. This finding is also supported by previous studies [12,16]. Indeed, the absence of mfs in the canine blood may depend on several factors, such as the long prepatent period of the parasite (i.e., about 7 months) and the low mfs concentration in the samples [1,16,28,29]. Dogs may become amicrofilaremic after 12 months post-infection due to the development of an immune response or to female HW senility in the absence of reinfections [30]. In contrast, the highest level of mfs in the blood is detected between 7 and 12 months post-infection and during the warmest seasons (i.e., between June and September), with circadian peaks occurring from 4:00 a.m. to 10:00 p.m. [30,31]. On the other hand, although the modified Knott's method allows the morphological identification of the mfs, no other filarial species were herein detected, probably due to the low pressure of proper vectors in the given environment.

This study has some limitations. Only two dog shelters were herein investigated, which could lead to sampling bias. In addition, the SNAP 4Dx Plus test was only performed on samples scored negative by the modified Knott's test. Thus, a full comparison between the modified Knott's test, the SNAP 4Dx Plus test and the microtiter plate ELISA methods was not possible. Although an increase in the number of positive samples after heating was more evident for *D. repens*-positive samples, the sample size for this conclusion was small.

The accuracy of the diagnostic methods for the detection of HW infection in dogs is essential for veterinary practitioners to decide which is the most suitable strategy for each dog, that is, curative or preventive treatments for positive and negative dogs, respectively [10,23]. Furthermore, chemoprophylaxis should be addressed in dog shelters, where dogs are confined and oftentimes much more exposed to mosquito vectors in the outside environment, as compared to single privately owned dogs. These results also suggest that veterinary practitioners dealing with dogs without clinical signs suggestive of HWD should interpret the results of qualitative serological tests with caution and, preferably, in combination with a modified Knott's test for circulating mfs. Both results should always be interpreted in conjunction with detailed clinical and anamnestic data. Considering the data above, the combination of more than one diagnostic tool is recommended to increase the probability of finding true positive dogs. Although the microtiter plate ELISA with preheating has been shown to detect more HW-antigen-positive dogs [26,27], this method is presently not recommended for routine HW screening (American Heartworm Society guidelines). Indeed, dogs receiving annual veterinary care and HW prevention were not found to have a high likelihood of false-negative HW antigen test results due to immune complexing [32]. Furthermore, the detection of HW antigens using the SNAP 4Dx Plus test remains one of the most common and reliable techniques to be used for a rapid POC diagnosis in veterinary clinics as well as in field studies [20,33]. From a public health perspective, these data highlight that the use of the modified Knott's test alone may result in many false-negative dogs, serving as a hidden source of infection to mosquito vectors, thus representing a risk for other dogs and humans sharing the same environment.

4. Materials and Methods

4.1. Study Design

Blood and serum samples (n = 363) used herein were collected during a previous study on *Dirofilaria* spp. infection in sheltered dogs [10]. All dogs were housed in two shelters in Southern Italy (40.608705N, 17.994495E, site 1; 40.419326N, 18.165582E, site 2), where 44.2% and 7% were infected by *D. immitis* and *D. repens*, respectively, as determined by the modified Knott's test [10]. The animals were handled and sampled following the approval of the Ethical Committee of the Department of Veterinary Medicine of the University of Bari, Italy (Prot. Uniba 8/19).

4.2. Diagnostic Procedures

Two HW antigen detection tests were used in this study: the rapid SNAP 4Dx Plus Test (IDEXX Laboratories, Inc., Westbrook, ME, USA) and the microtiter plate ELISA (i.e., Heartworm Antigen by ELISA, IDEXX Laboratories, Inc., Westbrook, ME, USA) with and without a heat pretreatment method for ICD of the serum sample. The microtiter plate ELISA, which is only available through the IDEXX Reference Laboratory, was performed as previously described [23]. Serum samples (100 µL) were added to each well of the microtiter plate, coated with the capture antibody and incubated for 30 min at room temperature. Each well was rinsed five times with wash solution and then 100 µL of a horseradish peroxidase-conjugated antibody solution was added to the well and incubated for 30 min at room temperature. The wells were washed and 50 µL of 3,3′,5,5′-tetramethylbenzidine (TMB) substrate solution was added and incubated for 10 min at room temperature. Following the addition of a 0.1% sodium dodecyl sulfate stop solution, the absorbance was measured at 650 nm. A positive result was recorded if the absorbance of the sample exceeded that of the negative control by an optical density (OD) of 0.05. For the microtiter plate ELISA with preheating, the serum sample was mixed with an equal volume of 0.1 M EDTA (pH 7.5) and the mixture was heated at 100 °C for 5 min [23]. Following centrifugation at 16,000× *g* for 5 min, the supernatant (100 µL) was transferred to the microtiter plate and the ELISA performed as described above. The microtiter plate ELISA for the heartworm antigen was used to test all samples. The SNAP 4Dx Plus Test, a bi-directional flow ELISA test designed for POC testing, was performed according to the manufacturer's instructions for any sample that scored negative on the modified Knott's test. Before testing, each serum sample was thawed at room temperature and then vortexed.

4.3. Data Handling and Statistical Analysis

In the absence of a gold standard (i.e., necropsy), percent positive and negative agreements were determined for the microtiter plate ELISA, both with and without preheating, by calculating the proportion of samples found to be positive by both methods over the total number of positives by the ICD method. Likewise, percent negative agreement was the proportion of negative samples found by both methods over the total number of negative samples by the ICD method. Concordance between ELISA with and without preheating was evaluated by McNemar's test for paired data. Statistical significance was defined as a *p*-value < 0.05. Exact binomial methods were used to calculate the 95% confidence intervals for both percent agreements. All analyses were performed using R (version 3.6.1). Visualization of data sets and test results employed Euler diagrams (created in Microsoft PowerPoint, 2016).

Author Contributions: Conceptualization, R.P. and D.O.; methodology, R.P., R.I., D.S., J.B., J.L., M.J.B. and D.O.; Software, D.S., J.B. and J.L.; validation, D.S., J.B. and J.L.; writing—original draft preparation, R.P., R.I., M.J.B. and D.O.; writing—review and editing, R.P., R.I., J.A.M.-R., D.S., J.B., J.L., F.B., F.D.-T., M.J.B. and D.O.; supervision, D.O. and M.J.B.; funding acquisition, F.B. and M.J.B. All authors have read and agreed to the published version of the manuscript.

Funding: This work was supported by Boehringer Ingelheim Animal Health (France, Europe) and IDEXX Laboratories, Inc., Westbrook, Maine, ME 04092, USA.

Acknowledgments: The authors would like to thank Phyllis Tyrrell and Jan Drexel for their assistance with this study.

Conflicts of Interest: The POC and microtiter plate ELISA tests were provided by IDEXX Laboratories, Inc. Donald Szlosek, Jennifer Braff, Joe Liu and Melissa J. Beall are all employees of IDEXX Laboratories, Inc., Westbrook, Maine, ME 04092, USA.

References

1. Simón, F.; Siles-Lucas, M.; Morchón, R.; González-Miguel, J.; Mellado, I.; Carretón, E.; Montoya-Alonso, J.A. Human and animal dirofilariasis: The emergence of a zoonotic mosaic. *Clin. Microbiol. Rev.* **2012**, *25*, 507–544. [CrossRef]

2. Dantas-Torres, F.; Otranto, D. Dirofilariosis in the Americas: A more virulent *Dirofilaria immitis? Parasit Vectors* **2013**, *6*, 288. [CrossRef]

3. Colella, V.; Nguyen, V.L.; Tan, D.Y.; Lu, N.; Fang, F.; Zhijuan, Y.; Wang, J.; Liu, X.; Chen, X.; Dong, J.; et al. Zoonotic vectorborne pathogens and ectoparasites of dogs and cats in Eastern and Southeast Asia. *Emerg. Infect. Dis.* **2020**, *26*, 1221–1233. [CrossRef] [PubMed]

4. Otranto, D.; Brianti, E.; Gaglio, G.; Dantas-Torres, F.; Azzaro, S.; Giannetto, S. Human ocular infection with *Dirofilaria repens* (Railliet and Henry, 1911) in an area endemic for canine dirofilariasis. *Am. J. Trop. Med. Hyg.* **2011**, *84*, 1002–1004. [CrossRef] [PubMed]

5. Otranto, D.; Diniz, D.G.; Dantas-Torres, F.; Casiraghi, M.; Almeida, I.N.F.; Almeida, L.N.F.; Santos, J.N.; Furtado, A.P.; Sobrinho, E.F.A.; Bain, O. Human intraocular filariasis caused by *Dirofilaria* sp. nematode, Brazil. *Emerg. Infect. Dis.* **2011**, *17*, 863–866. [CrossRef] [PubMed]

6. Capelli, G.; Genchi, C.; Baneth, G.; Bourdeau, P.; Brianti, E.; Cardoso, L.; Danesi, P.; Fuehrer, H.P.; Giannelli, A.; Ionică, A.M.; et al. Recent advances on *Dirofilaria repens* in dogs and humans in Europe. *Parasit Vectors* **2018**, *11*, 663. [CrossRef]

7. Dantas-Torres, F.; Otranto, D. Overview on *Dirofilaria immitis* in the Americas, with notes on other filarial worms infecting dogs. *Vet. Parasitol.* **2020**, *282*, 109113. [CrossRef]

8. Otranto, D.; Capelli, G.; Genchi, C. Changing distribution patterns of canine vector borne diseases in Italy: Leishmaniosis vs. dirofilariosis. *Parasit Vectors* **2009**, *26*, S1–S2. [CrossRef]

9. Mendoza-Roldan, J.; Benelli, G.; Panarese, R.; Iatta, R.; Furlanello, T.; Beugnet, F.; Zatelli, A.; Otranto, D. *Leishmania infantum* and *Dirofilaria immitis* infections in Italy, 2009–2019: Changing distribution patterns. *Parasit Vectors* **2020**, *13*, 193. [CrossRef]

10. Panarese, R.; Iatta, R.; Latrofa, M.S.; Zatelli, A.; Ćupina, A.I.; Montarsi, F.; Pombi, M.; Mendoza-Roldan, J.A.; Beugnet, F.; Otranto, D. Hyperendemic *Dirofilaria immitis* infection in a sheltered dog population: An expanding threat in the Mediterranean region. *Int. J. Parasitol.* **2020**, *S0020-7519(20)*, 30114-4.

11. Diosdado, A.; Gómez, P.; González-Miguel, J.; Simón, F.; Morchón, R. Current status of canine dirofilariosis in an endemic area of western Spain. *J. Helminthol.* **2018**, *92*, 520–523. [CrossRef] [PubMed]

12. Miterpáková, M.; Valentová, A.; Čabanová, V.; Berešíková, L. Heartworm on the rise—New insights into *Dirofilaria immitis* epidemiology. *Parasitol. Res.* **2018**, *117*, 2347–2350. [CrossRef] [PubMed]

13. Stoyanova, H.; Carretón, E.; Montoya-Alonso, J.A. Stray dogs of Sofia (Bulgaria) could be an important reservoir of heartworm (*Dirofilaria Immitis*). *Helminthologia* **2019**, *56*, 329–333. [CrossRef] [PubMed]

14. Watanabe, Y.; Yang, C.H.; Tung, K.C.; Ooi, H.K. Comparison of microfilaria concentration method for *Setaria digitata* infection in cattle and for *Dirofilaria immitis* infection in dogs. *J. Vet. Med. Sci.* **2004**, *66*, 543–545. [CrossRef]

15. Venco, L.; Genchi, C.; Simón, F. La filariosis cardiopulmonar (*Dirofilaria immitis*) en el perro. In *La Filariosis en las Especies Domésticas y en el Hombre*; Simón, F., Genchi, C., Venco, L., Montoya, M.N., Eds.; Merial Laboratorios: Barcelona, Spain, 2011; pp. 19–60.

16. Otto, G.F. The significance of microfilaremia in the diagnosis of heartworm infection. *Proc. Heartworm Symp.* **1978**, *77*, 22–30.

17. Newton, W.L.; Wright, W.H. The occurrence of a dog filariid other than *Dirofilaria immitis* in the United States. *J. Parasitol.* **1956**, *42*, 246–258. [CrossRef]

18. Lee, A.C.Y.; Bowman, D.D.; Lucio-Forster, A.; Beall, M.J.; Liotta, J.L.; Dillon, R. Evaluation of a new in-clinic method for the detection of canine heartworm antigen. *Vet. Parasitol.* **2011**, *177*, 387–391. [CrossRef]

19. Venco, L.; Manzocchi, S.; Genchi, M.; Kramer, L.H. Heat treatment and false-positive heartworm antigen testing in ex vivo parasites and dogs naturally infected by *Dirofilaria repens* and *Angiostrongylus vasorum*. *Parasit Vectors* **2017**, *10*, 476. [CrossRef]

20. Henry, L.G.; Brunson, K.J.; Walden, H.S.; Wenzlow, N.; Beachboard, S.E.; Barr, K.L.; Long, M.T. Comparison of six commercial antigen kits for detection of *Dirofilaria immitis* infections in canines with necropsy-confirmed heartworm status. *Vet. Parasitol.* **2018**, *254*, 178–182. [CrossRef]

21. Weil, G.J.; Malane, M.S.; Powers, K.G.; Blair, L.S. Monoclonal antibodies to parasite antigens found in the serum of *Dirofilaria immitis*-infected dogs. *J. Immunol.* **1985**, *134*, 1185–1191.

22. Little, S.E.; Munzing, C.; Heise, S.R.; Allen, K.E.; Starkey, L.A.; Johnson, E.M.; Meinkoth, J.; Reichard, M.V. Pretreatment with heat facilitates detection of antigen of *Dirofilaria immitis* in canine samples. *Vet. Parasitol.* **2014**, *20*, 250–252. [CrossRef] [PubMed]

23. Beall, M.J.; Arguello-Marin, A.; Drexel, J.; Liu, J.; Chandrashekar, R.; Alleman, A.R. Validation of immune complex dissociation methods for use with heartworm antigen tests. *Parasit Vectors* **2017**, *10*, 481. [CrossRef] [PubMed]

24. Courtney, C.H.; Zeng, Q.Y. Comparison of heartworm antigen test kit performance in dogs having low heartworm burdens. *Vet. Parasitol.* **2001**, *96*, 317–322. [CrossRef]

25. Starkey, L.A.; Bowles, J.V.; Payton, M.E.; Blagburn, B.L. Comparative evaluation of commercially available point-of-care heartworm antigen tests using well-characterized canine plasma samples. *Parasit Vectors* **2017**, *10*, 475. [CrossRef]

26. Velasquez, L.; Blagburn, B.L.; Duncan-Decoq, R.; Johnson, E.M.; Allen, K.E.; Meinkoth, J.; Gruntmeira, J.; Littlea, S.E. Increased prevalence of *Dirofilaria immitis* antigen in canine samples after heat treatment. *Vet. Parasitol.* **2014**, *206*, 67–70. [CrossRef]

27. Drake, J.; Gruntmeir, J.; Merritt, H.; Allen, L.; Little, S.E. False negative antigen tests in dogs infected with heartworm and placed on macrocyclic lactone preventives. *Parasit Vectors* **2015**, *8*, 68. [CrossRef]

28. Knott, J. A method for making microfilarial survey on day blood. *Med. Hyg.* **1939**, *33*, 191–196. [CrossRef]

29. Otranto, D.; Testini, G.; Dantas-Torres, F.; Latrofa, M.S.; Diniz, P.P.; de Caprariis, D.; Lia, R.P.; Mencke, N.; Stanneck, D.; Capelli, G.; et al. Diagnosis of canine vector-borne diseases in young dogs: A longitudinal study. *J. Clin. Microbiol.* **2010**, *48*, 3316–3324. [CrossRef]

30. Rawlings, C.A.; Dawe, D.L.; McCall, J.W.; Keith, J.C.; Prestwood, A.K. Four types of occult *Dirofilaria immitis* infection in dogs. *J. Am. Vet. Med. Assoc.* **1982**, *180*, 1323–1326.

31. Church, E.M.; Georgi, J.R.; Robson, D.S. Analysis of the microfilarial periodicity of *Dirofilaria immitis*. *Cornell Vet.* **1976**, *66*, 333–346.

32. Nafe, L. Prevalence of *Dirofilaria immitis* antigen in client-owned pet dogs before and after serum heat treatment. In Proceedings of the Annual ACVIM Forum, Denver, CO, USA, 9–11 June 2016; Volume 30, p. 1428.

33. Trancoso, T.A.L.; Lima, N.C.; Barbosa, A.S.; Leles, D.; Fonseca, A.B.M.; Labarthe, N.V.; Bastos, O.M.P.; Uchôa, C.M.A. Detection of *Dirofilaria immitis* using microscopic, serological and molecular techniques among dogs in Cabo Frio, RJ, Brazil. *Braz. J. Vet. Parasitol.* **2020**, *29*, e017219. [CrossRef] [PubMed]

pathogens

Review

Descriptive Comparison of ELISAs for the Detection of *Toxoplasma gondii* Antibodies in Animals: A Systematic Review

K. L. D. Tharaka D. Liyanage *, Anke Wiethoelter [iD], Jasmin Hufschmid [†][iD] and Abdul Jabbar [†][iD]

Department of Veterinary Biosciences, Melbourne Veterinary School, Faculty of Veterinary and Agricultural Sciences, The University of Melbourne, Werribee, VIC 3030, Australia; anke.wiethoelter@unimelb.edu.au (A.W.); huj@unimelb.edu.au (J.H.); jabbara@unimelb.edu.au (A.J.)
* Correspondence: tkoswaththal@student.unimelb.edu.au; Tel.: +61-397-312313
† These authors contributed equally.

Abstract: *Toxoplasma gondii* is the zoonotic parasite responsible for toxoplasmosis in warm-blooded vertebrates. This systematic review compares and evaluates the available knowledge on enzyme-linked immunosorbent assays (ELISAs), their components, and performance in detecting *T. gondii* antibodies in animals. Four databases were searched for published scientific studies on *T. gondii* and ELISA, and 57 articles were included. Overall, indirect (95%) and in-house (67%) ELISAs were the most used types of test among the studies examined, but the 'ID Screen® Toxoplasmosis Indirect Multi-species' was common among commercially available tests. Varying diagnostic performance (sensitivity and specificity) and Kappa agreements were observed depending on the type of sample (serum, meat juice, milk), antigen (native, recombinant, chimeric) and antibody-binding reagents used. Combinations of recombinant and chimeric antigens resulted in better performance than native or single recombinant antigens. Protein A/G appeared to be useful in detecting IgG antibodies in a wide range of animal species due to its non-species-specific binding. One study reported cross-reactivity, with *Hammondia hammondi* and *Eimeria* spp. This is the first systematic review to descriptively compare ELISAs for the detection of *T. gondii* antibodies across different animal species.

Keywords: toxoplasmosis; animals; ELISA; native antigens; recombinant antigens

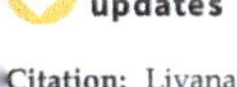

check for updates

Citation: Liyanage, K.L.D.T.D.; Wiethoelter, A.; Hufschmid, J.; Jabbar, A. Descriptive Comparison of ELISAs for the Detection of *Toxoplasma gondii* Antibodies in Animals: A Systematic Review. *Pathogens* **2021**, *10*, 605. https://doi.org/10.3390/pathogens10050605

Academic Editor: Geoff Hide

Received: 1 April 2021
Accepted: 12 May 2021
Published: 15 May 2021

Publisher's Note: MDPI stays neutral with regard to jurisdictional claims in published maps and institutional affiliations.

1. Introduction

Toxoplasma gondii (Apicomplexa: Sarcocystidae) is an intracellular parasite that can infect endothermic animal species, including mammals and birds, with a worldwide distribution [1,2]. Infection with *T. gondii* also affects nearly one-quarter of the human population, making it an important zoonotic problem globally [3]. In most immunocompetent individuals, toxoplasmosis remains asymptomatic and self-limiting [4,5]; however, the infection can lead to significant morbidity and even mortality [6,7]. In livestock, infection with *T. gondii* can cause serious reproductive complications, including abortion, congenital deformity, stillbirth and foetal mummification [7–9], leading to significant economic losses [6,7]. Furthermore, *T. gondii* is an emerging threat to the health and welfare of wildlife populations worldwide [10]. For example, New World monkeys and Australian marsupials are thought to be highly susceptible to toxoplasmosis, often resulting in clinical disease and even death [1,11]. A wide range of clinical signs, including sudden death, encephalitis, lymphadenopathy, respiratory distress, interstitial pneumonia, and neurological signs have been reported in wild animals [12–15].

Toxoplasma gondii has a complex life cycle, with the sexual phase occurring in the definitive host (i.e., cats and other felids) and the asexual phase in intermediate hosts (i.e., humans as well as virtually all warm-blooded animals) [16]. Three obvious parasitic stages can be identified in the life cycle: sporozoites within sporulated oocysts, tachyzoites and bradyzoites, all of which can infect both definitive and intermediate hosts [17]. Upon primary infection in felids by the ingestion of tissue cysts, bradyzoites invade intestinal cells, resulting in numerous asexual and sexual developmental stages, eventually forming millions of oocysts that are excreted in faeces [18]. Both definitive and intermediate hosts can be infected upon the ingestion of infective oocysts from contaminated water bodies, pasture, vegetation, or eating raw or undercooked meat including tissue cysts containing bradyzoites [5,18]. In both definitive and intermediate hosts, vertical transmission via transplacental and lactogenic routes has also been reported [1,5]. Upon the ingestion of oocysts by an intermediate host, bradyzoites transform into tachyzoites in the intestine, multiply rapidly and disseminate throughout the body, thereby infecting any kind of cell in the body eliciting a strong immune response in immunocompetent individuals [19,20]. Subsequently, tachyzoites transform into bradyzoites and form dormant tissue cysts in the intermediate host, with a preference for neural and muscular tissues [2].

The detection of *T. gondii* can be achieved through direct and indirect methods. Direct methods involve the identification of parasitic stages by microscopic examination, the detection of parasitic DNA in samples using polymerase chain reaction (PCR), or isolation of the parasite utilising bioassays [21]. However, while these direct tests can be highly specific, they generally have limited sensitivity because they rely on the presence of one of the three infective stages of the parasite in the tested sample [22]. The microscopic detection of oocysts in faecal, water or environmental samples has particularly low sensitivity and is time-consuming [23]. Molecular techniques such as PCR can be used for the detection of both acute infections [22,24], with parasitic DNA found in the bloodstream due to rapidly multiplying tachyzoites, and chronic infections, using tissue samples targeting cysts in muscle and nervous tissues, usually from deceased animals [25]. However, this, similar to histochemical techniques, can be limited by the low abundance and random distribution of tissue cysts and disseminating tachyzoites [25,26]. The cat bioassay is a highly sensitive and specific test and is considered the "gold standard" for the detection of *T. gondii*, because it relies on the shedding of oocysts in the faeces of cats who have been fed tissue cysts [19,27]. However, this method is expensive, time consuming, and poses ethical challenges, making it impractical for routine screening of larger samples [21]. Thus, indirect methods involving the serological detection of parasite-specific antibodies have become the routine test for the diagnosis of toxoplasmosis in both animals and humans [23,28].

Serological diagnosis of toxoplasmosis takes advantage of the persistent presence of specific antibodies in serum following exposure to the parasite. Upon primary exposure to the parasite, IgM antibodies are produced in immunocompetent animals, which are classically short-lived [29]. Subsequently, IgG antibodies appear and persist for years, providing a reliable serological marker for the detection of previous exposure to *T. gondii* [29,30]. Serological techniques are relatively inexpensive, require a small volume of the sample, and can be used in live animals [10,31,32]. Many serological techniques, including the Sabin–Feldman dye test (DT), modified agglutination test (MAT), direct agglutination test (DAT), indirect immunofluorescence test (IFAT), indirect hemagglutination assay (IHA), latex agglutination test (LAT), Western blot (WB), and enzyme-linked immunosorbent assay (ELISA) have been widely used to detect *T. gondii*-specific antibodies in animals and humans [23,33]. However, while each of these tests has certain advantages, they also have certain limitations. Among all these tests, ELISA appears to be the most reliable, practical, economical, and widely used test for the detection of exposure to *T. gondii* in animals [34–36]. Only a small volume of sample is required, and the assay can be semi-automated, thereby making it suitable for large-scale screening [37,38]. Moreover, ELISAs can differentiate between immunoglobulin classes and are, therefore, useful in determining the phase of infection [29]. ELISAs can be divided into four main types: direct, indirect, sandwich, and competitive. However, all types use a colorimetric technique to quantify the analyte of interest in a liquid sample based on an antigen–antibody reaction, with the antigen/antibody complex in an immobilised phase [39,40]. Different ELISAs use different types of antigen (native, recombinant, chimeric) and secondary antibodies/antibody binding reagents (species-specific conjugates, non-species-specific conjugates) to detect antibodies [7,23]. However, the choice of components may significantly influence test performance. Indicators of test performance include sensitivity, specificity, and overall agreement (usually indicated by the Kappa statistic) with a reference test [41–43]. Commercial ELISA kits for the detection of *T. gondii* antibodies in domesticated animals are available, making routine and large-scale screening more practical. Moreover, flexibility in adapting the ELISA technique to desired research interests (such as the evaluation of novel antigens/antibodies for the development of more accurate assays, and the comparison of the performance of different serological tests) have made "in-house" ELISAs widely popular [7,44,45].

This systematic review compares different ELISAs for the detection of *T. gondii* antibodies in animals and animal products, their individual components and protocols, and how these influence diagnostic performance. The review provides direction on how to overcome existing limitations to developing more reliable and accurate ELISAs for the detection of *T. gondii*-specific antibodies in a wide range of animals.

2. Results

2.1. Literature Search and Eligible Articles

During the literature search, 8736 studies were identified across four databases (Web of Science: $n = 3179$, Scopus: $n = 2993$, CAB Abstracts: $n = 2201$, and AGRICOLA: $n = 363$). Following the removal of duplicates ($n = 4292$), 4444 studies were subjected to title and abstract screening. Using exclusion and inclusion criteria, a total of 115 articles were selected for full-text evaluation, which resulted in the inclusion of 57 studies published between 1984 and 2020 (Figure 1).

2.2. General Characteristics of Studies Included in the Review

Out of 57 studies describing the evaluation of diagnostic performance of different ELISAs for the detection of *T. gondii* antibodies, the majority originated from Europe ($n = 21$), followed by Asia ($n = 16$), South America ($n = 9$), North America ($n = 6$), Africa ($n = 3$) and Australia ($n = 2$) (Figure 2). Out of the 20 animal species covered, most of the studies (91%) focused on domesticated animals [33,37,46–50], but 7% were on zoo and wild animals [28,44,51,52], and one study targeted both domestic and wild animals [43]. Serum was the most widely used (89% of studies) sample for the detection of *T. gondii*-specific antibodies in animals, followed by meat juice/tissue fluid (7%) [37,45,53,54] and milk (4%) [55,56]. Most of the studies applied an indirect ELISA (95% of studies) [35,43,44,57,58] as the primary technique, with a smaller number using competitive ELISAs (3%) [59,60] or a combination of different methods (2%) (indirect IgG and IgM, blocking ELISA, reverse IgM capture ELISA) [29].

2.3. Type of Antigen Used in ELISAs

Three main types of antigen, including native (66% of the 50 studies) [28,35,43,51,61], recombinant (30%) [24,44,50,57,62], and chimeric proteins (4%) [7,48], were used for the detection of *T. gondii*-specific antibodies in animals. Seven studies did not clearly indicate the type of antigen used, and hence were not included in the timeline below (Figure 3). Until approximately 2010, native antigens were most frequently used; however, the use of recombinant antigens became common in the last decade (Figure 3). Native antigens used in ELISAs consisted of either tachyzoite-based products or whole tachyzoites, whereas the six main types of recombinant antigens comprised surface antigens (SAG), dense granule proteins (GRA), microneme proteins (MIC), cyst matrix antigens (MAG), and rhoptry antigens (ROP), along with other peptide fragments.

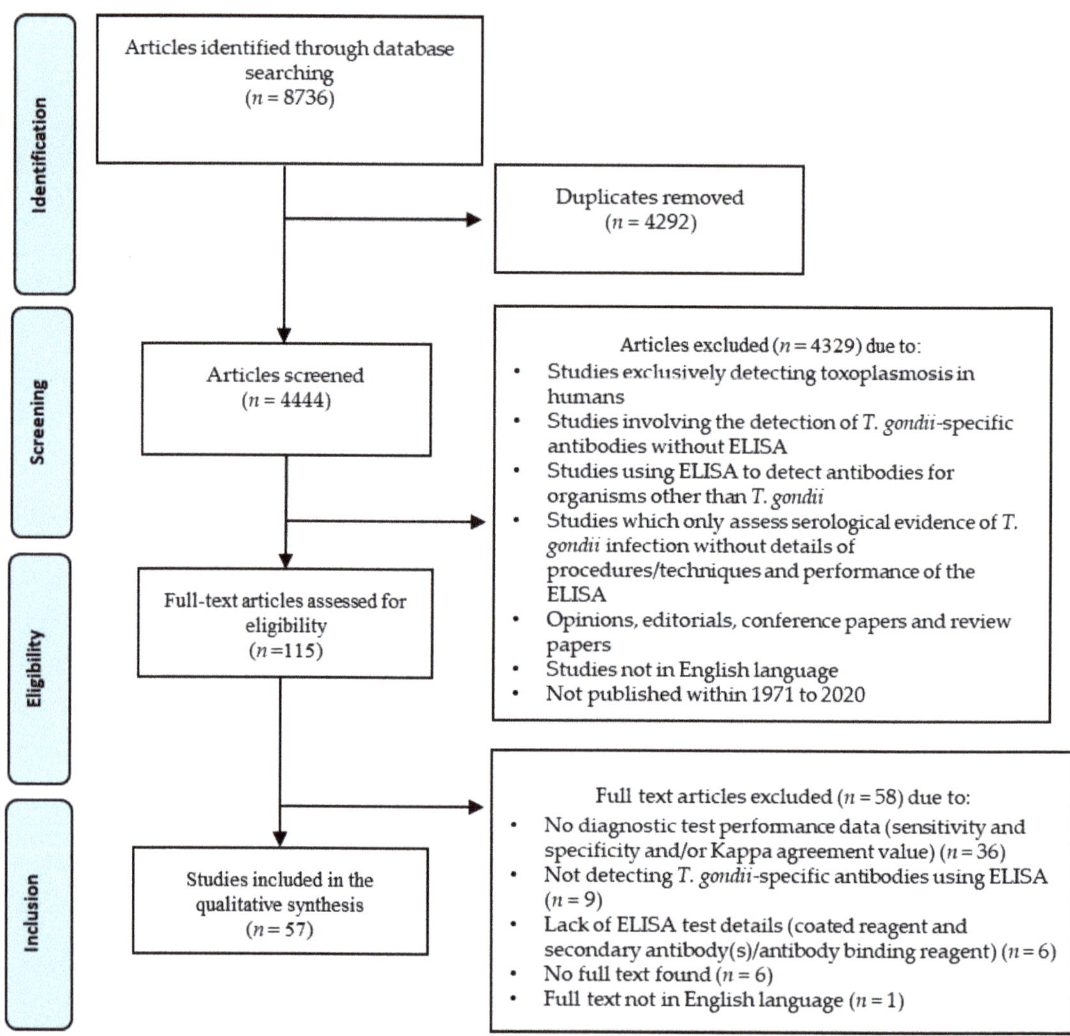

Figure 1. Preferred Reporting Items for Systematic Reviews and Meta-Analyses (PRISMA) flow diagram detailing the number of articles at each stage and the exclusion criteria applied.

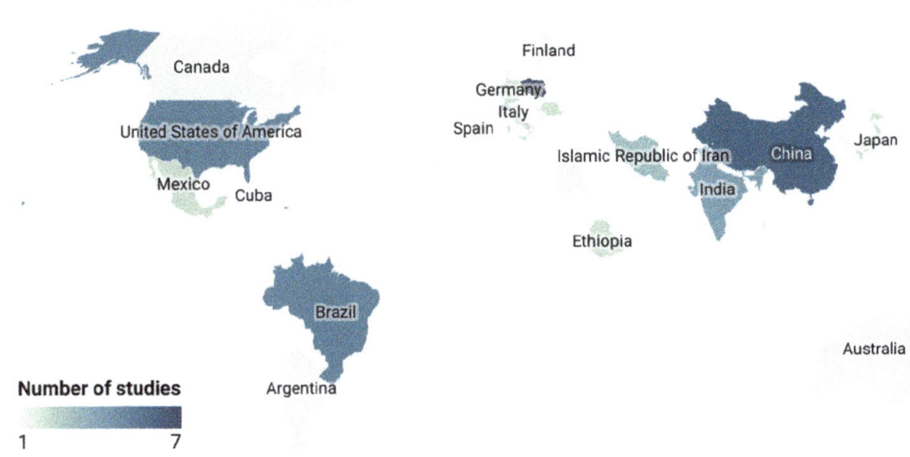

Figure 2. Geographical distribution of the studies (*n* = 57) included in the review. Detailed map can be accessed via https://datawrapper.dwcdn.net/MByRV/2/ (accessed on 30 March 2021).

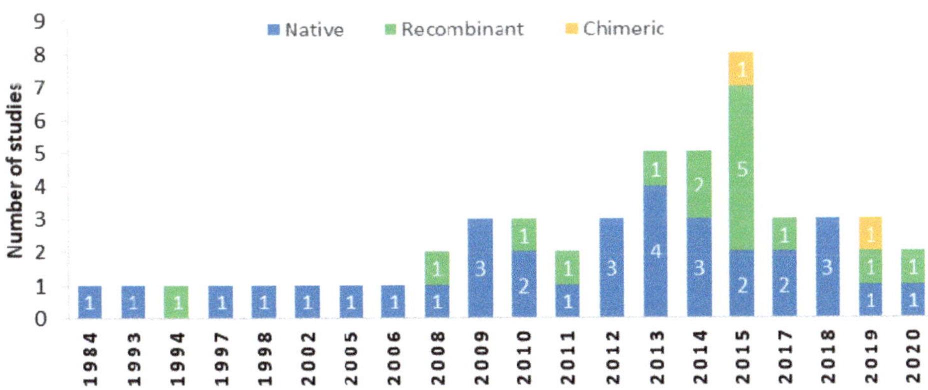

Figure 3. Frequency of antigen types used in ELISAs for the detection of *Toxoplasma gondii*-specific antibodies in animals over the study timeline. Native antigens (*n* = 33), recombinant (*n* = 15), and recombinant chimeric (*n* = 2).

2.4. Types of Antibodies Detected and Use of Secondary Antibodies/Antibody-Binding Reagents

Most of the studies (*n* = 52) focused on detecting *T. gondii*-specific IgG. Two studies detected specific IgM and IgG antibodies in pigs [29,63], where Lind et al. [29] used both an indirect IgM ELISA and a reverse IgM ELISA, while Terkawi et al. [63] only used an indirect IgM ELISA. Four studies tested *T. gondii*-specific IgY antibodies in birds, including three studies on chickens [33,47,64] and one in turkeys [62]. A variety of secondary antibodies or antibody-binding reagents were used. Based on their specificity for target species, these can be categorised into species-specific (74% of studies), multi-species (specific to a selected number of species) (10%), and non-species-specific (16%). A horseradish peroxidase (HRP) conjugate (not specified), which is a multi-species antibody-binding reagent, was used in both the ID Screen® Toxoplasmosis Indirect Multi-species (IDvet, Grabels, France) and Pigtype®Toxoplasma Ab (Qiagen, Leipzig, Germany) commercial ELISA kits. Protein A and G, either separately or combined, have also been used as non-species-specific reagents.

2.5. Types of ELISAs Used

The majority of studies used in-house ELISAs (67%) [28,43,64,65] followed by commercial ELISAs (24%) [21,37,66,67] and modified commercial ELISAs (9%) [44,47,52,68,69]. In modified commercial ELISAs, anti-human secondary antibodies were replaced by animal-specific secondary antibodies.

Six different commercial ELISA kits were used, including the ID Screen Toxoplasmosis Indirect Multi-species (IDvet) ($n = 6$), PrioCHECK *Toxoplasma* Ab porcine ELISA (Thermo Scientific, Zurich, Switzerland) ($n = 4$), Chekit-Toxotest (IDEXX Laboratories, Bern, Switzerland) ($n = 3$), *Toxoplasma gondii* Antibody Test Kit (SafePath Laboratories, Carlsbad, CA, USA) ($n = 2$), Pigtype *Toxoplasma* Ab (Qiagen, Leipzig, Germany) ($n = 1$) and Toxo SPA-ELISA Kit (Haitai Bio, Shenzhen, China) ($n = 1$). The ID Screen® Toxoplasmosis Indirect Multi-species (IDvet) ELISA kit uses the native P30 (SAG1) antigen and a multi-species HRP conjugate (not specified) to detect *T. gondii*-specific antibodies in multiple species. All other commercial kits use whole tachyzoite antigens, while no such information is available for the Chekit-Toxotest (IDEXX Laboratories). The PrioCHECK *Toxoplasma* Ab porcine ELISA (Thermo Scientific), Chekit-Toxotest (IDEXX Laboratories) and *Toxoplasma gondii* Antibody Test Kit (SafePath Laboratories) use species-specific secondary antibodies, whereas the Pigtype®*Toxoplasma* Ab (Qiagen) and Toxo SPA-ELISA kit (Haitai Bio) use a multi-species conjugate (not specified) and protein A, respectively, as secondary antibodies.

2.6. Diagnostic Performance

The diagnostic performance of different ELISAs was compared across different sample and antigen types, with the use of single vs. combinations of antigens and various antibody binding reagents. The following sections provide an overview of the diagnostic performance of ELISAs used for the detection of *T. gondii*-specific antibodies in animals.

2.6.1. Milk and Meat Juice ELISAs

Out of the five studies that utilised milk and meat juice samples, the ID Screen Toxoplasmosis Indirect Multi-species (IDvet) ELISA kit achieved more than 97% sensitivity and specificity and excellent agreement (Kappa value 0.949) for milk compared with serum, using the same ELISA kit. The reported sensitivity and specificity values for milk ranged from 88.7 to 97.55% and 97.42 to 97.83%, respectively. For meat juice, on the other hand, sensitivity and specificity values ranged from 3.6 to 96.7% and 83.9 to 100%, respectively (Table 1). Three commercial ELISA kits, including the PrioCHECK *Toxoplasma* Ab porcine ELISA (Thermo Scientific), Pigtype®*Toxoplasma* Ab (Qiagen), and ID Screen® Toxoplasmosis Indirect Multi-species (IDvet), reported better overall performance compared to other commercial kits using meat juice. The *Toxoplasma gondii* Antibody Test Kit (SafePath Laboratories) showed low sensitivity (3.6%), with slight agreement (Kappa value 0.05) with the MAT test [37]. However, the same ELISA kit reported a better sensitivity (88.6%) in another study using meat juice [45]. In-house ELISAs reported higher sensitivity and specificity values of 96.7% and 100%, respectively [54].

Table 1. Performance of different enzyme-linked immunoassays (ELISAs) in detecting *Toxcplasma gondii*-specific antibodies in milk and meat juice/tissue fluid samples.

Sample Type	ELISA	Host Species	Positive % (n/N)	Se (%)	Sp (%)	Agreement (Kappa Value)	Reference Test Used	Reference
Milk	In-house	Goat	20 (120/600)	88.7	97.4	ND	MAT (serum and milk)	[55]
	ID Screen® Toxoplasmosis Indirect Multi-species (IDvet)	Goat	59 (59/100)	97.55	97.8	0.949	Same commercial ELISA with serum	[56]
Meat Juice/ Tissue fluid	PrioCHECK *Toxoplasma* Ab porcine ELISA(Prionics)	Pig	41.1 (37/90)	96.4	83.9	0.74	MAT	[37]
	Pigtype®*Toxoplasma* Ab (Qiagen)	Pig	27.8 (25/90)	89.3	100	0.92	MAT	[37]
	ID Screen® Toxoplasmosis Indirect Multi-species (IDvet)	Pig	24.4 (22/90)	78.6	100	0.83	MAT	[37]
	Toxoplasma gondii Antibody Test Kit (SafePath Laboratories)	Pig	1.1 (1/90)	3.6	100	0.05	MAT	[37]
	Toxoplasma gondii Antibody Test Kit (SafePath Laboratories)	Pig	88.5 (62/70)	88.6	98	ND	Mouse bioassay	[45]
	In-house	Pig	6.2 (60/969)	96.7	100	ND	Commercial ELISA Kit	[54]

n—test positive; N—number of samples tested; Se—Sensitivity; Sp—Specificity; MAT—modified agglutination test; ND—no data.

2.6.2. Use of Single Recombinant Antigens

Recombinant antigens are specific immunogenic proteins produced in bacterial or appropriate eukaryotic systems using recombinant DNA technology, which are also used for the immunodetection of *T. gondii* infection [44,70]. Twelve different recombinant antigens across four major categories (surface granular antigens (SAG), dense granular proteins (GRA), microneme proteins (MIC) and peptide fragments) were used in a range of animal species (Table 2). Dense granular proteins (GRA) were the most frequently used recombinant antigens, with GRA7 being the most common (n = 8). Half of the studies using GRA7 reported excellent agreement with respective reference tests and high sensitivity (84.2–100%) and specificity (91.6–99.1%) (Table 2) [33,34,44,65]. Two studies claimed higher than 85% sensitivity and specificity, with substantial agreement [63,71], whereas one study reported a lower sensitivity and only fair agreement [46]. All studies using GRA1 (n = 4) reported higher than 75% sensitivity and specificity, and substantial agreement with the reference test. GRA2, GRA6, GRA14 and GRA15 were used less frequently (n = 1 per each antigen), the use of GRA6 and GRA14 resulted in higher than 80% sensitivity and substantial agreement with the reference tests, while GRA2 and GRA15 resulted in less than 30% sensitivity and fair to slight agreement with the reference tests, respectively.

Table 2. Comparison of sensitivity, specificity, and level of agreement with a reference test for different single recombinant antigen-based enzyme-linked immunoassays (ELISAs) to detect serum antibodies to *Toxoplasma gondii* in various animals.

Antigen Category	Antigen	Positive % (n/N)	Host Species	Se (%)	Sp (%)	Agreement (Kappa Value)	Reference Test Used for Comparison	Reference
Surface antigens (SAG)	SAG1	75 (39/52)	Jaguar	92.5	83.3	0.74	Commercial ELISA (TLA)	[44]
	SAG1	71.8 (181/252)	Cattle	84.38	87.9	0.73	IFAT	[24]
	SAG1	44.6 (25/56)	Goat	83.3	84.4	ND	MAT	[73]
	SAG2	ND	Cat	91.89	88.1	0.67	LAT	[46]
	SAG2	41.26 (26/63)	Goat	82.14	91.4	0.741	IFAT	[57]
	SAG2	50 (30/60)	Sheep	81.25	85.7	0.667	IFAT	[57]
	SAG2	64.44 (29/45)	Cattle	87.1	85.7	0.701	IFAT	[57]
	SAG2	61.5 (115/258)	Cattle	80	88.6	0.689	IFAT	[50]
Dense granule proteins (GRA)	GRA1	15.3 (20/131)	Mink	78.9	95.5	0.73	WB	[34]
	GRA1	75 (39/52)	Jaguar	92.5	83.3	0.74	Commercial ELISA (TLA)	[44]
	GRA1	16.4 (18/110)	Chicken	81.3	94.7	0.72	WB	[33]
	GRA1	16.2 (42/259)	Dog	81	95.4	0.66	ELISA (TLA)	[71]
	GRA2	ND	Cat	27.3	96.52	0.3	LAT	[46]
	GRA6	ND	Cat	82.43	88.7	0.62	LAT	[46]
	GRA7	ND	Cat	35.1	89.9	0.27	LAT	[46]
	GRA7	21.6 (40/185)	Cat	89.7	92.5	0.92	IFAT/MAT	[65]
	GRA7	13 (17/131)	Mink	84.2	99.1	0.83	WB	[34]
	GRA7	76.9 (40/52)	Jaguar	97.5	91.6	0.89	Commercial ELISA (TLA)	[44]
	GRA7	15.5 (17/110)	Chicken	100	98.9	0.96	WB	[33]
	GRA7	55.9 (33/59)	Pig	90.63	85.2	0.76	LAT	[63]
	GRA7	42.8 (24/56)	Goat	80	84.4	ND	MAT	[73]
	GRA7	16.2 (42/259)	Dog	91	97.7	0.8	ELISA (TLA)	[71]
	GRA14	47.4 (28/59)	Pig	81.25	92.6	0.73	LAT	[63]
	GRA15	ND	Cat	17.57	86.4	0.04	LAT	[46]
Microneme proteins (MIC)	MIC3	41.1 (81/197)	Pig	ND	ND	0.86	MAT	[58]
	MIC3	45.8 (11/24)	Dog	ND	ND	0.85	MAT	[58]
	MIC10	ND	Cat	16.21	85.8	0.02	LAT	[46]

Table 2. *Cont.*

Antigen Category	Antigen	Positive % (n/N)	Host Species	Se (%)	Sp (%)	Agreement (Kappa Value)	Reference Test Used for Comparison	Reference
Other peptide fragments	H4	25.81 (79/306)	Cat	93	100	ND	DAT, IFAT, DT	[72]
	H11	12.41 (38/306)	Cat	64	100	ND	DAT, IFAT, DT	[72]

n—test positive; N—number of samples tested; Se—Sensitivity; Sp—Specificity; MAT—modified agglutination test; IFAT—indirect fluorescent antibody test; WB—Western blot; LAT—latex agglutination test; DAT—direct agglutination test; TLA—*Toxoplasma gondii* lysate antigen; ND—no data.

Among SAG antigens, SAG 2 was used more frequently ($n = 5$) and with higher sensitivity (80–91.89%) and specificity (85.71–91.43%) than SAG 1 ($n = 3$; higher than 83% sensitivity and specificity). The least commonly used antigens were MIC and peptide fragments. Zhang et al. [58] used MIC3 in two host species and obtained excellent agreement. In contrast, the use of MIC10 achieved only low sensitivity and slight agreement [46]. Among recombinant polypeptide proteins used in cats, H4 performed better than H11 [72].

2.6.3. Use of Combinations of Recombinant Antigens

Seventeen different recombinant antigen combinations (M1–M17), containing between two and five antigens, were used (Table 3). Two additional categories of recombinant antigen, cyst matrix antigen (MAG), and rhoptry antigen (ROP), which were not used in single recombinant antigen ELISAs, were used in recombinant antigen combinations. Out of seventeen different combinations, nine combinations (M1, M2, M3 in jaguars, M4, M7, M8, M9, M14 in sheep and M15) resulted in the strongest performance, with greater than 90% sensitivity and specificity. Antigens from the SAG and/or GRA categories were widely used among combinations. Either SAG or GRA antigens were included in sixteen out of seventeen combinations (all combinations except M1). Moreover, combinations of both SAG and GRA antigens were used in seven out of seventeen instances (M2, M3, M4, M5, M6, M14, M17), which resulted in good sensitivity (77.8–100%) and specificity (84.4–100%). Furthermore, certain recombinant antigen mixtures, including M1 (H4 + H11), M2 (SAG1 + GRA7), M3 (SAG1 + GRA7), M5 (SAG2 + GRA6), M6 (SAG2 + GRA7), M10 (GRA2 + GRA7), M11 (GRA6 + GRA), M16 (GRA2 + GRA6+ GRA7 + GRA15) and M17 (SAG2 + GRA2 + GRA6 + GRA7 + GRA15), had better diagnostic performances (either sensitivity or specificity or both/Kappa values) than if being used as single recombinant antigens (Table 2) in the same host species. Moreover, variable diagnostic performance was observed when the same antigen combination was used in different animal species (M3, M13, M14, M15).

2.6.4. Recombinant Chimeric Antigens

Chimeric antigens are a new generation of recombinant antigens and have only been used in two studies. They are made by the fusion of two or more fragments of well-known antigens, hence containing multiple immunoreactive epitopes from each antigen. Nine different combinations of such recombinant chimeric antigens were reported for use in horses, sheep, goats, and pigs (Table 4), but none of the studies reported the level of Kappa agreement with their respective reference test. Of the nine chimeric antigens used, CM5 (SAG2-GRA1-ROP1L) and CM8 (AMA1-SAG2-GRA1-ROP1) were the most effective, with high sensitivity (93.8–100% and 95.56–97.92%, respectively) and specificity (100% in both). Both CM5 and CM8 comprised fragments from SAG, GRA, ROP antigens, and CM8 additionally included fragments of apical membrane antigen (AMA). The remaining chimeric antigens varied in test performance, with sensitivity ranging from 28.4 to 100% and specificity ranging from 95.12 to 100%, depending on the animal species tested. All chimeric antigens were reported to have high specificity of more than 95% across all animal species tested.

Table 3. Comparison of sensitivity, specificity, and level of agreement with a reference test for different recombinant antigen combinations based enzyme-linked immunoassays (ELISAs) to detect serum antibodies to *Toxoplasma gondii* in various animals.

Combination of Antigens	Antigens	Positive % (n/N)	Host Species	Se (%)	Sp (%)	Agreement (Kappa Value)	Reference Test Used for Comparison	Reference
M1	H4 + H11	31.37 (96/306)	Cat	95	100	ND	DAT, DT, IFAT	[72]
M2	SAG1 + GRA1	75 (39/52)	Jaguar	95	91.6	0.84	Commercial ELISA (TLA)	[44]
M3	SAG1 + GRA7	76.9 (40/52)	Jaguar	97.5	91.6	0.89	Commercial ELISA (TLA)	[44]
		46.4 (26/56)	Goat	86.6	84.4	ND	MAT	[73]
M4	SAG2 + GRA1	81.5 (88/108)	Sheep	100	95	ND	ND	[70]
M5	SAG2 + GRA6	ND	Cat	94.59	89.6	0.72	LAT	[46]
M6	SAG2 + GRA7	ND	Cat	90.54	85.5	0.62	LAT	[46]
M7	SAG2 + ROP1	81.5 (88/108)	Sheep	100	95	ND	ND	[70]
M8	GRA1 + ROP1	81.5 (88/108)	Sheep	100	100	ND	ND	[70]
M9	GRA1 + GRA7	76.9 (40/52)	Jaguar	97.5	91.6	0.89	Commercial ELISA (TLA)	[44]
M10	GRA2 + GRA7	ND	Cat	44.59	89.3	0.35	LAT	[46]
M11	GRA6 + GRA7	ND	Cat	74.32	89	0.58	LAT	[46]
M12	GRA7 + GRA8	20.2 (387/1913)	Turkey	92.6–100	78.1–100	ND	ND	[62]
M13	SAG1 + MIC1 + MAG1	37.21 (32/86)	Horse	88.9	100	ND	DAT, IFAT	[7]
		57.07 (109/191)	Sheep	77.9	92.2	ND	DAT, IFAT	[7]
		42.86 (72/168)	Pig	88.9	100	ND	DAT, IFAT	[7]
M14	SAG2 + GRA1 + ROP1	32.56 (28/86)	Horse	77.8	100	ND	DAT, IFAT	[7]
		73.30 (140/191)	Sheep	100	100	ND	DAT, IFAT	[7]
		39.29 (66/168)	Pig	81.5	100	ND	DAT, IFAT	[7]
		81.5 (88/108)	Sheep	100	100	ND	ND	[70]
M15	GRA1 + GRA2 + GRA6	36.05 (24/86)	Horse	66.7	100	ND	DAT, IFAT	[7]
		70.16 (129/191)	Sheep	92.1	100	ND	DAT, IFAT	[7]
		46.43 (44/168)	Pig	54.3	100	ND	DAT, IFAT	[7]

Table 3. *Cont.*

Combination of Antigens	Antigens	Positive % (n/N)	Host Species	Se (%)	Sp (%)	Agreement (Kappa Value)	Reference Test Used for Comparison	Reference
M16	GRA2 + GRA6+ GRA7 + GRA15	ND	Cat	70.27	86.1	0.5	LAT	[46]
M17	SAG2 + GRA2 + GRA6 + GRA7 + GRA15	ND	Cat	89.19	95.4	0.81	LAT	[46]

n—test positive; N—number of samples tested; Se—Sensitivity; Sp—Specificity; MAT—modified agglutination test; IFAT—indirect fluorescent antibody test; WB—Western blot; LAT—latex agglutination test; DAT—direct agglutination test; DT—Dye test; TLA—*Toxoplasma gondii* lysate antigen; ND—no data

Table 4. Comparison of sensitivity and specificity values for different recombinant chimeric antigen-based enzyme-linked immunoassays for the detection of antibodies to *Toxoplasma gondii* in various animals.

	Combination of Chimeric Recombinant Antigens	Species	Positive % (n/N)	Se (%)	Sp (%)	Reference Test	Reference
CM1	GRA1-GRA2-GRA6	Horse	36.1 (31/86)	86.1	100	DAT, IFAT	
		Sheep	70.2 (134/191)	95.7	100	DAT, IFAT	
		Pig	46.4 (78/168)	96.3	100	DAT, IFAT	
CM2	MIC1-MAG1-SAG1s	Horse	31.4 (27/86)	75	100	DAT, IFAT	
		Sheep	71.7 (137/191)	97.9	100	DAT, IFAT	
		Pig	22.0 (37/168)	45.7	100	DAT, IFAT	
CM3	SAG1L-MIC1-MAG1	Horse	32.6 (28/86)	77.8	100	DAT, IFAT	
		Sheep	73.3 (140/191)	100	100	DAT, IFAT	[7]
		Pig	43.5 (73/168)	90.1	100	DAT, IFAT	
CM4	SAG2-GRA1-ROP1S	Horse	21 (18/86)	50	100	DAT, IFAT	
		Sheep	73.3 (140/191)	100	100	DAT, IFAT	
		Pig	13.7 (23/168)	28.4	100	DAT, IFAT	
CM5	SAG2-GRA1-ROP1L	Horse	41.9 (36/86)	100	100	DAT, IFAT	
		Sheep	73.3 (140/191)	100	100	DAT, IFAT	
		Pig	45.2 (76/168)	93.8	100	DAT, IFAT	
CM6	AMA1N-SAG2-GRA1-ROP1	Sheep	Not clearly mentioned	97.9	97.62	LAT, IFAT	
		Goat	Not clearly mentioned	88.9	100	LAT, IFAT	
CM7	AMA1C-SAG2-GRA1-ROP1	Sheep	Not clearly mentioned	95.8	95.24	LAT, IFAT	
		Goat	Not clearly mentioned	95.6	97.56	LAT, IFAT	[48]
CM8	AMA1-SAG2-GRA1-ROP1	Sheep	Not clearly mentioned	97.9	100	LAT, IFAT	
		Goat	Not clearly mentioned	95.6	100	LAT, IFAT	
CM9	SAG2-GRA1-ROP1-GRA2	Sheep	Not clearly mentioned	97.9	97.62	LAT, IFAT	
		Goat	Not clearly mentioned	57.8	95.12	LAT, IFAT	

n—number of samples tested positive; N—total number of samples tested; Se—Sensitivity; Sp—Specificity; IFAT—indirect fluorescent antibody test; LAT—latex agglutination test; DAT—direct agglutination test.

2.6.5. Comparison of Native and Recombinant/Chimeric Antigens

Nine different studies compared the diagnostic performance of native and recombinant/chimeric antigens in eight species, resulting in twelve comparisons provided here (Table 5). Native antigens were either lysate antigens from whole tachyzoites (TLA) or soluble antigens (TSA) of *T. gondii*. Only the most effective recombinant antigen or antigen combination from each study (according to the authors of each study when multiple recombinant antigens or combinations were used) were included in this comparison. Most comparisons (8/12) reported similar or slightly higher sensitivity for recombinant and chimeric antigens (84.2–100%) compared to native antigens (68.4–100%). However, in four instances, native antigens produced slightly better sensitivity than a recombinant antigen combination: M17, M1 and two chimeric antigens, CM5 in pigs and CM8 in goats (Table 5). Recombinant and chimeric antigens reported overall better specificity than native antigens. Eleven out of 12 comparisons reported similar or higher specificity in recombinant/chimeric antigens, with values ranging from 95.36 to 100%. One study reported slightly higher specificity (99.3%) for the native antigen compared to GRA7 (92.5%) [65]; however, three other comparisons achieved higher specificity using GRA7 than with native antigens [33,34,71].

Table 5. Comparison of sensitivity and specificity between native and recombinant/chimeric antigen(s) in detecting *Toxoplasma gondii*-specific antibodies in multiple animal species.

Species	Antigen (s)	Se (%)	Sp (%)	Reference
Cat	M17 (SAG2 + GRA2 + GRA6 + GRA7 + GRA15) TLA	89.19 97.29	95.36 93.62	[46]
Cat	GRA7 TLA	89.7 84.6	92.5 99.3	[65]
Cat	M1 (H4 + H11) TSA	95 98	100 99	[72]
Pig	CM5 (SAG2-GRA1-ROP1L) TLA	93.8 100	100 100	[7]
Horse	CM5 (SAG2-GRA1-ROP1L) TLA	100 100	100 100	[7]
Mink	GRA7 TSA	84.2 68.4	99.1 96.4	[34]
Sheep	CM5 (SAG2-GRA1-ROP1L) TLA	100 100	100 100	[7]
Sheep	M14 (GRA1 + SAG2 + ROP1) TLA	100 100	100 100	[70]
Sheep	CM8 (AMA1-SAG2-GRA1-ROP1) TLA	97.92 97.92	100 100	[48]
Goat	CM8 (AMA1-SAG2-GRA1-ROP1) TLA	95.56 97.78	100 100	[48]
Chicken	GRA7 TSA	100 93.8	98.9 97.9	[33]
Dog	GRA7 TLA	91 88.1	97.7 96.8	[71]

TLA—Whole tachyzoites; TSA—*Toxoplasma gondii* soluble antigens; Se—Sensitivity; Sp—Specificity.

2.6.6. Diagnostic Performance of Non-Species-Specific Antibody Binding Reagents

Non-species-specific secondary antibody binding reagents have the advantage of being able to detect antibodies across a broad range of hosts without the need for species-specific conjugates. In this review, 16% of studies used the non-species-specific reagents protein A, protein G, and protein A/G, targeting eleven different species of animals and using indirect ELISAs (Table 6). Protein A/G was used in five different studies across 11 different mammalian species, with all of them (except one study where the Kappa value was not given) reporting substantial to excellent agreement with their reference test. Sensitivity and specificity of using protein A/G were 68.4–92% and 89–99.1%, respectively. Protein A was used in four different studies in three host species, and agreement with the reference test varied from moderate to excellent, reporting sensitivity and specificity values of 89.5–100% and 82–100%, respectively.

Table 6. Comparison of sensitivity, specificity, and level of agreement with a reference test for non-species-specific antibody binding reagents used in enzyme-linked immunoassays to detect antibodies to *Toxoplasma gondii* in various animals.

Conjugate	Species	Positive % (n/N)	Se (%)	Sp (%)	Agreement (Kappa Value)	Reference Test	Reference
Protein A/G	Pig	28.5 (4/14)–76.9 (10/12)	88.6	93.9	0.8	Commercial ELISA	[43]
			88.6	93.9	0.8	MAT	[43]
			84.8	96.8	0.8	WB	[43]
	Cat	100 (11/11)	ND	ND	1	MAT	[43]
	Mice	0 (0/3)–100% (3/3)	ND	ND	1	MAT	[43]
	Seal	0 (0/4)–(14/14)	ND	ND	0.8	MAT	[43]
	Mink	13 (17/131)–15.3(20/131)	68.4–84.2	95.5–99.1	0.7–0.8	WB	[34]
	White-tailed deer	42.2 (113/268)	92	89	ND	ND	[52]
	Alpaca Sheep Goat Horse Dog	57.1 (8/14) Alpaca 58.8 (10/17) Sheep 64.7 (11/17) Goat 13.3 (2/15) Horse 48.2 (27/56) Dog	92 (Overall value)	89 (Overall value)	0.81 (Overall value)	IHA	[68]
	Pig	Not clear	ND	ND	0.9	MAT	[58]
		Not clear	ND	ND	0.8	Commercial ELISA	[58]
	Dog	45.8 (11/24)	ND	ND	0.9	MAT	[58]
	Cat	38.5 (5/13)	ND	ND	0.9	MAT	[58]
Protein A	Goat	22 (132/600)	89.5	97.9	ND	MAT	[55]
	Dog	ND	93	82	0.8	IHA	[69]
	Cat	ND	100	100	1	IHA	[69]
	Dog	84 (178/212)	75-80	80-85	ND	WB	[74]
	Dog	34.7 (42/121)	ND	ND	0.6	MAT	[75]
	Cat	35.5 (16/45)	ND	ND	0.5	MAT	[75]
Protein G	Goat	ND	97	100	0.9	IHA	[69]
	Horse	ND	72	100	0.7	IHA	[69]
	Alpaca	ND	76	95	0.7	IHA	[69]
	Sheep	ND	91	100	0.9	IHA	[69]

n—number positive; N—number tested; Se—Sensitivity; Sp—Specificity; MAT—modified agglutination test; IFAT—indirect fluorescent antibody test; WB—Western blot; IHA—Indirect hemagglutination assay; ND—no data

Protein G was only used in one study, across four different species, producing substantial to excellent agreement with IHA as the reference test, and varying sensitivity (72–97%) and specificity (95–100%).

2.6.7. Cross-Reactivity of ELISAs Used for Animals

Eleven studies investigated, or mentioned, the potential cross-reactivity of antigens (Table 7). *Neospora caninum* was the most tested organism for cross-reactivity (*n* = 7), but none of the studies reported cross-reactions with that species. Several other studies tested other organisms, including *Sarcocystis* spp., *Besnoitia* spp., *Isospora suis*, *Trichinella* spp., *Ascaris suum*, *Salmonella* spp., *Yersinia* spp. and *Actinobacillus* spp., but did not detect cross-reactivity with *T. gondii*. Based on the studies included in this review, cross-reactivity was only observed with *Hammondia hammondi* and *Eimeria* spp. in experimentally infected turkeys when a recombinant antigen combination M12 (GRA7 and GRA8) was used [62].

Table 7. Studies testing or describing possible cross-reactivity in detecting *Toxoplasma gondii*-specific antibodies using enzyme-linked immunoassays in various animal species.

Possible Cross-Reactive Species	Host Species	Antigen(s) Used	Comments	Reference
Neospora caninum	Pig, Cat, Mice, Seal	Soluble tachyzoite	No cross-reactions were reported	[43]
	Chicken	Sonicated tachyzoite antigens	No cross-reactions were reported Dual infection with *T. gondii* and *N. caninum* was reported	[64]
	Sheep	M14 (GRA1 + SAG2 + ROP1)	No cross-reactions were reported	[70]
	Goat, Sheep	CM6 (AMA1N-SAG2-GRA1-ROP1) CM7 (AMA1C-SAG2-GRA1-ROP1) CM8 (AMA1-SAG2-GRA1-ROP1) CM9 (SAG2-GRA1-ROP1-GRA2)	No cross-reactions were reported	[48]
	Dog	Native purified SAG1 (From tachyzoites)	No cross-reactions were reported	[76]
	Turkey	M12 (GRA7 & GRA8)	No cross-reactions were reported with *N. caninum* positive turkeys Cross-reactions were observed in turkeys which were experimentally infected with *Hammondia hammondi* and turkey-specific *Eimeria* spp.	[62]
	White-tailed deer	Crude extract antigen	No cross-reactions were reported Dual infection with *T. gondii* and *N. caninum* was reported	[51]
Sarcocystis spp.	Cattle	Sonicated *T. gondii* antigens	Authors mentioned that some seropositive results for *T. gondii* in cattle could be due to cross-reactions with anti *Neospora* or anti *Sarcocystis* antibodies	[60]
	Pig	Crude rhoptries	No cross-reactions were reported	[77]
Besnoitia spp	Pig, Cat Mice, Seal	Soluble tachyzoite	No cross-reactions were reported	[43]
Isospora suis	Pig	Tachyzoite lysate	No cross-reactions were reported	[29]
Trichinella spp.	Pig, Cat Mice, Seal	Soluble tachyzoite	No cross-reactions were reported	[43]
	Pig	Tachyzoite lysate	No cross-reactions were reported	[29]
Ascaris suum	Pig	Tachyzoite lysate	No cross-reactions were reported	[29]

Table 7. *Cont.*

Possible Cross-Reactive Species	Host Species	Antigen(s) Used	Comments	Reference
Bacteria species *Salmonella* *Yersinia* *Actinobacillus*	Pig	Tachyzoite lysate	No cross-reactions were reported	[29]
No specific species	Pigs	Purified native SAG1	Authors mentioned the possibility of having low cross-reactions with purified native SAG 1	[49]

3. Discussion

ELISA is one of the most effective serological techniques used to detect exposure to *T. gondii* in animals [23,43]. This review evaluated 57 articles describing the performance of different ELISAs in detecting *T. gondii* antibodies in 20 different animal species. To the best of our knowledge, this is the first systematic review to provide a descriptive comparison on the performance of different ELISAs for the detection of *T. gondii* antibodies across different animal species. The results highlight the potential opportunities for refinements in ELISAs to be used in animals, including wildlife.

Researchers used different reference tests to estimate and compare the diagnostic performance of their ELISAs. Given that no perfect diagnostic test exists to detect toxoplasmosis in animals, each reference test is likely to differ in sensitivity and specificity [78], thereby affecting the calculated performance of the ELISA in question. Hence, comparison of diagnostic performance across different studies is complex and some seemingly perfect results (100% sensitivity and/or specificity) should be interpreted with care because perfect results may not necessarily mean the ELISA is perfect. In addition, sample sizes varied across studies and study power should be taken into consideration in serological comparison studies. The sample size is positively related to statistical power, and sufficiently large sample sizes are important in obtaining more accurate, valid, and reliable results [47,79].

There are clear, practical advantages to being able to use a single ELISA kit across multiple species and different types of sample. The most commonly used commercial kit, ID Screen® Toxoplasmosis Indirect Multi-species (IDvet, France), uses native P30 (SAG1) antigen and anti-multi-species conjugate as the secondary antibody, which makes it suitable for the detection of *T. gondii*-specific antibodies in ruminant, swine, dog and cat serum, but also in milk and meat juice [37]. It may subsequently be useful for the detection of *T. gondii* antibodies in wild ruminants, porcine species, canids, and felids. Moreover, the use of SAG 1 antigen in this kit might provide greater specificity than tests using whole tachyzoite antigen [49]. Thus, future validation of this test in a range of wildlife and other animal species suspected to be infected with *T. gondii* could be worthwhile.

The indirect ELISA was the most commonly used method in the studies included in the review. Indirect ELISAs involve two antibody-binding steps: first, primary antibodies in the sample bind to the immobilised coated antigen, and then labelled secondary antibodies bind to the primary antibodies, allowing signal amplification and identification of the primary antibody of interest [80]. A wide range of secondary antibodies are commercially available for most domesticated animal species, making the indirect ELISA method versatile for use in those species. An advantage of ELISAs over other serological methods is that they can be used for the detection of a range of immunoglobulin classes, including IgG, IgM, IgA, IgD and IgE [81–84]. However, most studies focused on animals only aim to detect mammalian IgG and IgM, and avian IgY antibodies (which resemble mammalian IgG). In fact, most studies (*n* = 52) focused on the detection of IgG or did not discriminate between the classes of antibodies. Only two studies specifically reported the detection of IgM [29,63]. IgM antibodies are often short-lived (2–4 weeks) and classically regarded as a marker for acute infection [85]. However, IgM antibodies can persist from a few months to a year, meaning that positive IgM results alone are not sufficient to discriminate between phases of infection [86]. However, testing IgM levels is useful when coupled with other diagnostic tests (e.g., IgG avidity testing, polymerase chain reaction) and widely used to determine the phase of infection in pregnancy-associated toxoplasmosis in humans [86,87]. IgG antibodies often persist for life in immunocompetent individuals, providing a reliable marker after primary infection, suggesting that a switch from isotype IgM to IgG has occurred, and they detect the chronic phase of *T. gondii* exposure in both animals and humans [29,30,85].

Based on the overall data reviewed, recombinant and chimeric antigens resulted in similar or better diagnostic performance than native *T. gondii* antigens. *Toxoplasma gondii* tachyzoite-based native antigens have commonly been used across many serological tests, including ELISA [23]. However, recombinant and chimeric antigens possess other advantages over native antigens, including the ease of production, reduced cost, and less exposure to biohazardous procedures [88]. Moreover, assay optimisation and standardisation are easier and researchers have the freedom to precisely construct the required antigen composition of interest [48,88]. However, there are some disadvantages, including inefficient expression and misfolding of proteins during the production process inside the prokaryotic systems, which could affect the affinity of the assay [89].

This review identified five well-defined categories of recombinant antigens (SAG, GRA, MIC, MAG, ROP), and members of each category were used as single antigens, or in a combination. Except for M1 (H4 + H11), all recombinant antigen combinations contained SAG and/or GRA antigens, indicating their wide use. Among single recombinant antigens, SAG and GRA were most frequently used. SAG1 and SAG2 are highly immunodominant and abundant antigens, present in the tachyzoite stage [90–92]. Moreover, SAG1 and SAG2 are highly conserved among different *T. gondii* strains and can be detected in both acute and chronic infections, thereby improving their value as diagnostic markers [57,91,93]. Furthermore, in human studies, both SAG1 [91,94] and SAG2 [93,95,96] have successfully been used to detect *T. gondii*-specific antibodies.

Dense granular proteins (GRA) are involved in *T. gondii*'s replication inside the host cells and are secreted by both tachyzoites and bradyzoites [97]. GRA7 is well-expressed on both the surface and within the cytoplasm of the infected host cell, resulting in direct exposure to the host's immune system and provoking a strong immune response in both acute and chronic toxoplasmosis [98,99]. Overall, this antigen performed better than several other GRAs across several host species, although one study reported poor sensitivity (35.1%) and only fair agreement with the reference test (LAT). However, while GRA7 was reported to have higher than 80% sensitivity and specificity, other members of the GRA family were not widely reported among studies.

Combinations of recombinant antigens tended to perform better than single recombinant antigens. The performance of single recombinant antigens is likely variable because they do not represent the same complex of epitopes as seen in native antigens [70]. Infected hosts mount varying humoral immune responses depending on the infective stage of the parasite; thus, a single recombinant antigen is probably not capable of binding with all stage-specific antibodies [70]. In contrast, a combination of different recombinant antigens may allow *T. gondii*-specific antibodies to recognise multiple epitopes from different parasitic stages [48,70]. Based on the diagnostic performance reported, the combination of SAG and GRA antigens, including SAG1, SAG2, and GRA7, may provide better diagnostic performance.

Only two studies reported the use of the novel generation of chimeric antigens. Similar to combinations of recombinant antigens, the enhanced epitope complexity of chimeric antigens probably results in recognition of different parasitic stages [7,48]. Thus, stage-specific *T. gondii* antibodies can be identified in the serum of the infected host, improving the sensitivity of the assay [7,48]. CM5 (SAG2-GRA1-ROP1L) and CM8 (AMA1-SAG2-GRA1-ROP1) performed better than other chimeric antigens. Despite the promising results reported to date, the diagnostic performance of chimeric antigens should be evaluated compared to combinations of recombinant antigens across many species to understand the broader value of this new, promising serological tool. Nevertheless, chimeric antigens have been successfully used as vaccine candidates [100] and to detect *T. gondii* infection in humans [101], and as a diagnostic tool to detect other pathogenic infections in humans [102–104].

In this review, three different categories of secondary antibodies/antibody binding reagents were identified based on their specificity to target species. Species-specific secondary antibodies were widely used in the studies [49,64,71]. This is likely a consequence of the fact that most of the studies focused on common domesticated animals, for which a wide range of species-specific or taxon-specific secondary antibodies are commercially available. However, the availability of specific-specific secondary antibodies is limited for other animals, such as wildlife species [10,43]. Subsequently, researchers have used multispecies and non-species-specific reagents to overcome this problem. Multispecies conjugates were used in two commercial ELISA kits, including the commercially available ID Screen® Toxoplasmosis Indirect Multi-species (IDvet) and Pigtype®Toxoplasma Ab (Qiagen) ELISA kits. However, data about the exact reagents present in those two kits were not available. Protein A/G combinations or protein A and G separately were used in several studies as non-species-specific antibody binding reagents. Both protein A and G are bacterial proteins that are capable of binding with the Fc region of the mammalian IgG [105], but appear to have poor binding capability for bird and reptilian antibodies [106]. Protein A/G combines the IgG binding capabilities of both protein A and G, and can therefore be used as a reliable serological tool for the detection of IgG antibodies across a wide range of mammal species [43,106]. Nonetheless, varying results in the binding capability of these non-species-specific reagents with IgG can be expected among different target host species. This might be due to slight variations in both binding ability and structure of the binding domains of the IgG as a result of genetic variation between species [43]. Hence, prior to use in an immunoassay, assessment of immunoglobulin binding capability of non-species-specific reagents with the target species is advisable [107,108].

Cross-reactivity between *T. gondii* antigens and antibodies against other organisms can reduce the specificity of serological assays [43]. Only one study, using the antigen M12 (GRA7 and GRA8), reported cross-reactivity (with *Hammondia hammondi* and *Eimeria* spp.) [62]. The apicomplexan parasites *T. gondii* and *H. hammondi* have structural similarities; cats act as a definitive host for both parasites [109]. Antigenic similarities between both parasites and the presence of cross immunity in infected hosts have been reported in other studies [109,110]. Thus, caution should be exercised when selecting the antigen, and ELISAs should be evaluated for possible cross-reactivity to optimise specificity [43].

Despite the effort to obtain and include all available literature related to this review, publication bias may have influenced the selection of studies, and some studies which were of low quality, unavailable, or in other languages were not included. Additionally, this review has provided a narrative synthesis of results, and statistical tests to compare results across various studies were not performed. Studies used different reference tests for comparison with ELISA performance; therefore, we did not evaluate the performance of such reference tests. Thus, future studies might consider statistical modelling approaches to standardise the performance of reference tests.

4. Materials and Methods

4.1. Review Protocol

This systematic review was reported according to the Preferred Reporting Items for Systematic Reviews and Meta-Analyses (PRISMA) guidelines (http://www.prisma-statement.org/, accessed on 30 March 2021). The review protocol was registered in an international prospective register of systematic reviews (PROSPERO) with the registration ID (CRD42020208925).

4.2. Search Strategy

A systematic search in four online scientific databases (Web of Science, Scopus, CAB Abstracts, Agricola) was conducted. The search strategy involved the use of Boolean operators with the following search terms: (*"Toxoplasma"* OR *"Toxoplasmosis"* OR *"T. gondii"* OR *"Toxoplasma gondii"* OR *"Toxoplasm*"*) AND (*"ELISA"* OR *"enzyme linked immunosorbent assay"* OR *"*ELISA"*). Only studies published in English and between 1971 and 2020 (since ELISA was developed, in 1971) were included [111].

4.3. Quality Assessment and Selection

Citations and abstracts were exported into EndNote X9 (Clarivate Analytics, Philadelphia, PA, USA) and all duplicates were removed. Subsequently, title and abstract screening was performed to select relevant articles based on the following exclusion criteria: (1) studies exclusively detecting toxoplasmosis in humans; (2) studies involving the detection of *T. gondii*-specific antibodies in animals using techniques other than ELISA; (3) studies using ELISA to detect antibodies in animals for organisms other than *T. gondii*; (4) studies which only assessed the serological evidence of toxoplasmosis without providing the details of procedures/techniques and diagnostic performance; and (5) opinions, editorials, conference papers, and review papers. Selected articles were then read in full to review their eligibility for inclusion. In addition to the above-mentioned exclusion criteria, studies without details of antigen-coated/immobilised agent and secondary antibody/antibody binding reagent were excluded. Studies providing information on diagnostic performance, using either sensitivity and specificity of the ELISA or Kappa agreement values, were included (Figure 1).

4.4. Data Extraction and Analyses

Data extraction was performed by the first author (K.L.D.T.D.L.). Any discrepancy was resolved by discussion with co-authors, who acted as secondary reviewers where required. The following data were extracted from eligible studies and documented in an Excel spreadsheet: species, sample type (serum, meat juice, milk) and size, animal category (domesticated, laboratory, wild and zoo), nature of the ELISA (in-house, commercial), ELISA type (direct, indirect, sandwich, competitive, other modifications, e.g., reverse), type of antigen(s) used (native, recombinant, chimeric), secondary antibody/immunoglobulins binding reagent, antibodies detected (IgG or IgM or both), the total number of samples tested and the number of positives, sensitivity and specificity of the ELISA, agreement (Kappa value) of the ELISA with the reference test, and cross-reactions. Animal types were categorised as follows: companion and livestock animals as domesticated animals; mice, hamsters, and guinea pigs as lab animals; free-ranging, undomesticated, and zoo animals as wild and zoo animals. All Kappa agreement values reported in the included studies were interpreted as follows: less than chance (≤ 0), slight (0.01–0.20), fair (0.21–0.40), moderate (0.41–0.60), substantial (0.61–0.80), and excellent (≥ 0.81) [41].

Relevant manufacturers were contacted to obtain technical data for their ELISA kits where required. Descriptive analyses of extracted data were performed, which are presented together with a narrative synthesis in this review. The bar chart and map were created using Microsoft Excel and Datawrapper.

5. Conclusions

In conclusion, most ELISAs used to detect exposure to *T. gondii* in animals were of the indirect type, and they generally used serum and targeted *T. gondii*-specific IgG. In-house ELISAs were most popular; however, among commercial kits, the ID Screen® Toxoplasmosis Indirect Multi-species (IDvet, France) kit appeared to be effective due to its better performance and utility across multiple species as well as the possibility of testing different types of samples (serum, milk, or meat juice). Recombinant antigen combinations and chimeric antigens overall provided better diagnostic performance than native antigens or single recombinant antigens. A wide range of secondary antibodies are commercially available for domestic animals, but for species where no secondary antibodies are available, protein A/G can provide an alternative solution. Cross-reactivity with *T. gondii*-related parasites should be considered to improve the diagnostic performance of the assay. The findings of this study can be used to overcome existing limitations and develop new and reliable serological assays for the detection of *T. gondii* antibodies in a range of animal species. In the future, updating this review including both animal and human studies with a combination of age, gender groups, and other diagnostic methods with statistical evaluation would provide a better understanding of the detection of *T. gondii* infection worldwide.

Author Contributions: Conceptualisation, A.J., J.H., K.L.D.T.D.L. and A.W.; methodology, J.H., A.J., K.L.D.T.D.L. and A.W.; formal analysis, K.L.D.T.D.L.; resources, J.H., A.J., K.L.D.T.D.L. and A.W.; data curation, K.L.D.T.D.L., J.H., and A.J.; writing—original draft preparation, K.L.D.T.D.L.; writing—review and editing, J.H., A.J., and A.W.; supervision, J.H. and A.J. All authors have read and agreed to the published version of the manuscript.

Funding: This work was supported by a University of Melbourne Graduate Research Scholarship for K.L.D.T.D.L.

Institutional Review Board Statement: Not applicable.

Informed Consent Statement: Not applicable.

Data Availability Statement: Only published data were included in this systematic review and their citations are provided in the reference list.

Conflicts of Interest: The authors declare no conflict of interest.

References

1. Dubey, J.P.; Beattie, C.P. *Toxoplasmosis of Animals and Man*; CRC Press: Boca Raton, FL, USA, 1988; ISBN 0849346185.
2. Tenter, A.M.; Heckeroth, A.R.; Weiss, L.M. *Toxoplasma gondii*: From animals to humans. *Int. J. Parasitol.* **2000**, *30*, 1217–1258. [CrossRef]
3. Petersen, E. Toxoplasmosis. *Semin. Fetal Neonatal Med.* **2007**, *12*, 214–223. [CrossRef]
4. Dubey, J.P. Toxoplasmosis—A waterborne zoonosis. *Vet. Parasitol.* **2004**, *126*, 57–72. [CrossRef] [PubMed]
5. Gajadhar, A.A.; Scandrett, W.B.; Forbes, L.B. Parásitos zoonóticos transmitidos por los alimentos y el agua en las granjas. *Rev. Sci. Tech. l'OIE* **2006**, *25*, 595–606. [CrossRef]
6. Stelzer, S.; Basso, W.; Benavides Silván, J.; Ortega-Mora, L.M.; Maksimov, P.; Gethmann, J.; Conraths, F.J.; Schares, G. *Toxoplasma gondii* infection and toxoplasmosis in farm animals: Risk factors and economic impact. *Food Waterborne Parasitol.* **2019**, *15*, e00037:00031–e00037:00032. [CrossRef]
7. Ferra, B.; Holec-Gasior, L.; Kur, J. Serodiagnosis of *Toxoplasma gondii* infection in farm animals (horses, swine, and sheep) by enzyme-linked immunosorbent assay using chimeric antigens. *Parasitol. Int.* **2015**, *64*, 288–294. [CrossRef] [PubMed]
8. Buxton, D.; Maley, S.W.; Wright, S.E.; Rodger, S.; Bartley, P.; Innes, E.A. *Toxoplasma gondii* and ovine toxoplasmosis: New aspects of an old story. *Vet. Parasitol.* **2007**, *149*, 25–28. [CrossRef]
9. Mainar, R.C.; De La Cruz, C.; Asensio, A.; Domínguez, L.; Vázquez-Boland, J.A. Prevalence of agglutinating antibodies to *Toxoplasma gondii* in small ruminants of the Madrid region, Spain, and identification of factors influencing seropositivity by multivariate analysis. *Vet. Res. Commun.* **1996**, *20*, 153–159. [CrossRef] [PubMed]
10. Elmore, S.A.; Samelius, G.; Al-Adhami, B.; Huyvaert, K.P.; Bailey, L.L.; Alisauskas, R.T.; Gajadhar, A.A.; Jenkins, E.J. Estimating *Toxoplasma gondii* exposure in Arctic foxes (*Vulpes lagopus*) while navigating the imperfect world of wildlife serology. *J. Wildl. Dis.* **2016**, *52*, 47–56. [CrossRef] [PubMed]
11. Dubey, J.; Jones, J. *Toxoplasma gondii* infection in humans and animals in the United States. *Int. J. Parasitol.* **2008**, *38*, 1257–1278. [CrossRef] [PubMed]
12. Dubey, J.P. A review of toxoplasmosis in wild birds. *Vet. Parasitol.* **2002**, *106*, 121–153. [CrossRef]
13. Bowater, R.O.; Norton, J.; Johnson, S.; Hill, B.; O'Donoghue, P.; Prior, H. Toxoplasmosis in Indo-Pacific humpbacked dolphins (*Sousa chinensis*), from Queensland. *Aust. Vet. J.* **2003**, *81*, 627–632. [CrossRef] [PubMed]
14. Sangster, C.; Gordon, A.; Hayes, D. Systemic toxoplasmosis in captive flying-foxes. *Aust. Vet. J.* **2012**, *90*, 140–142. [CrossRef] [PubMed]
15. Sørensen, K.K.; Mørk, T.; Sigurðardóttir, Ó.G.; Åsbakk, K.; Åkerstedt, J.; Bergsjø, B.; Fuglei, E. Acute toxoplasmosis in three wild Arctic foxes (*Alopex lagopus*) from Svalbard; one with co-infections of *Salmonella Enteritidis* PT1 and *Yersinia pseudotuberculosis* serotype 2b. *Res. Vet. Sci.* **2005**, *78*, 161–167. [CrossRef]
16. Frenkel, J.K.; Dubey, J.P.; Miller, N.L. *Toxoplasma gondii* in cats: Fecal stages identified as coccidian oocysts. *Science* **1970**, *167*, 893–896. [CrossRef] [PubMed]
17. Dubey, J.P.; Lindsay, D.S.; Speer, C.A. Structures of *Toxoplasma gondii* tachyzoites, bradyzoites, and sporozoites and biology and development of tissue cysts. *Clin. Microbiol. Rev.* **1998**, *11*, 267–299. [CrossRef]
18. Dubey, J.P. Advances in the life cycle of *Toxoplasma gondii*. *Int. J. Parasitol.* **1998**, *28*, 1019–1024. [CrossRef]

19. Hill, D.; Dubey, J.P. *Toxoplasma gondii*: Transmission, diagnosis and prevention. *Clin. Microbiol. Infect.* **2002**, *8*, 634–640. [CrossRef]
20. Kim, K.; Weiss, L.M. *Toxoplasma*: The next 100 years. *Microb. Infect.* **2008**, *10*, 978–984. [CrossRef] [PubMed]
21. Basso, W.; Bretislav, K.; Fraser, I.L.; Gereon, S.; Lais, P.; Maria, C.V.; Pavlo, M.; Peter, D.; Sonja, H.; Xaver, S. Assessment of diagnostic accuracy of a commercial ELISA for the detection of *Toxoplasma gondii* infection in pigs compared with IFAT, TgSAG1-ELISA and Western blot, using a Bayesian latent class approach. *Int. J. Parasitol.* **2013**, *43*, 565–570. [CrossRef]
22. Savva, D.; Morris, J.C.; Johnson, J.D.; Holliman, R.E. Polymerase chain reaction for detection of *Toxoplasma gondii*. *J. Med. Microbiol.* **1990**, *32*, 25–31. [CrossRef] [PubMed]
23. Liu, Q.; Wang, Z.-D.; Huang, S.-Y.; Zhu, X.-Q. Diagnosis of toxoplasmosis and typing of *Toxoplasma gondii*. *Parasites Vectors* **2015**, *8*. [CrossRef] [PubMed]
24. Sudan, V.; Tewari, A.K.; Singh, H. Serodiagnosis of *Toxoplasma gondii* infection in bovines from Kerala, India using a recombinant surface antigen 1 ELISA. *Biologicals* **2015**, *43*, 250–255. [CrossRef]
25. Garcia, J.L.; Gennari, S.M.; Machado, R.Z.; Navarro, I.T. *Toxoplasma gondii*: Detection by mouse bioassay, histopathology, and polymerase chain reaction in tissues from experimentally infected pigs. *Exp. Parasitol.* **2006**, *113*, 267–271. [CrossRef] [PubMed]
26. Dubey, J.P. Long-term persistence of *Toxoplasma gondii* in tissues of pigs inoculated with *T. gondii* oocysts and effect of freezing on viability of tissue cysts in pork. *Am. J. Vet. Res.* **1988**, *49*, 910–913. [PubMed]
27. Dubey, J.P. Oocyst shedding by cats fed isolated bradyzoites and comparison of infectivity of bradyzoites of the veg strain *Toxoplasma gondii* to cats and mice. *J. Parasitol.* **2001**, *87*, 215–219. [CrossRef]
28. Parameswaran, N.; O'Handley, R.M.; Grigg, M.E.; Fenwick, S.G.; Thompson, R.C.A. Seroprevalence of *Toxoplasma gondii* in wild kangaroos using an ELISA. *Parasitol. Int.* **2009**, *58*, 161–165. [CrossRef]
29. Lind, P.; Haugegaard, J.; Wingstrand, A.; Henriksen, S.A. The time course of the specific antibody response by various elisas in pigs experimentally infected with *Toxoplasma gondii*. *Vet. Parasitol.* **1997**, *71*, 1–15. [CrossRef]
30. Tekkesin, N.; Keskin, K.; Kılınc, C.; Orgen, N.; Molo, K. Detection of immunoglobulin G antibodies to *Toxoplasma gondii*: Evaluation of two commercial immunoassay systems. *J. Microbiol. Immunol. Infect.* **2011**, *44*, 21–26. [CrossRef] [PubMed]
31. Cakir-Koc, R. Production of anti-SAG1 IgY antibody against *Toxoplasma gondii* parasites and evaluation of antibody activity by ELISA method. *Parasitol. Res.* **2016**, *115*, 2947–2952. [CrossRef]
32. Villard, O.; Cimon, B.; L'Ollivier, C.; Fricker-Hidalgo, H.; Godineau, N.; Houze, S.; Paris, L.; Pelloux, H.; Villena, I.; Candolfi, E. Serological diagnosis of *Toxoplasma gondii* infection. *Diagn. Microbiol. Infect. Dis.* **2016**, *84*, 22–33. [CrossRef] [PubMed]
33. Sun, X.; Feng, W.; Jiping, L.; Quan, L.; Zedong, W. Evaluation of an indirect ELISA using recombinant granule antigen GRA1, GRA7 and soluble antigens for serodiagnosis of *Toxoplasma gondii* infection in chickens. *Res. Vet. Sci.* **2015**, *100*, 161–164. [CrossRef] [PubMed]
34. Gu, Y.; Wang, Z.; Cai, Y.; Li, X.; Wei, F.; Shang, L.; Li, J.; Liu, Q. A comparative study of *Toxoplasma gondii* seroprevalence in mink using a modified agglutination test, a Western blot, and enzyme-linked immunosorbent assays. *J. Vet. Diagn. Investig.* **2015**, *27*, 616–620. [CrossRef]
35. Titilincu, A.; Mircean, V.; Iovu, A.; Cozma, V. Development of an indirect ELISA test using tachyzoite crude antigen for sero-diagnosis of sheep *Toxoplasma gondii* infection. *Bull. Univ. Agric. Sci. Vet. Med. Cluj Napoca* **2009**, *66*, 137–141.
36. Tumurjav, B.; Terkawi, M.A.; Zhang, H.; Zhang, G.; Jia, H.; Goo, Y.-K.; Yamagishi, J.; Nishikawa, Y.; Igarashi, I.; Sugimoto, C.; et al. Serodiagnosis of ovine toxoplasmosis in Mongolia by an enzyme-linked immunosorbent assay with recombinant *Toxoplasma gondii* matrix antigen 1. *Jap. J. Vet. Res.* **2010**, *58*, 111–119.
37. Felin, E.; Anu, N.r.; Maria, F.-A. Comparison of commercial ELISA tests for the detection of *Toxoplasma* antibodies in the meat juice of naturally infected pigs. *Vet. Parasitol.* **2017**, *238*, 30–34. [CrossRef]
38. Payne, R.A.; Joynson, D.H.M.; Wilsmore, A.J. Enzyme-linked immunosorbent assays for the measurement of specific antibodies in experimentally induced ovine toxoplasmosis. *Epidemiol. Infect.* **1988**, *100*, 205–212. [CrossRef] [PubMed]
39. Engvall, E.; Perlmann, P. Enzyme-linked immunosorbent assay, ELISA III. Quantitation of specific antibodies by enzyme-labeled anti-immunoglobulin in antigen-coated tubes. *J. Immunol.* **1972**, *109*, 129–135.
40. Yolken, R.; Kim, H.; Clem, T.; Wyatt, R.; Kalica, A.; Chanock, R.; Kapikian, A. Enzyme-linked immunosorbent assay (ELISA) for detection of human reovirus-like agent of infantile gastroenteritis. *Lancet* **1977**, *310*, 263–267. [CrossRef]
41. Viera, A.; Garrett, J. Understanding interobserver agreement: The kappa statistic. *Fam. Med.* **2005**, *37*, 360–363. [PubMed]
42. Šimundić, A.-M. Measures of diagnostic accuracy: Basic definitions. *EJIFCC* **2009**, *19*, 203–211.
43. Al-Adhami, B.H.; Alvin, A.G. A new multi-host species indirect ELISA using protein A/G conjugate for detection of anti-*Toxoplasma gondii* IgG antibodies with comparison to ELISA-IgG, agglutination assay and Western blot. *Vet. Parasitol.* **2014**, *200*, 66–73. [CrossRef] [PubMed]
44. Reynoso-Palomar, A.; Moreno-Gálvez, D.; Villa-Mancera, A. Prevalence of *Toxoplasma gondii* parasite in captive Mexican jaguars determined by recombinant surface antigens (SAG1) and dense granular antigens (GRA1 and GRA7) in ELISA-based serodiagnosis. *Exp. Parasitol.* **2020**, *208*. [CrossRef] [PubMed]
45. Gamble, H.R.; Dubey, J.P.; Lambillotte, D.N. Comparison of a commercial ELISA with the modified agglutination test for detection of *Toxoplasma* infection in the domestic pig. *Vet. Parasitol.* **2005**, *128*, 177–181. [CrossRef] [PubMed]
46. Abdelbaset, A.E.; Doaa, S.; Hend, A.; Mahmoud Abd Ellah, R.; Makoto, I.; Mohamed Hassan, K.; Xuan, X. Evaluation of recombinant antigens in combination and single formula for diagnosis of feline toxoplasmosis. *Exp. Parasitol.* **2017**, *172*, 1–4. [CrossRef] [PubMed]

47. Casartelli-Alves, L.; Boechat, V.C.; Macedo-Couto, R.; Ferreira, L.C.; Nicolau, J.L.; Neves, L.B.; Millar, P.R.; Vicente, R.T.; Oliveira, R.V.C.; Muniz, A.G.; et al. Sensitivity and specificity of serological tests, histopathology and immunohistochemistry for detection of *Toxoplasma gondii* infection in domestic chickens. *Vet. Parasitol.* **2014**, *204*, 346–351. [CrossRef] [PubMed]

48. Holec-Gąsior, L.; Ferra, B.; Grąźlewska, W. *Toxoplasma gondii* tetravalent chimeric proteins as novel antigens for detection of specific immunoglobulin G in sera of small ruminants. *Animals* **2019**, *9*, 1146. [CrossRef]

49. Pardini, L.; Maksimov, P.; Herrmann, D.C.; Bacigalupe, D.; Rambeaud, M.; Machuca, M.; Moré, G.; Basso, W.; Schares, G.; Venturini, M.C. Evaluation of an in-house TgSAG1 (P30) IgG ELISA for diagnosis of naturally acquired *Toxoplasma gondii* infection in pigs. *Vet. Parasitol.* **2012**, *189*, 204–210. [CrossRef] [PubMed]

50. Sudan, V.; Tewari, A.K.; Singh, H. Detection of antibodies against *Toxoplasma gondii* in Indian cattle by recombinant SAG2 enzyme-linked immunosorbent assay. *Acta Parasitol.* **2019**, *64*, 148–151. [CrossRef]

51. Olamendi-Portugal, M.; Ortega-S, J.A.; Medina-Esparza, L.; García-Vázquez, Z.; Cantu, A.; Correa, D.; Caballero-Ortega, H.; Cruz-Vázquez, C.; Sánchez-Alemán, M.A. Serosurvey of antibodies against *Toxoplasma gondii* and *Neospora caninum* in white-tailed deer from Northern Mexico. *Vet. Parasitol.* **2012**, *189*, 369–373. [CrossRef]

52. Schaefer, J.J.; Kirchgessner, M.S.; Whipps, C.M.; Mohammed, H.O.; Bunting, E.M.; Wade, S.E. Prevalence of antibodies to *Toxoplasma Gondii* in white-tailed deer (*Odocoileus Virginianus*) in New York state, USA. *J. Wildl. Dis.* **2013**, *49*, 940–945. [CrossRef] [PubMed]

53. Glor, S.B.; Edelhofer, R.; Grimm, F.; Deplazes, P.; Basso, W. Evaluation of a commercial ELISA kit for detection of antibodies against *Toxoplasma gondii* in serum, plasma and meat juice from experimentally and naturally infected sheep. *Parasites Vectors* **2013**, *6*. [CrossRef] [PubMed]

54. Yoo, W.G.; Kim, S.-M.; Won, E.J.; Lee, J.-Y.; Dai, F.; Woo, H.C.; Nam, H.-W.; Kim, T.I.; Han, J.-H.; Kwak, D.; et al. Tissue fluid enzyme-linked immunosorbant assay for piglets experimentally infected with *Toxoplasma gondii* and survey on local and imported pork in Korean retail meat markets. *Korean J. Parasitol.* **2018**, *56*, 437–446. [CrossRef] [PubMed]

55. Attia, M.M.; Saad, M.F.; Abdel-Salam, A.B. Milk as a substitute for serum in diagnosis of toxoplasmosis in goats. *J. Egypt. Soc. Parasitol.* **2017**, *47*, 227–234. [CrossRef]

56. Gazzonis, A.L.; Olivieri, E.; Stradiotto, K.; Villa, L.; Manfredi, M.T.; Zanzani, S.A. *Toxoplasma gondii* antibodies in bulk tank milk samples of caprine dairy herds. *J. Parasitol.* **2018**, *104*, 560–565. [CrossRef] [PubMed]

57. Singh, H.; Tewari, A.K.; Mishra, A.K.; Maharana, B.; Sudan, V.; Raina, O.K.; Rao, J.R. Detection of antibodies to *Toxoplasma gondii* in domesticated ruminants by recombinant truncated SAG2 enzyme-linked immunosorbent assay. *Trop. Anim. Health Prod.* **2015**, *47*, 171–178. [CrossRef] [PubMed]

58. Zhang, D.; Wang, Z.; Fang, R.; Nie, H.; Feng, H.; Zhou, Y.; Zhao, J. Use of protein AG in an enzyme-linked immunosorbent assay for serodiagnosis of *Toxoplasma gondii* infection in four species of animals. *Clin. Vaccine Immunol.* **2010**, *17*, 485–486. [CrossRef] [PubMed]

59. Armas Valdes, Y.; Obregon Alvarez, D.; Grandia Guzman, R.; Mitat Valdes, A.; Roque Lopez, E.; Perez Ruano, M.; Entrena Garcia, A.A. Validation of an inhibition enzyme-linked immunosorbent assay system for the diagnosis of *Toxoplasma gondii* infection in buffaloes (Bubalus bubalis). *Rev. Sci. Tech. Off. Int. Epizoot.* **2018**, *37*, 917–924. [CrossRef]

60. Sroka, J.; Karamon, J.; Cencek, T.; Dutkiewicz, J. Preliminary assessment of usefulness of cELISA test for screening pig and cattle populations for presence of antibodies against *Toxoplasma gondii*. *Ann. Agric. Environ. Med.* **2011**, *18*, 335–339. [PubMed]

61. Opsteegh, M.; Teunis, P.; Mensink, M.; Züchner, L.; Titilincu, A.; Langelaar, M.; van der Giessen, J. Evaluation of ELISA test characteristics and estimation of *Toxoplasma gondii* seroprevalence in Dutch sheep using mixture models. *Prev. Vet. Med.* **2010**, *96*, 232–240. [CrossRef]

62. Koethe, M.; Bittame, A.; Spekker, K.; Mercier, C.; Straubinger, R.K.; Fehlhaber, K.; Tenter, A.M.; Ludewig, M.; Pott, S.; Bangoura, B.; et al. Prevalence of specific IgG-antibodies against *Toxoplasma gondii* in domestic turkeys determined by kinetic ELISA based on recombinant GRA7 and GRA8. *Vet. Parasitol.* **2011**, *180*, 179–190. [CrossRef] [PubMed]

63. Terkawi, M.A.; Kameyama, K.; Rasul, N.H.; Xuan, X.; Nishikawa, Y. Development of an immunochromatographic assay based on dense granule protein 7 for serological detection of *Toxoplasma gondii* infection. *Clin. Vaccine Immunol.* **2013**, *20*, 596–601. [CrossRef] [PubMed]

64. Hamidinejat, H.; Nabavi, L.; Mayahi, M.; Ghourbanpoor, M.; Pourmehdi Borojeni, M.; Norollahi Fard, S.; Shokrollahi, M. Comparison of three diagnostic methods for the detection of *Toxoplasma gondii* in free range chickens. *Trop. Biomed.* **2014**, *31*, 507–513. [PubMed]

65. Cai, Y.; Wang, Z.; Li, J.; Li, N.; Wei, F.; Liu, Q. Evaluation of an indirect elisa using recombinant granule antigen GRA 7 for serodiagnosis of *Toxoplasma gondii* infection in cats. *J. Parasitol.* **2015**, *101*, 37–40. [CrossRef] [PubMed]

66. Basso, W.; Sollberger, E.; Schares, G.; Küker, S.; Ardüser, F.; Moore-Jones, G.; Zanolari, P. *Toxoplasma gondii* and *Neospora caninum* infections in South American camelids in Switzerland and assessment of serological tests for diagnosis. *Parasites Vectors* **2020**, *13*. [CrossRef]

67. Endrias Zewdu, G.; Mukarim, A.; Tsehaye, H.; Tessema, T.S. Comparison between enzyme linked immunosorbent assay (ELISA) and modified agglutination test (MAT) for detection of Toxoplasma gondii infection in sheep and goats slaughtered in an export abattoir at Debre-Zeit, Ethiopia. *Glob. Vet.* **2013**, *11*, 747–752.

68. Schaefer, J.J.; Holly, A.W.; Hussni, O.M.; Stephanie, L.S.; Susan, E.W. Chimeric protein A/G conjugate for detection of anti–*Toxoplasma gondii* immunoglobulin G in multiple animal species. *J. Vet. Diagn. Investig.* **2012**, *24*, 572–575. [CrossRef] [PubMed]

69. Schaefer, J.J.; White, H.A.; Schaaf, S.L.; Mohammed, H.O.; Wade, S.E. Modification of a commercial *Toxoplasma gondii* immunoglobulin G enzyme-linked immunosorbent assay for use in multiple animal species. *J. Vet. Diagn. Investig.* **2011**, *23*, 297–301. [CrossRef] [PubMed]

70. Holec-Gąsior, L.; Ferra, B.; Hiszczyńska-Sawicka, E.; Kur, J. The optimal mixture of *Toxoplasma gondii* recombinant antigens (GRA1, P22, ROP1) for diagnosis of ovine toxoplasmosis. *Vet. Parasitol.* **2014**, *206*, 146–152. [CrossRef]

71. Wang, Z.; Ge, W.; Huang, S.-Y.; Li, J.; Zhu, X.-Q.; Liu, Q. Evaluation of recombinant granule antigens GRA1 and GRA7 for serodiagnosis of *Toxoplasma gondii* infection in dogs. *BMC Vet. Res.* **2014**, *10*. [CrossRef]

72. Tenter, A.M.; Vietmeyer, C.; Rommel, M. ELISAs based on recombinant antigens for sero-epidemiological studies on *Toxoplasma gondii* infections in cats. *Parasitology* **1994**, *109*, 29–36. [CrossRef] [PubMed]

73. Velmurugan, G.V.; Tewari, A.K.; Rao, J.R.; Surajit, B.; Kumar, M.U.; Mishra, A.K. High-level expression of SAG1 and GRA7 gene of *Toxoplasma gondii* (Izatnagar isolate) and their application in serodiagnosis of goat toxoplasmosis. *Vet. Parasitol.* **2008**, *154*, 185–192. [CrossRef]

74. Silva, N.M.; Lourenço, E.V.; Silva, D.A.O.; Mineo, J.R. Optimisation of cut-off titres in *Toxoplasma gondii* specific ELISA and IFAT in dog sera using immunoreactivity to SAG-1 antigen as a molecular marker of infection. *Vet. J.* **2002**, *163*, 94–98. [CrossRef] [PubMed]

75. Zhu, C.H.; Cui, L.L.; Zhang, L.S. Comparison of a commercial ELISA with the modified agglutination test for detection of *Toxoplasma gondii* antibodies in sera of naturally infected dogs and cats. *Iran. J. Parasitol.* **2012**, *7*, 89–95.

76. Hosseininejad, M.; Azizi, H.R.; Hosseini, F.; Schares, G. Development of an indirect ELISA test using a purified tachyzoite surface antigen SAG1 for sero-diagnosis of canine *Toxoplasma gondii* infection. *Vet. Parasitol.* **2009**, *164*, 315–319. [CrossRef] [PubMed]

77. Garcia, J.L.; Navarro, I.T.; Vidotto, O.; Gennari, S.M.; Machado, R.Z.; Pereira, B.D.; Sinhorini, I.L. *Toxoplasma gondii*: Comparison of a rhoptry-ELISA with IFAT and MAT for antibody detection in sera of experimentally infected pigs. *Exp. Parasitol.* **2006**, *113*, 100–105. [CrossRef] [PubMed]

78. Adriaanse, K.; Firestone, S.M.; Lynch, M.; Rendall, A.R.; Sutherland, D.R.; Hufschmid, J.; Traub, R. Comparison of the modified agglutination test and real-time PCR for detection of *Toxoplasma gondii* exposure in feral cats from Phillip Island, Australia, and risk factors associated with infection. *Int. J. Parasitol. Parasites Wildl.* **2020**, *12*, 126–133. [CrossRef]

79. Suresh, K.; Chandrashekara, S. Sample size estimation and power analysis for clinical research studies. *J. Hum. Reprod. Sci.* **2012**, *5*, 7–13. [CrossRef] [PubMed]

80. Aydin, S. A short history, principles, and types of ELISA, and our laboratory experience with peptide/protein analyses using ELISA. *Peptides* **2015**, *72*, 4–15. [CrossRef]

81. Halevy, B.; Sarov, I. Enzyme-linked immunosorbent assay (ELISA) for detection of specific IgA antibodies to mumps virus. *J. Clin. Pathol.* **1982**, *35*, 1129–1133. [CrossRef]

82. Klein-Schneegans, A.-S.; Gavériaux, C.; Fonteneau, P.; Loor, F. Indirect double sandwich ELISA for the specific and quantitative measurement of mouse IgM, IgA and IgG subclasses. *J. Immunol. Methods* **1989**, *119*, 117–125. [CrossRef]

83. Plebani, A.; Avanzini, M.A.; Massa, M.; Ugazio, A.G. An avidin-biotin ELISA for the measurement of serum and secretory IgD. *J. Immunol. Methods* **1984**, *71*, 133–140. [CrossRef]

84. Zhang, Z.; Cai, Z.; Hou, Y.; Hu, J.; He, Y.; Chen, J.; Ji, K. Enhanced sensitivity of capture IgE-ELISA based on a recombinant Der f 1/2 fusion protein for the detection of IgE antibodies targeting house dust mite allergens. *Mol. Med. Rep.* **2019**, *19*, 3497–3504. [CrossRef] [PubMed]

85. Sroka, J.; Cencek, T.; Ziomko, I.; Karamon, J.; Zwoliński, J. Preliminary assessment of ELISA, MAT, and LAT for detecting *Toxoplasma gondii* antibodies in pigs. *Bull. Vet. Inst. Pulawy* **2008**, *52*, 545–549.

86. Dhakal, R.; Gajurel, K.; Pomares, C.; Talucod, J.; Press, C.J.; Montoya, J.G. Significance of a positive *Toxoplasma* Immunoglobulin M test result in the United States. *J. Clin. Microbiol.* **2015**, *53*, 3601–3605. [CrossRef]

87. Sharifi, K.; Hosseini Farash, B.R.; Tara, F.; Khaledi, A.; Sharifi, K.; Shamsian, S.A.A. Diagnosis of acute toxoplasmosis by IgG and IgM antibodies and IgG avidity in pregnant women from Mashhad, Eastern Iran. *Iran. J. Parasitol.* **2019**, *14*, 639–645. [CrossRef] [PubMed]

88. Dai, J.; Jiang, M.; Wang, Y.; Qu, L.; Gong, R.; Si, J. Evaluation of a recombinant multiepitope peptide for serodiagnosis of *Toxoplasma gondii* infection. *Clin. Vaccine Immunol.* **2012**, *19*, 338–342. [CrossRef]

89. Pietkiewicz, H.; Hiszczyńska-Sawicka, E.; Kur, J.; Petersen, E.; Nielsen, H.V.; Paul, M.; Stankiewicz, M.; Myjak, P. Usefulness of *Toxoplasma gondii* recombinant antigens (GRA1, GRA7 and SAG1) in an immunoglobulin G avidity test for the serodiagnosis of toxoplasmosis. *Parasitol. Res.* **2007**, *100*, 333–337. [CrossRef]

90. Wang, Y.; Yin, H. Research progress on surface antigen 1 (SAG1) of *Toxoplasma gondii*. *Parasites Vectors* **2014**, *7*, 180. [CrossRef] [PubMed]

91. Chahed Bel-Ochi, N.; Bouratbine, A.; Mousli, M. Enzyme-linked immunosorbent assay using recombinant SAG1 antigen to detect *Toxoplasma gondii*-specific immunoglobulin G antibodies in human sera and saliva. *Clin. Vaccine Immunol.* **2013**, *20*, 468–473. [CrossRef] [PubMed]

92. Cong, H.; Zhang, M.; Zhang, Q.; Gong, J.; Cong, H.; Xin, Q.; He, S. Analysis of structures and epitopes of surface antigen glycoproteins expressed in bradyzoites of *Toxoplasma gondii*. *BioMed Res. Int.* **2013**, *2013*, 1–9. [CrossRef]

93. Li, S.; Galvan, G.; Araujo, F.G.; Suzuki, Y.; Remington, J.S.; Parmley, S. Serodiagnosis of recently acquired *Toxoplasma gondii* infection using an enzyme-linked immunosorbent assay with a combination of recombinant antigens. *Clin. Diagn. Lab. Immunol.* **2000**, *7*, 781–787. [CrossRef] [PubMed]
94. Kasper, L.H.; Bradley, M.S.; Pfefferkorn, E.R. Identification of stage-specific sporozoite antigens of *Toxoplasma gondii* by monoclonal antibodies. *J. Immunol.* **1984**, *132*, 443–449. [PubMed]
95. Hiszczyńska-Sawicka, E.; Kur, J.; Pietkiewicz, H.; Holec-Gasior, L.; Gąsior, A.; Myjak, P. Efficient production of the *Toxoplasma gondii* GRA6, p35 and SAG2 recombinant antigens and their applications in the serodiagnosis of toxoplasmosis. *Acta Parasitol.* **2005**, *50*, 249–254.
96. Crawford, J.; Lamb, E.; Wasmuth, J.; Grujic, O.; Grigg, M.E.; Boulanger, M.J. Structural and functional characterization of sporoSAG: A SAG2-related surface antigen from *Toxoplasma gondii*. *J. Biol. Chem.* **2010**, *285*, 12063–12070. [CrossRef] [PubMed]
97. Ahn, H.-J.; Kim, S.; Kim, H.-E.; Nam, H.-W. Interactions between secreted GRA proteins and host cell proteins across the parasitophorous vacuolar membrane in the parasitism of *Toxoplasma gondii*. *Korean J. Parasitol.* **2006**, *44*, 303. [CrossRef]
98. Jacobs, D.; Vercammen, M.; Saman, E. Evaluation of recombinant dense granule antigen 7 (GRA7) of *Toxoplasma gondii* for detection of immunoglobulin G antibodies and analysis of a major antigenic domain. *Clin. Diagn. Lab. Immunol.* **1999**, *6*, 24–29. [CrossRef] [PubMed]
99. Pfrepper, K.-I.; Enders, G.; Gohl, M.; Krczal, D.; Hlobil, H.; Wassenberg, D.; Soutschek, E. Seroreactivity to and avidity for recombinant antigens in toxoplasmosis. *Clin. Diagn. Lab. Immunol.* **2005**, *12*, 977–982. [CrossRef] [PubMed]
100. Dodangeh, S.; Fasihi-Ramandi, M.; Daryani, A.; Valadan, R.; Sarvi, S. In silico analysis and expression of a novel chimeric antigen as a vaccine candidate against *Toxoplasma gondii*. *Microb. Pathog.* **2019**, *132*, 275–281. [CrossRef] [PubMed]
101. Hajissa, K.; Zakaria, R.; Suppian, R.; Mohamed, Z. An evaluation of a recombinant multiepitope based antigen for detection of *Toxoplasma gondii* specific antibodies. *BMC Infect. Dis.* **2017**, *17*. [CrossRef]
102. Camussone, C.; Gonzalez, V.N.; Belluzo, M.A.S.; Pujato, N.; Ribone, M.A.E.; Lagier, C.M.; Marcipar, I.N.S. Comparison of recombinant *Trypanosoma cruzi* peptide mixtures versus multiepitope chimeric proteins as sensitizing antigens for immunodiagnosis. *Clin. Vaccine Immunol.* **2009**, *16*, 899–905. [CrossRef] [PubMed]
103. Cheng, Z.; Zhao, J.-W.; Sun, Z.-Q.; Song, Y.-Z.; Sun, Q.-W.; Zhang, X.-Y.; Zhang, X.-L.; Wang, H.-H.; Guo, X.-K.; Liu, Y.-F.; et al. Evaluation of a novel fusion protein antigen for rapid serodiagnosis of tuberculosis. *J. Clin. Lab. Anal.* **2011**, *25*, 344–349. [CrossRef]
104. Dipti, C.A.; Jain, S.K.; Navin, K. A novel recombinant multiepitope protein as a hepatitis C diagnostic intermediate of high sensitivity and specificity. *Protein Expr. Purif.* **2006**, *47*, 319–328. [CrossRef] [PubMed]
105. Rispens, T.; Vidarsson, G. Chapter 9—Human IgG subclasses. In *Antibody Fc*; Ackerman, M.E., Nimmerjahn, F., Eds.; Academic Press: Boston, MA, USA, 2014; pp. 159–177. ISBN 978-0-12-394802-1.
106. Kelly, P.J.; Tagwira, M.; Matthewman, L.; Mason, P.R.; Wright, E.P. Reactions of sera from laboratory, domestic and wild animals in Africa with protein a and a recombinant chimeric protein AG. *Comp. Immunol. Microbiol. Infect. Dis.* **1993**, *16*, 299–305. [CrossRef]
107. Vaz, P.K.; Hartley, C.A.; Browning, G.F.; Devlin, J.M. Marsupial and monotreme serum immunoglobulin binding by proteins A, G and L and anti-kangaroo antibody. *J. Immunol. Methods* **2015**, *427*, 94–99. [CrossRef]
108. Stöbel, K.; Schönberg, A.; Staak, C. A new non-species dependent ELISA for detection of antibodies to *Borrelia burgdorferi* s. l. in zoo animals. *Int. J. Med. Microbiol.* **2002**, *291*, 88–99. [CrossRef]
109. Araujo, F.G.; Dubey, J.P.; Remington, J.S. Antigenic similarity between the coccidian parasites *Toxoplasma gondii* and *Hammondia hammondi* 1. *J. Protozool.* **1984**, *31*, 145–147. [CrossRef] [PubMed]
110. Frenkel, J.K.; Dubey, J.P. *Hammondia hammondi* gen. nov., sp.nov., from domestic cats, a new coccidian related to *Toxoplasma* and *Sarcocystis*. *Z. Parasitenkd.* **1975**, *46*, 3–12. [CrossRef] [PubMed]
111. Engvall, E.; Perlmann, P. Enzyme-linked immunosorbent assay (ELISA) quantitative assay of immunoglobulin G. *Immunochemistry* **1971**, *8*, 871–874. [CrossRef]

Article

LAMP Assay for the Detection of *Echinococcus multilocularis* Eggs Isolated from Canine Faeces by a Cost-Effective NaOH-Based DNA Extraction Method

Barbara J. Bucher [1], Gillian Muchaamba [1], Tim Kamber [1], Philipp A. Kronenberg [1,2], Kubanychbek K. Abdykerimov [3,4], Myktybek Isaev [5], Peter Deplazes [1] and Cristian A. Alvarez Rojas [1,*]

[1]   Institute of Parasitology, Vetsuisse and Medical Faculty, University of Zurich, 8057 Zurich, Switzerland; barbara.bucher@uzh.ch (B.J.B.); gillian.muchaamba@uzh.ch (G.M.); timpeter.kamber@uzh.ch (T.K.); philipp.kronenberg2@uzh.ch (P.A.K.); deplazesp@access.uzh.ch (P.D.)

[2]   Graduate School for Cellular and Biomedical Sciences, University of Bern, 3012 Bern, Switzerland

[3]   Section of Epidemiology, Vetsuisse Faculty, University of Zurich, 8057 Zurich, Switzerland; kubanychbek.abdykerimov@uzh.ch

[4]   Life Science Zurich Graduate School, University of Zürich, 8057 Zurich, Switzerland

[5]   Department of Parasitology, Kyrgyz Research Institute of Veterinary Medicine Arstanbek Duisheev, Ministry of Education and Science of the Kyrgyz Republic, Bishkek 720033, Kyrgyzstan; isaev-ww-1988@mail.ru

*   Correspondence: cristian.alvarezrojas@uzh.ch

Citation: Bucher, B.J.; Muchaamba, G.; Kamber, T.; Kronenberg, P.A.; Abdykerimov, K.K.; Isaev, M.; Deplazes, P.; Alvarez Rojas, C.A. LAMP Assay for the Detection of *Echinococcus multilocularis* Eggs Isolated from Canine Faeces by a Cost-Effective NaOH-Based DNA Extraction Method. *Pathogens* **2021**, *10*, 847. https://doi.org/10.3390/pathogens10070847

Academic Editor: Vito Colella

Received: 20 May 2021
Accepted: 29 June 2021
Published: 5 July 2021

Publisher's Note: MDPI stays neutral with regard to jurisdictional claims in published maps and institutional affiliations.

Abstract: The detection of *Echinococcus multilocularis* in infected canids and the environment is pivotal for a better understanding of the epidemiology of alveolar echinococcosis in endemic areas. Necropsy/sedimentation and counting technique remain the gold standard for the detection of canid infection. PCR-based detection methods have shown high sensitivity and specificity, but they have been hardly used in large scale prevalence studies. Loop-mediated isothermal amplification (LAMP) is a fast and simple method to detect DNA with a high sensitivity and specificity, having the potential for field-application. A specific LAMP assay for the detection of *E. multilocularis* was developed targeting the mitochondrial *nad1* gene. A crucial step for amplification-based detection methods is DNA extraction, usually achieved utilising silica-gel membrane spin columns from commercial kits which are expensive. We propose two cost-effective and straightforward methods for DNA extraction, using NaOH (method 1A) and InstaGene^{TM} Matrix (method 1B), from isolated eggs circumventing the need for commercial kits. The sensitivity of both assays with fox samples was similar (72.7%) with multiplex-PCR using protocol 1A and LAMP using protocol 1B. Sensitivity increased up to 100% when testing faeces from 12 foxes infected with more than 100 intestinal stages of *E. multilocularis*. For dogs, sensitivity was similar (95.4%) for LAMP and multiplex-PCR using protocol 1B and for both methods when DNA was extracted using protocol 1A (90.9%). The DNA extraction methods used here are fast, cheap, and do not require a DNA purification step, making them suitable for field studies in low-income countries for the prevalence study of *E. multilocularis*.

Keywords: *Echinococcus*; NaOH; LAMP; PCR; DNA extraction; taeniid egg isolation

1. Introduction

Echinococcus multilocularis is the canine intestinal cestode responsible for alveolar echinococcosis (AE) in humans [1]. Primarily foxes, and to a lesser extent, other wild canids and domestic dogs play a role as definitive hosts for *E. multilocularis* [2]. Humans act as dead-end hosts, acquiring the infection by oral intake of viable eggs of *E. multilocularis* from infected food, soil, water, or hand-to-mouth contact after a direct interaction with infected dogs and contaminated matrices [3]. Establishing the exact link between AE and its origin (source attribution) is challenging due to the long incubation time of the disease of up to 15 years. Therefore, it is essential to have tools for the accurate determination of the environmental contamination with *E. multilocularis* eggs and the detection of canine intestinal infections.

Different methodologies for the diagnosis of *E. multilocularis* in canids have been standardised and published over the years. Post-mortem examination (i.e., sedimentation and counting technique, SCT) remains the gold standard [4]. However, a post-mortem examination has significant disadvantages, requiring either expensive infrastructure, strict biosafety measures and involves the killing of canids. Several methods based on the detection of *E. multilocularis* antigens [5–8] and DNA [9–11] in faeces have been developed (reviewed in [12,13]). Coproantigen detection (copro-ELISA) has the advantage of detecting prepatent infections [5,14]. Although some coproantigen tests are commercially available, they are not regularly used for massive screening in canid faeces from endemic areas. On the other hand, DNA of *E. multilocularis* can be potentially detected from the prepatent period as free DNA or from cells of the growing worms; and mostly from proglottids/eggs during patency. DNA can be isolated from total faeces and used, for example, in nested PCR (copro-DNA-PCR) [9] or from isolated taeniid eggs through floatation and sieving [10] and used in multiplex-PCR (egg-DNA-PCR) [15]. Although the egg-DNA-PCR focuses only on patent infections, it has the advantage of concentrating eggs and decreasing the presence of PCR inhibitors.

PCR-based methods for identifying parasites in faecal samples rely on isolation and DNA concentration using silica-gel membrane in spin columns from commercial kits removing PCR inhibitors (proteinases, bile salts, polyphenols, and acids). However, some publications report that the total removal of inhibitors from faeces is not always possible [16,17]. Furthermore, the usage of commercial kits requires trained staff performing several liquid handling steps, and the commercial kit itself, which can be a costly item when large numbers of samples need to be analysed. The cost of such kits ranges between 3 and 11 USD per sample, which is unreasonable for governments and researchers in low-income countries. Besides the PCR inhibitors, the successful disruption of the typical taeniid eggshell keratin-like layer by homogenisation or alkaline lysis is essential for DNA detection. Unlike polymerases used in PCR, the in silico designed *Bst* 2.0 DNA Polymerase used in LAMP (loop-mediated isothermal amplification) displays improved tolerance for inhibitors [18–20]. LAMP assays have been developed to detect *E. multilocularis* [21,22] and *E. granulosus* [23–26] in faecal samples and also several other parasites, including *Clonorchis sinensis* [27], *Trichuris muris* [28], *Strongyloides* spp. [29], *Necator americanus* [30], and *Taenia* spp. [31]. LAMP also offers a robust tool for DNA amplification on the presence of reagents like NaOH used in alternative DNA extraction methods suggesting that commercial kits for DNA isolation could be circumvented. Methods for DNA extraction without silica-gel membranes have been described in the literature since the 1990s [32–35]. A reappraisal of such methods for developing cost-effective DNA extraction (i.e., NaOH-based) has provided evidence to produce a template of sufficient quality for subsequent PCR from different tissues [36–38]. Furthermore, single *Echinococcus* eggs have been used for PCR- and LAMP-based methods directly after treatment with NaOH [26,39–42]. In the present study, we propose two cost-effective DNA extraction methods (without purification) from isolated taeniid eggs, based on NaOH and InstaGene™ Matrix (Chelex 6%) to be used directly as a template for multiplex-PCR [15] and LAMP (developed in this study) for the detection of *E. multilocularis*. We also labelled two of the primers used in LAMP to combine them with lateral flow dipstick for the same diagnostic purpose. The proposed cost-effective DNA extractions can be used in large prevalence studies to detect *E. multilocularis* eggs and potentially *E. granulosus* in canine faecal samples in low-income endemic countries.

2. Results

2.1. Optimisation, Specificity, Analytical Sensitivity, Limit of Detection, and Stability of LAMP

After optimisation of the LAMP assay, the reaction was set up with 6 mM MgSO₄, 0.8 M betaine, 0.004% malachite green, and 1× primer mix (1.6 µM of each FIP and BIP primer, 0.2 µM of each F3 and B3 primer and 0.4 µM of the loop primer); amplification occurred at 65 °C for 60 min. The selected LAMP primer set showed successful amplification when using *E. multilocularis* DNA rendering a clear blue-green colour, a ladder-like structure on agarose gel after electrophoresis and a specific band on the lateral-flow-dipstick (when using the labelled BIP and FIP primers). Conversely, a negative result followed by testing DNA of other *Echinococcus* spp., *Taenia* spp., and other cestodes as a template (see material and methods). The limit of detection (analytical sensitivity) of LAMP with the selected primer set was 1 pg/µL of *E. multilocularis* DNA in all serial dilutions performed in triplicate. On the other hand, the limit of detection of the multiplex-PCR [15] was 10 fg/µL of *E. multilocularis* DNA. Serial dilutions of samples derived from protocol 1A were performed to exclude inhibition of LAMP reactions. The 1:50 dilution showed to perform best, as no inhibition could be detected. The lowest amount of eggs possible to detect with LAMP and multiplex-PCR was one egg treated with the protocol 1B in 4/4 replicates and two eggs in 3/4 replicates with protocol 1A. In order to investigate if free DNA of the parasite was present in the supernatant of the egg suspensions used for the limit of detection we used it as a template for LAMP and multiplex-PCR [15], all reactions were negative. In this way, we can confirm that DNA originated from the eggs in suspension. Aliquots of a prepared master mix for LAMP containing all the reagents except DNA were tested weekly over six weeks. A positive LAMP reaction was achieved in all the period when adding *E. multilocularis* DNA (2 ng).

2.2. Examination of Field Samples with Protocols 1A and 1B

A schematic representation of the sample processing is shown in Figure 1. In the case of negative samples, we collected the sediment from the 21 µm filter, in the same way as in positive samples, to perform the DNA extraction.

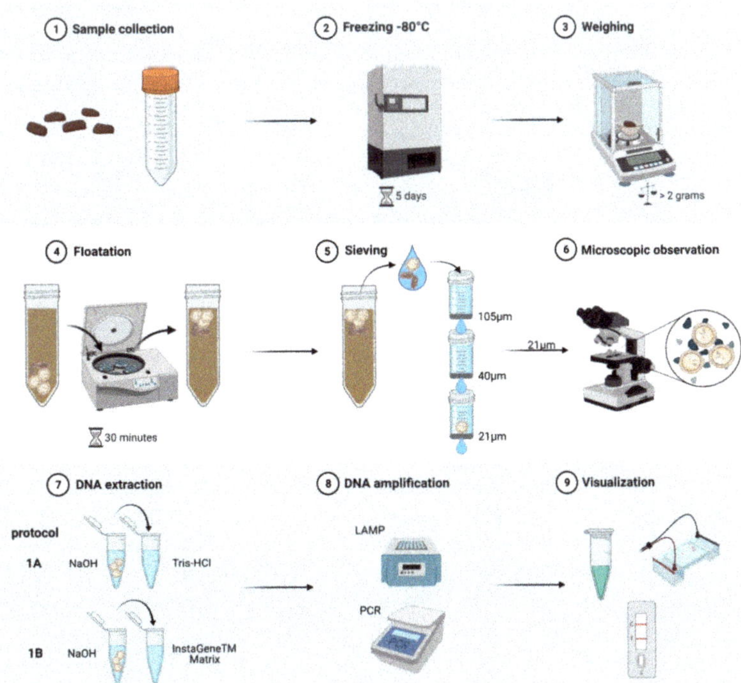

Figure 1. Schematic representation of the process of field samples analysed in the present study. Step 7 shows the two methods used for DNA extraction in this study (protocols 1A and 1B). The visualisation of multiplex-PCR occurred after electrophoresis and in the case of LAMP results were assessed after change of colour of the reaction, electrophoresis, and lateral flow dipsticks as shown in step 9.

2.2.1. Foxes

Intestinal stages of *E. multilocularis* (between 2 to >100) were detected in 44/62 foxes after necropsy/SCT. Microscopic detection of taeniid eggs was successful for 30 samples from the 44 foxes documented to be infected with *E. multilocularis*. On the other hand, 18 faecal samples were collected from foxes without intestinal stages of *E. multilocularis* at necropsy/SCT; from them, it was possible to observe taeniid eggs in five samples (Table 1); 8/18 foxes mentioned before were infected with *Taenia* spp. When using protocol 1A, LAMP showed a positive reaction in 26/44 samples from animals harbouring *E. multilocularis* in their intestine and 1/18 samples from foxes without intestinal stages of *E. multilocularis*. Conversely, multiplex-PCR was positive for *E. multilocularis* in 32/44 samples from foxes infected with *E. multilocularis* and 1/18 samples from foxes without intestinal stages of *E. multilocularis*. Using multiplex-PCR, other cestodes and *Taenia* spp. were identified, producing an amplicon (267 bp) in five samples from the 18 animals without intestinal stages of *E. multilocularis* (protocol 1A and 1B), suggesting that if PCR inhibitors were present, they did not impede the amplification of DNA (See detailed list of results in. Detailed list of results is shown in Table S1.

The second aliquot of faeces from the same foxes was used to isolate eggs for treatment with protocol 1B. In this case, taeniid eggs were observed in 31/44 faeces from animals with *E. multilocularis* in their intestine and in 5/18 faecal samples from negative animals. After treating eggs with protocol 1B, we found positive LAMP reactions in 32/44 foxes infected with *E. multilocularis* and in 1/18 foxes with no intestinal stages of *E. multilocularis*. Conversely, multiplex-PCR was positive for *E. multilocularis* in 30/44 positive and 1/18 negative foxes, respectively. A positive result for the amplification of 267 bp corresponding to other cestodes including *Taenia* spp. was found in 10 samples with a negative result for *E. multilocularis* using the DNA from protocol 1B (Table S1).

Furthermore, multiplex-PCR using protocols 1A and 1B showed a positive result for the amplification of 267 bp corresponding to other cestodes, including *Taenia* spp. in all the eight foxes identified to be infected with *Taenia* spp. at necropsy/SCT.

2.2.2. Dogs

The detection of *E. multilocularis* in dogs was performed by a multiplex-PCR in a previous prevalence study. In the case of the dog group (Table 1), taeniid eggs were observed in 20/22 samples from dogs previously identified to be infected with *E. multilocularis*. No eggs were observed in samples that were considered negative to *E. multilocularis*. When the isolated eggs were treated with protocol 1A and used for DNA amplification, we found a positive result for *E. multilocularis* in 20/22 samples using LAMP and multiplex-PCR. No positive reactions occurred with LAMP and multiplex-PCR in the ten samples previously identified as negative to *E. multilocularis*. When isolating taeniid eggs to be treated with protocol 1B, we observed eggs in 21/22 samples from dogs previously identified positive to *E. multilocularis*. In this group, LAMP and multiplex-PCR showed a positive result in 21/22 samples and a negative result in 10/10 samples identified as negative to *E. multilocularis*.

In the case of fox samples, sensitivity in LAMP and multiplex-PCR was 72.7% (C.I. 95%: 57.2–85) using protocol 1B and 1A, respectively (Table 2). Specificity values were not 100% for fox samples, the highest value was 94.4% (72.7–99.8) achieved using both protocols for LAMP and multiplex-PCR (Table 2). In the dog group, sensitivity for LAMP and multiplex-PCR was 95.4% (77.1–99.8) achieved with protocol 1B. The specificity in LAMP and multiplex-PCR with both protocols was 100% (Table 2).

Table 1. Results of LAMP and multiplex-PCR [1] amplification of *E. multilocularis* using the product of two DNA extraction methods (protocols 1A and 1B) as a template from isolated taeniid eggs from 62 fox faecal samples and 32 dog faecal samples. In the case of foxes, the status of *E. multilocularis* infection was established at necropsy with the Sedimentation and Counting Test (SCT); for dogs, *E. multilocularis* infections status was established based on egg isolation/multiplex-PCR [15].

Host	Number of Animals Confirmed Positive (+) or Without (–) Detected *E. multilocularis* Infection (Method for Diagnosis)	Protocol 1A: NaOH/Dilution 1:50 with Tris-HCl			Protocol 1B: NaOH + InstaGene™ Matrix		
		Microscopic Detection of Taeniid Eggs [2]	LAMP Positive	Multiplex-PCR [1] Positive	Microscopic Detection of Taeniid Eggs [2]	LAMP Positive	Multiplex-PCR [1] Positive
Foxes	44 + (necropsy/SCT)	30 positive	25	29	31 positive	28	27
		14 negative	1	3	13 negative	4	3
	18 − (necropsy/SCT)	5 positive	1	1	5 positive	1	1
		13 negative	0	0	13 negative	0	0
Dogs	22 + (multiplex-PCR [1])	20 positive	20	20	21 positive	21	21
		2 negative	0	0	1 negative	0	0
	10 − (multiplex-PCR [1])	10 negative	0	0	10 negative	0	0

[1] as described in [15]; [2] independent egg isolation from two aliquots of two grams of faeces performed as previously described [10].

Table 2. Sensitivity and specificity values including confidence intervals in brackets for LAMP and multiplex-PCR for the detection of *E. multilocularis* using as template material treated with protocols 1A and 1B (see Table 1 for details) from foxes and dogs. Foxes were considered truly infected if intestinal stages of *E. multilocularis* were found in the intestine at necropsy/SCT. Dogs were considered as truly infected based on a positive multiplex-PCR [15] performed previously as part of a prevalence study in Kyrgyzstan.

	Foxes Protocol 1A		Foxes Protocol 1B		Dogs Protocol 1A		Dogs Protocol 1B	
	LAMP	Multiplex-PCR	LAMP	Multiplex-PCR	LAMP	Multiplex-PCR	LAMP	Multiplex-PCR
Sensitivity (IC 95%)	59 (43.2–73.6)	72.7 (57.2–85)	72.7 (57.2–85)	68.1 (52.4–81.3)	90.9 (70.8–98.8)	90.9 (70.8–98.8)	95.4 (77.1–99.8)	95.4 (77.1–99.8)
Specificity (IC 95%)	94.4 (72.7–99.8)	94.4 (72.7–99.8)	94.4 (72.7–99.8)	94.4 (72.7–99.8)	100 (69.1–100)	100 (69.1–100)	100 (69.1–100)	100 (69.1–100)

2.3. Examination of Field Samples with Protocols 1A and 1B According to Worm Burden

From 28 fox samples (positive at necropsy/SCT) collected in 2020, we assessed the worm burden for *E. multilocularis* after a thorough examination of the intestinal mucosa. Worm burden ranged between 2 and >100 *E. multilocularis* intestinal stages (Table 3). We divided the samples based on worm burden from foxes with two to 20 intestinal stages of *E. multilocularis* (9 foxes), 21–100 (7) and samples from foxes with >100 worms (12). We also included 17 samples from foxes that were negative at necropsy/SCT for *E. multilocularis*. LAMP and multiplex-PCR showed a positive result in 1/9 samples with worm burden between two and 20 parasites. When worm burden was between 21–100 worms, LAMP could detect 4/7 of the samples with protocol 1A and 6/7 were positive with multiplex-PCR. With protocol 1B, LAMP detected 6/7, and multiplex-PCR found 7/7 to be positive. Finally, when the worm burden was higher than 100 worms LAMP and multiplex-PCR detected 12/12 positive samples with protocol 1B. Multiplex-PCR detected 12/12 samples positives with protocol 1A (Table 3). Finally, from the 17 samples with no *E. multilocularis* found at necropsy/SCT, LAMP and multiplex-PCR rendered a positive result in one sample treated with both protocols.

Table 3. Number of positive results for *E. multilocularis* in LAMP and multiplex-PCR using protocols 1A and 1B in faecal samples from foxes necropsied in 2020 related to the total worm burden for *E. multilocularis*.

		Protocol 1A: NaOH/Dilution 1:50 Tris-HCl		Protocol 1B: NaOH + InstaGene™ Matrix	
# *E. multilocularis* at SCT	# Examined	LAMP Positive (Sensitivity)	Multiplex-PCR Positive (Sensitivity)	LAMP Positive (Sensitivity)	Multiplex-PCR Positive (Sensitivity)
0	17	1	1	1	1
2–20	9	1 (11.1%)	1 (11.1%)	2 (18.1%)	2 (18.1%)
21–100	7	4 (57.1%)	6 (85.7%)	6 (85.7%)	7 (100%)
>100	12	10 (83.3%)	12 (100%)	12 (100%)	12 (100%)

2.4. DNA Extraction from Whole Faeces (Protocol 2)

We investigated if it was possible to use NaOH+InstaGene™ Matrix (called protocol 2A) to treat a faecal sample to be used directly as a template for LAMP reaction (from this study) and a multiplex-PCR [15]. For comparison, we isolated DNA from the same samples using the Qiagen mini stool kit (named here protocol 2B). Using protocol 2A, we found a positive LAMP reaction for *E. multilocularis* in 10/30 faeces from foxes positive to *E. multilocularis* at necropsy/SCT; and a positive result in 17/30 in multiplex-PCR (Table 4). When an aliquot of the same samples was used for DNA isolation with the commercial kit, LAMP was positive in 16/30 and multiplex-PCR in 20/30 samples. Furthermore, a positive amplification corresponding to other cestodes, including *Taenia* spp. [15] was found in seven and five samples which were negative to *E. multilocularis* using a template from protocol 2A and the QIAGEN stool kit, respectively, with the multiplex-PCR. Detailed record of the multiplex-PCR results showing amplification of other cestodes, including *Taenia* spp. can be found in Table S1.

In the case of dogs, LAMP was positive in 2/18 samples and 7/18 samples in multiplex-PCR with protocol 2A. Using DNA isolated with a commercial kit, we found a positive LAMP reaction in 7/18 samples and 6/18 using the multiplex-PCR. Overall, the sensitivity achieved with the multiplex PCR using protocol 2B was 66% (CI: 47.1–82.7) followed by multiplex-PCR using protocol 2A (56.6%, 37.4–74.5) both in fox faeces. For dog faeces, the sensitivity achieved was 38.8% (17.3–64.2) with multiplex-PCR using protocol 2A and LAMP using protocol 2B.

3. Discussion

Loop-mediated isothermal amplification (LAMP) [43] has been portrayed as an affordable alternative to PCR for the detection of different pathogens, including parasites in faeces [18,19,44,45]. *Bst* polymerase used in LAMP offers high robustness being able to withstand harsh chemicals like NaOH and to overcome the presence of PCR inhibitors. In this study, we used the *Bst* 2.0 DNA polymerase which has improved performances regarding amplification speed, yield, tolerance to salt, and inhibitors present in different matrices including faeces [18–20]. Nevertheless, in the present study, a dilution of samples was necessary for protocol 1A, either to dilute NaOH or inhibitors present in samples. We developed a simple, cost-efficient methodology for DNA extraction involving lysis of isolated taeniid eggs with NaOH alone or adding Chelex resin (InstaGene™ Matrix). Then, we coupled the DNA extraction with a LAMP test developed in this study. Additionally, we included the multiplex-PCR [15] to compare the results for the amplification of the *E. multilocularis* target.

Necropsy/SCT remains the gold standard for the detection of *Echinococcus* spp. in canids but it is not rational to perform. We had access to freshly hunted foxes from which a necropsy and SCT could be performed. Therefore, we were able to compare the performance of the methods for DNA extraction using LAMP and multiplex-PCR [15] to detect *E. multilocularis* in fox faecal samples against the gold standard. As seen in Table 2, the sensitivity of both assays with fox samples was 72.7 (57.2–85) with multiplex-PCR using protocol 1A and LAMP using protocol 1B. We performed a thorough worm count in the intestinal content of foxes necropsied in 2020, showing that the sensitivity of the tests increased up to 100% in multiplex-PCR using protocol 1A for DNA extraction in faeces from 12 foxes with more than 100 worms. Protocol 1B LAMP and multiplex-PCR also detected 12/12 positive samples. It is possible to suggest that sensitivity could be improved further using a higher amount of faeces as starting material to increase the chances of finding eggs in animals with low worm burden.

In samples from dogs, the infection status was assessed with multiplex-PCR from isolated eggs from faeces collected from the ground in a highly endemic area for AE in Kyrgyzstan. In this case, sensitivity was 95.4% for LAMP and multiplex-PCR using protocol 1B (Table 2) and 90.9% for both methods when DNA was extracted using protocol 1A. In the present study, we were able to detect other cestodes, including *Taenia* spp. and *Mesocestoides* spp. in some of the samples which were negative for *E. multilocularis* using the multiplex PCR developed by Trachsel et al. [15], suggesting that no inhibition was present in these samples. Furthermore, we detected *Taenia* in all the samples from naturally infected foxes harbouring adult *Taenia* specimens in their intestine. By using whole faeces (400–500 mg) as starting material, multiplex-PCR was more sensitive than LAMP detecting *E. multilocularis* infection in foxes and dogs that using DNA isolated with a commercial kit. We used a large amount of faecal sample (500 mg) as input for the QIAamp FAST DNA Stool Mini Kit. Using the same commercial kit, Skrzypek et al. [46] used 1 g of faeces to detect *E. multilocularis* reporting positive results in 45.7% and 48.6% of the faeces from infected foxes (diagnosed at necropsy/SCT) with nested and multiplex-PCR, respectively, which is similar to the results found by us.

Table 4. Results of LAMP for *E. multilocularis* (from this study) and multiplex-PCR [15] using as a template the supernatant of 500 mg of fox (from Switzerland) and dog faeces (from Kyrgyzstan) treated with protocol 2A and from the DNA isolated with the QIAGEN stool kit (protocol 2B).

Host	Number of Animals Confirmed Positive (+) or Without (–) Detected *E. multilocularis infection* (Method for Diagnosis)	Protocol 2A: NaOH + InstaGene™ Matrix		Protocol 2B: QIAamp DNA Stool Mini Kit	
		LAMP Positive	Multiplex-PCR [1] Positive	LAMP Positive	Multiplex-PCR [1] Positive
Foxes	30 + (necropsy/SCT)	10^2	17	16^2	20
	5 – (necropsy/SCT)	0	0	0	0
Dogs	18 + (multiplex-PCR)	2^2	7	7	6
	10 – (multiplex-PCR)	0	0	0	0

[1] as described in [15]; [2] all samples found to be positive in LAMP were also positive in multiplex-PCR.

Previously, a LAMP assay has been used to detect *E. multilocularis* in canine faeces in China, reporting higher sensitivity than PCR in experimentally and naturally infected dogs, detecting as low as 1 pg DNA [21] which is the same limit of detection in the LAMP from the present study. The LAMP assay developed by Ni et al. has been used for detecting *E. multilocularis* [21] in wastewater in China [47]. Interestingly, in the work by Ni et al. [48], DNA was isolated with the QIAamp DNA Stool Mini Kit which in our hands did not produce satisfactory results. However, the results using the commercial kit were better than using NaOH-based method directly in faeces. The same kit was used in another report of a LAMP assay for the detection of *E. multilocularis* and other taeniids in a Tibetan rural area [22]. The LAMP assay detected 1 pg of DNA from *E. multilocularis* and a minimum of two eggs from this parasite [22]. They found that multiplex-PCR was able to detect more samples infected with *E. granulosus* and *T. hydatigena* than LAMP.

Finally, LAMP has been portrayed as a simple and easy to implement method that could be used in field surveillance of pathogens, specifically for *Echinococcus* species. However, so far it has not been used in large epidemiological studies in endemic areas. One of the advantages of LAMP is that there is no need for a thermocycler; however, it still requires some equipment including a heating block or water bath, a centrifuge, gel electrophoresis system (if results need to be confirmed in case of ambiguous colour change), and a clean lab with trained staff. In LAMP reactions, large quantities of DNA are amplified and pose a high risk of contamination when tubes are opened. To minimise this risk, a typical lab where PCR is routinely used, needs different separate environments for DNA extraction (pre-LAMP), setup of LAMP reactions, and visualisation of LAMP results (post-LAMP). The irregular shedding of eggs by the definitive hosts hampers the correct diagnosis of patent *E. multilocularis* infections. Furthermore, there are differences in the egg shedding between hosts; the highest egg output occurs 37 to 42 days post experimental infection in foxes and between 43 and 45 days in dogs in the same conditions [49]. Additionally, the patent period lasts one month for 98% of the worm burden, with a residual worm burden that lasts several months [49]. Therefore, detecting eggs from naturally infected canids in a single sample from naturally infected animals is challenging, as we have shown here since the presence of eggs will depend on the stage of the infection. Furthermore, the taeniid egg isolation has some disadvantages which is important to keep in mind, it can be time consuming especially analysing large number of samples, it requires the purchase of nylon filters which need to be inserted in the lid of the tubes or PET bottles [50] for sieving and it involves the use of a floatation solution which in the case of zinc chloride can be toxic for the manipulator and the environment [51]. A sugar solution [52] could potentially replace the zinc chloride avoiding the toxicity. We also tested the feasibility of using lateral flow dipstick coupled with LAMP reaction, and the results are promising; however, the high cost of the dipsticks precluded us from suggesting them as a cost-effective tool for the diagnosis of *E. multilocularis*.

Methods based on DNA detection from *E. multilocularis* have been published and extensively reviewed [12,13,53]. In general, they can be divided into those using whole faeces and the ones based on isolated taeniid eggs as starting material for DNA extraction. We focus in this study on the latter method which allows the enrichment of helminth eggs from faecal samples; in doing so, we aim to decrease the presence of PCR inhibitors commonly found in faecal samples based on the thorough wash of the filters used in the sieving process [10,15]. The taeniid egg isolation based on floatation/sieving allows the use of a relatively large amount of faeces as starting material, increasing the chances of finding eggs in canines with low worm burden for example. Up to 20 mL of faeces suspended in ethanol were used in the original publication describing the method [10] reporting a positive PCR in 33 out of 35 samples from foxes infected with *E. multilocularis* (diagnosed at necropsy). On the other hand, protocols for DNA extraction from whole faeces using commercial kits generally accept from 100 up to 500 mg of faeces as starting material and have the advantage of acquiring high yield purified DNA, in theory without PCR inhibitors. But they are expensive, costing between 3.1 and 10.2 USD per sample, and require between 40 to 90 min to be completed for 10 samples [16], making them unsuitable for use in large-scale prevalence studies. Therefore, there is a need to investigate and standardise cost-effective and straightforward methods for DNA extraction to be used in such studies. The use of NaOH-based DNA extraction methods from different tissues can produce a template of sufficient quality for PCR [36–38,54]. Furthermore, DNA has been extracted from *Echinococcus* spp. eggs, and also protoscoleces and cyst tissue, using NaOH for direct PCR and LAMP for genotyping studies without a purification step in several publications [26,39–41,55,56]. Similarly, the use of Chelex as a DNA extraction method has also been proposed since the 1990s [33] and used in recent publications as an alternative to producing input for PCR-based diagnostics [57,58] and mammal identification of scats [58] for example. NaOH and Chelex are components of the method used in seminal publications reporting on DNA isolation from taeniid eggs in which a commercial kit was also included [10,11,15,59]. The DNA extraction methods used in the present study offer a cost-effective alternative to the commercial kits for this purpose. If we consider the cost of a NaOH 0.2 M commercially available, the cost per sample is 0.014 USD for protocol 1A; in the case of protocol 1B the cost of InstaGene™ Matrix per sample is 0.95 USD. Furthermore, the time of DNA extraction is reduced in protocol 1A to 20 and 40 min with protocol 1B for processing ten samples, offering advantages over the 90 min required for DNA extraction described [59]. However, the time needed to isolate eggs needs to be considered as 3 h are required for 10–20 samples (Figure 1).

To conclude, we show two cost-effective methods for DNA extraction from isolated taeniid eggs for direct use in multiplex-PCR and LAMP developed in this study. These procedures circumvent the use of commercial kits for DNA extraction. Compared with the gold standard, the sensitivity of the test remains lower mainly because our methodology is focusing on the detection of eggs, therefore being unable to detect prepatent or low worm burden infections. Nevertheless, considering the reduction in costs and time, we propose the methods for DNA extraction as a valuable tool that can be used in extensive prevalence studies investigating the presence of eggs in canine environmental faecal samples in endemic areas of *E. multilocularis*.

4. Materials and Methods

4.1. Parasites

Eggs of *E. multilocularis* were isolated from the faeces of a naturally infected fox with no *Taenia* or *Mesocestoides* spp., detected at necropsy. Faeces were stored at −80 °C for five days for biosafety reasons and then subjected to floatation with zinc chloride (specific gravity 1.45) and sieving protocol with different nylon filters [10]. Eggs were stored at 4 °C in PBS with penicillin-streptomycin until further use. Multiplex-PCR [15] confirmed that only eggs of *E. multilocularis* and no *Taenia* spp. or *Mesocestoides* spp. were present in the suspension.

DNA was isolated from *E. multilocularis* adult worms and metacestode tissue cultured in vitro [60] using the QIAamp DNA Mini Kit (Qiagen, Hilden, Germany). The same procedure was used to isolate DNA from metacestode or adult stages of different cestodes identified morphologically including *E. granulosus sensu stricto*, *E. equinus*, *E. ortleppi*, *E. intermedius* (G6 and G7), *E. vogeli*, *E. shiquicus*, *Taenia polyacantha*, *T. multiceps*, *T. ovis*, *T. saginata*, *T. solium*, *T. crassiceps*, *T. hydatigena*, *Hydatigera taeniaeformis*, *T. pisiformis*, *T. krabbei*, *Dipylidium caninum*, *Diphyllobothrium latum*, *Mesocestoides litteratus*, and *M. lineatus*. Parasite identification was confirmed with PCR/sequencing of a section of the *cox1* gene [61].

4.2. LAMP

4.2.1. Primer Design

Multiple alignments of the mitochondrial genome of *E. multilocularis* and other *Echinococcus* and *Taenia* species were performed to select a region for designing LAMP primers using Genious R10 V10.1.3. The mitochondrial *nad1* gene of *E. multilocularis* (accession number in GenBank AB668376) was selected as it showed sequence variation between species, and primers were created using Primer Explorer V5 (EikenChemicalCoLtd.) [62]

and Primer Designer 1.16 (Premier) [63]. Primers were tested in silico for specificity using BLAST [64] and Primer-BLAST [65]. In total, seven primer sets, four primers each, were synthesised by Microsynth (Balgach, Switzerland). Primer sets were tested, assessing their specificity and analytical sensitivity to detect DNA from *E. multilocularis*.

4.2.2. Specificity and Analytical Sensitivity

Initially, the primer sets were tested in LAMP reactions using DNA isolated from *E. multilocularis*. Serial dilutions (1:10) of *E. multilocularis* DNA were prepared in triplicate starting at 10 ng/µL to 0.1 fg/µL. The primer set showing the highest consistency, specifically amplifying only *E. multilocularis* DNA with the lowest concentration as a template, was chosen for standardisation of the LAMP reaction (Table 5). Additionally, we tested all primer sets with DNA from the different cestodes mentioned above. To allow the visualization of samples positive to *E. multilocularis* DNA with the HybriDetect-Universal Lateral Flow Assay Kit (Milenia Biotec, Giessen, Germany), the FIP and BIP primers were labelled at the 5′ end with FAM (6-fluorescein amidite) and DIG (digoxigenin) as shown in Table 5.

Table 5. Primer set selected for the detection of *Echinococcus multilocularis* in a LAMP assay. For detection with the HybriDetect-Universal Lateral Flow Assay Kit (LFD), a FAM (6-fluorescein amidite) and DIG (digoxigenin) modifications were included in the Em-FIP and Em-BIP primers.

Primer	Sequence (5′–3′)	5′-Modification (LFD)
Em-F3	GCTTGTTGTTGTTTCCATTGA	-
Em-B3	ACAAAACCACCACCAACC	-
Em-FIP	TCCCTTTCAGACTCCCCATAATCA-TTTTTGGTGTGTGTGCTATG	FAM
Em-BIP	AGCGGTATATACTTTACGTGTTTGT-TCATTACAACAATCAACCATGA	DIG
Em-LB	TTGCTTGTGAGTATATAGTTGTATATGTGT	-

4.2.3. LAMP Assay

For the optimisation of LAMP assays, different concentrations of $MgSO_4$ (4–8 mM); primer mix (2×, 1.5×, 1×, 0.75×, 0.5×, 0.25×); betaine (0.4, 0.8, 1 and 1.2 M); malachite green (0.004%, 0.008% and 0.016%) were tested. Bovine serum albumin (0.1% w/v) was added to the reaction mixture to improve the performance of the *Bst* Polymerase during amplification. Various amplification times (30–90 min) and temperatures (61–70 °C) were tested. The reagent concentrations showing the strongest amplification (see the visualisation of the LAMP reaction) without unspecific amplification were chosen for the field study. Finally, the LAMP reaction was set up in 25 µL containing the isothermal amplification buffer from New England Biolabs (Ipswich, MA, USA) [20 mM Tris-HCl, 10 mM $(NH_4)_2SO_4$, 50 mM KCl, 2 mM $MgSO_4$, and 0.1% Tween 20], 6 mM $MgSO_4$, 1.4 mM of each dNTP, 0.8 M betaine, 0.004% malachite green, 8 U/mL *Bst* 2.0 DNA Polymerase Warmstart (New England Biolabs), 1.6 µM of each FIP(-FAM) and BIP(-DIG) primer, 0.2 µM of each F3 and B3 primer, and 0.4 µM of LB primer (loop primer). Finally, 2 µL DNA were added into each tube as a template. The amplification occurred at 65 °C for 60 min in a heating block. In every experiment, positive control (DNA of *E. multilocularis*) and negative control (water) were included.

4.2.4. Visualisation of the LAMP Reaction

Three different visualisation methods were used to assess the results of LAMP reactions. First, directly after the incubation time, visual judgment of the colour change was performed and documented. A positive reaction was indicated by different intensities of blue colour and in a negative reaction by a colour change from blue or green to colourless. Secondly, 3 µL of each LAMP product were subjected to gel electrophoresis in 1.5% agarose gels stained with Gel Red (Biotium, Fremont, CA, USA) and visualised with a UV transilluminator (BioRad, Hercules, CA, USA). A DNA ladder-like pattern represented a positive reaction while no ladder was a negative result. Finally, we used a lateral flow dipstick (HybriDetect 2T, Milenia Biotec, Giessen, Germany) to assess the LAMP reactions with the FIP and BIP labelled primers (Table 5). Briefly, 100 µL of citrate-phosphate-buffer (Milenia Biotec) were mixed with 10 µL of the LAMP reaction in a 1.5 mL tube. Then, the dipstick was inserted into each tube and the results were assessed after five to ten minutes. LAMP amplicons labelled with FAM and DIG bind to the anti-digoxigenin antibodies in the strip. The test band will appear on the dipstick, along with the control band, whereas in a negative LAMP reaction, only the control band is visible.

4.2.5. Stability of the LAMP Master Mix

The stability of the LAMP master mix was assessed by preparing aliquots of 23 µL of a LAMP master mix (as described above) without DNA template in 0.2 mL tubes stored at 4 °C, over a period of six weeks [66]. Every week, two tubes were taken from the fridge and 2 µL of *E. multilocularis* DNA (1 ng/µL) were added to one tube and 2 µL of water to the second tube as negative control. The amplification and visualisation of the reaction occurred as explained above.

4.2.6. Limit of Detection of LAMP

The limit of detection (No. of eggs) of the LAMP reaction with the primer set from Table 5 was established using eggs of *E. multilocularis* isolated from faeces of a naturally infected fox (as explained above). Eggs were aspirated individually with a micropipette under the microscope into 1.5 mL tubes. Four replicates of tubes containing ten, five, four, three, two, and one egg(s) were prepared in 30 µL of water. Egg suspensions were subjected to protocols 1A and 1B as explained below for DNA extraction.

4.3. DNA Extraction (Protocols 1A and 1B)

The methods presented basically aim to break the eggs of *E. multilocularis*, and subsequently lyse cells of the embryo to release DNA to be used directly in LAMP and multiplex-PCR without a further purification step. We use here the term "DNA extraction" to refer to these methods for treating taeniid eggs. For protocol 1A, egg suspensions were treated with 0.2 M NaOH (ratio 1:1) and incubated at 95 °C for 10 min in a heating block [26,41]. Tubes were centrifuged (quick spin) and then dilutions with Tris-HCl pH 8.3 (100 mM) were prepared (ratios 1:2, 1:10, 1:50, and 1:100). Finally, 2 µL of each dilution were used as a template for LAMP (developed in this study using primers from Table 5) and multiplex-PCR [15]; the dilution offering the most consistent result was chosen for the DNA extraction in the field study. For protocol 1B, egg suspensions were treated as in protocol 1A without dilution with Tris-HCl. Subsequently, 100 µL of Instagene Matrix (Bio-Rad) were added to each tube and incubated for another 15 min at 56 °C. After vortexing and centrifugation (12,000× g for 3 min), 2 µL of the supernatant were used directly as a template for LAMP (developed in this study) and multiplex-PCR [15].

4.4. Application of LAMP and Multiplex-PCR in Field Samples
4.4.1. Fox Faecal Samples

In total, 63 faecal samples from foxes shot during the official hunting season (January–February) in the surroundings of Zurich (Switzerland) from 2018 until 2020 were used in this study. At necropsy, fox intestines were examined for the presence of *E. multilocularis* and other parasites based on the SCT method [4]. Faecal samples were collected from the rectum, deposited in 50 mL tubes, and stored at −80 °C for five days to inactivate taeniid eggs. Worm burden was assessed in 45 foxes necropsied in 2020 including 28 positive foxes and 17 negative animals to the presence of *E. multilocularis*. Two aliquots of two grams of faeces per sample were taken for independent isolation of taeniid eggs as previously described [10]. The material retained in the 21 µm filter was collected and carefully screened with an inverted microscope for the presence of taeniid eggs in 10 mL tubes with a flat side. Then, all tubes were centrifuged at 200× g for 10 min, the supernatant discarded and the sediment was transferred to a 1.5 mL tube for treatments with protocols 1A or 1B and subsequently use as a template for LAMP (from this study) and multiplex-PCR [15].

4.4.2. Dog Faecal Samples

In total, we used 32 faecal samples from dogs collected as part of a large prevalence study for *Echinococcus* spp. in Kyrgyzstan between 2017 and 2018 [including a genetic characterisation of *E. multilocularis* [67] and identification by egg isolation + multiplex-PCR [15]]. In the study mentioned above, taeniid eggs were concentrated from dog faecal samples as previously described and DNA was isolated combining alkaline lysis and the QIAamp Kit [10,50]. Briefly, eggs were resuspended in 200 µL of distilled water and 25 µL KOH (1M) and 7 µL of DTT (1M) were added and incubated at 65 °C for 15 min. Afterwards, 60 µL 2M Tris-HCl pH 8.4 and 2 µL HCl (12.4N/≥37%) were pipetted into the tubes. Finally, 200 µL of Buffer AL (QIAamp Kit) and 20 µL of Proteinase K were added and incubated at 56 °C for 10 min. Then, 50 µL of Chelex solution (50%) were added and tubes were mixed in a rotator for 1 h at room temperature. After centrifugation, the supernatant (approximately 400 µL) was transferred to a new tube with 200 µL of ethanol (100%) and mixed; the content of each tube was then transferred into a Qiagen spin column. The protocol continues following the manufacturer instructions for the QIAamp DNA with two washes with buffer AW1 instead of one. DNA was eluted in 100 µL of 10 mM Tris-HCl, pH 8.3, and stored at −20 °C until use as a template in multiplex-PCR [15].

4.5. Application of LAMP and Multiplex-PCR in Whole Faeces (Protocols 2A and 2B)

Two aliquots of 400–500 mg were taken from faecal samples of 30 foxes infected with *E. multilocularis* and five from negative foxes from the same group of animals used for protocols 1A and 1B (diagnosed at necropsy/SCT). Dog faecal samples included 18 *E. multilocularis* positive and ten negative (diagnosed using egg isolation + multiplex-PCR). One aliquot was treated with the protocol 2A: 400 µL of NaOH solution (0.2 M) were added, mixed, and incubated at 95 °C for 10 min, then 1 mL of Instagene Matrix (Bio-Rad) was added and the tubes were kept at 56 °C for 15 min. The samples were vortexed, centrifuged at 12,000× g for 3 min, and 500 µL of the supernatant transferred to a new tube. From the supernatant, a 1:50 dilution with water was prepared and 2 µL were used as a template for LAMP and multiplex-PCR. The second aliquot was used for DNA isolation following the QIAamp Fast DNA Stool Mini Kit® instructions for large stool volumes (protocol 2B).

4.6. Sensitivity and Specificity

The diagnostic sensitivities and specificities were calculated for protocol 1A and 1B by comparing the results of LAMP and multiplex-PCR with the SCT result in foxes and with the initial multiplex-PCR (from the prevalence study) result in the case of dogs.

Supplementary Materials: The following are available online at https://www.mdpi.com/article/10.339 0/pathogens10070847/s1, Table S1. List with all the results from foxes positive and negative to intestinal stages of *E. multilocularis* diagnosed at necropsy and used in the present study. Isolated eggs were treated with protocol 1A and 1B. Results for multiplex PCR also shows if an amplicon corresponding to other cestodes (oc) including *Taenia* spp. was also amplified. Em: *Echinococcus multilocularis*.

Author Contributions: Conceptualization, C.A.A.R., P.D.; methodology, B.J.B., T.K., C.A.A.R.; software, B.J.B., T.K.; validation, B.J.B., G.M., C.A.A.R.; formal analysis, B.J.B., C.A.A.R.; investigation, B.J.B., G.M., C.A.A.R.; resources, K.K.A., M.I., P.A.K., P.D., C.A.A.R.; writing—original draft preparation, B.J.B., P.D., C.A.A.R.; writing—review and editing B.J.B., G.M., T.K., P.A.K., P.D., C.A.A.R.; visualization B.J.B., C.A.A.R., P.A.K.; supervision, P.D., C.A.A.R.; project administration, P.D. All authors have read and agreed to the published version of the manuscript.

Funding: This research received no external funding.

Institutional Review Board Statement: Not applicable.

Informed Consent Statement: Not applicable.

Data Availability Statement: Not applicable.

Acknowledgments: Figure 1 was created using BioRender.com.

Conflicts of Interest: The authors declare no conflict of interest.

References

1. Eckert, J.; Deplazes, P. Biological, epidemiological, and clinical aspects of echinococcosis, a zoonosis of increasing concern. *Clin. Microbiol. Rev.* **2004**, *17*, 107–135. [CrossRef]
2. Romig, T.; Deplazes, P.; Jenkins, D.; Giraudoux, P.; Massolo, A.; Craig, P.S.; Wassermann, M.; Takahashi, K.; De La Rue, M. Ecology and life cycle patterns of *Echinococcus* species. *Adv. Parasitol.* **2017**, *95*, 213–314. [CrossRef]
3. Alvarez Rojas, C.A.; Mathis, A.; Deplazes, P. Assessing the contamination of food and the environment with *Taenia* and *Echinococcus* eggs and their zoonotic transmission. *Curr. Clin. Microbiol. Rep.* **2018**, *5*, 154–163. [CrossRef]
4. Eckert, J.; Deplazes, P.; Craig, P.S.; Gemmell, M.A.; Gottstein, B.; Heath, D.; Jenkins, D.J.; Kamiya, M.; Lightowlers, M. Chapter 3 Echinococcosis in animals: Clinical aspects, diagnosis and treatment. In *WHO/OIE Manual on Echinococcosis in Humans and Animals: A Public Health Problem of Global Concern*; Eckert, J., Gemmell, M.A., Meslin, F.X., Pawlowski, Z.S., Eds.; World Organisation for Animal Health: Paris, France, 2001.
5. Deplazes, P.; Gottstein, B.; Eckert, J.; Jenkins, D.J.; Ewald, D.; Jimenez-Palacios, S. Detection of *Echinococcus* coproantigens by enzyme-linked immunosorbent assay in dogs, dingoes and foxes. *Parasitol. Res.* **1992**, *78*, 303–308. [CrossRef]
6. Deplazes, P.; Alther, P.; Tanner, I.; Thompson, R.C.A.; Eckert, J. *Echinococcus multilocularis* coproantigen detection by enzyme-linked immunosorbent assay in fox, dog, and cat populations. *J. Parasitol.* **1999**, *85*, 115–121. [CrossRef]
7. Sakai, H.; Nonaka, N.; Yagi, K.; Oku, Y.; Kamiya, M. Coproantigen detection in a survey of *Echinococcus multilocularis* infection among red foxes, *Vulpes vulpes schrencki*, in Hokkaido, Japan. *J. Vet. Med. Sci.* **1998**, *60*, 639–641. [CrossRef] [PubMed]
8. Machnicka, B.; Dziemian, E.; Rocki, B.; Kołodziej-Sobocińska, M. Detection of *Echinococcus multilocularis* antigens in faeces by ELISA. *Parasitol. Res.* **2003**, *91*, 491–496. [CrossRef] [PubMed]
9. Dinkel, A.; Von Nickisch-Rosenegk, M.; Bilger, B.; Merli, M.; Lucius, R.; Romig, T. Detection of *Echinococcus multilocularis* in the definitive host: Coprodiagnosis by PCR as an alternative to necropsy. *J. Clin. Microbiol.* **1998**, *36*, 1871–1876. [CrossRef]
10. Mathis, A.; Deplazes, P.; Eckert, J. An improved test system for PCR-based specific detection of *Echinococcus multilocularis* eggs. *J. Helminthol.* **1996**, *70*, 219–222. [CrossRef] [PubMed]
11. Bretagne, S.; Guillou, J.P.; Morand, M.; Houin, R. Detection of *Echinococcus multilocularis* DNA in fox faeces using DNA amplification. *Parasitology* **1993**, *106*, 193–199. [CrossRef] [PubMed]
12. Deplazes, P.; Dinkel, A.; Mathis, A. Molecular tools for studies on the transmission biology of *Echinococcus multilocularis*. *Parasitology* **2003**, *127*, S53–S61. [CrossRef]
13. Siles-Lucas, M.; Casulli, A.; Conraths, F.J.; Müller, N. Chapter Three—Laboratory diagnosis of *Echinococcus* spp. in human patients and infected animals. *Adv. Parasitol.* **2017**, *96*, 159–257. [PubMed]
14. Yimam, A.E.; Nonaka, N.; Oku, Y.; Kamiya, M. Prevalence and intensity of *Echinococcus multilocularis* in red foxes (*Vulpes vulpes schrencki*) and raccoon dogs (*Nyctereutes procyonoides albus*) in Otaru City, Hokkaido, Japan. *Jpn. J. Vet. Res.* **2002**, *49*, 287–296.
15. Trachsel, D.; Deplazes, P.; Mathis, A. Identification of taeniid eggs in the faeces from carnivores based on multiplex PCR using targets in mitochondrial DNA. *Parasitology* **2007**, *134*, 911–920. [CrossRef] [PubMed]
16. Maksimov, P.; Schares, G.; Press, S.; Fröhlich, A.; Basso, W.; Herzig, M.; Conraths, F.J. Comparison of different commercial DNA extraction kits and PCR protocols for the detection of *Echinococcus multilocularis* eggs in faecal samples from foxes. *Vet. Parasitol.* **2017**, *237*, 83–93. [CrossRef] [PubMed]
17. Videnska, P.; Smerkova, K.; Zwinsova, B.; Popovici, V.; Micenkova, L.; Sedlar, K.; Budinska, E. Stool sampling and DNA isolation kits affect DNA quality and bacterial composition following 16S rRNA gene sequencing using MiSeq Illumina platform. *Sci. Rep.* **2019**, *9*, 3837. [CrossRef] [PubMed]
18. Kaneko, H.; Kawana, T.; Fukushima, E.; Suzutani, T. Tolerance of loop-mediated isothermal amplification to a culture medium and biological substances. *J. Biochem. Biophys. Methods* **2007**, *70*, 499–501. [CrossRef] [PubMed]
19. Grab, D.J.; Lonsdale-Eccles, J.; Inoue, N. Lamp for tadpoles. *Nat. Methods* **2005**, *2*, 635. [CrossRef] [PubMed]
20. Francois, P.; Tangomo, M.; Hibbs, J.; Bonetti, E.-J.; Boehme, C.C.; Notomi, T.; Perkins, M.D.; Schrenzel, J. Robustness of a loop-mediated isothermal amplification reaction for diagnostic applications. *FEMS Immunol. Med. Microbiol.* **2011**, *62*, 41–48. [CrossRef]
21. Ni, X.; Mcmanus, D.P.; Yan, H.; Yang, J.; Lou, Z.; Li, H.; Li, L.; Lei, M.; Cai, J.; Fan, Y.; et al. Loop-mediated isothermal amplification (LAMP) assay for the identification of *Echinococcus multilocularis* infections in canine definitive hosts. *Parasites Vectors* **2014**, *7*, 254. [CrossRef]
22. Feng, K.; Li, W.; Guo, Z.; Duo, H.; Fu, Y.; Shen, X.; Tie, C.; Xiao, C.; Luo, Y.; Qi, G.; et al. Development of LAMP assays for the molecular detection of taeniid infection in canine in Tibetan rural area. *J. Vet. Med. Sci.* **2017**, *79*, 1986–1993. [CrossRef]
23. Avila, H.G.; Mozzoni, C.; Trangoni, M.D.; Cravero, S.L.P.; Pérez, V.M.; Valenzuela, F.; Gertiser, M.L.; Butti, M.J.; Kamenetzky, L.; Jensen, O.; et al. Development of a copro-LAMP assay for detection of several species of *Echinococcus granulosus sensu lato* complex. *Vet. Parasitol.* **2020**, *277*, 109017. [CrossRef] [PubMed]
24. Ni, X.-W.; Mcmanus, D.P.; Lou, Z.-Z.; Yang, J.-F.; Yan, H.-B.; Li, L.; Li, H.-M.; Liu, Q.-Y.; Li, C.-H.; Shi, W.-G.; et al. A comparison of loop-mediated isothermal amplification (LAMP) with other surveillance tools for *Echinococcus granulosus* diagnosis in canine definitive hosts. *PLoS ONE* **2014**, *9*, e100877. [CrossRef] [PubMed]
25. Salant, H.; Abbasi, I.; Hamburger, J. The development of a loop-mediated isothermal amplification method (LAMP) for *Echinococcus granulosus* coprodetection. *Am. J. Trop. Med. Hyg.* **2012**, *87*, 883–887. [CrossRef] [PubMed]

26. Wassermann, M.; Mackenstedt, U.; Romig, T. A loop-mediated isothermal amplification (LAMP) method for the identification of species within the *Echinococcus granulosus* complex. *Vet. Parasitol.* **2014**, *200*, 97–103. [CrossRef] [PubMed]

27. Rahman, S.M.M.; Song, H.B.; Jin, Y.; Oh, J.-K.; Lim, M.K.; Hong, S.-T.; Choi, M.-H. Application of a loop-mediated isothermal amplification (LAMP) assay targeting cox1 gene for the detection of *Clonorchis sinensis* in human fecal samples. *PLoS Negl. Trop. Dis.* **2017**, *11*, e0005995. [CrossRef]

28. Fernández-Soto, P.; Fernández-Medina, C.; Cruz-Fernández, S.; Crego-Vicente, B.; Febrer-Sendra, B.; García-Bernalt Diego, J.; Gorgojo-Galindo, Ó.; López-Abán, J.; Vicente Santiago, B.; Muro Álvarez, A. Whip-LAMP: A novel LAMP assay for the detection of *Trichuris muris*-derived DNA in stool and urine samples in a murine experimental infection model. *Parasites Vectors* **2020**, *13*, 552. [CrossRef]

29. Fernández-Soto, P.; Sánchez-Hernández, A.; Gandasegui, J.; Bajo Santos, C.; López-Abán, J.; Saugar, J.M.; Rodríguez, E.; Vicente, B.; Muro, A. Strong-LAMP: A LAMP Assay for *Strongyloides* spp. Detection in Stool and Urine Samples. Towards the Diagnosis of Human Strongyloidiasis Starting from a Rodent Model. *PLoS Negl. Trop. Dis.* **2016**, *10*, e0004836. [CrossRef]

30. Mugambi, R.M.; Agola, E.L.; Mwangi, I.N.; Kinyua, J.; Shiraho, E.A.; Mkoji, G.M. Development and evaluation of a Loop Mediated Isothermal Amplification (LAMP) technique for the detection of hookworm (*Necator americanus*) infection in fecal samples. *Parasites Vectors* **2015**, *8*, 574. [CrossRef] [PubMed]

31. Nkouawa, A.; Sako, Y.; Nakao, M.; Nakaya, K.; Ito, A. Loop-mediated isothermal amplification method for differentiation and rapid detection of *Taenia* species. *J. Clin. Microbiol.* **2009**, *47*, 168–174. [CrossRef]

32. Steiner, J.J.; Polemba, C.J.; Fjellstrom, R.G.; Elliott, L.F. A rapid one-tube genomic DNA extraction process for PCR and RAPD analyses. *Nucleic Acids Res.* **1995**, *23*, 2569–2570. [CrossRef]

33. De Lamballerie, X.; Zandotti, C.; Vignoli, C.; Bollet, C.; De Micco, P. A one-step microbial DNA extraction method using "Chelex 100" suitable for gene amplification. *Res. Microbiol.* **1992**, *143*, 785–790. [CrossRef]

34. Kawasaki, E.S. Sample preparation from blood, cells and other fluids. In *PCR Protocols*; Innis, M.A., Gelfand, D.H., Sninsky, J.J., White, T.J., Eds.; Academic Press: San Diego, CA, USA, 1990; pp. 146–152.

35. Wang, H.; Qi, M.; Cutler, A.J. A simple method of preparing plant samples for PCR. *Nucleic Acids Res.* **1993**, *21*, 4153–4154. [CrossRef] [PubMed]

36. Osmundson, T.W.; Eyre, C.A.; Hayden, K.M.; Dhillon, J.; Garbelotto, M.M. Back to basics: An evaluation of NaOH and alternative rapid DNA extraction protocols for DNA barcoding, genotyping, and disease diagnostics from fungal and oomycete samples. *Mol. Ecol. Resour.* **2013**, *13*, 66–74. [CrossRef]

37. Satya, P.; Mitra, S.; Ray, D.P.; Mahapatra, B.S.; Karan, M.; Jana, S.; Sharma, A.K. Rapid and inexpensive NaOH based direct PCR for amplification of nuclear and organelle DNA from ramie (Boehmeria nivea), a bast fibre crop containing complex polysaccharides. *Ind. Crop. Prod.* **2013**, *50*, 532–536. [CrossRef]

38. Werner, O.; Ros, R.M.; Guerra, J. Direct amplification and NaOH extraction: Two rapid and simple methods for preparing bryophyte DNA for polymerase chain reaction (PCR). *J. Bryol.* **2002**, *24*, 127–131. [CrossRef]

39. Hüttner, M.; Nakao, M.; Wassermann, T.; Siefert, L.; Boomker, J.D.F.; Dinkel, A.; Sako, Y.; Mackenstedt, U.; Romig, T.; Ito, A. Genetic characterization and phylogenetic position of *Echinococcus felidis* Ortlepp, 1937 (Cestoda: Taeniidae) from the African lion. *Int. J. Parasitol.* **2008**, *38*, 861–868. [CrossRef]

40. Mulinge, E.; Magambo, J.; Odongo, D.; Njenga, S.; Zeyhle, E.; Mbae, C.; Kagendo, D.; Addy, F.; Ebi, D.; Wassermann, M.; et al. Molecular characterization of *Echinococcus* species in dogs from four regions of Kenya. *Vet. Parasitol.* **2018**, *255*, 49–57. [CrossRef] [PubMed]

41. Nakao, M.; Sako, Y.; Ito, A. Isolation of polymorphic microsatellite loci from the tapeworm *Echinococcus multilocularis*. *Infect. Genet. Evol.* **2003**, *3*, 159–163. [CrossRef]

42. Wassermann, M.; Aschenborn, O.; Aschenborn, J.; Mackenstedt, U.; Romig, T. A sylvatic lifecycle of *Echinococcus equinus* in the Etosha National Park, Namibia. *Int. J. Parasitol. Parasites Wildl.* **2015**, *4*, 97–103. [CrossRef] [PubMed]

43. Notomi, T.; Okayama, H.; Masubuchi, H.; Yonekawa, T.; Watanabe, K.; Amino, N.; Hase, T. Loop-mediated isothermal amplification of DNA. *Nucleic Acids Res.* **2000**, *28*, E63. [CrossRef] [PubMed]

44. Xu, J.; Rong, R.; Zhang, H.Q.; Shi, C.J.; Zhu, X.Q.; Xia, C.M. Sensitive and rapid detection of *Schistosoma japonicum* DNA by loop-mediated isothermal amplification (LAMP). *Int. J. Parasitol.* **2010**, *40*, 327–331. [CrossRef] [PubMed]

45. Cai, X.Q.; Xu, M.J.; Wang, Y.H.; Qiu, D.Y.; Liu, G.X.; Lin, A.; Tang, J.D.; Zhang, R.L.; Zhu, X.Q. Sensitive and rapid detection of *Clonorchis sinensis* infection in fish by loop-mediated isothermal amplification (LAMP). *Parasitol. Res.* **2010**, *106*, 1379–1383. [CrossRef] [PubMed]

46. Skrzypek, K.; Karamon, J.; Samorek-Pieróg, M.; Dąbrowska, J.; Kochanowski, M.; Sroka, J.; Bilska-Zając, E.; Cencek, T. Comparison of two DNA extraction methods and two PCRs for detection of *Echinococcus multilocularis* in the stool samples of naturally infected red foxes. *Animals* **2020**, *10*, 2381. [CrossRef] [PubMed]

47. Lass, A.; Ma, L.; Kontogeorgos, I.; Xueyong, Z.; Li, X.; Karanis, P. Contamination of wastewater with *Echinococcus multilocularis*—Possible implications for drinking water resources in the QTP China. *Water Res.* **2020**, *170*, 115334. [CrossRef]

48. Chamai, M.; Omadang, L.; Erume, J.; Ocaido, M.; Oba, P.; Othieno, E.; Bonaventure, S.; Kitibwa, A. Identification of *Echinococcus granulosus* strains using polymerase chain reaction–restriction fragment length polymorphism amongst livestock in Moroto district, Uganda. *Onderstepoort J. Vet. Res.* **2016**, *83*, 1–7. [CrossRef] [PubMed]

49. Kapel, C.M.O.; Torgerson, P.R.; Thompson, R.C.A.; Deplazes, P. Reproductive potential of *Echinococcus multilocularis* in experimentally infected foxes, dogs, raccoon dogs and cats. *Int. J. Parasitol.* **2006**, *36*, 79–86. [CrossRef]

50. Guggisberg, A.R.; Alvarez Rojas, C.A.; Kronenberg, P.A.; Miranda, N.; Deplazes, P. A sensitive, one-way sequential sieving method to isolate helminths' eggs and protozoal oocysts from lettuce for genetic identification. *Pathogens* **2020**, *9*, 624. [CrossRef]
51. Bodar, C.W.M.; Pronk, M.E.J.; Sijm, D.T.H.M. The European Union risk assessment on zinc and zinc compounds: The process and the facts. *Integr. Environ. Assess. Manag.* **2005**, *1*, 301–319. [CrossRef]
52. Matsuo, K.; Kamiya, H. Modified sugar centrifugal flotation technique for recovering *Echinococcus multilocularis* eggs from soil. *J. Parasitol.* **2005**, *91*, 208–209. [CrossRef]
53. Mathis, A.; Deplazes, P. Copro-DNA tests for diagnosis of animal taeniid cestodes. *Parasitol. Int.* **2006**, *55*, S87–S90. [CrossRef]
54. Meeker, N.D.; Hutchinson, S.A.; Ho, L.; Trede, N.S. Method for isolation of PCR-ready genomic DNA from zebrafish tissues. *Biotechniques* **2007**, *43*, 610–614. [CrossRef]
55. Boubaker, G.; Macchiaroli, N.; Prada, L.; Cucher, M.A.; Rosenzvit, M.C.; Ziadinov, I.; Deplazes, P.; Saarma, U.; Babba, H.; Gottstein, B.; et al. A Multiplex PCR for the simultaneous detection and genotyping of the *Echinococcus granulosus* complex. *PLoS Negl. Trop. Dis.* **2013**, *7*, e2017. [CrossRef] [PubMed]
56. Sean Walsh, P.; Metzger, D.A.; Higuchi, R. Chelex 100 as a medium for simple extraction of DNA for PCR-based typing from forensic material. *Biotechniques* **2013**, *54*, 134–139. [CrossRef]
57. Singh, U.A.; Kumari, M.; Iyengar, S. Method for improving the quality of genomic DNA obtained from minute quantities of tissue and blood samples using Chelex 100 resin. *Biol. Proced. Online* **2018**, *20*, 12. [CrossRef] [PubMed]
58. Paxinos, E.; Mcintosh, C.; Ralls, K.; Fleischer, R. A noninvasive method for distinguishing among canid species: Amplification and enzyme restriction of DNA from dung. *Mol. Ecol.* **1997**, *6*, 483–486. [CrossRef] [PubMed]
59. Stefanic, S.; Shaikenov, B.S.; Deplazes, P.; Dinkel, A.; Torgerson, P.R.; Mathis, A. Polymerase chain reaction for detection of patent infections of *Echinococcus granulosus* ("sheep strain") in naturally infected dogs. *Parasitol. Res.* **2004**, *92*, 347–351. [CrossRef] [PubMed]
60. Laurimäe, T.; Kronenberg, P.A.; Alvarez Rojas, C.A.; Ramp, T.W.; Eckert, J.; Deplazes, P. Long-term (35 years) cryopreservation of *Echinococcus multilocularis* metacestodes. *Parasitology* **2020**, *147*, 1048–1054. [CrossRef]
61. Bowles, J.; Blair, D.; Mcmanus, D.P. Genetic variants within the genus *Echinococcus* identified by mitochondrial DNA sequencing. *Mol. Biochem. Parasitol.* **1992**, *54*, 165–173. [CrossRef]
62. ECC Ltd. PrimerExplorer V5. Available online: http://primerexplorer.jp/lampv5e/index.html (accessed on 24 August 2020).
63. PREMIER Biosoft. LAMP Designer. Available online: http://www.premierbiosoft.com/isothermal/index.html (accessed on 24 August 2020).
64. Altschul, S.F.; Gish, W.; Miller, W.; Myers, E.W.; Lipman, D.J. Basic local alignment search tool. *J. Mol. Biol.* **1990**, *215*, 403–410. [CrossRef]
65. Ye, J.; Coulouris, G.; Zaretskaya, I.; Cutcutache, I.; Rozen, S.; Madden, T.L. Primer-BLAST: A tool to design target-specific primers for polymerase chain reaction. *BMC Bioinform.* **2012**, *13*, 134. [CrossRef]
66. Kamber, T.; Koekemoer, L.L.; Mathis, A. Loop-mediated isothermal amplification (LAMP) assays for *Anopheles funestus* group and *Anopheles gambiae* complex species. *Med. Vet. Entomol.* **2020**, *34*, 295–301. [CrossRef] [PubMed]
67. Alvarez Rojas, C.A.; Kronenberg, P.A.; Aitbaev, S.; Omorov, R.A.; Abdykerimov, K.K.; Paternoster, G.; Mullhaupt, B.; Torgerson, P.; Deplazes, P. Genetic diversity of *Echinococcus multilocularis* and *Echinococcus granulosus sensu lato* in Kyrgyzstan: The A2 haplotype of *E. multilocularis* is the predominant variant infecting humans. *PLoS Negl. Trop. Dis.* **2020**, *14*, e0008242. [CrossRef] [PubMed]

 *pathogens*

Article

Non-Invasive Molecular Survey of Sarcoptic Mange in Wildlife: Diagnostic Performance in Wolf Faecal Samples Evaluated by Multi-Event Capture–Recapture Models

Julieta Rousseau [1,2,*], Mónia Nakamura [2,3], Helena Rio-Maior [2], Francisco Álvares [2], Rémi Choquet [4], Luís Madeira de Carvalho [1,*], Raquel Godinho [2,3] and Nuno Santos [2]

1 CIISA—Centro de Investigação Interdisciplinar em Sanidade Animal, Faculty of Veterinary Medicine, University of Lisbon, Avenida da Universidade Técnica, 1300-477 Lisbon, Portugal
2 CIBIO/InBIO—Centro de Investigação em Biodiversidade e Recursos Genéticos, Universidade do Porto, Campus de Vairão, 4485-661 Vairão, Portugal; moniayui@gmail.com (M.N.); helenariomaior@gmail.com (H.R.-M.); falvares@cibio.up.pt (F.Á.); rgodinho@cibio.up.pt (R.G.); nuno.santos@cibio.up.pt (N.S.)
3 Department of Biology, Faculty of Sciences, University of Porto, 4169-007 Porto, Portugal
4 CEFE—Centre d'Écologie Fonctionnelle et Évolutive, UMR 5175, CNRS, University of Montpellier, 1919 Route de Mende, CEDEX 5, FR-34293 Montpellier, France; Remi.CHOQUET@cefe.cnrs.fr
* Correspondence: julieta_rousseau@hotmail.com (J.R.); madeiradecarvalho@fmv.ulisboa.pt (L.M.d.C.)

Citation: Rousseau, J.; Nakamura, M.; Rio-Maior, H.; Álvares, F.; Choquet, R.; Madeira de Carvalho, L.; Godinho, R.; Santos, N. Non-Invasive Molecular Survey of Sarcoptic Mange in Wildlife: Diagnostic Performance in Wolf Faecal Samples Evaluated by Multi-Event Capture–Recapture Models. *Pathogens* **2021**, *10*, 243. https://doi.org/10.3390/pathogens 10020243

Academic Editors: Bonto Faburay and Jörg Jores

Received: 18 January 2021
Accepted: 17 February 2021
Published: 20 February 2021

Publisher's Note: MDPI stays neutral with regard to jurisdictional claims in published maps and institutional affiliations.

Abstract: Sarcoptic mange is globally enzootic, and non-invasive methods with high diagnostic specificity for its surveillance in wildlife are lacking. We describe the molecular detection of *Sarcoptes scabiei* in non-invasively collected faecal samples, targeting the 16S rDNA gene. We applied this method to 843 Iberian wolf *Canis lupus signatus* faecal samples collected in north-western Portugal (2006–2018). We further integrated this with serological data (61 samples from wolf and 20 from red fox *Vulpes vulpes*, 1997–2019) in multi-event capture–recapture models. The mean predicted prevalence by the molecular analysis of wolf faecal samples from 2006–2018 was 7.2% (CI_95 5.0–9.4%; range: 2.6–11.7%), highest in 2009. The mean predicted seroprevalence in wolves was 24.5% (CI_95 18.5–30.6%; range: 13.0–55.0%), peaking in 2006–2009. Multi-event capture–recapture models estimated 100% diagnostic specificity and moderate diagnostic sensitivity (30.0%, CI_95 14.0–53.0%) for the molecular method. Mange-infected individually identified wolves showed a tendency for higher mortality versus uninfected wolves (ΔMortality 0.150, CI_95 −0.165–0.458). Long-term serology data highlights the endemicity of sarcoptic mange in wild canids but uncovers multi-year epidemics. This study developed and evaluated a novel method for surveying sarcoptic mange in wildlife populations by the molecular detection of *S. scabiei* in faecal samples, which stands out for its high specificity and non-invasive character.

Keywords: *Canis lupus*; *Vulpes vulpes*; *Sarcoptes scabiei*; PCR; serology; epidemiology; Iberian Peninsula

1. Introduction

Sarcoptic mange is a highly contagious and globally widespread skin disease caused by the burrowing mite *Sarcoptes scabiei*, affecting more than 100 wild and domestic mammal species [1]. Its transmission occurs mainly by intra- or interspecific direct contact, but indirect transmission can also occur if mites contaminate the environment, such as dens [2,3]. Sarcoptic mange causes severe pruritus, accompanied by erythematous eruptions, papules, alopecia, and crusts [4]. As the disease progresses, it can give rise to a complex cascade of interacting physiological and behavioural effects on the host, which can lead to death [5]. These effects are related to compromised thermoregulatory capacity, increased metabolic rates, and altered activity patterns [5,6].

Sarcoptic mange is enzootic in several wildlife populations throughout the world, but may become epizootic [7,8]. Epizootics can occur due to the introduction of a new virulent strain of *S. scabiei*, increased pathogenicity of an already existing strain, and/or increased susceptibility of the host population [9]. The impact of the disease is potentially more severe in small, genetically compromised, or fragmented populations, mediated by demographic stochasticity, weakened immune responses, and lack of metapopulation dynamics [1,10,11].

Surveillance of sarcoptic mange in wildlife can be accomplished by invasive or non-invasive methods. In non-invasive sampling, the source sample is left behind by the animal and can be recovered without having to capture or disturb it [12]. To date, mange surveillance tends to rely on invasive methods, such as the detection of *S. scabiei* in skin samples by molecular methods, microscopy, or the recovery of live mites [13,14], or by serology resorting to blood samples, body fluids, or lung tissue [15–17]. These methods are challenging due to the elusiveness of wildlife and require the capture or death of individuals, raising welfare issues [18,19]. The availability of non-invasive surveillance methods is thus of utmost relevance [20]. Camera-trapping is a non-invasive method that has been providing insights into the epidemiology of alopecic lesions in wildlife populations [13,21]. However, it is constrained by low diagnostic specificity, because it only detects mange-compatible dermatological lesions, without confirmation of the aetiology [1,21].

The genetic analysis of non-invasive samples, such as faeces, can be a valuable tool for the surveillance of mange [20]. The associated pruritus causes the affected animals to spend much time grooming, potentially ingesting mites and passing them in their faeces [20]. Besides sarcoptic mange, the same occurs in notoedric mange, caused by the sarcoptiform mite *Notoedres* sp., which is most common in felids [22]. Taking advantage of this, Stephenson et al. [20] developed a PCR assay targeting the internal transcribed spacer 2 (ITS-2) gene for the detection of *Notoedres cati* DNA in faecal samples of bobcats *Lynx rufus*, providing insights into the distribution of the disease in northern California, U.S.A. However, PCR assays targeting ITS-2 and the cytochrome c oxidase subunit 1 (COX-1) genes were unsuccessful in detecting the DNA of *S. scabiei* in faecal samples of black bears *Ursus americanus* [23].

Longitudinal surveillance of wildlife diseases can be presented as capture–recapture (CR) data, which correspond to an individual's history of encounters [24], whenever samples can be assigned to individually identified animals. Capture–recapture data can be analysed through multi-event CR models, which consider states (usually "infected"/"uninfected"/"dead") and observations generated from the underlying state of an individual, while accounting for the uncertainty in state assignment and imperfect detection [25]. These states, often not observable in the field, are linked by probabilistic matrices to observable events [25]. At each encounter occasion, an event is observed and recorded in an encounter history. Multi-event CR models were initially developed for ecological studies; however, they have proven to be powerful tools to study infectious and parasitic diseases in wildlife, allowing the estimation of epidemiological parameters related to infection rates and survival [26,27]. These models can be used to evaluate the performance of diagnostic tests, because they estimate the uncertainty in state assignment, which corresponds to the diagnostic sensitivity and specificity of the test.

Regarding wild canids, the study of sarcoptic mange by conventional methods is hampered by their elusive behaviour and low population densities [20,28]. The red fox *Vulpes vulpes* is often considered as a reservoir host [8,13], and sarcoptic mange has caused local declines of this species in Europe [7,29]. While wolves, *Canis lupus*, seem to usually mount an effective immune response to sarcoptic mange [4,13], the disease caused average mortality estimated at 27–34% in the Midwestern U.S.A. [1]. Such estimates are lacking for Europe [6], where wolf populations are more fragmented than in North America [28]. Particularly, the Iberian Peninsula maintains an isolated, stable to slightly increasing population of >2000 Iberian wolves *Canis lupus signatus* [28], where 20% seroprevalence and mortality caused by sarcoptic mange has been reported [13,30].

Given the potential impact of sarcoptic mange on populations of wild canids and the lack of specific non-invasive surveillance techniques, this study aims to: (i) optimize a molecular method for the surveillance of sarcoptic mange in non-invasive faecal samples of wolves; (ii) use multi-event capture–recapture models, integrating serology and molecular data, to evaluate the performance of the non-invasive method and estimate epidemiological parameters of sarcoptic mange in this wolf population.

2. Results

2.1. Non-Invasive Molecular Method

A 16S rDNA PCR fragment of the same size of the positive control (132 bp) was amplified in 39/843 faecal samples, while in another 32 samples the amplified fragment was 1–2 bp larger or smaller than the positive control. We assumed these differences as the result of polymorphisms (insertions or deletions) in the hypervariable regions of the 16S rDNA. We thus classified the 71 samples as positive to *S. scabiei* DNA. In 48 of the 71 (68%) samples, the results of the two replicas were concordant.

The generalized linear mixed model (GLMM) showed non-significant differences in prevalence between sexes ($p = 0.405$) and sampling at wolf homesites or in transects ($p = 0.471$) (Table 1).

Table 1. Results of the binomial generalized linear model of non-invasive molecular data in wolves.

Variable	Samples (n)	β	Standard Error (β)	z-Score	p-Value
Fixed effects					
Intercept		−2.710	0.399	−6.801	<0.001
Sex					
Male	237	0.309	0.371	0.832	0.405
Type of sampling					
Homesites	286	−0.317	0.440	−0.721	0.471
Random effect					
Variance			0.519		
Standard deviation			0.721		
Years (N)			11		
Samples (n)			442		

n sample size; N total number of units.

The overall prevalence of sarcoptic mange in the non-invasive wolf faecal samples, estimated by GLMM from 2006 to 2018, was 7.2% (CI$_{95}$ 5.0–9.4%). The predicted prevalence was highest in 2009 (11.7% CI$_{95}$ 5.1–26.9%) and lowest in 2017 (2.6% CI$_{95}$ 0.9–8.3%) (Figure 1).

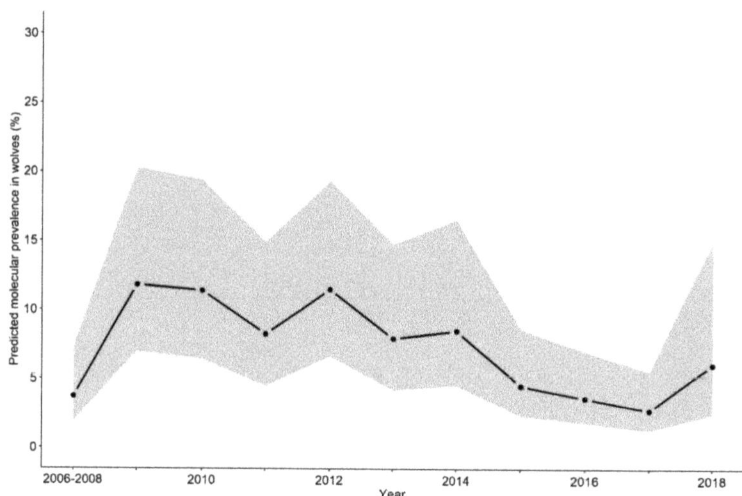

Figure 1. Predicted prevalence of sarcoptic mange in wolves by the non-invasive molecular method, from 2006 to 2018. An 80% confidence interval of the fixed and random effects from the binomial generalized linear mixed model is shaded in grey.

2.2. Serology

The GLMM showed non-significant tendencies for higher seroprevalence in foxes compared to wolves (p = 0.437) and lower in lung tissue extract (LTE) compared to serum (p = 0.076) (Table 2).

Table 2. Results of the binomial generalized linear mixed model of serology data in wolves and red foxes.

Variable	Samples (n)	β	Standard Error (β)	z-Score	p-Value
Fixed effects					
Intercept		−1.199	0.600	−1.999	0.046
Test matrix					
Lung tissue extract	54	−1.238	0.698	−1.775	0.076
Species					
Red fox	20	0.582	0.749	0.777	0.437
Random effect					
Variance		1.023			
Standard deviation		1.012			
Years (N)		17			
Samples (n)		80			

n sample size; N total number of units.

The overall predicted seroprevalence of sarcoptic mange in wolves from 1997 to 2019 was 24.5% (CI$_{95}$ 18.5–30.6%). The predicted seroprevalence in wolves ranged from 13.0% (CI$_{95}$ 2.4–46.4%) in 2018, to 55.0% (CI$_{95}$ 23.0–84.2%) in 2008, and highlights an outbreak peaking between 2006 and 2009 (Figure 2).

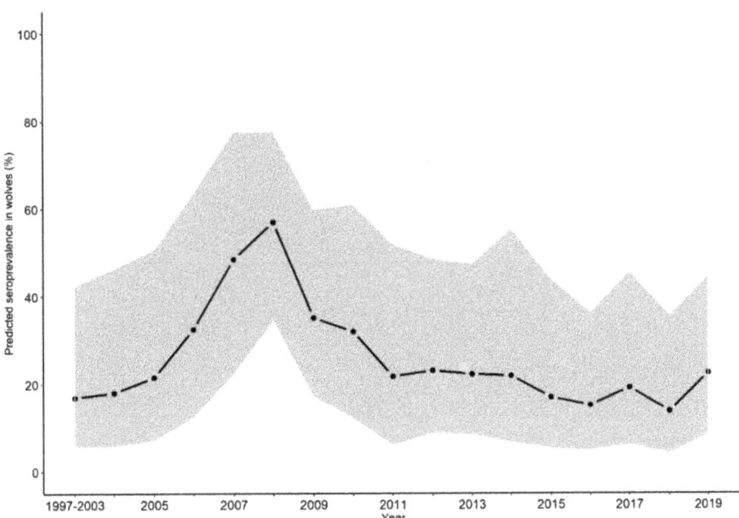

Figure 2. Predicted seroprevalence of sarcoptic mange in wolves considering serum as the test matrix, from 1997 to 2019. The 80% confidence interval of the fixed and random effects from the binomial generalized linear mixed model is shaded in grey.

2.3. Multi-Event Capture–Recapture Models

The most supported model (Akaike information criterion corrected for small sample size (AICc) = 1115.46, Model 1, Supplementary Table S1) assumed no difference in mortality between mange-infected and uninfected wolves. This model estimated that 30.0% (CI_{95} 14.0–53.0%) of the infected wolves would test positive, which corresponds to the diagnostic sensitivity of the non-invasive molecular method. It also estimated that 100% of the uninfected wolves test negative, which corresponds to the diagnostic specificity of the non-invasive molecular method. This last estimate was consistent across models, even in those where this parameter was estimated instead of fixed (Model 4, Supplementary Table S1).

While the most supported model does not include an effect of sarcoptic mange on survival, there is good support for the model including such an effect ($\Delta AICc$ = 0.94, model 2, Supplementary Table S1). Model 2 estimated the mortality of individually identified infected wolves to be higher by 0.150 (CI_{95} −0.165–0.458) than for uninfected wolves (Figure 3). Furthermore, there is moderate support for a differential effect of sarcoptic mange on survival between sexes; model 3 ($\Delta AICc$ = 2.02, Supplementary Table S1) estimated the mortality of individually identified infected male wolves to be higher by 0.199 (CI_{95} −0.377–0.671) than for infected females (Figure 3).

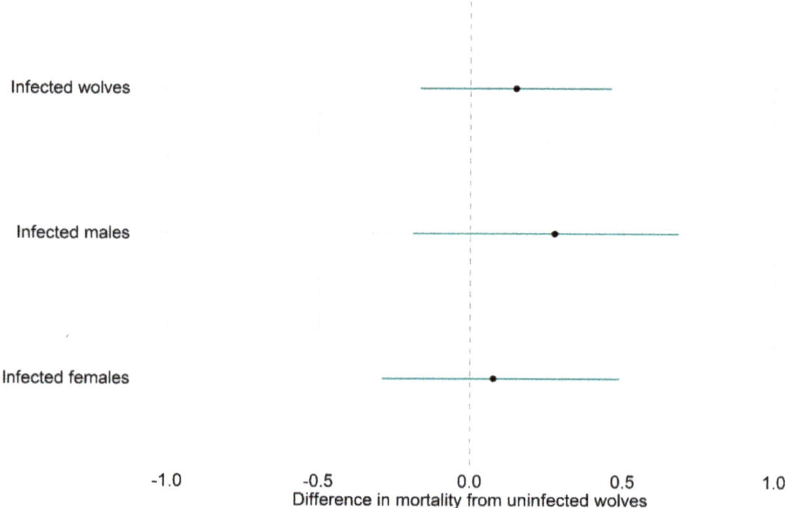

Figure 3. Difference in the estimated mortality between sarcoptic mange-infected and uninfected wolves. Estimates from the multi-event capture–recapture models 2 and 3 (Supplementary Table S1). The 95% confidence intervals of the difference in the mortality estimates are shown as horizontal error bars.

3. Discussion

Sarcoptic mange in wildlife populations is notoriously difficult to monitor because of the elusive nature of the hosts and the difficulty in obtaining invasive biological samples, which also raises animal welfare concerns [13,18,21]. Here, we describe a novel method for the surveillance of sarcoptic mange in wildlife, relying on PCR assays applied to non-invasively collected faecal samples. The present study is, to the best of our knowledge, the first to successfully detect *S. scabiei* DNA in wildlife using non-invasive samples. This method has implications for wildlife management because it allows disease surveillance avoiding capture and immobilization, which require considerable time and resources and have implications for animal welfare [18,19]. It also allows individual genetic identification and facilitates achieving larger sample sizes and repeated sampling over time, essential for longitudinal epidemiological studies.

However, non-invasive genetic samples have potential drawbacks such as the possibility of false-negative results due to the poor quality of the DNA in aged samples, the presence of PCR inhibitors, inconsistent grooming and ingestion of mites, the dilution of low numbers of ingested mites in large volumes of faeces, or intermittent mite excretion [20,23]. These issues likely contribute to the moderate diagnostic sensitivity achieved by the method (30.0%), as estimated by multi-event CR models. The study by Stephenson et al. [20] on notoedric mange in bobcats estimated a sensitivity of 52.6%, somewhat higher than in our study on sarcoptic mange in wolves. The number of mites ingested is probably influenced by the different pathology of mange in the two species. Sarcoptic mange in wolves is usually characterized by low parasite loads [4], however notoedric mange in bobcats presents numerous parasites [22]. Additionally, grooming behaviour is more pronounced in felids than in canids [31]. We thus hypothesize that the diagnostic sensitivity of molecular methods based on faecal samples depends on the pathology of the disease in the host species, as well as on its grooming behaviour. Furthermore, the poor quality of the DNA in aged faecal samples (the majority of our sample) likely contributed to the rate of false negatives, although we were able to amplify *Sarcoptes* spp. DNA from samples without individual genetic identification (i.e., presumably with more degraded DNA).

While multi-event CR models provided relevant information on the diagnostic performance of the non-invasive method, more research is warranted. Particularly relevant would be to compare the estimates of the diagnostic sensitivity and specificity with those obtained from animals with known infections status and varying parasite loads, although these are difficult to obtain in wild species. False-positive results arising from pseudo-parasitism from the consumption of mange-infected prey or allogrooming of infected conspecifics could be expectable [32]. However, our results suggest that such events are infrequent; in all the most supported multi-event CR models, the probability of amplifying *S. scabiei* DNA from faecal samples of uninfected wolves was estimated as null.

The predicted prevalence of *S. scabiei* DNA in faecal samples of wolves in our study region was 7.2% from 2006–2018. No differences were found in prevalence by sex, as previously reported in Iberian wolves [13]. Significant temporal variation in prevalence was identified, with a maximum predicted by GLMM in 2009 (11.7%) and minimum in 2017 (2.6%). This temporal trend is mostly concordant with the pattern we found using serological methods.

The detection of antibodies against *S. scabiei* in wolves and foxes using a commercial enzyme-linked immunosorbent assay (ELISA) kit for dogs is warranted by the antigenic similarity between varieties of *S. scabiei* [23,33,34]. Although the test employed uses HRP-conjugated anti-dog antibodies, the immunological similarity between canid species makes it possible to detect antibodies against *S. scabiei* in fox and wolf samples, as shown in other studies [15,23].

The overall seroprevalence estimated by GLMM from 1997 to 2019 was 24.5%, peaking during 2006–2009, up to 55.0% in 2008. Interestingly, an outbreak of sarcoptic mange in wolves and red foxes was also recorded in Asturias, northern Spain (~200 km from our study area) in 2007–2008 [13]. In the Iberian Peninsula, in 2004–2008, seroprevalence in Asturias averaged 20.5%, similar to our estimate [13]. The seroprevalence detected in our study was higher than that reported for Scandinavian wolves (10.1%) during 1998–2013, which was hypothesized to be related to higher densities of wild carnivores and greater contact of wolves with domestic animals in the Iberian Peninsula [35].

Finite mixture models allow characterizing the distributions of the seropositive and seronegative subgroups within bimodal datasets [36], thus being an alternative tool to estimate the probability of each sample being positive or negative to serological tests [37]. These models have been increasingly used to evaluate the performance of diagnostic tests in the absence of reference tests in humans [38–40] and livestock [41,42]. They are particularly relevant in epidemiology studies in wildlife, where reference tests or reference infected/uninfected animals are seldom available [37]. The multi-species nature and limited size of our dataset, including wolves and red foxes sampled post-mortem, might have affected the diagnostic performance of the serological test. However, we must emphasize the use of lung tissue extract, with the advantage of the ease of collection (simple, cheap and without the need for specialized personnel), the possibility of freezing cadavers or organs until the test is performed and the serological evaluation of animals submitted to necropsy, as being particularly useful when studying elusive wild species [38–40]. The representativeness of our serological study could be affected by the limited size and opportunistic nature of the sample collection. We strived to control for these confounding factors by using mixed models with "biological matrix" and "species" as independent variables and present the seroprevalence data predicted by the model accounting for these confounding factors (Figure 2). Nevertheless, this limitation of our study is evident in the wide confidence intervals of the predicted seroprevalence.

The integration of *S. scabiei* DNA detection in individually identified faecal samples with invasive serological data in multi-event CR models is a powerful tool to estimate epidemiological parameters. Our data support that wolves infected with sarcoptic mange might have higher mortality compared to uninfected wolves (0.15). The mortality was also elevated (0.27–0.34) in mange-infected wolves in North America [1], mediated by pack size, because diseased wolves could obtain nutritional support from the pack's predation [32].

Furthermore, individual wolves' genomic variation was shown to be negatively correlated with sarcoptic mange severity [43].

Interestingly, the excess mortality of mange-infected wolves was estimated in our study to be larger for males than for females. Although there is a general trend for higher parasite loads in males than females in many wildlife species, as assessed by prevalence and intensity of infection, evidence for differential mortality between sexes is scant [44,45]. Sex was not a significant determinant of sarcoptic mange severity in Yellowstone wolves [43]. The apparently elevated susceptibility of male Iberian wolves to sarcoptic mange could be due to immunological, hormonal, or behavioural sexual differences [44].

In the Iberian Peninsula, despite wolf mortality from sarcoptic mange being occasionally recorded [30], it was suggested that the disease had a limited demographic effect on wolf populations [13]. Although the confidence interval of our estimates does not exclude null differences in mortality (Figure 3), it suggests higher mortality in Iberian wolves infected by sarcoptic mange, particularly in males, which may raise conservation issues and requires further research.

The Iberian wolf population is the largest in western Europe, and it diverged from other European wolves approximately ten thousand years ago [46]. Most of Europe's wolf populations have been increasing, however the Iberian population is, in general, stable [28]. Inhabiting a largely human-dominated landscape, such as our study area [47], frequent contact with humans and domestic animals may contribute to the occurrence of sarcoptic mange outbreaks [1,35]. Additionally, the Iberian wolf population is highly structured, presents low connectivity [48], and exhibits low genetic diversity in the major histocompatibility complex class II locus [49]. This low diversity reported at a locus involved in the immune response to extracellular parasites [50] could lead to increased susceptibility to sarcoptic mange.

We highlight the particular case of the small and threatened subpopulation of Iberian wolves located south of the Douro river in Portugal, consisting of fewer than 10 packs showing reproductive instability, low genetic diversity, and isolation [48,51], and one of the few in Europe considered on the verge of extinction [52]. The impact of infectious and parasitic diseases, with the foremost example of sarcoptic mange, in isolated populations can lead to population declines and even local extinctions [7,29,32]. It is therefore essential to implement disease surveillance schemes, preferentially using non-invasive methods such as the one we have established and proposed here as a novel diagnostic tool for this zoonotic disease in wildlife.

4. Materials and Methods

4.1. Study Area

This study was conducted in an ~8000 km^2 region located in north-western Portugal (south-western Europe), comprising a mountainous area up to 1430 m above sea level (a.s.l.) with a temperate Atlantic–Mediterranean climate characterised by hot summers and rainy winters, with annual average precipitation of 1200 mm with little snow (<30 days/year of snow cover) and average annual temperature of 14 °C [53,54]. The study area encompasses a human-dominated landscape [47] inhabited by a largely stable population of 15–20 wolf packs, which prey mostly on livestock [51,55]. The red fox is an abundant and widespread wild canid in the study area [56].

4.2. Molecular Screening of Faecal Samples

The Iberian wolf faecal samples were collected between 2006 and 2018 in a subset of the area surveyed by serology (~2000 km^2 inhabited by 5–8 wolf packs) [47]. DNA was available for a total of 843 scats collected using preventive measures to avoid sample contamination and stored at room temperature in sterile tubes in 96% ethanol [57]. Faecal samples were systematically collected either along transects (n = 557) or at wolf homesites (n = 286), including both fresh and aged samples [47,57]. All samples were previously confirmed as belonging to wolves by genetic analysis, and information on the sex and individual identification (genotype) was available for most of the samples ($n = 445$, from 219 individuals), following the procedures described in Nakamura et al. [57]. Briefly, DNA was extracted from the scat outer layer following the protocol of Frantz et al. [58] using the GuSCN/silica approach [59]. As a final step, DNA was purified using Microcon YM-30 columns (Millipore, Billerica, MA, USA) [60]. Negative controls were included throughout the process to monitor for potential DNA contamination, and all pre-PCR procedures were performed in laboratories dedicated to low-quality DNA samples [57]. Until further processing, DNA was stored at $-20\,^{\circ}$C.

In the present study, DNA was extracted from a skin sample of a *S. scabiei*-infected red fox (as confirmed by microscopy) and used as a positive control in the PCRs. DNA was extracted using the EasySpin Genomic DNA Tissue Kit (Citomed, Lisbon, Portugal), following the manufacturer's instructions. The presence of a 132 bp fragment from the 16S rDNA mitochondrial gene of *S. scabiei* was tested in the 843 non-invasive samples using primers SSUDF and SSUDR, as described by Angelone-Alasaad et al. [14]. An M13-tail fluorescence labelling protocol was implemented to allow size screening in an automated sequencer [61]. The PCR reaction was prepared with a final volume of 10.5 µL, consisting of 5 µL of MyTaq HS Mix (Bioline, London, UK), 0.04 µL of primer forward, 0.4 µL of primer reverse, 0.4 µL of fluorescent M13-tail, and 2.5 µL of water. Primers were used at 10 µM. PCRs were prepared in plates, each including negative and positive controls, in rooms dedicated to low-quality DNA. Each sample was assayed twice to increase the probability of amplification. Samples were considered positive if one of the replicas amplified the target DNA sequence.

DNA amplification was performed in a T100 thermocycler (Bio-Rad, Hercules, CA, USA) following the protocol: initial denaturation of 10 min at 95 °C and 45 cycles of 30 s at 95 °C, 45 s at 53 °C, and 20 s at 72 °C, with a final extension of 10 min at 72 °C. The amplified products were separated by capillary electrophoresis in an ABI3130xl Genetic Analyser (Applied Biosystems, Foster City, CA, USA), and fragment size scoring was performed against the GeneScan 500 LIZ molecular size standard (Applied Biosystems) using GeneMapper 5 (Applied Biosystems). Results were checked manually.

4.3. Collection of Invasive Samples for Serology

Between 1997 and 2019, whole blood samples were opportunistically collected from Iberian wolves ($n = 22$) and red foxes ($n = 5$) whenever they were captured for scientific purposes, as described in Santos et al. [19]. Briefly, wild canids were captured with Belisle leg-hold snares (Edouard Belisle, Saint Veronique, QC, Canada) and chemically immobilised with an intramuscular injection of a mixture of ketamine (4.71 ± 1.17 mg/kg) (Imalgene, Boehringer Ingelheim, Lyon, France) and medetomidine (0.10 ± 0.03 mg/kg) (Domitor, Boehringer Ingelheim, Lyon, France). Whole blood was collected by venepuncture of the cephalic or saphenous veins, and the serum obtained by centrifugation at $1430\times g$ for 10 min was stored at $-20\,^{\circ}$C until its use. The chemical immobilization was reversed with an intramuscular injection of atipamezole (0.40 ± 0.01 mg/kg) (Revertor, Boehringer Ingelheim, Lyon, France) [19]. Trapping was conducted under permits issued by the Portuguese nature conservation authority (*Instituto da Conservação da Natureza e das Florestas:* 338/2007/CAPT, 258/2008/CAPT, 286/2008/CAPT, 260/2009/CAPT, 332/2010/MANU, 333/2010/CAPT, 336/2010/MANU, 26/2012/MANU, and 72/2014/CAPT) and according to Portuguese (Decreto-Lei 113/2013) and European legislation (Directive 2010/63/EU) on animal experimentation and international wildlife standards [62,63].

In the same period, lung samples were collected upon standard necropsy of opportunistically-found dead wolves ($n = 39$) and red foxes ($n = 15$) and stored at $-20\,^{\circ}$C until processing. Lung tissue extract (LTE) was obtained following the protocol by Ferroglio et al. [64] with minor adaptations. Briefly, ~1 g of lung tissue was cut into 4–5 pieces and placed in a Falcon tube, where 1 mL of phosphate-buffered saline was added and shaken manually for 4 min. Samples were then centrifuged at $800\times g$ for 10 min and the supernatant was collected and stored at $-20\,^{\circ}$C until use.

4.4. Serology

Antibodies against *S. scabiei* antigens were detected with a commercial indirect ELISA kit (*Sarcoptes*-ELISA 2001 Dog Kit, AFOSA GmbH, Germany) following the manufacturer's instructions with minor adaptations. The antibody concentration in LTE was shown in other species to be 1–3-fold lower compared to serum [64,65]. To minimize this expected difference between the two biological matrices used in this study, sera were tested at a dilution of 1:100 in sample dilution solution, according to the manufacturer's instructions, and LTE at 1:50. Sera, LTE samples, and controls (dog sera provided with the kit) were tested in duplicate. The test result (TR) was calculated according to the equation:

$$TR = \frac{ODsample - ODnegative\_control}{ODpositive\_control - ODnegative\_control} \times 100 \qquad (1)$$

The ELISA was applied to types of samples (serum/LTE) for which it was not originally validated; therefore, it was essential to estimate the cut-off for positivity. The cut-off was estimated by finite mixture models, which enable characterizing of the distributions of the subgroups (seropositive and seronegative in this study) within bimodal datasets [36]. These models are powerful tools in the scope of probabilistic diagnostics of serological tests in the absence of reference tests [37,39].

Finite mixture modelling of the TR data was performed separately for sera and LTE, because differences in antibody concentration may occur between these biological matrices [40]. A non-parametric stochastic model for independent data was implemented, which does not assume any type of distribution for the seropositive and seronegative subsets of the dataset [36]. The distribution of the seropositive and seronegative subsets is thus solely derived from the data. The probability of each sample belonging to the seropositive or seronegative subsets was estimated from 2000 model iterations, and each sample was assigned to one of those subsets when the estimated probability was >95%. Finite mixture models were implemented using the package "mixtools" [36] in R.

4.5. Statistical Analysis

Binomial generalized linear mixed models (GLMM) with a logit link were used to identify the variables related to serology and to non-invasive molecular results, as well as to correct for potential confounding factors related to the sampling process. The categorical independent variables included in the serology GLMM were the test matrix (serum/LTE) and species (wolf/fox), as fixed effects, and year (1997–2019), as a random effect. The reference classes were set based on their sample size and proportion of positives: serum and wolf.

In the non-invasive molecular GLMM, the categorical independent variables were the type of sample collection (transect/homesites) and sex, as fixed effects, and year (2006–2018), as a random effect. The reference classes were samples collected in transects, and female, for the abovementioned reasons.

Some of the years were pooled (1997–2003 in the serology GLMM and 2006–2008 in the molecular GLMM) to achieve adequate sample sizes in each of the models (Tables 1 and 2). Correlation between fixed effects was always <0.600. The serological and molecular prevalence of mange in wolves for each year, the overall prevalence, and the corresponding 80% and 95% confidence intervals for the random effects were predicted from the GLMM using the package "merTools" [66]. *p*-values were calculated following Satterthwaite's degrees of freedom method using the package "lmerTest" [67]. All statistical analyses were performed in R 3.6.1 [68].

4.6. Multi-Event Capture–Recapture Models

Multi-event CR models were applied to a subset of the non-invasive molecular data for which individual identification was already available (445 samples, assigned to 219 individual wolves between 2006 and 2018). The following states were considered in the model: wolves infected with *S. scabiei* (M+), uninfected wolves (M−), and dead wolves

(D). In any sampling occasion (year), an individual wolf may be alive in classes M+ or M−
or may be dead. In each sampling occasion, the possible observations were: "individual
wolf not detected", "individual wolf detected and *S. scabiei* PCR-negative", "individual
wolf detected and *S. scabiei* PCR-positive", or "individual wolf detected but not tested for
S. scabiei".

Models were implemented with the software E-SURGE [24], which uses a maximum
likelihood approach to estimate the parameters. The model included the following matrices:
initial state, survival, transitions between M+ and M− conditional on survival, detection,
probability of being tested, and uncertainty in state assignment (Supplementary File S1).
An EM (20) + Quasi-Newton nonlinear maximum likelihood solver was used to obtain the
maximum likelihood estimator, and 50 model runs using a different set of random initial
values were applied to avoid local minima.

A set of candidate models were defined, incorporating biologically relevant combinations of effects on survival (constant, sarcoptic mange, and sex effects). The standardized
annual predicted seroprevalence of sarcoptic mange estimated from the serology GLMM
was included as a temporal covariate to estimate the transition between the M- and M+
states (equivalent to the incidence of the infection). For the events, the detection probability
at the first encounter was fixed at 1 because the encounter history is conditional on being
caught in the first period, and the following detection probabilities depend on the state
and the time occasion. The probability of being assigned the observation "individual wolf
detected but not tested for *S. scabiei*" was fixed at the proportion of individually identified
samples that were not tested for *S. scabiei* DNA (0.247). Regarding the uncertainty in state assignment, the probability of a non-infected wolf testing positive was consistently estimated
as close to null, and was thus fixed at 0 in models 1–3 and 5 (Supplementary File S1).

No multi-event goodness-of-fit test exists; therefore, we applied a single-state goodness-
of-fit testing approach implemented in U-CARE [69]. Test 3.SR indicated significant transience effects ($\chi^2 = 20.91$, $p = 0.034$, 11 df). Transience can be defined as individuals
captured for the first time having a lower probability of being re-captured, as compared
to individuals that had been captured previously [70]. Survival was thus estimated in all
models separately for wolves captured once or more than once [70]. Different models were
selected under an information–theoretical approach by their Akaike information criterion
corrected for small sample size (AICc) [71].

5. Conclusions

We describe a novel non-invasive method for monitoring sarcoptic mange in wildlife,
based on the detection of *S. scabiei* DNA in faecal samples. Although the method was
developed and validated using wolf samples, it should be useful in other wildlife hosts,
and stands out for its high specificity and non-invasive character. The application of multi-
event CR models to datasets with individually identified samples was fundamental for
estimating epidemiological parameters in mange-infected wolves and for evaluating the
performance of the new surveillance method.

Supplementary Materials: The following are available online at https://www.mdpi.com/2076-081
7/10/2/243/s1. File S1: Methodological approach for the multi-event capture–recapture models;
Table S1: Parametrizations of the most supported model and models with ΔAICc < 4. Differences to
the most supported model highlighted in bold.

Author Contributions: Conceptualization, L.M.d.C., N.S., and R.G.; methodology: N.S. and R.G.;
software: J.R. and N.S.; validation: F.Á., H.R.-M., J.R., L.M.d.C., M.N., N.S., R.G., and R.C.; formal
analysis: J.R. and N.S.; investigation: J.R. and N.S.; resources: F.Á., H.R-M., M.N., N.S., and R.G.;
data curation: F.Á., H.R.-M., M.N., N.S., and R.G.; writing—original draft preparation: J.R. and N.S.;
writing—review and editing: F.Á., H.R.-M., J.R., L.M.d.C., M.N., N.S., R.G., and R.C.; visualization:
F.Á., H.R.-M., J.R., L.M.d.C., M.N., N.S., R.G., and R.C.; supervision: L.M.d.C., N.S., and R.G.; project
administration: N.S. and R.G.; funding acquisition: L.M.d.C., F.Á., N.S., and R.G. All authors have
read and agreed to the published version of the manuscript.

Funding: This research was funded by Fundação para a Ciência e Tecnologia (FCT; grant numbers SFRH/BPD/116596/2016 and SFRH/BD/144087/2019 and contract under DL57/2016); CIISA—Centro de Investigação Interdisciplinar em Sanidade Animal, Faculdade de Medicina Veterinária, Universidade de Lisboa, Lisboa, Portugal (project number UIDB/00276/2020—funded by FCT). Fieldwork for sample collection was funded by Empreendimentos Eólicos do Vale do Minho S.A. and ACHLI, while the molecular work was partially funded by CIBIO private funds.

Institutional Review Board Statement: Live animals were captured under permits issued by the Portuguese nature conservation authority (*Instituto da Conservação da Natureza e das Florestas*: 338/2007/CAPT, 258/2008/CAPT, 286/2008/CAPT, 260/2009/CAPT, 332/2010/MANU, 333/2010/CAPT, 336/2010/MANU, 26/2012/MANU, and 72/2014/CAPT) and according to Portuguese (Decreto-Lei 113/2013) and European legislation (Directive 2010/63/EU) on animal experimentation and international wildlife standards [62,63].

Informed Consent Statement: Not applicable.

Data Availability Statement: The data that support the findings of this study are available from the corresponding author upon reasonable request.

Acknowledgments: We are grateful to S. Lopes, D. Castro, P. Ribeiro, and S. Mourão for lab assistance, and the Sistema de Monitorização de Lobos Mortos /Instituto de Conservação da Natureza e das Florestas I.P. for providing part of the wolf tissue sample. This is paper number 26 from the Iberian Wolf Research Team (IWRT).

Conflicts of Interest: The authors declare no conflict of interest.

References

1. Niedringhaus, K.D.; Brown, J.D.; Sweeley, K.M.; Yabsley, M.J. A review of sarcoptic mange in North American wildlife. *Int. J. Parasitol. Parasites Wildl.* **2019**, *9*, 285–297. [CrossRef] [PubMed]
2. Kołodziej-Sobocińska, M.; Zalewski, A.; Kowalczyk, R. Sarcoptic mange vulnerability in carnivores of the Białowieża Primeval Forest, Poland: Underlying determinant factors. *Ecol. Res.* **2014**, *29*, 237–244. [CrossRef]
3. Montecino-Latorre, D.; Cypher, B.L.; Rudd, J.L.; Clifford, D.L.; Mazet, J.A.K.; Foley, J.E. Assessing the role of dens in the spread, establishment and persistence of sarcoptic mange in an endangered canid. *Epidemics* **2019**, *27*, 28–40. [CrossRef]
4. Oleaga, A.; Casais, R.; Prieto, J.M.; Gortázar, C.; Balseiro, A. Comparative pathological and immunohistochemical features of sarcoptic mange in five sympatric wildlife species in Northern Spain. *Eur. J. Wildl. Res.* **2012**, *58*, 997–1000. [CrossRef]
5. Martin, A.M.; Fraser, T.A.; Lesku, J.A.; Simpson, K.; Roberts, G.L.; Garvey, J.; Polkinghorne, A.; Burridge, C.P.; Carver, S. The cascading pathogenic consequences of *Sarcoptes scabiei* infection that manifest in host disease. *R. Soc. Open Sci.* **2018**, *5*, 180018. [CrossRef] [PubMed]
6. Astorga, F.; Carver, S.; Almberg, E.S.; Sousa, G.R.; Wingfield, K.; Niedringhaus, K.D.; Van Wick, P.; Rossi, L.; Xie, Y.; Cross, P.; et al. International meeting on sarcoptic mange in wildlife, June 2018, Blacksburg, Virginia, USA. *Parasites Vectors* **2018**, *11*, 449. [CrossRef]
7. Soulsbury, C.D.; Iossa, G.; Baker, P.J.; Cole, N.C.; Funk, S.M.; Harris, S. The impact of sarcoptic mange *Sarcoptes scabiei* on the British fox *Vulpes vulpes* population. *Mamm. Rev.* **2007**, *37*, 278–296. [CrossRef]
8. Davidson, R.K.; Bornstein, S.; Handeland, K. Long-term study of *Sarcoptes scabiei* infection in Norwegian red foxes (*Vulpes vulpes*) indicating host/parasite adaptation. *Vet. Parasitol.* **2008**, *156*, 277–283. [CrossRef]
9. Pence, D.B.; Windberg, L.A.; Pence, B.C.; Sprowls, R. The epizootiology and pathology of sarcoptic mange in coyotes, *Canis latrans*, from South Texas. *J. Parasitol.* **1983**, *69*, 1100–1115. [CrossRef] [PubMed]
10. Pence, D.B.; Ueckermann, E. Sarcoptic mange in wildlife. *OIE Rev. Sci. Tech.* **2002**, *21*, 385–398. [CrossRef]
11. Benson, J.F.; Mahoney, P.J.; Vickers, T.W.; Sikich, J.A.; Beier, P.; Riley, S.P.D.; Ernest, H.B.; Boyce, W.M. Extinction vortex dynamics of top predators isolated by urbanization. *Ecol. Appl.* **2019**, *29*, e01868. [CrossRef]
12. Taberlet, P.; Waits, L.P.; Luikart, G. Noninvasive genetic sampling: Look before you leap. *Trends Ecol. Evol.* **1999**, *14*, 323–327. [CrossRef]
13. Oleaga, A.; Casais, R.; Balseiro, A.; Espí, A.; Llaneza, L.; Hartasánchez, A.; Gortázar, C. New techniques for an old disease: Sarcoptic mange in the Iberian wolf. *Vet. Parasitol.* **2011**, *181*, 255–266. [CrossRef] [PubMed]
14. Angelone-Alasaad, S.; Molinar Min, A.R.; Pasquetti, M.; Alagaili, A.N.; D'Amelio, S.; Berrilli, F.; Obanda, V.; Gebely, M.A.; Soriguer, R.C.; Rossi, L. Universal conventional and real-time PCR diagnosis tools for *Sarcoptes scabiei*. *Parasites Vectors* **2015**, *8*, 587. [CrossRef] [PubMed]
15. Bornstein, S.; Frössling, J.; Näslund, K.; Zakrisson, G.; Mörner, T. Evaluation of a serological test (indirect ELISA) for the diagnosis of sarcoptic mange in red foxes (*Vulpes vulpes*). *Vet. Dermatol.* **2006**, *17*, 411–416. [CrossRef]

16. Jakubek, E.B.; Mattsson, R.; Mörner, T.; Mattsson, J.G.; Gavier-Widén, D. Potential application of serological tests on fluids from carcasses: Detection of antibodies against *Toxoplasma gondii* and *Sarcoptes scabiei* in red foxes (*Vulpes vulpes*). *Acta Vet. Scand.* **2012**, *54*, 13. [CrossRef] [PubMed]
17. Haas, C.; Rossi, S.; Meier, R.; Ryser-Degiorgis, M.-P. Evaluation of a commercial ELISA for the detection of antibodies to *Sarcoptes scabiei* in wild boar (*Sus scrofa*). *J. Wildl. Dis.* **2015**, *51*, 729–733. [CrossRef] [PubMed]
18. Lindsjö, J.; Fahlman, Å.; Törnqvist, E. Animal welfare from mouse to moose—Implementing the principles of the 3Rs in wildlife research. *J. Wildl. Dis.* **2016**, *52*, S65–S77. [CrossRef] [PubMed]
19. Santos, N.; Rio-Maior, H.; Nakamura, M.; Roque, S.; Brandão, R.; Álvares, F. Characterization and minimization of the stress response to trapping in free-ranging wolves (*Canis lupus*): Insights from physiology and behavior. *Stress* **2017**, *20*, 513–522. [CrossRef]
20. Stephenson, N.; Clifford, D.; Worth, S.J.; Serieys, L.E.K.; Foley, J. Development and validation of a fecal PCR assay for *Notoedres cati* and application to notoedric mange cases in bobcats (*Lynx rufus*) in Northern California, USA. *J. Wildl. Dis.* **2013**, *49*, 303–311. [CrossRef]
21. Carricondo-Sanchez, D.; Odden, M.; Linnell, J.D.C.; Odden, J. The range of the mange: Spatiotemporal patterns of sarcoptic mange in red foxes (*Vulpes vulpes*) as revealed by camera trapping. *PLoS ONE* **2017**, *12*, e0176200. [CrossRef] [PubMed]
22. Foley, J.; Serieys, L.E.K.; Stephenson, N.; Riley, S.; Foley, C.; Jennings, M.; Wengert, G.; Vickers, W.; Boydston, E.; Lyren, L.; et al. A synthetic review of *Notoedres* species mites and mange. *Parasitology* **2016**, *143*, 1847–1861. [CrossRef]
23. Peltier, S.K.; Brown, J.D.; Ternent, M.A.; Fenton, H.; Niedringhaus, K.D.; Yabsley, M.J. Assays for detection and identification of the causative agent of mange in free-ranging black bears (*Ursus americanus*). *J. Wildl. Dis.* **2018**, *54*, 471–479. [CrossRef] [PubMed]
24. Choquet, R.; Rouan, L.; Pradel, R. Program E-SURGE: A software application for fitting multievent models. In *Modeling Demographic Processes in Marked Populations. Environmental and Ecological Statistics Series*; Thomson, D.L., Cooch, E.G., Conroy, M.J., Eds.; Springer: New York, NY, USA, 2009; Volume 3, pp. 845–865.
25. Pradel, R. Multievent: An extension of multistate capture-recapture models to uncertain states. *Biometrics* **2005**, *61*, 442–447. [CrossRef]
26. Lachish, S.; Knowles, S.C.L.; Alves, R.; Wood, M.J.; Sheldon, B.C. Infection dynamics of endemic malaria in a wild bird population: Parasite species-dependent drivers of spatial and temporal variation in transmission rates. *J. Anim. Ecol.* **2011**, *80*, 1207–1216. [CrossRef] [PubMed]
27. Chambert, T.; Staszewski, V.; Lobato, E.; Choquet, R.; Carrie, C.; Mccoy, K.D.; Tveraa, T.; Boulinier, T. Exposure of black-legged kittiwakes to Lyme disease spirochetes: Dynamics of the immune status of adult hosts and effects on their survival. *J. Anim. Ecol.* **2012**, *81*, 986–995. [CrossRef]
28. Chapron, G.; Kaczensky, P.; Linnell, J.D.C.; von Arx, M.; Huber, D.; Andrén, H.; López-bao, J.V.; Adamec, M.; Álvares, F.; Anders, O.; et al. Recovery of large carnivores in Europe's modern human-dominated landscapes. *Science* **2014**, *346*, 1517–1519. [CrossRef]
29. Forchhammer, M.C.; Asferg, T. Invading parasites cause a structural shift in red fox dynamics. *Proc. R. Soc. B Biol. Sci.* **2000**, *267*, 779–786. [CrossRef] [PubMed]
30. Domínguez, G.; Espí, A.; Prieto, J.M.; De La Torre, J.A. Sarcoptic mange in Iberian wolves (*Canis lupus signatus*) in northern Spain. *Vet. Rec.* **2008**, *162*, 754–755. [CrossRef]
31. Read, J.; Gigliotti, F.; Darby, S.; Lapidge, S. Dying to be clean: Pen trials of novel cat and fox control devices. *Int. J. Pest Manag.* **2014**, *60*, 166–172. [CrossRef]
32. Almberg, E.S.; Cross, P.C.; Dobson, A.P.; Smith, D.W.; Metz, M.C.; Stahler, D.R.; Hudson, P.J. Social living mitigates the costs of a chronic illness in a cooperative carnivore. *Ecol. Lett.* **2015**, *18*, 660–667. [CrossRef]
33. Lower, K.S.; Medleau, L.M.; Hnilica, K.; Bigler, B. Evaluation of an enzyme-linked immunosorbant assay (ELISA) for the serological diagnosis of sarcoptic mange in dogs. *Vet. Dermatol.* **2001**, *12*, 315–320. [CrossRef] [PubMed]
34. Arlian, L.G.; Morgan, M.S. A review of *Sarcoptes scabiei*: Past, present and future. *Parasites Vectors* **2017**, *10*, 297. [CrossRef]
35. Fuchs, B.; Zimmermann, B.; Wabakken, P.; Bornstein, S.; Månsson, J.; Evans, A.L.; Liberg, O.; Sand, H.; Kindberg, J.; Ågren, E.O.; et al. Sarcoptic mange in the Scandinavian wolf *Canis lupus* population. *BMC Vet. Res.* **2016**, *12*, 156. [CrossRef]
36. Benaglia, T.; Chauveau, D.; Hunter, D.R.; Young, D.S. Mixtools: An R package for analyzing finite mixture models. *J. Stat. Softw.* **2009**, *32*, 1–29. [CrossRef]
37. Peel, A.J.; McKinley, T.J.; Baker, K.S.; Barr, J.A.; Crameri, G.; Hayman, D.T.S.; Feng, Y.R.; Broder, C.C.; Wang, L.-F.; Cunningham, A.A.; et al. Use of cross-reactive serological assays for detecting novel pathogens in wildlife: Assessing an appropriate cutoff for henipavirus assays in African bats. *J. Virol. Methods* **2013**, *193*, 295–303. [CrossRef]
38. Ades, A.E.; Price, M.J.; Kounali, D.; Akande, V.A.; Wills, G.S.; McClure, M.O.; Muir, P.; Horner, P.J. Proportion of tubal factor infertility due to *Chlamydia*: Finite mixture modeling of serum antibody titers. *Am. J. Epidemiol.* **2017**, *185*, 124–134. [CrossRef]
39. Migchelsen, S.J.; Martin, D.L.; Southisombath, K.; Turyaguma, P.; Heggen, A.; Rubangakene, P.P.; Joof, H.; Makalo, P.; Cooley, G.; Gwyn, S.; et al. Defining seropositivity thresholds for use in trachoma elimination studies. *PLoS Negl. Trop. Dis.* **2017**, *11*, e0005230. [CrossRef]
40. Seck, M.C.; Badiane, A.S.; Thwing, J.; Moss, D.; Fall, F.B.; Gomis, J.F.; Deme, A.B.; Diongue, K.; Sy, M.; Mbaye, A.; et al. Serological data shows low levels of chikungunya exposure in Senegalese nomadic pastoralists. *Pathogens* **2019**, *8*, 113. [CrossRef]

41. Charlier, J.; Ghebretinsae, A.; Meyns, T.; Czaplicki, G.; Vercruysse, J.; Claerebout, E. Antibodies against *Dictyocaulus viviparus* major sperm protein in bulk tank milk: Association with clinical appearance, herd management and milk production. *Vet. Parasitol.* **2016**, *232*, 36–42. [CrossRef]

42. Deng, H.; Dam-Deisz, C.; Luttikholt, S.; Maas, M.; Nielen, M.; Swart, A.; Vellema, P.; van der Giessen, J.; Opsteegh, M. Risk factors related to *Toxoplasma gondii* seroprevalence in indoor-housed Dutch dairy goats. *Prev. Vet. Med.* **2016**, *124*, 45–51. [CrossRef] [PubMed]

43. DeCandia, A.L.; Schrom, E.C.; Brandell, E.E.; Stahler, D.R.; vonHoldt, B.M. Sarcoptic mange severity is associated with reduced genomic variation and evidence of selection in Yellowstone National Park wolves (*Canis lupus*). *Evol. Appl.* **2020**, 1–17. [CrossRef]

44. Klein, S.L. Hormonal and immunological mechanisms mediating sex differences in parasite infection. *Parasite Immunol.* **2004**, *26*, 247–264. [CrossRef] [PubMed]

45. Zuk, M. The sicker sex. *PLoS Pathog.* **2009**, *5*, e1000267. [CrossRef]

46. Silva, P.; Galaverni, M.; Ortega-Del Vecchyo, D.; Fan, Z.; Caniglia, R.; Fabbri, E.; Randi, E.; Wayne, R.K.; Godinho, R. Genomic evidence for the old divergence of Southern European wolf populations. *Proc. R. Soc. B* **2020**, *287*, 20201206. [CrossRef]

47. Rio-Maior, H.; Nakamura, M.; Álvares, F.; Beja, P. Designing the landscape of coexistence: Integrating risk avoidance, habitat selection and functional connectivity to inform large carnivore conservation. *Biol. Conserv.* **2019**, *235*, 178–188. [CrossRef]

48. Silva, P.; López-Bao, J.V.; Llaneza, L.; Álvares, F.; Lopes, S.; Blanco, J.C.; Cortés, Y.; García, E.; Palacios, V.; Rio-Maior, H.; et al. Cryptic population structure reveals low dispersal in Iberian wolves. *Sci. Rep.* **2018**, *8*, 14108. [CrossRef] [PubMed]

49. Rocha, R.G.; Magalhães, V.; López-Bao, J.V.; Van Der Loo, W.; Llaneza, L.; Alvares, F.; Esteves, P.J.; Godinho, R. Alternated selection mechanisms maintain adaptive diversity in different demographic scenarios of a large carnivore. *BMC Evol. Biol.* **2019**, *19*, 90. [CrossRef]

50. Schwensow, N.; Fietz, J.; Dausmann, K.H.; Sommer, S. Neutral versus adaptive genetic variation in parasite resistance: Importance of major histocompatibility complex supertypes in a free-ranging primate. *Heredity* **2007**, *99*, 265–277. [CrossRef] [PubMed]

51. Pimenta, V.; Barroso, I.; Álvares, F.; Correia, J.; Ferrão da Costa, G.; Moreira, L.; Nascimento, J.; Petrucci-Fonseca, F.; Roque, S.; Santos, E. *Situação populacional do Lobo em Portugal: Resultados do Censo Nacional 2002/2003*; Techincal report; Instituto de Conservação da Natureza/Grupo Lobo: Lisboa, Portugal, 2005; 158p, Available online: http://www2.icnf.pt/portal/pn/biodiversidade/patrinatur/resource/docs/Mam/rel-lobo (accessed on 15 July 2020). (In Portuguese)

52. Boitani, L.; Ciucci, P. Wolf management across Europe: Species conservation without boundaries. In *A new Era for Wolves and People: Wolf Recovery, Human Attitudes and Policy*; Musiani, M., Boitani, L., Paquet, P., Eds.; University of Calgary Press: Calgary, AB, Canada, 2009; pp. 15–39.

53. IPMA Normas Climatológicas 1981–2010. Instituto Português do Mar e da Atmosfera. Available online: http://www.ipma.pt/pt/oclima/normais.clima/ (accessed on 15 July 2020).

54. INE Instituto Nacional de Estatística-Statistics Portugal. Available online: http://www.ine.pt (accessed on 15 July 2020).

55. Pimenta, V.; Barroso, I.; Boitani, L.; Beja, P. Risks *a la carte*: Modelling the occurrence and intensity of wolf predation on multiple livestock species. *Biol. Conserv.* **2018**, *228*, 331–342. [CrossRef]

56. Álvares, F.; Ferreira, C.C.; Barbosa, A.M.; Rosalino, L.M.; Pedroso, N.M.; Bencatel, J. Carnívoros. In *Atlas de Mamíferos de Portugal*, 2nd ed.; Bencatel, J., Sabino-Marques, H., Álvares, F., Moura, A.E., Barbosa, A.M., Eds.; Universidade de Évora: Évora, Portugal, 2019; pp. 66–99.

57. Nakamura, M.; Godinho, R.; Rio-Maior, H.; Roque, S.; Kaliontzopoulou, A.; Bernardo, J.; Castro, D.; Lopes, S.; Petrucci-Fonseca, F.; Álvares, F. Evaluating the predictive power of field variables for species and individual molecular identification on wolf noninvasive samples. *Eur. J. Wildl. Res.* **2017**, *63*, 1–10. [CrossRef]

58. Frantz, A.C.; Pope, L.C.; Carpenter, P.J.; Roper, T.J.; Wilson, G.J.; Delahay, R.J.; Burke, T. Reliable microsatellite genotyping of the Eurasian badger (*Meles meles*) using faecal DNA. *Mol. Ecol.* **2003**, *12*, 1649–1661. [CrossRef] [PubMed]

59. Boom, R.; Sol, C.J.A.; Salimans, M.M.M.; Jansen, C.L.; Wertheim-Van Dillen, P.M.E.; Van der Noordaa, J. Rapid and simple method for purification of nucleic acids. *Cinical Microbiol.* **1990**, *28*, 495–503. [CrossRef]

60. Godinho, R.; López-Bao, J.V.; Castro, D.; Llaneza, L.; Lopes, S.; Silva, P.; Ferrand, N. Real-time assessment of hybridization between wolves and dogs: Combining noninvasive samples with ancestry informative markers. *Mol. Ecol. Resour.* **2015**, *15*, 317–328. [CrossRef] [PubMed]

61. Blacket, M.J.; Robin, C.; Good, R.T.; Lee, S.F.; Miller, A.D. Universal primers for fluorescent labelling of PCR fragments—An efficient and cost-effective approach to genotyping by fluorescence. *Mol. Ecol. Resour.* **2012**, *12*, 456–463. [CrossRef]

62. Sikes, R.S.; Gannon, W.L. Guidelines of the American Society of Mammalogists for the use of wild mammals in research. *J. Mammal.* **2011**, *92*, 235–253. [CrossRef]

63. Chinnadurai, S.K.; Strahl-Heldreth, D.; Fiorello, C.V.; Harms, C.A. Best-practice guidelines for field-based surgery and anesthesia of free-ranging wildlife. I. Anesthesia and analgesia. *J. Wildl. Dis.* **2016**, *52*, S14–S27. [CrossRef]

64. Ferroglio, E.; Rossi, L.; Gennero, S. Lung-tissue extract as an alternative to serum for surveillance for brucellosis in chamois. *Prev. Vet. Med.* **2000**, *43*, 117–122. [CrossRef]

65. Mörner, T.; Sandström, G.; Mattsson, R. Comparison animals of serum and lung extracts for surveys of wild animals for antibodies to *Francisella tularensis* biovar *palaearctica*. *J. Wildl. Dis.* **1988**, *24*, 10–14. [CrossRef] [PubMed]

66. Knowles, J.E.; Frederick, C. merTools: Tools for Analyzing Mixed Effect Regression Models. R Package Version 0.2.1. Available online: https://cran.r-project.org/package=merTools (accessed on 17 July 2020).

67. Kuznetsova, A.; Brockhoff, P.B.; Christensen, R.H. lmerTest package: Tests in linear mixed effects models. *J. Stat. Softw.* **2017**, *82*, 1–26. [CrossRef]
68. R Core Team. R: A Language and Environment for Statistical Computing. 2019. Available online: https://www.r-project.org/ (accessed on 17 July 2020).
69. Choquet, R.; Lebreton, J.D.; Gimenez, O.; Reboulet, A.M.; Pradel, R. U-CARE: Utilities for performing goodness of fit tests and manipulating CApture-REcapture data. *Ecography* **2009**, *32*, 1071–1074. [CrossRef]
70. Genovart, M.; Pradel, R. Transience effect in capture-recapture studies: The importance of its biological meaning. *PLoS ONE* **2019**, *14*, e0222241. [CrossRef] [PubMed]
71. Burnham, K.P.; Anderson, D.R. *Model. Selection and Multimodel Inference: A Practical Information-Theoretic Approach*, 2nd ed.; Springer: New York, NY, USA, 2002.

Article

Opportunistic Mapping of *Strongyloides stercoralis* and Hookworm in Dogs in Remote Australian Communities

Meruyert Beknazarova [1,*], **Harriet Whiley** [1], **Rebecca Traub** [2] and **Kirstin Ross** [1]

1 Faculty of Science and Engineering, Flinders University, Bedford Park, SA 5042, Australia; harriet.whiley@flinders.edu.au (H.W.); kirstin.ross@flinders.edu.au (K.R.)

2 Faculty of Veterinary and Agricultural Sciences, University of Melbourne, Parkville, VIC 3052, Australia; rebecca.traub@unimelb.edu.au

* Correspondence: meruyert.cooper@flinders.edu.au or mirabeknazarova@gmail.com

Received: 27 April 2020; Accepted: 19 May 2020; Published: 21 May 2020

Abstract: Both *Strongyloides stercoralis* and hookworms are common soil-transmitted helminths in remote Australian communities. In addition to infecting humans, *S. stercoralis* and some species of hookworms infect canids and therefore present both environmental and zoonotic sources of transmission to humans. Currently, there is limited information available on the prevalence of hookworms and *S. stercoralis* infections in dogs living in communities across the Northern Territory in Australia. In this study, 274 dog faecal samples and 11 faecal samples of unknown origin were collected from the environment and directly from animals across 27 remote communities in Northern and Central Australia. Samples were examined using real-time polymerase chain reaction (PCR) analysis for the presence of *S. stercoralis* and four hookworm species: *Ancylostoma caninum, Ancylostoma ceylanicum, Ancylostoma braziliense* and *Uncinaria stenocephala*. The prevalence of *S. stercoralis* in dogs was found to be 21.9% (60/274). *A. caninum* was the only hookworm detected in the dog samples, with a prevalence of 31.4% (86/274). This study provides an insight into the prevalence of *S. stercoralis* and hookworms in dogs and informs future intervention and prevention strategies aimed at controlling these parasites in both dogs and humans. A "One Health" approach is crucial for the prevention of these diseases in Australia.

Keywords: *Strongyloides stercoralis*; soil-transmitted helminths; hookworms; zoonotic parasites; Australian remote communities; One Health

---

1. Introduction

Soil-transmitted helminths (STHs) are estimated to infect up to 2 billion people worldwide, with a high prevalence recorded in Southeast Asia [1–3]. Australia as a whole has a relatively low prevalence of STHs due to widespread access to adequate hygiene, sanitation and clean water [4]. *Strongyloides stercoralis*, distributed throughout the tropics, is estimated to infect up to 370 million people worldwide, predominantly in socioeconomically disadvantaged communities [5,6]. Strongyloidiasis is a major health concern in remote Australian communities with up to 60% of indigenous populations found to be seropositive for the disease [4,7,8]. *Strongyloides stercoralis* can infect humans chronically and, in the case of immunocompromised patients, can develop into severe hyperinfective or disseminated strongyloidiasis, which has a mortality rate of up to 90% [9].

Genetic studies worldwide and in Australia have shown that there are at least two genetically different strains of *S. stercoralis*—one that is zoonotic, infecting both humans and dogs, and one that only infects dogs [10–12]. There is sufficient evidence to suggest that dogs can act as potential reservoirs for human strongyloidiasis and that controlling the parasite in dogs may play a role in preventing the disease in humans.

Hookworms infect up to half a billion people worldwide [13]. The most prevalent hookworms in humans in Southeast Asia and the Pacific are *Necator americanus, Ancylostoma ceylanicum,* and *Ancylostoma duodenale* [14,15]. Hookworms in humans can contribute to iron deficiency anaemia and can have an impact on maternal and child health [16]. Hookworm infection in humans was considered a widespread public health problem in parts of Australia until intervention campaigns successfully eradicated it from the mainstream population [17–21]. Only a single autochthonous case of *A. ceylanicum* in humans was reported in Western Australia and an imported case was reported in an Australian soldier returning from the Solomon Islands [22,23]. More recent studies found that hookworms, specifically *A. duodenale* [24], remain sporadically reported in remote communities in far north Queensland, northern parts of New South Wales, Western Australia and the Northern Territory (NT). In the Northern Territory, hookworm prevalence in humans is reported to be significantly lower than that of *S. stercoralis* [18,21,22,25]. Overall, a reduction has been seen in both *S. stercoralis* and hookworm infections in humans in the remote communities in the NT, and this has been attributed to deworming programs [20]. However, neither strongyloidiasis nor hookworm infection has been eradicated completely from remote communities, despite various intervention programs.

In Australia, as in other countries of the Asia-Pacific region, dogs are considered a potential zoonotic reservoir for STH infections, including strongyloidiasis and hookworms. Within indigenous Australian communities, the risk of transmission may be increased by the fact that dogs tend to live in close contact with humans [26].

In Australia, the most common hookworms in dogs are *Ancylostoma caninum, A. ceylanicum, Ancylostoma braziliense* and *Uncinaria stenocephala* [15]. These hookworm species are zoonotic and all are capable of causing cutaneous larva migrans in humans [27]. *A. ceylanicum* and *A. caninum* are of particular interest, as *A. ceylanicum* larvae can develop into the adult stage in humans, and *A. ceylanicum* is now recognised as the second most common species of hookworm infecting humans in the Asia-Pacific [28–30]. *A. caninum* infection in humans is non-patent and is strongly associated with eosinophilic enteritis [31,32]. Recent data show a high prevalence of both *A. caninum* and *A. ceylanicum* in dogs, dingoes and soil in remote communities in Western Australia and North-East Queensland. [33,34]. Both *A. ceylanicum* and *A. caninum* are considered neglected zoonotic parasites and accurate data on their prevalence in dogs and humans residing in the Indigenous communities of northern Australia are largely lacking [15,24,28,32,35].

In this study, we aimed to map the distribution of zoonotic *S. stercoralis* and hookworm species in dogs in remote communities in northern Australia. To the best of our knowledge, this is the first large-scale molecular study of dogs in these remote communities for the presence of *S. stercoralis* and hookworms.

2. Results

2.1. Dog DNA Origin

We tested 285 fresh faecal samples, presumed to be from dogs, which had been collected from communities across the Northern Territory, Central Australia, northern areas of Western Australia and the north-west of South Australia. These samples were screened for *Canis lupus familiaris* and *Canis lupus dingo* DNA. We confirmed that 274 out of 285 DNA samples extracted from the faeces were of dog origin (*Canis lupus familiaris* or *Canis lupus dingo*) through the use of polymerase chain reaction (PCR)-based amplification of the partial mitochondrial DNA (mtDNA).

2.2. Prevalence of Strongyloides stercoralis *and Hookworms*

The prevalence of *Strongyloides* species (spp.) among the 285 environmental faecal samples was 21.1% (60/285) as determined by PCR-based amplification of the partial 18 Svedberg unit ribosomal

RNA (18S rRNA). The prevalence of *S. stercoralis* among the 274 dog faecal samples was 21.9% (60/274) (Figure 1).

Out of four hookworm species tested, only *A. caninum* was detected. The prevalence of hookworm infection (*A. caninum*) among the 285 environmental faecal samples was 30.2% (86/285) by PCR-based amplification of the partial internal transcribed spacer (ITS) gene. The prevalence of hookworm infection (*A. caninum*) among the 274 dog samples was 31.4% (86/274) (Figure 1).

Maps showing sample locations and *S. stercoralis* and hookworm prevalence in dogs are shown in Figures 2 and 3.

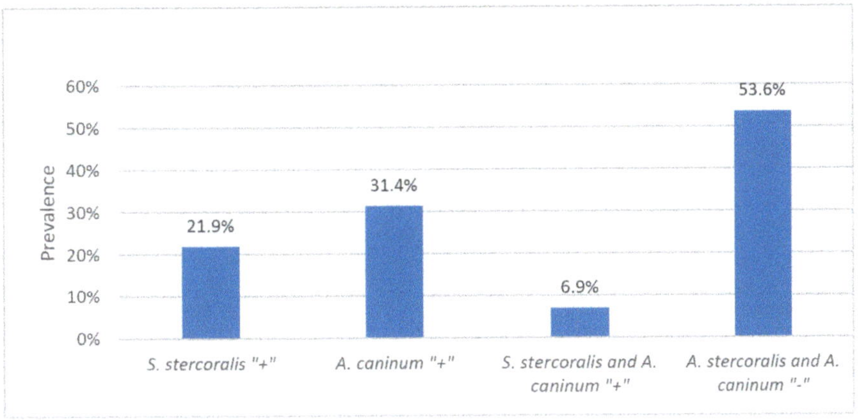

Figure 1. The percentage of dog faecal samples positive for *Strongyloides. stercoralis*, *Ancylostoma caninum*, *S. stercoralis* and *A. caninum* and the percentage of dog samples negative for both *S. stercoralis* and *A. caninum*.

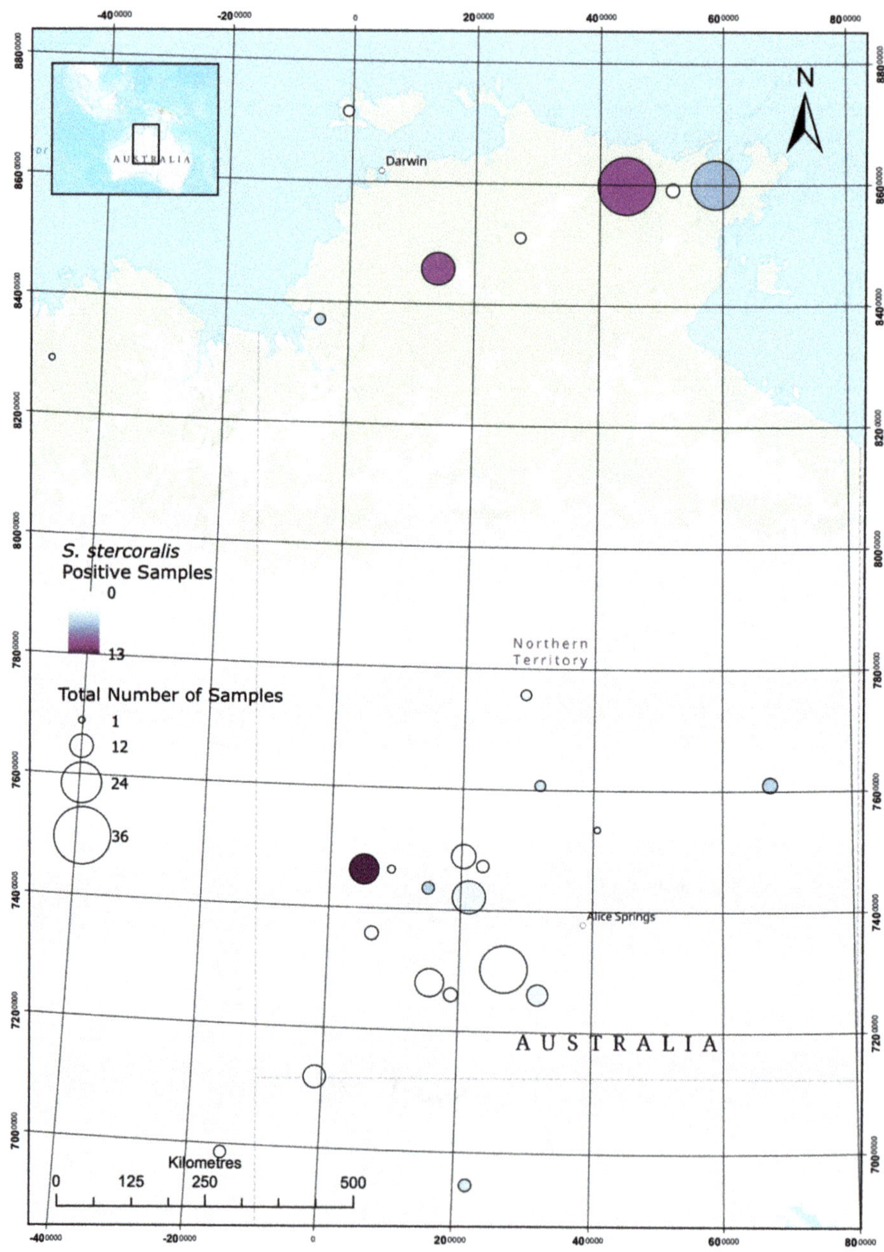

Figure 2. Opportunistic mapping of *S. stercoralis* in dogs in remote communities.

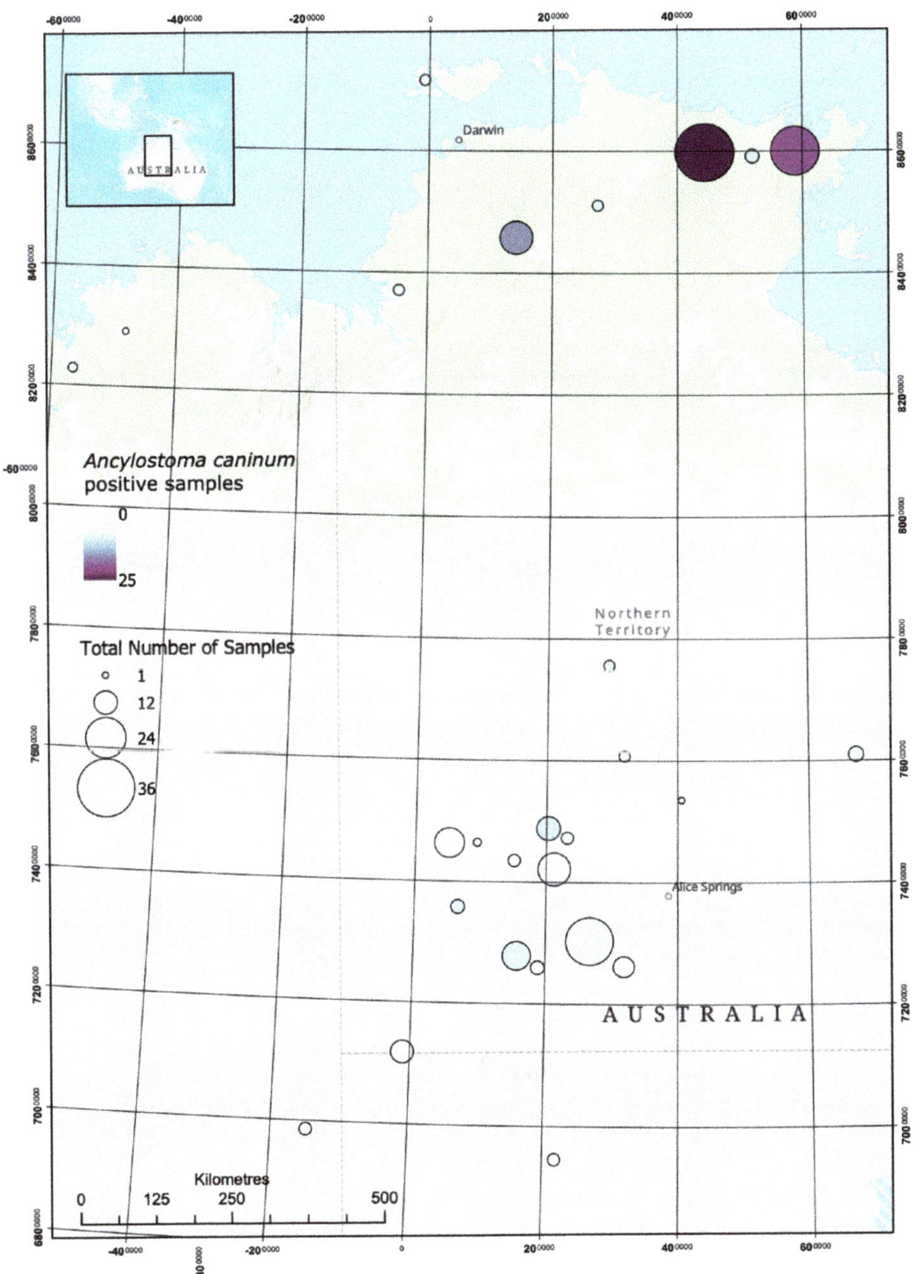

Figure 3. Opportunistic mapping of *A. cacinum* in dogs in remote communities.

2.3. Association of Hookworms with Strongyloidiasis

Chi-squared analysis did not identify a statistically significant association between *S. stercoralis* and *A. caninum* (x^2 (1) = 0.003, p = 0.958, n = 274). Of the 274 dog faecal samples, 6.9% (19/274) tested positive for both *S. stercoralis* and *A. caninum* and 53.6% tested negative for both parasites (Figure 1).

None of the non-dog faecal samples were infected with *S. stercoralis* or *A. caninum*.

3. Discussion

In this study, we used the quantitative polymerase chain reaction (qPCR) technique to detect potentially zoonotic *S. stercoralis* and zoonotic hookworms in dog faecal samples collected from remote communities in Northern and Central Australia. The prevalence of *S. stercoralis* and *A. caninum* in dogs was found to be high. All samples were negative for *A. ceylanicum*, *A. braziliense* and *U. stenocephala*, which supports previous studies demonstrating that *A. caninum* is the most common hookworm in dogs living in remote communities in Australia [15,33,34].

Ancylostoma caninum is known to cause eosinophilic enteritis in humans. Although infection is asymptomatic in most cases, symptoms can include strong abdominal pain with or without peripheral eosinophilia, nausea, diarrhoea, anorexia and allergic reactions [34]. In cases in which the infection is patent, the impact of a hookworm infection on nutritional status and immunocompetence may be associated with other health problems, including increased susceptibility to other helminth infections [36].

Although *A. ceylanicum* is the predominant hookworm affecting dogs and cats in Asia, it was reported only recently in dogs in Australia [28]. However, its presence in a cat from far north Queensland was retrospectively dated back to 1994 [37]. *Ancylostoma ceylanicum* was detected for the first time in Australia in 6.5% of dogs from rural and urban areas in Broome, Brisbane, the Sunshine Coast, Melbourne and Alice Springs [15]. More recently, *A. caninum* and *A. ceylanicum* infections were reported for the first time at a prevalence of 98.4% (62/63) and 1.6% (1/64), respectively, in domestic dogs in far north Queensland [24]. The same study discovered prevalence of *A. ceylanicum* ranging from 25% to 100% in the soil in different communities in far north Queensland [24]. A study of dingoes and dogs in Northeast Queensland reported 100% (35/35) and 11% (4/35) prevalence of *A. caninum* and *A. ceylanicum*, respectively, in dingoes, and a 92% (78/85) prevalence of *A. caninum* in dogs, based on both necropsy and faecal examination [34]. A more recent study found 66% (93/141) of camp dogs in remote communities in Western Australia to be infected with *A. caninum*, based on molecular examination [33]. The absence of *A. ceylanicum* in this study was likely due to the climatic conditions, such as dry weather, of the study area at the time of sampling. To date, there has been no evidence that *A. ceylanicum* possesses the biological advantage of undergoing arrested development, a process in which larvae undergo a period of hypobiosis in host tissue and then resume development in the intestinal tract when climatic conditions favour transmission [38].

The absence of *U. stenocephala* in the samples is supported by its association with lower temperatures [39]. *U. stenocephala* is predominately found in the southern regions of Australia, because the optimum temperature conditions for *U. stenocephala* larvae development up to the infective stage is between 7.5 and 27 °C and the ideal temperature for the free-living stages is 20 °C [40]. Likewise, previous studies exclusively detected *A. braziliense* in dogs located in North Queensland [15,41].

Molecular detection methods have been shown to be highly effective in the detection of *S. stercoralis* and hookworms in faecal samples [42–46]. However, the sensitivity of the PCR technique for the detection of *S. stercoralis* is lower when there is a low number of larvae in the faeces [47]. Although the *S. stercoralis* primers and probe used in this study have been described as *S. stercoralis*-specific [43,47], they can also amplify *S. ratti*, as previously demonstrated [47], meaning that for environmental samples we can only assume that positive samples contain *Strongyloides* spp. As for the dogs, we know from the previous genotyping study on dogs living in remote communities in Australia that dogs have been found to be infected with *S. stercoralis* strains [12]. However, due to the possibility that dogs engaged in hunting and coprophagia, we cannot rule out the possibility of mechanical ingestion of other species of *Strongyloides*, including human-sourced species.

Increased humidity and temperature are typically associated with the presence of *Ancylostoma* spp. and *S. stercoralis*. Tropical climates have been shown to be associated with multiple parasite infections in humans [36,48]. Hookworm infection intensity has also been associated with multiparasitism because co-infection with hookworms weakens the immune system of the host. The intensity of a strongyloidiasis infection is in turn highly dependent on the immune status of the host [36,48]. This illustrates the importance of detecting and differentiating parasite infections. Moreover, the indiscriminate use of anthelmintic drugs may cause the development of anthelmintic resistance [49]. A study conducted in Brazil showed a strong association between hookworm and other helminth infections (but not *S. stercoralis* infections) in humans [36]. In the present study, we did not find any significant association between *Strongyloides* spp. and *A. caninum* in dogs. Non-infected dogs might be a result of dog health programs targeted at desexing and deworming dogs in remote communities, which are administered by the Animal Management in Rural and Remote Indigenous Communities (AMRRIC) organisation.

Infections of both *S. stercoralis* and hookworm occur through exposure to soil contaminated with free-living infective stages of a parasite [1]. In the studied locations, dogs live in close proximity to their owners. Climate, sanitation, hygiene, environmental contamination with human or dog faeces and lack of knowledge of STH diseases are the main factors influencing the persistence of the disease, and can also influence its transmission [5,32,50]. Our findings demonstrate the importance of the "One Health" initiative, an approach which considers veterinary and public health interventions together. The One Health approach should be central to the development of methods of eliminating *S. stercoralis* and hookworms. To maintain the health of both dogs and humans, veterinarians and pet owners are encouraged to coordinate their efforts and to work in partnership [51].

The findings of this study need to be interpreted in light of its limitations. The faeces samples were collected from the ground, rather than directly from the rectums of dogs. Therefore, some of the samples that were collected were found not to be from dogs. Samples collected from the environment might have been contaminated with DNA from extraneous environmental organisms, which could have caused further inhibition of the DNA of the target organisms (*Strongyloides* spp. and hookworms) [52,53]. Researchers could also have accidentally collected faeces that were old enough for parasites' DNA to have degraded. Both these limitations could have resulted in false negatives. Furthermore, the possibility that the dogs had engaged in hunting and coprophagia could lead to false positives. The opportunistic sampling method did not allow us to consider risk factors associated with parasite prevalence, such as seasonal variation, climate conditions or the use of anthelmintic drugs. Furthermore, there was significant variation in the number of samples from each geographical area.

The aim of this study was to map the prevalence of *S. stercoralis* and hookworm infection in dogs in remote communities in Australia based on the molecular screening of dog faeces. The objective was to develop and optimise detection methods that can be applied in similar environmental settings without laboratory facilities and in a respectful and non-intrusive manner. We detected high levels of *S. stercoralis* and *A. caninum* in dog faecal samples collected from remote communities. Future research is needed to examine parasite prevalence in both dogs and humans from the same communities to determine whether there is an association between them, and thus to assess the zoonotic potential of dogs to transmit the diseases. Given the zoonotic nature of these parasitic species, the findings of this study can be used to develop control measures to maintain dog and human health.

4. Materials and Methods

4.1. Ethical Considerations

The project was registered with the Flinders University Animal Welfare Committee, part of the Research Development and Support division. The research was approved by the Social and Behavioural Research Ethics Committee (SBREC) (No. 6852, dated 1 June, 2015). For dog faeces collected from residential or private land, consent was obtained from the owners of the dogs or from the local managers of the communities.

4.2. Study Area and Population

Two hundred and eighty-five faecal samples presumed to be from dogs were collected from remote communities across the Northern Territory, Central Australia, Western Australia, and the northwest of South Australia during 2016 and 2019. The samples were collected from 27 locations in total, including 23 communities in the Northern Territory, two communities in the northern parts of Western Australia, one community in the northwest of South Australia and in the vicinity of Alice Springs.

4.3. Specimen Collection and DNA Extraction

Faeces were collected either by the Flinders University researchers, Northern Territory Department of Health environmental health officers (EHOs) or veterinarians primarily from the AMRRIC organisation.

In the cases where samples were collected by EHOs or representatives of AMRRIC, they would do so during their routine inspections or dog treatments. A sampling package containing the project's information sheet, risk assessment and consent forms, sampling instructions and sampling equipment was provided to them in advance.

Permission from the community elders, Traditional Owners or community managers was obtained prior to collecting samples from private or residential land. Approximately 2–3 g of faeces were collected and preserved immediately in 6 mL DESS (dimethyl sulfoxide, disodium EDTA, and saturated NaCl) and kept at room temperature [54]. The samples were shipped to the Environmental Health laboratory, Flinders University, within 30 days after collection for further sample processing. The genomic DNA was extracted using the PowerSoil DNA Isolation Kit (QIAGEN, Hilden, Germany) as described previously [12,47].

4.4. Real-Time PCR Assays

The real-time PCR assay was adopted from Verweij et al. [43] using *S. stercoralis*-specific primers (Stro18S-1530F and Stro18S-1630R) and a probe (Stro18S-1586T) targeting the 101 base pair (bp) region of the 18S rRNA, and conducted as described previously [12]. All qPCR reactions were performed in triplicate on the two-channel Corbett Rotor-Gene 6000 machine (QIAGEN, Hilden, Germany). The primers, probes and qPCR conditions are shown in Table 1. It should be noted that although this primer/probe set is considered specific for *S. stercoralis*, it can also amplify other species of *Strongyloides*, including *Strongyloides ratti*.

Positive, non-template and negative control samples were included in each qPCR run. The cycle quantification (Cq) value for *S. stercoralis* was 0.02 to 0.03. A sample was considered positive when the cycle threshold (Ct) value was lower than the mean negative Ct value minus 2.6 standard deviations of a mean negative control Ct value [54]. Positive samples were amplified in every qPCR reaction.

Multiplex qPCR assays for detection of *A. ceylanicum*, *A. caninum*, *A. braziliense* and *U. stenocephala* using primers and probes targeting the internal transcribed spacer 1 (*ITS1*) gene were adopted and performed as described by Massetti et al. [46].

Synthetic block gene fragments (IDT Technologies, Skokie, Illinois, USA) of *ITS1* genes targeted by the PCR primers and probes for *A. ceylanicum*, *A. caninum*, *A. braziliense* and *U. stenocephala* were used as positive controls in the PCR runs (Table 2). Nuclease-free water was used as the non-template or negative control. Synthetic block gene fragments (IDT Technologies, Skokie, Illinois, USA) of a herpes virus (Equine herpesvirus type 4, accession number KT324745.1) was used as an internal control. Primers and a probe to amplify a region of the dog mtDNA (*Canis lupus familiaris* or *Canis lupus dingo*, accession numbers MH 105047.1 and MH035676.1) were used as DNA extraction controls in all runs. Primers, probes and qPCR conditions are shown in Table 1. The GenBank Accession numbers and sequences of the synthetic block gene fragments used as controls in this study are presented in Table 2. All hookworm qPCR reactions were performed in duplicate on the multiplex channel Corbett Rotor-Gene 6000 machine (QIAGEN, Hilden, Germany).

Table 1. Primers and probes and PCR conditions.

Primer/Probe	Amplicon	Sequence	Reaction Conditions
Stro18S-1530F Stro18S-1630R Stro18S-1586T FAM	rDNA 101 bp	5′-GAATTCCAAGTAAACGTAAGTCATTAGC-3′ 5′-TGCCTCTGGATATTGCTCAGTTC-3′ 5′-FAM-ACACACCGGCCGTCGCTGC-3′-BHQ1	**Step 1:** 95 °C for 15 min, **Step 2:** 95 °C for 15 s, **Step 3:** 60 °C for 30 s. Repeat steps two and three 40 times.
A. cancey F *A. cancey* R Ahumanceylanicum probe Acantub probe	ITS1 region	5′- GGGAAGGTTGGGAGTATCG-3′ 5′- CGAACTTCGCACAGCAATC-3′ 5′- Cy5/CCGTTC+CTGGGTGGC/3IABkRQSp/-3′ 5′-HEX/ AG+T+CGT+T+A+C+TGG/3IABkRFQ/-3′	**Step 1:** 95 °C for 2 min, **Step 2:** 95 °C for 15 s, **Step 3:** 60°C for 60 s. Repeat steps two and three 40 times.
Uncbraz F Uncbraz R Unc Probe Abra probe	ITS1 region	5′- GAG CTT TAG ACT TGA TGA GCA TTG-3′ 5′- GCA GAT CAT TAA GGT TTC CTG AC-3′ 5′-/5HEX/CAT TAG GCG /ZEN/GCA ACG TCT GGT G/3IABkFQ/-3′ 5′-/56FAM/TGA GCG CTA /ZEN/GGC TAA CGC CT/3IABkFQ/-3′	**Step 1:** 95 °C for 2 min, **Step 2:** 95 °C for 15 s, **Step 3:** 64 °C for 60 s. Repeat steps two and three 40 times.
EMV F ENV R ENV probe	Equine herpesvirus type 4	5′-GATGACACTAGCG-ACTTCGA-3′ 5′-CAGGGCAGAAACC-ATAGACA-3′ 5′-TEX-TTTCGCGTGC-CTCCTCCAG-IBRQ-3′	**Step 1:** 95 °C for 2 min, **Step 2:** 95 °C for 15 s, **Step 3:** 60 °C for 60 s. Repeat steps two and three 40 times.
Dog F Dog R Dog probe	mtDNA	5′-CGACCTCGATGTTGGATCAG-3′ 5′-GAACTCAGATCACGTAGGACTTT-3′ 5′-FAM/ CCTAATGGT/ ZEN/ GCAGCAGCTATTAA/ LABKFQ-3′	**Step 1:** 95 °C for 2 min, **Step 2:** 95 °C for 15 s, **Step 3:** 60 °C for 60 s. Repeat steps two and three 40 times.

Table 2. Synthetic block gene fragments used for positive controls.

Species	GenBank Accession Number	Sequence
Ancylostoma ceylanicum	DQ780009.1	CGTGCTAGTCTTCAGGACTTTGTCGGGAAGGTTGGGAGTATCGCCCCCCGTTACAGCCCTACGTGAGGTGTCTATGTGCAGCAAGAGCCGTTCCTGGGTGGCGGCAGTGATTGCTGTGCGAAGTTCGCGTTTCGCTGAGCTTTAGACTTGAG
Ancylostoma duodenale/Ancylostoma caninum	EU344797.1	CGTGCTAGTCTTCACGACTTTGTCGGGAAGGTTGGGAGTATCGCCCCCCGTTATAGCCCTACGTAAGGTGTCTATGTGCAGCAAGAGTCGTTACTGGGTGACGGCAGTGATTGCTGTGCGAAGTTCGCGTTTCGCTGAGCTTTAGACTTGAT
Ancylostoma braziliense	JQ812692.1	TGTACGAAGCTCGCGGTTTCGTCAGAGCTTTAGACTTGATGAGCATTGCTAGAATGCCGCCTTACCTGCTTGTGTTGGTGGTTGAGCGCTAGGCTAACGCCTGGTGCGGCACCTGTCTGTCAGGAAACCTTAATGATCTGCTAACGCGGACGCCAGCACAGCAAT
Uncinaria stenocephala	HQ262054.1	GCTGTGCGAAGTTCGCGTTTCGCTGAGCTTTAGACTTGATGAGCATTGCTGGAATGCCGCCTTACTGTTTGTGTTGGTGGTTGGGCATTAGGCGGCAACGTCTGGTGCGACACCTGTTTGTCAGGAAACCTTAATGATCTGCTCACGTGGACGCCAATACAGCACT
Equid herpesvirus	KT324745.1	ATGAAAGCTCTATACCCAATAACAACCAGGAGCCTTAAAAACAAAGCCAAAGCCTCATACGGCCAAAACGACGATGATGACACTAGCGACTTCGATGAAGCCAAGCTGGAGGAGGCACGCGAAATGATCAAATATATGTCTATGGTTTCTGCCCTGGAAAAACAGGAAAAAAAGGCAATGAAGAAAAACAAGGGGGTTGGACTTATTGCC

The Cq value for *A. ceylanicum* and *A. caninum* was 0.05 and a Ct value of 32 was established. The Cq values for *A. braziliense* and *U. stenocephala* were 0.08 and 0.1, respectively, and the Ct value was set to 32.

Synthetic block gene fragments of hookworms were also spiked with negative dog DNA and analysed with qPCR to check for any inhibitors that might be contained in dog DNA. All spiked hookworm synthetic block gene fragments were amplified by means of qPCR.

4.5. Statistical Analysis

A chi-square independence test was performed to determine whether there was an association between hookworms and *S. stercoralis* infection. Data were analysed using Statistical Package for Social Sciences (SPSS) software (SPSS for Windows, Version 23, IBM) and Excel 2016 (Microsoft).

Author Contributions: Conceptualisation, M.B., H.W. and K.R.; methodology, M.B., H.W., R.T.; validation, M.B. and R.T.; formal analysis, M.B.; investigation, M.B.; resources, H.W., R.T. and K.R.; data curation, M.B.; writing—original draft preparation, M.B.; writing—review and editing, M.B., H.W., R.T. and K.R.; supervision, H.W., R.T. and K.R.; project administration, M.B., H.W. and K.R.; funding acquisition, H.W. and K.R.; All authors have read and agreed to the published version of the manuscript.

Funding: No funding was received to publish this work.

Acknowledgments: The authors would like to thank Jan Allen and Madeleine Kelso from the Animal Management in Rural and Remote Indigenous Communities; Ted Donelan from the West Arnhem Regional Council; Fiona Smith, Aaron Clifford, Kiri Gould, Russel Spargo and Ryan McLean from the Environmental Health Branch at the Department of Health, NT for helping us collecting dog faeces. We wish to thank Rogan Lee, Matthew Watts, John Clancy and Vishal Ahuja at the Westmead Hospital, NSW for sending us *S. ratti* infected rat faeces. The authors would also like to thank Patsy Zanjedes from the Melbourne Veterinary School, University of Melbourne for her assistance in the lab. The work has been supported by the Australian Government Research Training Program Scholarship.

Conflicts of Interest: The authors declare no conflict of interest.

References

1. Bethony, J.; Brooker, S.; Albonico, M.; Geiger, S.M.; Loukas, A.; Diemert, D.; Hotez, P.J. Soil-transmitted helminth infections: Ascariasis, trichuriasis, and hookworm. *Lancet* **2006**, *367*, 1521–1532. [CrossRef]
2. Pullan, R.L.; Smith, J.L.; Jasrasaria, R.; Brooker, S.J. Global numbers of infection and disease burden of soil transmitted helminth infections in 2010. *Parasites Vectors* **2014**, *7*, 37. [CrossRef]
3. Jex, A.R.; Lim, Y.A.; Bethony, J.M.; Hotez, P.J.; Young, N.D.; Gasser, R.B. Soil-transmitted helminths of humans in Southeast Asia—Towards integrated control. In *Advances in Parasitology*; Elsevier: Amsterdam, The Netherlands, 2011; Volume 74, pp. 231–265.
4. Gordon, C.A.; Kurscheid, J.; Jones, M.K.; Gray, D.J.; McManus, D.P. Soil-transmitted helminths in Tropical Australia and Asia. *Trop. Med. Infect. Dis.* **2017**, *2*, 56. [CrossRef]
5. Olsen, A.; van Lieshout, L.; Marti, H.; Polderman, T.; Polman, K.; Steinmann, P.; Stothard, R.; Thybo, S.; Verweij, J.J.; Magnussen, P. Strongyloidiasis–the most neglected of the neglected tropical diseases? *Trans. R. Soc. Trop. Med. Hyg.* **2009**, *103*, 967–972. [CrossRef]
6. Beknazarova, M.; Whiley, H.; Ross, K. Strongyloidiasis: A disease of socioeconomic disadvantage. *Int. J. Environ. Res. Public Health* **2016**, *13*, 517. [CrossRef]
7. Adams, M.; Page, W.; Speare, R. Strongyloidiasis: An issue in Aboriginal communities. *Rural Remote Health* **2003**, *3*, 152.
8. Johnston, F.H.; Morris, P.S.; Speare, R.; McCarthy, J.; Currie, B.; Ewald, D.; Page, W.; Dempsey, K. Strongyloidiasis: A review of the evidence for Australian practitioners. *Aust. J. Rural Health* **2005**, *13*, 247–254. [CrossRef]
9. Geri, G.; Rabbat, A.; Mayaux, J.; Zafrani, L.; Chalumeau-Lemoine, L.; Guidet, B.; Azoulay, E.; Pène, F. Strongyloides stercoralis hyperinfection syndrome: A case series and a review of the literature. *Infection* **2015**, *43*, 691–698. [CrossRef]
10. Jaleta, T.G.; Zhou, S.; Bemm, F.M.; Schär, F.; Khieu, V.; Muth, S.; Odermatt, P.; Lok, J.B.; Streit, A. Different but overlapping populations of Strongyloides stercoralis in dogs and humans—Dogs as a possible source for zoonotic strongyloidiasis. *PLoS Negl. Trop. Dis.* **2017**, *11*, e0005752. [CrossRef]

11. Nagayasu, E.; Htwe, M.P.P.T.H.; Hortiwakul, T.; Hino, A.; Tanaka, T.; Higashiarakawa, M.; Olia, A.; Taniguchi, T.; Win, S.M.T.; Ohashi, I. A possible origin population of pathogenic intestinal nematodes, Strongyloides stercoralis, unveiled by molecular phylogeny. *Sci. Rep.* **2017**, *7*, 4844. [CrossRef]

12. Beknazarova, M.; Barratt, J.L.; Bradbury, R.S.; Lane, M.; Whiley, H.; Ross, K. Detection of classic and cryptic Strongyloides genotypesby deep amplicon sequencing: A preliminary survey of dog and human specimens collected from remote Australian communities. *PLoS Negl. Trop. Dis.* **2019**, *13*, e0007241. [CrossRef]

13. Forouzanfar, M.H.; Afshin, A.; Alexander, L.T.; Anderson, H.R.; Bhutta, Z.A.; Biryukov, S.; Brauer, M.; Burnett, R.; Cercy, K.; Charlson, F.J. Global, regional, and national comparative risk assessment of 79 behavioural, environmental and occupational, and metabolic risks or clusters of risks, 1990–2015: A systematic analysis for the global burden of disease study 2015. *Lancet* **2016**, *388*, 1659–1724. [CrossRef]

14. Holt, D.C.; McCarthy, J.S.; Carapetis, J.R. Parasitic diseases of remote Indigenous communities in Australia. *Int. J. Parasitol.* **2010**, *40*, 1119–1126. [CrossRef]

15. Palmer, C.S.; Traub, R.J.; Robertson, I.D.; Hobbs, R.P.; Elliot, A.; While, L.; Rees, R.; Thompson, R.A. The veterinary and public health significance of hookworm in dogs and cats in Australia and the status of A. ceylanicum. *Vet. Parasitol.* **2007**, *145*, 304–313. [CrossRef]

16. Hotez, P.; Whitham, M. Helminth infections: A new global women's health agenda. *Obstet. Gynecol.* **2014**, *123*, 155–160. [CrossRef]

17. Bearup, A. The intensity and type of hookworm infestation in the ingham district of North Queensland. *Med. J. Aust.* **1931**, *2*, 65–74. [CrossRef]

18. Bradbury, R.; Traub, R.J. Hookworm infection in oceania. In *Neglected Tropical Diseases-Oceania*; Springer: Berlin/Heidelberg, Germany, 2016; pp. 33–68.

19. Prociv, P.; Luke, R.A. The changing epidemiology of human hookworm infection in Australia. *Med. J. Aust.* **1995**, *162*, 150–154. [CrossRef]

20. Holt, D.C.; Shield, J.; Harris, T.M.; Mounsey, K.E.; Aland, K.; McCarthy, J.S.; Currie, B.J.; Kearns, T.M. Soil-Transmitted Helminths in Children in a Remote Aboriginal Community in the Northern Territory: Hookworm is Rare but Strongyloides stercoralis and Trichuris trichiura Persist. *Trop. Med. Infect. Dis.* **2017**, *2*, 51. [CrossRef]

21. Davies, J.; Majumdar, S.S.; Forbes, R.; Smith, P.; Currie, B.J.; Baird, R.W. Hookworm in the Northern Territory: Down but not out. *Med. J. Aust.* **2013**, *198*, 278–281. [CrossRef]

22. Koehler, A.V.; Bradbury, R.S.; Stevens, M.A.; Haydon, S.R.; Jex, A.R.; Gasser, R.B. Genetic characterization of selected parasites from people with histories of gastrointestinal disorders using a mutation scanning-coupled approach. *Electrophoresis* **2013**, *34*, 1720–1728. [CrossRef]

23. Speare, R.; Bradbury, R.S.; Croese, J. A case of Ancylostoma ceylanicum infection occurring in an Australian soldier returned from Solomon Islands. *Korean J. Parasitol.* **2016**, *54*, 533. [CrossRef] [PubMed]

24. Smout, F.A.; Skerratt, L.F.; Butler, J.R.; Johnson, C.N.; Congdon, B.C.; Thompson, R.A. The hookworm Ancylostoma ceylanicum: An emerging public health risk in Australian tropical rainforests and Indigenous communities. *One Health* **2017**, *3*, 66–69. [CrossRef] [PubMed]

25. Hopkins, R.M.; Hobbs, R.P.; Thompson, R.A.; Gracey, M.S.; Spargo, R.M.; Yates, M. The prevalence of hookworm infection, iron deficiency and anaemia in an Aboriginal community in north-west Australia. *Med. J. Aust.* **1997**, *166*, 241–244. [CrossRef]

26. Constable, S.; Dixon, R.; Dixon, R. For the love of dog: The human–dog bond in rural and remote Australian indigenous communities. *Anthrozoös* **2010**, *23*, 337–349. [CrossRef]

27. Traub, R.J.; Robertson, I.D.; Irwin, P.; Mencke, N.; Thompson, R.A. Application of a species-specific PCR-RFLP to identify Ancylostoma eggs directly from canine faeces. *Vet. Parasitol.* **2004**, *123*, 245–255. [CrossRef]

28. Traub, R.J. Ancylostoma ceylanicum, a re-emerging but neglected parasitic zoonosis. *Int. J. Parasitol.* **2013**, *43*, 1009–1015. [CrossRef]

29. Bradbury, R.S.; Hii, S.F.; Harrington, H.; Speare, R.; Traub, R. Ancylostoma ceylanicum hookworm in the Solomon Islands. *Emerg. Infect. Dis.* **2017**, *23*, 252. [CrossRef]

30. Inpankaew, T.; Schär, F.; Dalsgaard, A.; Khieu, V.; Chimnoi, W.; Chhoun, C.; Sok, D.; Marti, H.; Muth, S.; Odermatt, P. High prevalence of Ancylostoma ceylanicum hookworm infections in humans, Cambodia, 2012. *Emerg. Infect. Dis.* **2014**, *20*, 976. [CrossRef]

31. Prociv, P.; Croese, J. Human eosinophilic enteritis caused by dog hookworm Ancylostoma caninum. *Lancet* **1990**, *335*, 1299–1302. [CrossRef]

32. McCarthy, J.; Moore, T.A. Emerging helminth zoonoses. *Int. J. Parasitol.* **2000**, *30*, 1351–1359. [CrossRef]
33. Rusdi, B.; Laird, T.; Abraham, R.; Ash, A.; Robertson, I.D.; Mukerji, S.; Coombs, G.W.; Abraham, S.; O'Dea, M.A. Carriage of critically important antimicrobial resistant bacteria and zoonotic parasites amongst camp dogs in remote Western Australian indigenous communities. *Sci. Rep.* **2018**, *8*, 8725. [CrossRef]
34. Smout, F.; Skerratt, L.; Johnson, C.; Butler, J.; Congdon, B. Zoonotic helminth diseases in dogs and dingoes utilising shared resources in an australian aboriginal community. *Trop. Med. Infect. Dis.* **2018**, *3*, 110. [CrossRef]
35. Walker, N.I.; Croese, J.; Clouston, A.D.; Parry, M.; Loukas, A.; Prociv, P. Eosinophilic enteritis in northeastern Australia. Pathology, association with Ancylostoma caninum, and implications. *Am. J. Surg. Pathol.* **1995**, *19*, 328–337. [CrossRef]
36. Fleming, F.M.; Brooker, S.; Geiger, S.M.; Caldas, I.R.; Correa-Oliveira, R.; Hotez, P.J.; Bethony, J.M. Synergistic associations between hookworm and other helminth species in a rural community in Brazil. *Trop. Med. Infect. Dis.* **2006**, *11*, 56–64. [CrossRef]
37. Traub, R.J.; Inpankaew, T.; Sutthikornchai, C.; Sukthana, Y.; Thompson, R.A. PCR-based coprodiagnostic tools reveal dogs as reservoirs of zoonotic ancylostomiasis caused by Ancylostoma ceylanicum in temple communities in Bangkok. *Vet. Parasitol.* **2008**, *155*, 67–73. [CrossRef]
38. Schad, G.A.; Page, M.R. Ancylostoma caninum: Adult worm removal, corticosteroid treatment, and resumed development of arrested larvae in dogs. *Exp. Parasitol.* **1982**, *54*, 303–309. [CrossRef]
39. Beveridge, I. Australian hookworms (Ancylostomatoidea): A review of the species present, their distributions and biogeographical origins. *Parassitologia* **2002**, *44*, 83–88.
40. Gibbs, H.; Gibbs, K. The effects of temperature on the development of the free-living stages of Dochmoides stenocephala (Railliet, 1884)(Ancylostomidae: Nematoda). *Can. J. Zool.* **1959**, *37*, 247–257. [CrossRef]
41. Stewart, L. The Taxonomy of Ancylostoma Species in the Townsville Region of North Queensland. Master's Thesis, James Cook University, Townsville, Australia, 1994.
42. Schär, F.; Odermatt, P.; Khieu, V.; Panning, M.; Duong, S.; Muth, S.; Marti, H.; Kramme, S. Evaluation of real-time PCR for Strongyloides stercoralis and hookworm as diagnostic tool in asymptomatic schoolchildren in Cambodia. *Acta Trop.* **2013**, *126*, 89–92. [CrossRef]
43. Verweij, J.J.; Canales, M.; Polman, K.; Ziem, J.; Brienen, E.A.; Polderman, A.M.; van Lieshout, L. Molecular diagnosis of *Strongyloides stercoralis* in faecal samples using real-time PCR. *Trans. R. Soc. Trop. Med. Hyg.* **2009**, *103*, 342–346. [CrossRef]
44. Gasser, R.; Cantacessi, C.; Loukas, A. DNA technological progress toward advanced diagnostic tools to support human hookworm control. *Biotechnol. Adv.* **2008**, *26*, 35–45. [CrossRef]
45. Hii, S.F.; Senevirathna, D.; Llewellyn, S.; Inpankaew, T.; Odermatt, P.; Khieu, V.; Muth, S.; McCarthy, J.; Traub, R.J. Development and evaluation of a multiplex quantitative real-time polymerase chain reaction for hookworm species in human stool. *Am. J. Trop. Med. Hyg.* **2018**, *99*, 1186–1193. [CrossRef]
46. Massetti, L.; Colella, V.; Zandejas, P.A.; Ng-Nguyen, D.; Harriott, L.; Marwedel, L.; Wiethoelter, A.; Traub, R.J. High-throughput multiplex qPCRs for the surveillance of zoonotic species of canine hookworms. *PLoS Negl. Trop. Dis.* **2020**. [CrossRef]
47. Sultana, Y.; Jeoffreys, N.; Watts, M.R.; Gilbert, G.L.; Lee, R. Real-time polymerase chain reaction for detection of *Strongyloides stercoralis* in stool. *Am. J. Trop. Med. Hyg.* **2013**, *88*, 1048–1051. [CrossRef]
48. Brooker, S.; Miguel, E.A.; Moulin, S.; Louba, A.I.; Bundy, D.A.; Kremer, M. Epidemiology of single and multiple species of helminth infections among school children in Busia District, Kenya. *East Afr. Med. J.* **2000**, *77*, 77. [CrossRef]
49. Thompson, R.A.; Roberts, M.G. Does pet helminth prophylaxis increase the rate of selection for drug resistance? *Trends Parasitol.* **2001**, *17*, 576–578. [CrossRef]
50. Traub, R.J.; Robertson, I.D.; Irwin, P.; Mencke, N.; Thompson, R.A. The prevalence, intensities and risk factors associated with geohelminth infection in tea-growing communities of Assam, India. *Trop. Med. Int. Health* **2004**, *9*, 688–701. [CrossRef]
51. Willis, E.; Ross, K. Review of principles governing dog health education in remote Aboriginal communities. *Aust. Veter. J.* **2019**, *97*, 4–9. [CrossRef]
52. Alaeddini, R. Forensic implications of PCR inhibition—A review. *Forensic Sci. Int. Genet.* **2012**, *6*, 297–305. [CrossRef]

53. Schrader, C.; Schielke, A.; Ellerbroek, L.; Johne, R. PCR inhibitors–occurrence, properties and removal. *J. Appl. Microbiol.* **2012**, *113*, 1014–1026. [CrossRef]

54. Beknazarova, M.; Millsteed, S.; Robertson, G.; Whiley, H.; Ross, K. Validation of DESS as a DNA Preservation method for the detection of strongyloides spp. in canine feces. *Int. J. Environ. Res. Public Health* **2017**, *14*, 624. [CrossRef]

 *pathogens*

Article

"Begging the Question"—Does *Toxocara* Infection/Exposure Associate with Multiple Sclerosis-Risk?

Ali Taghipour [1], Ali Rostami [2,3,*], Sahar Esfandyari [4], Saeed Aghapour [5], Alessandra Nicoletti [6] and Robin B. Gasser [7,*]

[1] Department of Parasitology, Faculty of Medical Sciences, Tarbiat Modares University, Tehran, Iran; alitaghipor71@yahoo.com
[2] Infectious Diseases and Tropical Medicine Research Center, Health Research Institute, Babol University of Medical Sciences, Babol, Iran
[3] Immunoregulation Research Center, Health Research Institute, Babol University of Medical Sciences, Babol, Iran
[4] Department of Anatomy, School of Medicine, Tehran University of Medical Sciences, Tehran, Iran; saharesfandyari1366@gmail.com
[5] Department of Neurosurgery, Faculty of Medicine, Mazandaran University of Medical Sciences, Sari, Iran; aghapour.saeed@yahoo.com
[6] Department G.F. Ingrassia, Section of Neurosciences, University of Catania, 95123 Catania, Italy; anicolet@unict.it
[7] Department of Veterinary Biosciences, Melbourne Veterinary School, The University of Melbourne, Parkville 3010, VIC, Australia
* Correspondence: alirostami1984@gmail.com (A.R.); robinbg@unimelb.edu.au (R.B.G.); Tel.: +98-11-32190557 (A.R.)

Received: 27 September 2020; Accepted: 4 November 2020; Published: 11 November 2020

Abstract: Although the cause of multiple sclerosis (MS) is unclear, infectious agents, including some parasitic roundworms (nematodes), have been proposed as possible risk factors or contributors. Here, we conducted a systematic review and meta-analysis of published observational studies to evaluate whether there is a possible association between infection with, or exposure to, one or more members of the genus *Toxocara* (phylum Nematoda; superfamily Ascaridoidea) and MS. We undertook a search of public literature databases to identify relevant studies and then used a random-effects meta-analysis model to generate the pooled odds ratio (OR) and 95% confidence intervals (CIs). This search identified six of a total of 1371 articles that were relevant to the topic; these published studies involved totals of 473 MS patients and 647 control subjects. Anti-*Toxocara* IgG serum antibodies were detected in 62 MS patients and 37 controls, resulting in respective seroprevalences of 13.1% (95% CI: 8.2–20.3) and 4.8% (95% CI: 2.5–9.2), indicating an association (pooled OR, 3.01; 95% CI: 1.46–6.21). Because of the publication bias identified (six eligible studies), well-designed and -controlled studies are required in the future to rigorously test the hypothesis that *Toxocara* infection/exposure has an association with MS.

Keywords: *Toxocara*; multiple sclerosis; association; meta-analysis

---

1. Introduction

Over the past decade, the World Health Organization (WHO) has emphasized the major importance of investigating neurological disorders in humans [1]. Key disorders, such as multiple sclerosis (MS) as well as Alzheimer's and Parkinson's diseases, cause major morbidity and mortality worldwide [1]. MS is a prevalent, chronic, and immune-mediated disease of the central nervous system (CNS), causing significant neurological disability worldwide [2,3]. According to the Global Burden of Disease Study [4], MS contributes to >1.151 million disability-adjusted life-years (DALYs) annually, and was linked to 18,932 deaths in 2016. While the exact cause(s) of MS is (are) not yet understood, the underlying mechanism is thought to be autoimmunity and/or a failure of particular cells to produce myelin [5,6]. Multiple factors, including genetic background, immune dysregulation, and environment, are proposed to contribute to the aetio-pathogenesis of this disease [7–9].

Studies have investigated the possible roles of infectious agents, mainly viruses, of which Epstein-Barr virus (anti-EBNA IgG sero-positivity) and infectious mononucleosis were reported to have a positive association with MS [10]. On the other hand, according to the 'hygiene hypothesis', multiple infectious exposures in early childhood reduce the risk of autoimmune and allergic diseases, as is commonly observed in tropical and subtropical areas [11]. The hygiene hypothesis has been proposed to explain an apparent increase in MS in Western countries, with an imbalance between Th1 and Th2 immune responses being the immunological reasoning [12]. In accord with this hypothesis, the modulation of autoimmune responses by some helminths has been shown to associate with several autoimmune diseases [13–15], and it has been proposed by some workers that helminths could have a protective effect against MS via a down-modulation of inflammatory responses and an enhancement of immune regulation [16]. Although some experimental evidence supports a protective effect, the relationship between helminths and MS is still a matter of major contention [17]. Indeed, some helminths (e.g., *Necator americanus* and *Trichuris suis*) appear to reduce the risk of MS [18,19], while others, such as *Toxocara* spp., may contribute to MS development [20,21].

Toxocara infection of humans is caused by the larval stages of members of the genus *Toxocara*, principally *T. canis*, but sometimes *T. cati* or related species from canids (dogs) or felids (cats) [22–25]. It is estimated that >1.4 billion people worldwide are infected with, or exposed to, *Toxocara* spp., indicating that infections are widespread, and thus, likely being responsible for human toxocariasis [25,26]. Human infection occurs via the ingestion of *Toxocara* eggs (containing infective third-stage larvae) from contaminated soil or raw vegetables, or sometimes via eating undercooked or raw meat (e.g., chicken and lamb) from paratenic hosts carrying encysted *Toxocara* larvae [27,28]. Following the ingestion of eggs or larvae, individual larvae emerge/activate in the small intestine, penetrate the gut wall and then migrate to different organs via the systemic circulation, but do not develop to mature adult worms in the gut [29]. The migrating larvae can cause significant damage to multiple organ systems in the accidental human host, such as the viscera and nervous system (including eyes), which can lead to disease and permanent damage [28,30].

Toxocara larvae can cross the blood-brain barrier to invade the CNS and cause neurotoxocariasis [31], although the detection of such larvae in brain tissues can be challenging [32,33]. Larval migration induces a host response, characterized mainly by a T-helper cell (Th2) response, cellular infiltration around larvae, and increased production of cytokines (interleukins-4, -5, -10 and -13), peripheral eosinophilia and/or specific serum IgG and IgE antibodies [29,34]. The neurological manifestation of neurotoxocariasis is variable and can include encephalitis, meningitis, myelitis, and/or cerebral vasculitis, but asymptomatic CNS infection is common [31,35,36]; MRI findings of neurotoxocariasis include subcortical, cortical or deep white matter lesions with variable enhancement, which can associate with hydrocephalus and leptomeningeal enhancement, as well as spinal cord involvement [37]. Human neurotoxocariasis is thought to be rare, even if, in many animal models (e.g., rodents, pigs, and primates), *Toxocara* larvae usually migrate to the brain [38]. Although toxocariasis is a prevalent helminthiasis worldwide, characterized by a pronounced Th2 host response [29], few studies have evaluated its possible role on the risk of MS [20,21,39,40], and no study has yet been carried out to systematically review existing data and information on this topic. This study scrutinizes all publicly available peer-reviewed literature to critically evaluate whether human *Toxocara* infection/exposure associates with the risk of MS, or not.

2. Materials and Methods

The Preferred Reporting Items for Systematic Reviews and Meta-analyses (PRISMA) guidelines [41] were followed for the present study design, as well as for the analysis and interpretation of results.

2.1. Search Strategy and Study Selection

A comprehensive search of the literature was performed using five international databases, including PubMed, Scopus, Science Direct, Web of Science, and Google Scholar, from inception to 15 June 2020. The search terms followed medical subject headings (MeSH): "*Toxocara* infection", "toxocariasis", "*Toxocara canis*", "*Toxocara cati*", "multiple sclerosis", "association", and "relationship", alone or in combination with "OR" and/or "AND". The reference lists in the eligible studies were used to access relevant, related studies and to optimize data acquisition. After completing the search, the articles selected were independently reviewed by the two researchers (A.T. and A.R.). Following an appraisal of the title, abstract, and full text of individual articles, all duplicates or publications unrelated to the topic were excluded. The full texts of relevant publications were read and scrutinized to identify articles eligible for inclusion, and any conflicts of opinion or uncertainties were resolved through detailed discussions, and a consensus was reached. Included were peer-reviewed, original or conference papers describing: (1) Cross-sectional, cohort or case–control studies; (2) prevalence studies of humans for *Toxocara* infection or exposure, with both MS and suitable control groups; (3) serological or histopathological investigations of human *Toxocara* infection or toxocariasis; (4) published (up to 15 June 2020), without applying a language, time or geographic limitation. Excluded were articles: (1) Without one or more appropriate control groups of healthy people; (2) with a sample size of ≤30 in each group; (3) experimental studies; and (4) reviews, case reports, and letters without original research results or data sets.

2.2. Data Extraction and Quality Assessment

Eligible articles were individually scrutinized, and data/information extracted. The data recorded were: First author, publication year, country, diagnostic methodology, age (mean or range) and gender of human subjects, total numbers of MS patients and healthy control subjects, as well as the prevalence of anti-*Toxocara* serum antibodies in individuals of each of the subject groups. The quality of each eligible article was assessed using the Newcastle–Ottawa Scale (NOS), as recommended by the Cochrane network [42,43]. The scoring system was: Subject selection criteria (0–4 points), comparability of subjects (0–2 points), and exposure (0–3 points)—with a nine-point maximum. An article was given one star for each numbered item meeting the selection and exposure criteria, and two stars were given for comparability. Using the sum of all points, the quality of each article was rated as high (7–9 points), moderate (4–6), or poor (0–3).

2.3. Data Synthesis and Statistical Analysis

All statistical analyses were conducted using comprehensive meta-analysis software (version 2, BIOSTAT, Englewood, NJ, USA). First, we estimated the pooled prevalence of *Toxocara* infection/exposure with a corresponding 95% confidence interval (CI) in each case and control groups employing the DerSimonian-Laird random-effects model, and the difference was calculated by χ^2 test. We used the exact binomial method of Hamza et al. to model within-study variability by binomial distribution and Freeman-Tukey Double Arcsine Transformation to stabilize the variances in the meta-analysis [44]. Then, the respective odds ratio (OR) and 95% CI was calculated for each article/study. To assess an association between *Toxocara* exposure/infection and MS in humans, ORs from individual studies were combined to produce a pooled OR and 95% CI, employing the random-effects model with a restricted maximum-likelihood estimator. Heterogeneity among studies was assessed using I^2 and Cochran's Q statistics [45]. The effects of a small study and of publication bias on results were inferred using the Egger's regression test [46]. A *p*-value of < 0.05 was considered statistically significant.

3. Results

The systematic search identified 1371 articles of possible relevance, 8 of which remained and were pertinent, following duplicate-removal, title- and abstract-screening, and implementation of all eight inclusion and exclusion criteria (Figure 1). Of the eight articles selected, two [21,47] were excluded. The first article [21] was removed because of its small sample size (<30 in both the case- and control-groups) and sero-negative results for all participants in both groups; the second [47] was eliminated, due to the recruitment of inappropriate control subjects (i.e., patients with clinically isolated syndrome [CIS]—a first neurological episode of MS). In this latter study, serum IgG-antibodies were detected in ELISA using against *Toxocara* excretory/secretory (TES) antigens in one of 62 (1.6%) MS patients, whereas none of the CIS patients had measurable titers.

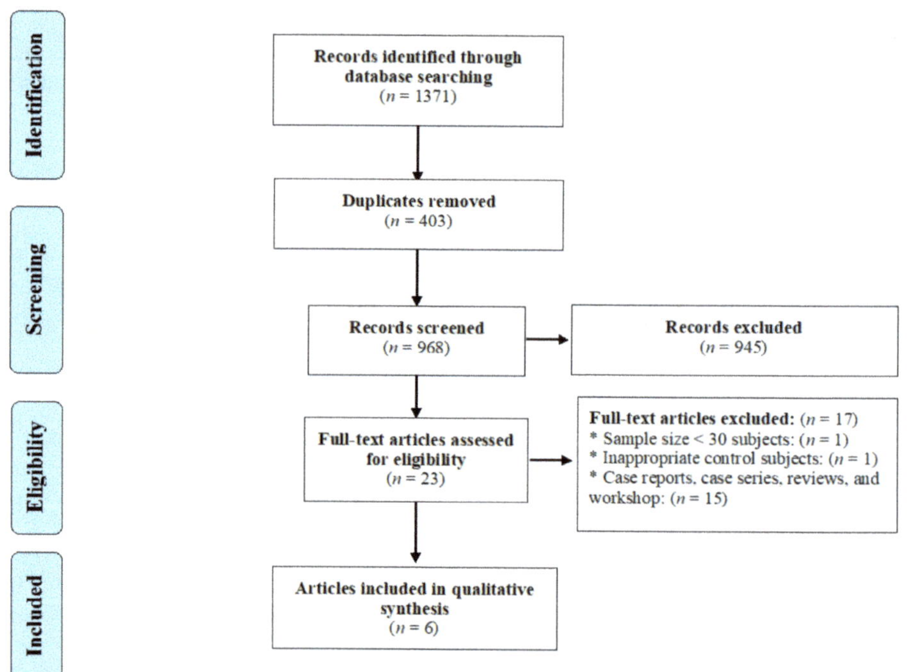

Figure 1. PRISMA flow diagram of the search strategy and study selection process.

Therefore, six studies [20,39,40,48–50] qualified and were, ultimately, included in the meta-analysis (Figure 1). Five studies were original papers, published in peer-reviewed journals, and one [49] was a conference paper. These studies had a case–control design, were from Iran (n = 4), Italy (n = 1) and Turkey (n = 1), and were published between 2006 and 2020 (Table 1). The key characteristics of each of these studies are listed in Table 1, and the results of the quality assessment of case–control studies are given in Table 2.

Table 1. Key characteristics of studies used to investigate an association between human *Toxocara* infection/exposure and multiple sclerosis (MS).

Reference	Country	Age Range, or Mean Age ± Standard Deviation (Years)		Subjects with MS		Subjects without MS (Controls)		*p*-Value
		Subjects with MS	Subjects without MS Controls	Number Tested	Positive for Anti-*Toxocara* IgG Serum Antibodies (%)	Number Tested	Positive for Anti-*Toxocara* IgG Serum Antibodies (%)	
Kuk et al. [47]	Turkey	20–54	20–50	37	4 (10.8)	50	1 (2.0)	0.08
Zibaei and Ghorbani. [20]	Iran	3–49	3–52	68	10 (14.7)	70	1 (1.4)	0.004
Khalilidehkordi et al. [49]	Iran	not recorded	not recorded	70	8 (11.4)	70	1 (1.4)	<0.05
Cicero et al. [39]	Italy	44.6 ± 11.1	48.1 ± 15.6	132	12 (9.1)	287	23 (8.0)	0.7
Khalili et al. [48]	Iran	41.2 ± 9.5	38.8 ± 7.6	70	20 (28.5)	70	8 (11.4)	0.02
Esfandiari et al. [40]	Iran	11–60	11–60	96	8 (8.3)	100	3 (3.0)	0.1

Table 2. Newcastle–Ottawa Scale for assessing the quality of the six case–control studies included to assess an association between human *Toxocara* infection/exposure and multiple sclerosis (MS).

	Selection				Comparability		Exposure		
Reference	Adequate Case Definition	Representat-Iveness of MS Cases	Selection of Controls	Definition of Controls	Comparability of Cases and Controls on the Basis of Design or Analysis	Ascertainment of Exposure	Same Method of Ascertainment for Cases and Controls	Non-Response Rate	Score
Kuk et al. [47]	*	*	*	na	*	*	*	na	6
Zibaei and Ghorbani. [20]	*	*	*	*	**	*	*	na	8
Khalilidehkordi et al. [49]	*	*	*	na	na	*	*	na	5
Cicero et al. [39]	*	*	*	*	**	*	*	*	9
Khalili et al. [48]	*	*	*	na	**	*	*	na	7
Esfandiari et al. [40]	*	*	*	*	**	*	*	na	8

In this table, one star was given to each article for each item meeting the selection and exposure criteria, and two stars were given for comparability. Using the sum of all points, the quality of each article was rated as high (7–9 points), moderate (4–6), or poor (0–3); not applicable (na).

The six articles included in the present meta-analysis showed an acceptable quality (i.e., had a high or moderate quality-score; cf. Table 2). All of them reported prevalences of anti-*Toxocara* IgG serum antibodies, established by enzyme-linked immunosorbent assay (ELISA), although results were divergent between studies; three reported a non-significant, while the three others reported a significant, positive association between *Toxocara* infection/exposure and MS (Table 1).

In total, 473 MS patients were recruited to all six studies selected; 62 subjects had anti-*Toxocara* IgG serum antibodies. Of the 647 control subjects, 37 people had anti-*Toxocara* IgG serum antibodies. The pooled anti-*Toxocara* sero-prevalence rate in MS patients (13.1%; 95% CI, 8.2–20.3; I^2 = 71.1; 95% CI, 32.7 to 87.5; Q-value = 17.325) was higher (*p* value < 0.01) than in the control group (4.8%; 95% CI, 2.5–9.2; I^2 = 57.4; −5.3 to 82.6; Q-value = 11.83) (Figure 2).

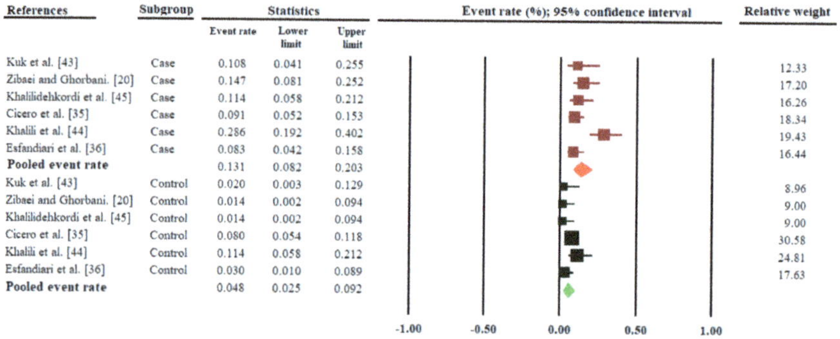

References	Subgroup	Event rate	Lower limit	Upper limit	Event rate (%); 95% confidence interval	Relative weight
Kuk et al. [43]	Case	0.108	0.041	0.255		12.33
Zibaei and Ghorbani. [20]	Case	0.147	0.081	0.252		17.20
Khalilidehkordi et al. [45]	Case	0.114	0.058	0.212		16.26
Cicero et al. [35]	Case	0.091	0.052	0.153		18.34
Khalili et al. [44]	Case	0.286	0.192	0.402		19.43
Esfandiari et al. [36]	Case	0.083	0.042	0.158		16.44
Pooled event rate		0.131	0.082	0.203		
Kuk et al. [43]	Control	0.020	0.003	0.129		8.96
Zibaei and Ghorbani. [20]	Control	0.014	0.002	0.094		9.00
Khalilidehkordi et al. [45]	Control	0.014	0.002	0.094		9.00
Cicero et al. [35]	Control	0.080	0.054	0.118		30.58
Khalili et al. [44]	Control	0.114	0.058	0.212		24.81
Esfandiari et al. [36]	Control	0.030	0.010	0.089		17.63
Pooled event rate		0.048	0.025	0.092		

−1.00   −0.50   0.00   0.50   1.00

Figure 2. Forest plots for random-effects meta-analysis of the prevalence rates of *Toxocara* infection/exposure (established by anti-*Toxocara* IgG serum antibody detection) in MS patients (cases) and in healthy control subjects (controls). Relative weight: Weight of each study by comparison with all six studies—in percent.

The results of the meta-analysis of all six studies showed a pooled OR of 3.01 (95% CI, 1.46–6.21), suggesting that *Toxocara* infection/exposure could be significantly associated with an increased risk of MS (Figure 3). Some heterogeneity (I^2 = 42.4; 95% CI, −45.2 to 76.1; Q-value = 8.68) was detected among studies. Sensitivity analysis was performed to determine the effect of one study (reported in a conference paper) on our estimated OR; this analysis showed that, after removing this study [49], the association was still significant (OR, 2.6; 95% CI, 1.2–5.6; I^2 = 43.2; 95% CI, −53.8 to 78.40; Q-value = 7.05) (Figure S1). In one of the six studies included [49], based on the proportion of anti-*Toxocara* sero-positive patients, women were significantly associated with developing MS. For two [20,40] of the six studies, there was no significant difference in gender between the MS patient- and healthy control groups. Employing the Egger's regression test, a significant publication bias was found in the studies included here (*t*-value = 3.53, *p*-value = 0.02; Figure S2).

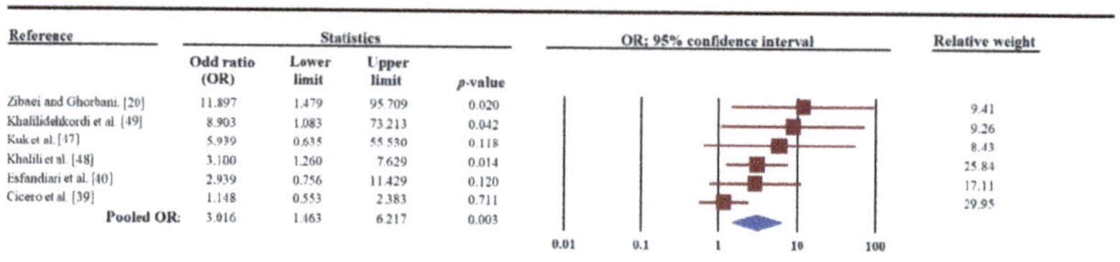

Reference	Odd ratio (OR)	Lower limit	Upper limit	*p*-value	OR; 95% confidence interval	Relative weight
Zibaei and Ghorbani. [20]	11.897	1.479	95.709	0.020		9.41
Khalilidehkordi et al. [49]	8.903	1.083	73.213	0.042		9.26
Kuk et al. [17]	5.939	0.635	55.530	0.118		8.43
Khalili et al. [48]	3.100	1.260	7.629	0.014		25.84
Esfandiari et al. [40]	2.939	0.756	11.429	0.120		17.11
Cicero et al. [39]	1.148	0.553	2.383	0.711		29.95
Pooled OR:	3.016	1.463	6.217	0.003		

0.01   0.1   1   10   100

Figure 3. Forest plot, pooled with random-effects regarding the association between *Toxocara* infection/exposure (assessed by anti-*Toxocara* IgG serum antibody detection) and multiple sclerosis (MS), showing the odd ratio (OR) and a 95% confidence interval (CI). The *p*-value referred to the significance of OR.

4. Discussion

The hygiene hypothesis has been proposed as a possible explanation for the increased incidence of allergy and autoimmune diseases, including MS, in the Western world, with an imbalance between Th1 and Th2 responses being promoted as an immunological explanation [12]. Epidemiological studies have indicated that the prevalence of MS has been increasing over time, particularly in countries whose socioeconomic and sanitation levels have improved, probably through a progressive decrease in the prevalence of infections [51]. Some helminth species have been reported or proposed to have a role in preventing autoimmune diseases. Supporting this hypothesis are some experimental studies showing that parasites, such as *Schistosoma mansoni*, exert a protective effect on the development of experimental allergic encephalomyelitis (EAE) in infected mice, dampening the classical Th1 response through an immunological switch to a Th2 response [52]. However, clearly, the possible relationship between helminths and MS is still controversial. On the one hand, for instance, *Necator americanus* larvae [18] and *Trichuris suis* eggs [19] are proposed to play a role in protecting people against MS. On the other hand, some other worms, such as *Toxocara* species [20] and *Onchocerca volvulus* [53], have been suggested to contribute to autoimmune diseases [21].

The present meta-analysis provides epidemiological evidence of a significant association between *Toxocara* infection/exposure and MS, with a pooled overall OR of 3.01 (95% CI, 1.46–6.21). The results of this study accord with those reported in some systematic reviews and meta-analyses examining the role of *Toxocara* infection as a potential risk factor for other neurological syndromes, such as epilepsy [54,55].

Toxocara larvae can migrate in the tissues of the central nervous system (CNS) and induce the neural larva migrans (NLM) syndrome [35,56]. At the end of the visceral phase of migration, *Toxocara* larvae commence the myotropic-neurotropic phase of migration, and reach the brain within 28 days of infection [56]. An experimental study of NLM in mice [57] showed that larvae penetrate arteries, near the brain surface, and assume a predilection for the cerebellum, rather than the cerebrum or brainstem. *Toxocara* larvae might survive for up to two years after infection in the brains of experimentally infected mice [58]. Furthermore, CNS invasion of *Toxocara* larvae results in parenchymal damage, hemorrhagic lesions, demyelination, focal malacia, and neuronal necrosis in the brain [59–63].

Of the animal/toxocariasis models employed, the pig model has been particularly useful because of physiological and biochemical similarities between pigs and humans, especially in relation to immune responses. In pigs, *Toxocara* larvae can be recovered from the brain between days 10–21 after infection, but disease in the pig is self-limiting, and larvae become undetectable in the brain after a period of 120 days. Pathological changes associated with porcine cerebral toxocariasis include congestion, oedema, shrinkage of nerve cells, vacuolization, gliosis, satellitosis, neurophagia, and liquefactive necrosis [38]. However, it should be noted that, according to a recent study in pigs [64], the passage of *Toxoxara* larvae through the brain does not always induce lesions detectable by magnetic resonance imaging (MRI), suggesting that they do not cause structural lesions, thus leaving no detectable damage.

Despite this evidence, there is a 'missing link' between *Toxocara* infection/exposure and the pathogenesis of MS in humans, which warrants serious attention [65]. Some pathogenic and immunologic mechanisms have been suggested to explain the possible implication(s) of toxocariasis or *Toxocara* infection in MS: (1) While demyelination and neurodegeneration processes are key characteristics of MS [5], they are also frequently-observed histopathological hallmarks of NLM and neurotoxocariasis in mice with experimental *T. canis* infection [60,61,63,66,67]. Here, demyelination might relate to a reduced cholesterol concentration or a down-regulation of key genes involved in cholesterol synthesis or transport via alterations in signal transduction induced by the presence of *Toxocara* [56,63]. (2) Experimental studies of mice have shown that migrating *Toxocara* larvae can be responsible for increased permeability of the blood-brain barrier, and an elevated expression of nitric oxide synthase (iNOS) and pro-inflammatory cytokines, such as interleukin–1β, interleukin–6 and tumor necrosis factor-α, which are potentially neurotoxic substances, and could lead to neurodegeneration or neuronal damage [35,68]. Although these pathological and immunological observations attempt to explain a possible association of *Toxocara* infection/exposure and/or toxocariasis with MS, it is important to point out: (i) That many observations have been made in mice experimentally infected with *T. canis* by comparison with matching, uninfected control mice, and (ii) that, although both humans and mice are accidental hosts, the immunological responses in experimentally infected mice could be distinct from those in people who contract accidental *Toxocara* infection. Other factors to consider (in relation to both host species) would be the intensity of the larval infection/burden and different immune responses to these larvae among individuals (i.e., high responders *versus* low responders). As mouse models exist for human toxocariasis [30,56,69], it would be informative to conduct systematic, comparative studies to critically assess experimentally whether *Toxocara* infection/exposure contributes to MS, but it will be important to evaluate which experimental model should be used to answer which specific question, and ensure cautious interpretation of experimental findings for mice with respect to MS in humans.

This study is the first systematic review and meta-analysis of all publicly accessible, published data/information to assess an association between human *Toxocara* infection/exposure, and MS risk. Although the present literature search was comprehensive and the methodology rigorous, the results of this meta-analysis need to be interpreted with caution, mainly for the following reasons: (1) All of the studies included here are retrospective case–control studies, such that one cannot be confident that *Toxocara* exposure or infection occurred before the outcome (i.e., MS), and consequently, a possible 'reverse causality' cannot be excluded; (2) most studies selected were hospital-based, such that selection bias could not be ruled out; (3) the numbers of studies ($n = 6$) and participants were limited, and from a small number of countries; (4) some studies published in local journals (not indexed in global databases) might have been missed in the literature search; (5) human case- and control-groups were not adjusted for different factors, such as age or gender; (6) potential risk factors (pica or exposure to pets) were not systematically evaluated; (7) in most articles, the definition of a 'control group' was unclear; (8) there was a publication bias, such that that studies with a significant difference between MS patients and controls were probably more likely to be published; and (9) eligible studies included in the meta-analysis had used only ELISA to detect anti-*Toxocara* antibodies, and did not employ a confirmatory method (e.g., immunoblotting) to exclude cross-reactivity with serum antibodies against other nematodes, such as *Ascaris*.

5. Conclusions

Although the small number of published studies ($n = 6$) investigating whether *Toxocara* infection/exposure could be associated with MS limit the interpretations of the findings and the conclusions that could be drawn from the meta-analysis, the present study emphasizes the need for well-designed and well-controlled longitudinal (cohort) studies in the future, to rigorously test the hypothesis that *Toxocara* infection/exposure has an association with MS, and to assess whether such infection/exposure is a co-factor contributing to the development of MS.

Supplementary Materials: The following are available online at http://www.mdpi.com/2076-0817/9/11/938/s1. Figure S1: Forest plot for sensitivity analysis (after removing of study by Khalilidehkordi et al. [49], which was a conference paper) pooled with random-effects regarding the association between *Toxocara* infection/exposure (assessed by anti-*Toxocara* IgG serum antibody detection) and multiple sclerosis (MS), showing the odd ratio (OR) and a 95% confidence interval (CI). Figure S2: Publication bias, calculated using an Egger's plot.

Author Contributions: Conceptualization, A.T., A.R. and R.B.G.; methodology, A.T., A.R. and S.A.; software, A.T. and S.A.; validation, A.R., and S.A.; formal analysis, A.T. and S.E.; investigation, A.R.; resources, A.T.; data curation, A.T. and S.E.; writing—original draft preparation, A.R. and R.B.G.; writing—review and editing, A.R., R.B.G. and A.N.; visualization, A.R.; supervision, R.B.G.; project administration, A.R. and R.B.G.; funding acquisition, R.B.G. All authors have read and agreed to the published version of the manuscript.

Funding: Funding from the Australian Research Council (ARC), Yourgene Health Singapore and Melbourne Water is gratefully acknowledged (R.B.G.).

Acknowledgments: This research was supported by the Health Research Institute at the Babol University of Medical Sciences, Babol, Iran (A.R.).

Conflicts of Interest: The authors declare no conflict of interest. Funders had no role in the design of the study; in the collection, analyses or interpretation of data; in the writing of the manuscript; or in the decision to publish the results.

References

1. Feigin, V.L.; Abajobir, A.A.; Abate, K.H.; Abd-Allah, F.; Abdulle, A.M.; Abera, S.F.; Abyu, G.Y.; Ahmed, M.B.; Aichour, A.N.; Aichour, I. Global, regional, and national burden of neurological disorders during 1990–2015: A systematic analysis for the Global Burden of Disease Study 2015. *Lancet Neurol.* **2017**, *16*, 877–897. [CrossRef]
2. Goldenberg, M.M. Multiple sclerosis review. *Pharm. Ther.* **2012**, *37*, 175–184.
3. Wingerchuk, D.M.; Weinshenker, B.G. Disease modifying therapies for relapsing multiple sclerosis. *BMJ* **2016**, *354*, i3518. [CrossRef] [PubMed]
4. Wallin, M.T.; Culpepper, W.J.; Nichols, E.; Bhutta, Z.A.; Gebrehiwot, T.T.; Hay, S.I.; Khalil, I.A.; Krohn, K.J.; Liang, X.; Naghavi, M. Global, regional, and national burden of multiple sclerosis 1990–2016: A systematic analysis for the Global Burden of Disease Study 2016. *Lancet Neurol.* **2019**, *18*, 269–285. [CrossRef]
5. Brownlee, W.J.; Hardy, T.A.; Fazekas, F.; Miller, D.H. Diagnosis of multiple sclerosis: Progress and challenges. *Lancet* **2017**, *389*, 1336–1346. [CrossRef]
6. Baecher-Allan, C.; Kaskow, B.J.; Weiner, H.L. Multiple sclerosis: Mechanisms and immunotherapy. *Neuron* **2018**, *97*, 742–768. [CrossRef]
7. Lauer, K. Environmental risk factors in multiple sclerosis. *Expert Rev. Neurother.* **2010**, *10*, 421–440. [CrossRef]
8. Guerau-de-Arellano, M.; Smith, K.M.; Godlewski, J.; Liu, Y.; Winger, R.; Lawler, S.E.; Whitacre, C.C.; Racke, M.K.; Lovett-Racke, A.E. Micro-RNA dysregulation in multiple sclerosis favours pro-inflammatory T-cell-mediated autoimmunity. *Brain* **2011**, *134*, 3578–3589. [CrossRef]
9. Berrih-Aknin, S.; Le Panse, R. Myasthenia gravis: A comprehensive review of immune dysregulation and etiological mechanisms. *J. Autoimmun.* **2014**, *52*, 90–100. [CrossRef]
10. Belbasis, L.; Bellou, V.; Evangelou, E.; Ioannidis, J.P.; Tzoulaki, I. Environmental risk factors and multiple sclerosis: An umbrella review of systematic reviews and meta-analyses. *Lancet Neurol.* **2015**, *14*, 263–273. [CrossRef]
11. Thompson, A.J.; Baranzini, S.E.; Geurts, J.; Hemmer, B.; Ciccarelli, O. Multiple sclerosis. *Lancet* **2018**, *391*, 1622–1636. [CrossRef]
12. Stiemsma, L.T.; Reynolds, L.A.; Turvey, S.E.; Finlay, B.B. The hygiene hypothesis: Current perspectives and future therapies. *Immunotargets Ther.* **2015**, *4*, 143–157. [CrossRef] [PubMed]
13. Zaccone, P.; Hall, S.W. Helminth infection and type 1 diabetes. *Rev. Diabet. Stud.* **2012**, *9*, 272–286. [CrossRef] [PubMed]
14. Elliott, D.E.; Weinstock, J.V. Helminth–host immunological interactions: Prevention and control of immune-mediated diseases. *Ann. N. Y. Acad. Sci.* **2012**, *1247*, 83–96. [CrossRef]
15. Versini, M.; Jeandel, P.-Y.; Bashi, T.; Bizzaro, G.; Blank, M.; Shoenfeld, Y. Unraveling the hygiene hypothesis of helminthes and autoimmunity: Origins, pathophysiology, and clinical applications. *BMC Med.* **2015**, *13*, 81. [CrossRef]
16. Correale, J.; Farez, M.F.; Gaitán, M.I. Environmental factors influencing multiple sclerosis in Latin America. *Mult. Scler. J. Exp. Transl. Clin.* **2017**, *3*, 2055217317715049. [CrossRef]
17. Sewell, D.; Qing, Z.; Reinke, E.; Elliot, D.; Weinstock, J.; Sandor, M.; Fabry, Z. Immunomodulation of experimental autoimmune encephalomyelitis by helminth ova immunization. *Int. Immunol.* **2003**, *15*, 59–69. [CrossRef]
18. Hansen, C.S.; Hasseldam, H.; Bacher, I.H.; Thamsborg, S.M.; Johansen, F.F.; Kringel, H. *Trichuris suis* secrete products that reduce disease severity in a multiple sclerosis model. *Acta Parasitol.* **2017**, *62*, 22–28. [CrossRef]
19. Tanasescu, R.; Tench, C.R.; Constantinescu, C.S.; Telford, G.; Singh, S.; Frakich, N.; Onion, D.; Auer, D.P.; Gran, B.; Evangelou, N. Hookworm treatment for relapsing multiple sclerosis: A randomized double-blinded placebo-controlled trial. *JAMA Neurol.* **2020**, *77*, 1089–1098. [CrossRef]
20. Zibaei, M.; Ghorbani, B. Toxocariasis and multiple sclerosis: A case-control study in Iran. *Neurol. Asia* **2014**, *109*, 283–286.

21. Söndergaard, H.P.; Theorell, T. A putative role for *Toxocara* species in the aetiology of multiple sclerosis. *Med. Hypotheses* **2004**, *63*, 59–61. [CrossRef] [PubMed]

22. Rostami, A.; Riahi, S.M.; Hofmann, A.; Ma, G.; Wang, T.; Behniafar, H.; Taghipour, A.; Fakhri, Y.; Spotin, A.; Chang, B.C. Global prevalence of *Toxocara* infection in dogs. *Adv. Parasitol.* **2020**, *109*, 561–583. [PubMed]

23. Rostami, A.; Sepidarkish, M.; Ma, G.; Wang, T.; Ebrahimi, M.; Fakhri, Y.; Mirjalali, H.; Hofmann, A.; Macpherson, C.N.; Hotez, P.J. Global prevalence of *Toxocara* infection in cats. *Adv. Parasitol.* **2020**, *109*, 615–639. [PubMed]

24. Ma, G.; Rostami, A.; Wang, T.; Hofmann, A.; Hotez, P.J.; Gasser, R.B. Global and regional seroprevalence estimates for human toxocariasis: A call for action. *Adv. Parasitol.* **2020**, *108*, 273–288.

25. Rostami, A.; Riahi, S.M.; Holland, C.V.; Taghipour, A.; Khalili-Fomeshi, M.; Fakhri, Y.; Omrani, V.F.; Hotez, P.J.; Gasser, R.B. Seroprevalence estimates for toxocariasis in people worldwide: A systematic review and meta-analysis. *PLoS Negl. Trop. Dis.* **2019**, *13*, e0007809. [CrossRef]

26. Rostami, A.; Ma, G.; Wang, T.; Koehler, A.V.; Hofmann, A.; Chang, B.C.; Macpherson, C.N.; Gasser, R.B. Human toxocariasis—A look at a neglected disease through an epidemiological 'prism'. *Infect. Genet. Evol.* **2019**, *74*, 104002. [CrossRef]

27. Siyadatpanah, A.; Tabatabaei, F.; Emami, Z.A.; Spotin, A.; Fallah, O.V.; Assadi, M.; Moradi, S.; Rostami, A.; Memari, F.; Hajialiani, F. Parasitic contamination of raw vegetables in Amol, North of Iran. *Arch. Clin. Infect. Dis.* **2013**, *8*, e15983. [CrossRef]

28. Carlin, E.P.; Tyungu, D.L. *Toxocara*: Protecting pets and improving the lives of people. *Adv. Parasitol.* **2020**, *109*, 3–16.

29. Ma, G.; Holland, C.V.; Wang, T.; Hofmann, A.; Fan, C.-K.; Maizels, R.M.; Hotez, P.J.; Gasser, R.B. Human toxocariasis. *Lancet Infect. Dis.* **2018**, *18*, e14–e24. [CrossRef]

30. Wu, T.; Bowman, D.D. Visceral larval migrans of *Toxocara canis* and *Toxocara cati* in non-canid and non-felid hosts. *Adv. Parasitol.* **2020**, *109*, 63–88.

31. Deshayes, S.; Bonhomme, J.; de La Blanchardière, A. Neurotoxocariasis: A systematic literature review. *Infection* **2016**, *44*, 565–574. [CrossRef] [PubMed]

32. Murrell, T.; Harbige, L.; Robinson, I. A review of the aetiology of multiple sclerosis: An ecological approach. *Ann. Hum. Biol.* **1991**, *18*, 95–112. [CrossRef] [PubMed]

33. Woodruff, A. Toxocariasis. *BMJ* **1970**, *3*, 663. [CrossRef]

34. Maizels, R.M. *Toxocara* canis: Molecular basis of immune recognition and evasion. *Vet. Parasitol.* **2013**, *193*, 365–374. [CrossRef] [PubMed]

35. Fan, C.-K.; Holland, C.V.; Loxton, K.; Barghouth, U. Cerebral toxocariasis: Silent progression to neurodegenerative disorders? *Clin. Microbiol. Rev.* **2015**, *28*, 663–686. [CrossRef]

36. Sánchez, S.; Garcıa, H.; Nicoletti, A. Clinical and Magnetic Resonance Imaging Findings of Neurotoxocariasis. *Front. Neurol.* **2018**, *9*, 53. [CrossRef]

37. Jewells, V.L.; Latchaw, R.E. What Can Mimic Multiple Sclerosis? In *Seminars in Ultrasound, CT and MRI*; WB Saunders: Philadelphia, PA, USA, 2020; Volume 41, pp. 284–295.

38. Strube, C.; Heuer, L.; Janecek, E. *Toxocara* spp. infections in paratenic hosts. *Vet. Parasitol.* **2013**, *193*, 375–389. [CrossRef]

39. Cicero, C.E.; Patti, F.; Lo Fermo, S.; Giuliano, L.; Rascunà, C.; Chisari, C.G.; D'Amico, E.; Paradisi, V.; Marin, B.; Preux, P.-M. Lack of association between *Toxocara canis* and multiple sclerosis: A population-based case–control study. *Mult. Scler. J.* **2020**, *26*, 258–259. [CrossRef]

40. Esfandiari, F.; Mikaeili, F.; Rahmanian, R.; Asgari, Q. Seroprevalence of toxocariasis in multiple sclerosis and rheumatoid arthritis patients in Shiraz city, southern Iran. *Clin. Epidemiol. Glob. Health* **2019**, *8*, 158–160. [CrossRef]

41. Moher, D.; Shamseer, L.; Clarke, M.; Ghersi, D.; Liberati, A.; Petticrew, M.; Shekelle, P.; Stewart, L.A. Preferred reporting items for systematic review and meta-analysis protocols (PRISMA-P) 2015 statement. *Syst. Rev.* **2015**, *4*, 1. [CrossRef]

42. Higgins, J.P.; Thomas, J.; Chandler, J.; Cumpston, M.; Li, T.; Page, M.J.; Welch, V.A. *Cochrane Handbook for Systematic Reviews of Interventions*; John Wiley & Sons: Hoboken, NJ, USA, 2019; ISBN 978-1-119-53662-8.

43. Stang, A. Critical evaluation of the Newcastle-Ottawa scale for the assessment of the quality of nonrandomized studies in meta-analyses. *Eur. J. Epidemiol.* **2010**, *25*, 603–605. [CrossRef] [PubMed]

44. Hamza, T.H.; van Houwelingen, H.C.; Stijnen, T. The binomial distribution of meta-analysis was preferred to model within-study variability. *J. Clin. Epidemiol.* **2008**, *61*, 41–51. [CrossRef] [PubMed]

45. Higgins, J.P.; Thompson, S.G.; Deeks, J.J.; Altman, D.G. Measuring inconsistency in meta-analyses. *BMJ* **2003**, *327*, 557–560. [CrossRef] [PubMed]

46. Egger, M.; Smith, G.D.; Schneider, M.; Minder, C. Bias in meta-analysis detected by a simple, graphical test. *BMJ* **1997**, *315*, 629–634. [CrossRef] [PubMed]

47. Posová, H.; Hrušková, Z.; Havrdová, E.; Kolářová, L. *Toxocara* spp. seronegativity in Czech patients with early form of multiple sclerosis-clinically isolated syndrome. *Epidemiol. Mikrobiol. Imunol.* **2017**, *66*, 124–127.

48. Kuk, S.; Ozgocmen, S.; Bulut, S. Seroprevalance of *Toxocara* antibodies in multiple sclerosis and ankylosing spondylitis. *Indian J. Med. Sci.* **2006**, *60*, 297–299. [CrossRef] [PubMed]

49. Khalili, N.; Khalili, N.; Nickhah, A.; Khalili, B. Seroprevalence of anti-*Toxocara* antibody among multiple sclerosis patients: A case–control study. *J. Parasit. Dis.* **2020**, *44*, 145–150. [CrossRef]

50. Khalilidehkordi, B. *Toxocara* infection in multiple sclerosis patients in Shahrekord district. In Proceedings of the 9th International Congress in Clinical Laboratories, Tehran, Iran, 21–24 February 2017.

51. Fleming, J.O.; Cook, T.D. Multiple sclerosis and the hygiene hypothesis. *Neurology* **2006**, *67*, 2085–2086. [CrossRef]

52. La Flamme, A.C.; Ruddenklau, K.; Bäckström, B.T. Schistosomiasis decreases central nervous system inflammation and alters the progression of experimental autoimmune encephalomyelitis. *Infect. Immun.* **2003**, *71*, 4996–5004. [CrossRef]

53. Kaiser, C.; Pion, S.D.; Boussinesq, M. Case-control studies on the relationship between onchocerciasis and epilepsy: Systematic review and meta-analysis. *PLoS Negl. Trop. Dis.* **2013**, *7*, e2147. [CrossRef]

54. Quattrocchi, G.; Nicoletti, A.; Marin, B.; Bruno, E.; Druet-Cabanac, M.; Preux, P.-M. Toxocariasis and epilepsy: Systematic review and meta-analysis. *PLoS Negl. Trop. Dis.* **2012**, *6*, e1775. [CrossRef] [PubMed]

55. Luna, J.; Cicero, C.E.; Rateau, G.; Quattrocchi, G.; Marin, B.; Bruno, E.; Dalmay, F.; Druet-Cabanac, M.; Nicoletti, A.; Preux, P.-M. Updated evidence of the association between toxocariasis and epilepsy: Systematic review and meta-analysis. *PLoS Negl. Trop. Dis.* **2018**, *12*, e0006665. [CrossRef] [PubMed]

56. Strube, C.; Waindok, P.; Raulf, M.-K.; Springer, A. *Toxocara*-induced neural larva migrans (neurotoxocarosis) in rodent model hosts. *Adv. Parasitol.* **2020**, *109*, 189–218. [PubMed]

57. Burren, C. The distribution of *Toxocara* larvae in the central nervous system of the mouse. *Trans. R. Soc. Trop. Med. Hyg.* **1971**, *65*, 450–453. [CrossRef]

58. Sprent, J. On the migratory behavior of the larvae of various *Ascaris* species in white mice: II. Longevity. *J. Infect. Dis.* **1953**, *92*, 114–117. [CrossRef]

59. Cardillo, N.; Rosa, A.; Ribicich, M.; López, C.; Sommerfelt, I. Experimental infection with *Toxocara cati* in BALB/c mice, migratory behaviour and pathological changes. *Zoonoses Public Health* **2009**, *56*, 198–205. [CrossRef]

60. Janecek, E.; Beineke, A.; Schnieder, T.; Strube, C. Neurotoxocarosis: Marked preference of *Toxocara canis* for the cerebrum and *T. cati* for the cerebellum in the paratenic model host mouse. *Parasit. Vectors* **2014**, *7*, 194. [CrossRef]

61. Heuer, L.; Beyerbach, M.; Lühder, F.; Beineke, A.; Strube, C. Neurotoxocarosis alters myelin protein gene transcription and expression. *Parasitol. Res.* **2015**, *114*, 2175–2186. [CrossRef]

62. Fonseca, G.R.e.; Dos Santos, S.V.; Chieffi, P.P.; Paula, F.M.d.; Gryschek, R.C.B.; Lescano, S.A.Z. Experimental toxocariasis in BALB/c mice: Relationship between parasite inoculum and the IgG immune response. *Mem. Inst. Oswaldo Cruz* **2017**, *112*, 382–386. [CrossRef]

63. Epe, C.; Sabel, T.; Schnieder, T.; Stoye, M. The behavior and pathogenicity of *Toxacara canis* larvae in mice of different strains. *Parasitol. Res.* **1994**, *80*, 691–695. [CrossRef]

64. Nicoletti, A.; Gomez-Puerta, L.A.; Arroyo, G.; Bustos, J.; Gonzalez, A.E.; Garcia, H.H.; Peru, C.W.G.I. *Toxocara* brain infection in pigs is not associated with visible lesions on brain magnetic resonance imaging. *Am. J. Trop. Med. Hyg.* **2020**, *103*, 273–275. [CrossRef] [PubMed]

65. Nicoletti, A. Toxocariasis. In *Handbook of Clinical Neurology*; Elsevier: Amsterdam, The Netherlands, 2013; Volume 114, pp. 217–228.

66. Dolinsky, Z.S.; Hardy, C.A.; Burright, R.G.; Donovick, P.J. The progression of behavioral and pathological effects of the parasite *Toxocara canis* in the mouse. *Physiol. Behav.* **1985**, *35*, 33–42. [CrossRef]

67. Springer, A.; Heuer, L.; Janecek-Erfurth, E.; Beineke, A.; Strube, C. Histopathological characterization of *Toxocara canis*-and *T. cati*-induced neurotoxocarosis in the mouse model. *Parasitol. Res.* **2019**, *118*, 2591–2600. [CrossRef] [PubMed]
68. Othman, A.A.; Abdel-Aleem, G.A.; Saied, E.M.; Mayah, W.W.; Elatrash, A.M. Biochemical and immunopathological changes in experimental neurotoxocariasis. *Mol. Biochem. Parasitol.* **2010**, *172*, 1–8. [CrossRef]
69. Lassmann, H.; Bradl, M. Multiple sclerosis: Experimental models and reality. *Acta Neuropathol.* **2017**, *133*, 223–244. [CrossRef]

Publisher's Note: MDPI stays neutral with regard to jurisdictional claims in published maps and institutional affiliations.

pathogens

Article

First Evidence of Function for *Schistosoma japonicum riok-1* and RIOK-1

Mudassar N. Mughal [1,2], Qing Ye [1], Lu Zhao [1], Christoph G. Grevelding [2], Ying Li [1], Wenda Di [3], Xin He [1], Xuesong Li [1], Robin B. Gasser [4] and Min Hu [1,*]

1   State Key Laboratory of Agricultural Microbiology, College of Veterinary Medicine, Huazhong Agricultural University, Wuhan 430070, China; Mudassar.N.Mughal@vetmed.uni-giessen.de (M.N.M.); Qingye198173@163.com (Q.Y.); bentengzhilu@163.com (L.Z.); lyhappycool@163.com (Y.L.); anniehe1991@gmail.com (X.H.); xuesong.li84@gmail.com (X.L.)
2   Biomedical Research Center Seltersberg, Institute of Parasitology, Justus Liebig University Giessen, D-35392 Giessen, Germany; christoph.grevelding@vetmed.uni-giessen.de
3   College of Animal Science and Technology, Guangxi University, Nanning 530005, China; diwenda@gxu.edu.cn
4   Department of Veterinary Biosciences, Faculty of Veterinary and Agricultural Sciences, Melbourne Veterinary School, The University of Melbourne, Parkville, VIC 3010, Australia; robinbg@unimelb.edu.au
*   Correspondence: mhu@mail.hzau.edu.cn

Abstract: Protein kinases are known as key molecules that regulate many biological processes in animals. The right open reading frame protein kinase (*riok*) genes are known to be essential regulators in model organisms such as the free-living nematode *Caenorhabditis elegans*. However, very little is known about their function in parasitic trematodes (flukes). In the present study, we characterized the *riok-1* gene (*Sj-riok-1*) and the inferred protein (*Sj*-RIOK-1) in the parasitic blood fluke, *Schistosoma japonicum*. We gained a first insight into function of this gene/protein through double-stranded RNA interference (RNAi) and chemical inhibition. RNAi significantly reduced *Sj-riok-1* transcription in both female and male worms compared with untreated control worms, and subtle morphological alterations were detected in the ovaries of female worms. Chemical knockdown of *Sj*-RIOK-1 with toyocamycin (a specific RIOK-1 inhibitor/probe) caused a substantial reduction in worm viability and a major accumulation of mature oocytes in the seminal receptacle (female worms), and of spermatozoa in the sperm vesicle (male worms). These phenotypic alterations indicate that the function of *Sj-riok-1* is linked to developmental and/or reproductive processes in *S. japonicum*.

Keywords: schistosomiasis; *Schistosoma japonicum*; right open reading frame protein kinase (*riok*) genes; *riok-1*; RIOK-1; double-stranded RNA interference (RNAi); chemical inhibition; toyocamycin; developmental and reproductive biology

Citation: Mughal, M.N.; Ye, Q.; Zhao, L.; Grevelding, C.G.; Li, Y.; Di, W.; He, X.; Li, X.; Gasser, R.B.; Hu, M. First Evidence of Function for *Schistosoma japonicum riok-1* and RIOK-1. *Pathogens* **2021**, *10*, 862. https://doi.org/10.3390/pathogens10070862

Academic Editor: Vito Colella

Received: 13 June 2021
Accepted: 5 July 2021
Published: 8 July 2021

Publisher's Note: MDPI stays neutral with regard to jurisdictional claims in published maps and institutional affiliations.

1. Introduction

In multicellular organisms, protein kinases (PKs) are encoded by large gene families, and regulate cellular processes, including DNA transcription, DNA replication, cell-cycle progression as well as metabolism [1,2]. PKs function to activate/inactivate proteins by catalyzing the transfer of phosphate groups to specific amino acid residues (i.e., Arg, His/Asp and Ser/Thr/Tyr) on their target proteins and, thus, play a regulatory role in many cell signaling pathways [1]. PKs can be classified as eukaryotic protein (ePKs), atypical protein (aPKs) kinases and protein kinase-like (PKL) [3]. For example, of >500 human PKs, <10% are PKL proteins, many of which were known as aPKs. There are 19 families of PKL kinases, one of which is called right open reading-frame kinases (RIOKs) [3].

Multicellular (metazoan) organisms usually have three *riok* genes (called *riok-1*, *riok-2*, and *riok-3*) [4]. However, it has been shown that flatworms (i.e., trematodes and cestodes) lack *riok-3* [5]. Structural information for RIOK proteins in metazoans is limited to the partially-solved crystal structure for RIOK-1 of humans [4], and functional domains of RIOK proteins have been modeled using three-dimensional (3D) structural modeling to assist in predicting and prioritizing kinase inhibitors that might target RIOK-1 of parasitic worms [6,7]. Functional information is available for metazoan model organisms, including the free-living nematode, *C. elegans*, for which investigations have demonstrated that *riok-1* and *riok-2* genes and their products play essential part in the process of ribosome biosynthesis, cell cycle progression, and/or chromosome stability [8–10], whereas *riok-3* has received limited attention, likely because it is not an essential gene [11–13].

There is scant information on the structure and function of *riok-1* and *riok-2* genes of parasitic flatworms, such as representative of the genus *Schistosoma* (blood flukes)—which are dioecious trematodes. Key species include *Schistosoma japonicum, S. haematobium*, and *S. mansoni*, which cause the neglected tropical disease (NTD) complex "schistosomiasis", affecting ~200 million people worldwide [14,15]. In parts of Asia, *S. japonicum* is particularly important because it can be transmitted (via the snail intermediate host) from water buffaloes, cattle, pigs, dogs, or rats to humans [16]. In the human host, infection is initiated by cercariae, free-living larvae released from the snail intermediate host; upon contact with water, cercariae penetrate the skin, transform into schistosomula, enter the blood circulation, and reach the hepatic portal and mesenteric veins, where the adult female and male worms pair up and reproduce. Eggs produced by female worms either penetrate the vessel walls and enter the intestinal lumen, being released via feces into the environment, or are passively transported via blood to the liver and spleen where they become entrapped, inducing the formation of granulomata [15].

Patients suffering from schistosomiasis are usually treated with praziquantel (PZQ), as no anti-schistosome vaccine is available. However, due to the widespread and regular use of PZQ in mass treatment programs, there is major concern that schistosomes develop resistance to this compound [17]. Moreover, PZQ fails to affect the juvenile stage of the parasite [18,19] and does not prevent reinfection [17]. These issues motivate efforts to functionally characterize essential genes or their products, particularly those involved in growth, developmental, and reproductive processes in schistosomes [20–22], in search for new interventions. In this context, we explored the *riok-1* gene (*Sj-riok-1*) and the function of RIOK-1 (*Sj*-RIOK-1) in *S. japonicum*, and provided the first evidence that this gene is involved in reproductive processes.

2. Results

2.1. Sj-riok-1 Encodes a Protein with Features Characteristic of RIOK-1

The genomic DNA sequence of *Sj-riok-1* was assembled using data from Worm-Base (PRJEA34885, scaffold SJC_S002310). The sequence was 6761 bp in length, and it had three exons which were 139, 488, and 642 bp long, respectively, and two introns (3036 and 2456 bp). The coding region of *Sj-riok-1* is 1269 bp long, representing 422 amino acids. A comparison of the inferred amino acid sequence (*Sj*-RIOK-1) with select RIOK-1 orthologs/homologs revealed high sequence identities with those from congeners *S. haematobium* (76.7%) and *S. mansoni* (84.8%), and a lower identity (56.4%) to that from *Clonorchis sinensis*—a carcinogenic liver fluke (Figure 1a). A phylogenetic analysis showed that the sequences of all trematodes formed a clade (with absolute nodal support) to the exclusion of orthologs from other invertebrate and vertebrate species (Figure 1b).

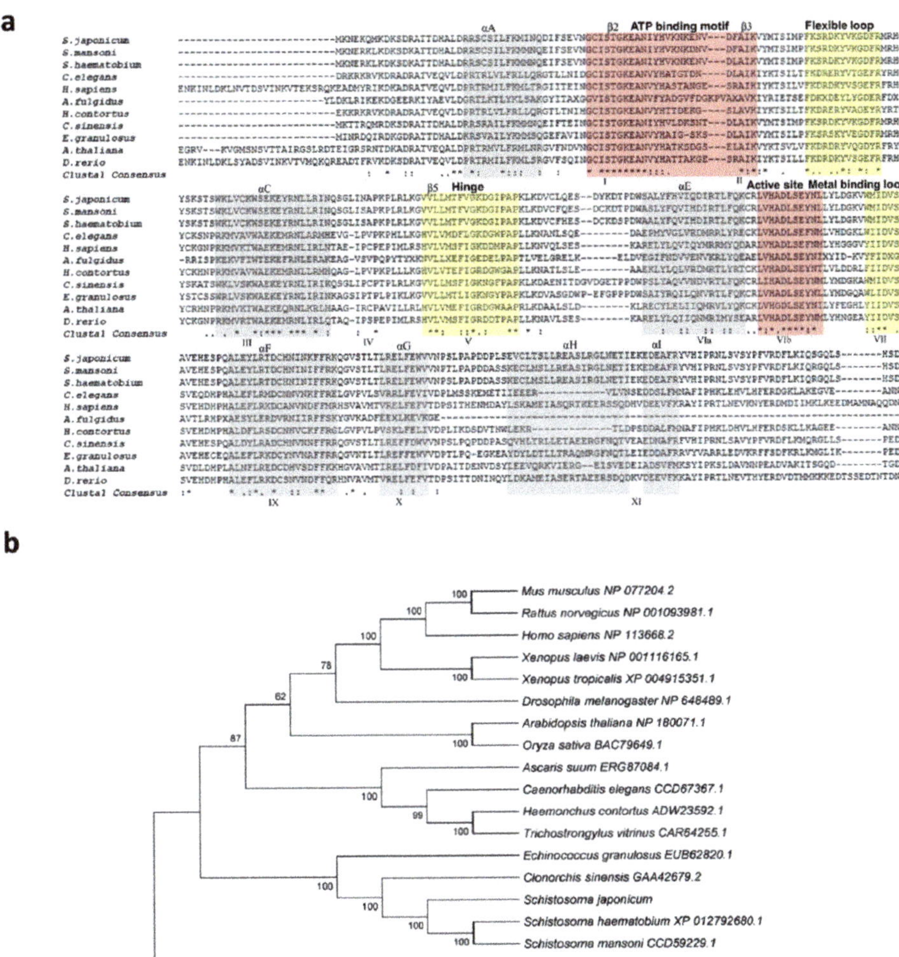

Figure 1. Comparison of *Schistosoma japonicum* Sj-RIOK-1 with RIOK-1s of other organisms. (**a**) Multiple sequence alignment among predicted RIOK-1 proteins from 11 organisms (*Schistosoma japonicum*, *Schistosoma mansoni*, *Schistosoma haematobium*, *Caenorhabditis elegans*, *Homo sapiens*, *Archaeoglobus fulgidus*, *Haemonchus contortus*, *Clonorchis sinensis*, *Echinococcus granulosus*, *Arabidopsis thaliana*, and *Danio rerio*). Alpha helices A–I or beta-sheet structures are colored light gray and labeled above the alignment. The subdomains I–XI are marked below the alignment. Functional domains, including the ATP binding motif and active site (red), flexible loop, hinge, and metal binding loop (yellow), are highlighted and marked above the alignment. Identical residue (*); high similarity (:); limited similarity (.); no similarity (no symbol). (**b**) The neighbor-joining (NJ) tree of RIOK-1 amino acid sequences from a range of organisms. The RIOK-1 of *Saccharomyces cerevisiae* (CAA99317.1) was used as an outgroup, and the bootstrap values are given above or below the branches. GenBank accession numbers are listed besides the species name. Accession numbers and related references of RIOK-1 amino acid sequences used for the multiple alignments and phylogeny are given in Table 1.

2.2. Transcription in Different Developmental Stages

Sj-riok-1 was found to be transcribed throughout all developmental stages and sexes assessed (Figure 2a), with transcription being highest in adult females and schistosomules from the lung. Transcript levels of *Sj-riok-1* in single, cultured females for 3 days was 7.2-fold ($F_{(1,8)}$ = 7.578, p = 0.0295; $t_{(6)}$ = 3.866, $p < 0.01$) higher than in couples cultured for the same time (Figure 2b). There was a significant reduction of *Sj-riok-1* transcript levels following re-pairing of females and males in vitro (6 days) which had been separated for 3 days prior to re-paring (Figure 2c; $t_{(6)} = 2.255$, $p < 0.05$ for females; $t_{(6)} = 4.910$, $p < 0.001$ for males). The results suggest that *Sj-riok-1* is involved in pairing-dependent developmental and/or reproductive processes.

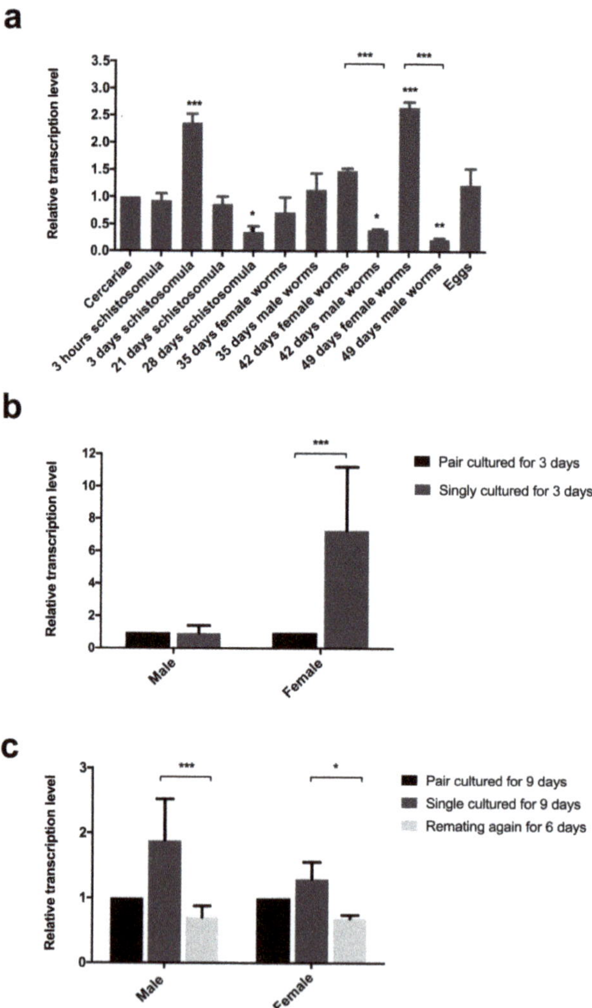

Figure 2. Relative transcription levels of *Sj-riok-1* in various developmental stages, and the effect of pairing on the transcription of this gene in adult males and females of *Schistosoma japonicum*. (**a**) Transcriptional level of *Sj-riok-1* in various developmental stages of *S. japonicum* (female worms from mixed cultures were used and the data normalized relative to the cercarial stage). (**b,c**) The effect of pairing on the transcription of *Sj-riok-1* in adult males and females. Data for separated and re-paired worms were normalized relative to the paired worms. Data given are representatives of the mean ± SD of three independent experiments, and statistically significant differences are indicated as * ($p < 0.05$), ** ($p < 0.01$), and *** ($p < 0.001$).

2.3. Significant Knockdown of Transcription by RNAi in Both Sexes, and Subtle Morphological Change in the Ovary

As morphological changes are seldomly seen in adult parasitic helminths following short-term RNAi by soaking [23], we elected to assess the specific reduction in gene transcription as the phenotype. Since the *Sj-riok-1* transcript level was highest in the adult stage during the sexual life-phase of *S. japonicum* (Figure 2a), we used paired adult worms for RNAi for a period of 14 days. Results showed that *Sj-riok-1* mRNA transcription was significantly and consistently reduced (Figure 3a) by 75–84% ($F_{(2,12)}$ = 90.24, $p < 0.0001$) in both male and female worms (35- or 28-days old, do) as compared with controls (i.e., worms treated with an irrelevant-dsRNA or without dsRNA) (Figure 3a,b). Although subtle, a significant decrease in ovary size (reduced length and width) was seen in treated worms (n = 28) compared with control worms (n = 28) (Figure 3c–f).

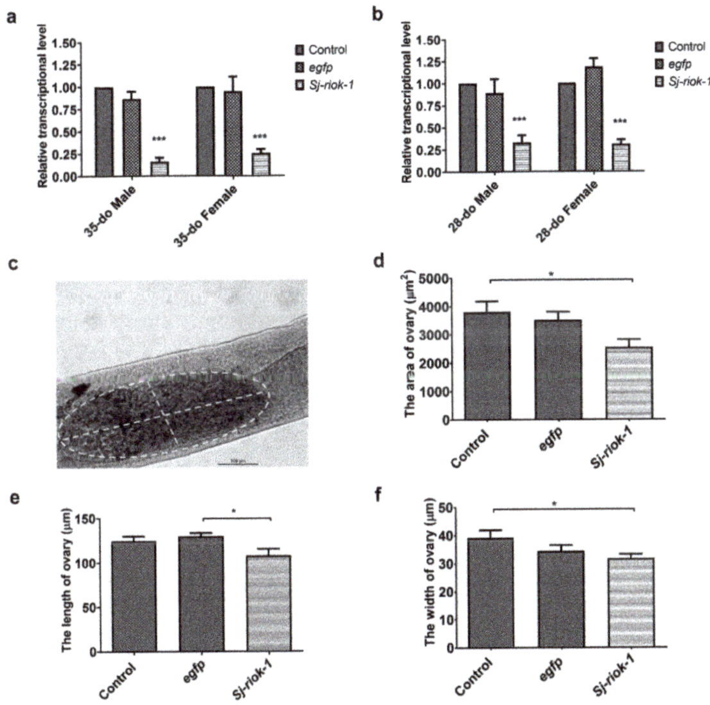

Figure 3. RNAi-mediated knockdown of *Sj-riok-1* gene decreased transcription in 35- or 28-do adult male and female worms and the ovarian dimension of the 35-do adult females of *Schistosoma japonicum*. (**a,b**) Relative *Sj-riok-1* transcript levels determined by qRT-PCR in 35- or 28-do worms treated with *Sj-riok-1* dsRNA, non-specific (egfp) dsRNA, and no dsRNA (control). (**c–f**) The area, length, and width of the ovary in 35-do females after 2 weeks of treatment with *Sj-riok-1* dsRNA or irrelevant (egfp) dsRNA, including a no-dsRNA (control) were measured using a bright-field microscope. Data are given as the mean ± SD of three independent experiments with 8–10 females in each experiment. * indicates $p < 0.05$; *** indicates $p < 0.001$.

2.4. Toyocamycin Affects Viability and Induces Pathological Changes in the Reproductive Tracts

Toyocamycin is a competitive, small molecule inhibitor of RIOK1 kinase activity [24]. In vitro treatment of paired adult worms (35- or 28-do) with toyocamycin (1 µM) significantly affected their viability; all worms died after 4–5 days (Figure 4a), whereas untreated controls lived for 12–14 days in culture. Worm pairs started to separate at 24 h, were completely separated at 48 h, and then curled and had a marked swelling in the gut at

$\geq$72 h. At this time point, worm motility, gut peristalsis, and egg production (females) ceased, as compared with untreated controls (Figure 4a–d). Toyocamycin treatment at a lower concentration (0.5 μM) had a similar, but a less intense effect compared with 1 μM (Figure 4a–d). At 1 μM, the effect of toyocamycin on viability ($F_{(2,24)}$ = 246.2, $p < 0.0001$) and egg production ($F_{(2,44)}$ = 71.66, $p < 0.0001$) of 28-do worms (Figure 4c,d) was greater than on older worms (35-do) (viability: $F_{(2,24)}$ = 163.1, $p < 0.0001$; egg production: $F_{(2,24)}$ = 86.42, $p < 0.0001$) (Figure 4a,b).

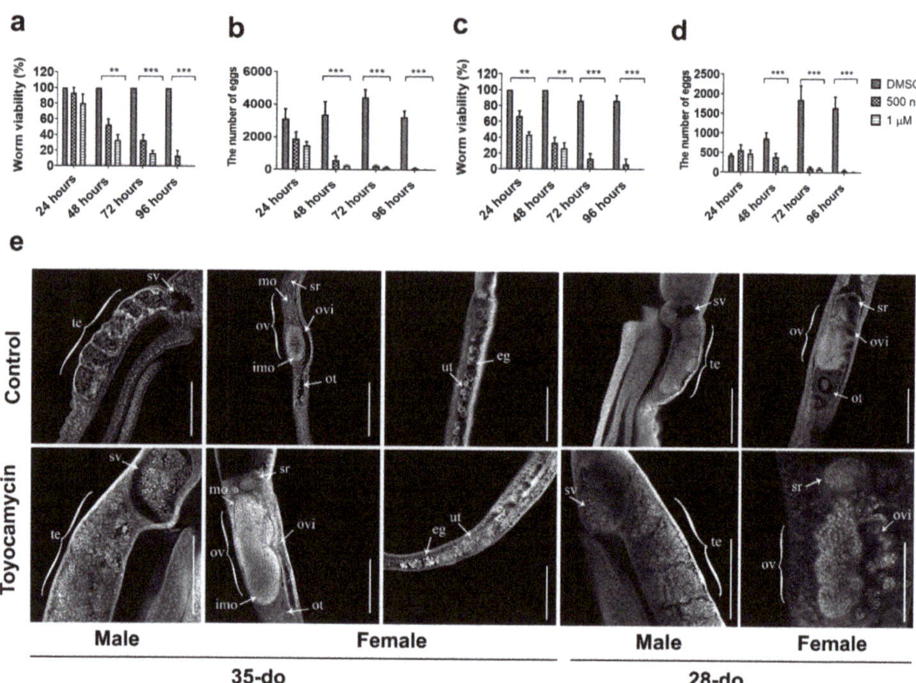

Figure 4. The effect of toyocamycin on viability and egg production of 35- or 28-do adult worms of *Schistosoma japonicum* and CLSM (confocal laser scanning microscopy) study of gonads of 35- or 28-do adult worms after toyocamycin treatment. (**a**) Effect of toyocamycin treatment between 24 and 96 h on viability of 35-do worms. (**b**) Effect of toyocamycin treatment between 24 and 96 h on egg production in 35-do worms. (**c**) Effect of toyocamycin treatment between 24 and 96 h on viability in 28-do worms. (**d**) Effect of toyocamycin treatment between 24 and 96 h on egg production in 28-do worms. (**e**) CLSM analyses of paired worms (35- or 28-do) incubated with DMSO (control) or 1 μM toyocamycin for 96 h. Abbreviations: te, testes; sv, sperm vesicle; ov, ovary; mo, mature oocytes; imo, immature oocytes; ovi, oviduct; ot, ootype; ut, uterus; sr, seminal receptacle; eg, egg. Scale bars: 200 μm. Data are representatives of the mean ± SD of three independent experiments. ** indicates $p < 0.01$; *** indicates $p < 0.001$.

Although not detected at 24, 48, or 72 h, some morphological changes were evident in the reproductive tracts at 96 h (Figure 4e). The changes initially detected at 96 h in the reproductive tracts of female and male worms exposed to 1 μM toyocamycin (Figure 4) were investigated in more detail. Toyocamycin treatment of paired adult worms (28- and 35-do) for 96 h led to a cessation of egg-release in female worms (Figure 4b,d). CLSM examination of gonads revealed a significant accumulation of mature oocytes in the oviduct close to the seminal receptacle of females and of spermatozoa in the sperm vesicle of males (28- and 35-do, Figure 4e). An accumulation of oocytes in the uterus was also seen in toyocamycin-treated 35-do females (Figure 4e).

3. Discussion

This study showed a significant and (relatively) consistent reduction in *Sj-riok-1* transcript levels in both female and male *S. japonicum* using RNAi compared with well-defined control worms (worms treated with dsRNA from an irrelevant-gene, and untreated worms) as well as subtle ovarian alterations. Chemical knockdown of *Sj*-RIOK-1 with toyocamycin—a competitive inhibitor and biochemical probe [24]—led to a substantial reduction in worm viability and pathological changes in the reproductive tracts, including a significant accumulation of mature oocytes in the area of the seminal receptacle that is part of the oviduct in females, and of spermatozoa in the sperm vesicle in males. These alterations suggest that *Sj-riok-1* is linked to developmental and/or reproductive processes in *S. japonicum*.

The functional genomic studies revealed that RIOK-1 of *Strongyloides stercoralis* is essential for the development and survival of *S. stercoralis* larvae [25,26], suggesting that RIOK-1 might be an anthelmintic target. Published studies show that RIO kinases homologs are receiving attention as possible targets for the development of anti-cancer treatments and anti-infectives [5–7,13,27]. This applies particularly for *riok-1* and *riok-2*, for which genetic manipulation studies have indicated their involvement in fundamental biological mechanisms such as ribosomal biosynthesis, chromosome stability, and/or cell cycle advancement [13,28–37]. Breugelmans et al. [5] showed that the *riok-1* and *riok-2* genes were present and transcribed in ≥ 52 species of metazoans, including 25 flatworms species (together with trematodes and cestodes). We observed high similarity in sequence and gene structures among RIOK-1 homologs of four trematode and four cestode species (with exon numbers ranging from 3 to 7, and gene lengths ranging from 1618 to 14,971 bp) than reported for nine nematode species [7], which is consistent with previous results [5], indicating relative conservation in *riok* structure in flatworms. For instance, the conserved RIOK-1 signature sequence "S-T-G-K-E-A" in the ATP binding motif is analogous to the signature sequence "G-x-G-K-E-S" of RIOK-2. However, the active site of RIOK-1 "L-V-H-x-D-L-S-E-Y-N" is different to that of "I-H-x-D-o-N-E-F-N" in RIOK-2 [38]. As identified in our alignment, *Sj*-RIOK-1 has Ser165 as part of conserved dipeptide motif present within the flexible loop capable of phosphorylation and autophosphorylation, indicating that *Sj*-RIOK-1 shares common features with the RIOK-1 family.

This apparent conservation of RIOK-1 for flatworms and distinctiveness from orthologs in mammals and other eukaryote groups (Figure 1b) suggests that this PKL kinase, or elements thereof, may represent a selective target. As most current kinase inhibitors for therapeutic use [39] target nucleotide binding sites, future work could focus on assessing the selectivity of these sites between schistosomes (and other flatworms) and mammalian hosts. However, it needs to be considered that the topologies of these binding domains could be similar in kinases other than RIOK, so that challenges regarding selectivity might arise when designing inhibitors specifically against a pathogen's RIOK-1. Selectivity not only relates to a discrepancy between pathogen and host RIOK sites, but probably also nucleotide binding domains in any of the many other kinases (n = 500) in the kinome of the human host. Therefore, it would be useful to critically appraise the nature and extent of evolutionary diversity in the nucleotide binding domains in RIOKs between flatworms and the principal mammalian host (human). A promising candidate might be the residue at position L289 in the sequence of human RIOK-1, which relates to P197 in *Schistosoma* species (*S. haematobium*). This distinction in the nucleotide-binding domain between these trematodes and the host could be relevant for designing ligands that particularly target RIOK-1 proteins of *S. japonicum* and its congeners, although such work should be assisted by investigating the crystal structures of these flukes.

As PZQ (a pyrazinoisoquinoline) is used to treat schistosomiasis of humans, this compound is not effective against all developmental stages of schistosomes [40,41]. Therefore, there is an urgency to work toward novel and improved drug targets against flatworms build on deep insights of the molecular biology and development of these worms. In this perspective, it is critical to target gene products that are transcribed or expressed in suitably "druggable" developmental stages of these worms, as is true of RIOKs. *S. japonicum* appears to transcribe *riok-1* in all developmental stages and both sexes, which is similar to findings for *S. haematobium* [42] and *S. mansoni* [20,43], and also other parasitic worms including *Ascaris suum*, *Brugia malayi*, and *H. contortus* [7]. Such constitutive transcription for *riok-1* in parasitic trematodes and nematodes (studied to date) indicates that RIOK-1 performs essential house-keeping functions pertaining to development and reproduction, which concords with essential roles in ribosome biosynthesis and cell-cycle progression in the well-studied model organisms *C. elegans* and *D. melanogaster* [8,13].

4. Materials and Methods

4.1. Procurement of the Parasite

The collection, processing, and storage of the different developmental stages (i.e., cercariae, skin-stage and lung-stage schistosomula, juvenile and adult-stage male and female worms, and eggs) of *Schistosoma japonicum* were conducted using established protocols [44].

4.2. Cloning of Sj-Riok-1 cDNA from S. japonicum, and Informatic Analyses

Total 3′-end cDNA (using *Sj-riok-1*-specific internal primers Sj-Riok1-F and Sj-Riok1-R, designed from expressed sequence tag (EST) sequence—GenBank accession no. AY810901.1, Supplemental Table S1) was prepared as previously described [44]. After cloning and sequencing, the sequence amplified by 3′ RACE-PCR was merged to the known sequence region to generate a full-length *Sj-riok-1* sequence, which was subsequently used in designing two additional primers Sjriok1-ORF-F and Sjriok1-ORF-R (Supplemental Table S1) for obtaining the full-length *Sj-riok-1* coding sequence (accession no. MN335243) using the conditions described previously [44].

The *Sj*-RIOK-1 protein sequence was inferred and its protein domains, motifs, and/or functional sites inferred using PROSITE [45] and Pfam [46]. RIOK-1 homologs representing 20 species other than *S. japonicum* (Table 1) were extracted from GenBank for sequence alignment with *Sj*-RIOK-1 and phylogenetic analysis employing the neighbor joining (NJ), maximum likelihood (ML), and maximum parsimony (MP) methods in MEGA v.5.0 using the same parameters as previously described [44].

Table 1. Accession numbers of RIOK-1 genes used for the multiple alignments and phylogeny in Figure 1. The RIOK-1 of the yeast *Saccharomyces cerevisiae* (CAA99317.1) was used as outgroup.

Species	Accession Numbers	References
Saccharomyces cerevisiae [1]	CAA99317.1	[47]
Xenopus laevis	NP_001116165.1	[48]
Xenopus tropicalis	XP_004915351.1	[49]
Homo sapiens [2]	NP_113668.2	[50]
Archaeoglobus fulgidus [2]	NP_73535983	[51]
Arabidopsis thaliana [2]	NP_180071.1	[52]
Danio rerio [2]	NP_998160.1	[53]
Rattus norvegicus	NP_001092981.1	[54]
Mus musculus	NP_077204.2	[55]
Oryza sativa	BAC79649.1	[56]
Arabidopsis thaliana [2]	NP_180071.1	[52]
Drosophila melanogaster	NP_648489.1	[57]
Ascaris suum	ERG87084.1	[58]

Table 1. *Cont.*

Species	Accession Numbers	References
Trichostrongylus vitrinus	CAR64255.1	[59]
Haemonchus contortus [2]	ADW23592.1	[6]
Caenorhabditis elegans [2]	CCD67367.1	[8]
Clonorchis sinensis [2]	GAA42679.2	[60]
Echinococcus granulosus [2]	EUB62820.1	[61]
Schistosoma haematobium [2]	XP_012792680.1	[42]
Schistosoma mansoni [2]	CCD59229.1	[62]

[1] Sequence used as an outgroup in phylogenetic analysis. [2] Sequence used in the alignment.

4.3. qPCR to Assess Transcript Levels of Different Developmental Stages and from In Vitro Paring Experiment

Quantitative real-time PCR (qPCR) was performed to assess transcript levels in distinct developmental stages of *S. japonicum* as described previously [44] using the primer pair Riok1-qPCR-F/Riok1-qPCR-R for *Sj-riok-1* and the primer pair β-Tubulin-qPCR-F/β-Tubulin-qPCR-R for the *β-tubulin* gene (accession no. AY220457.2) as the reference (Supplemental Table S1). The $2^{-\Delta\Delta Ct}$ method [63] was used for relative quantification. The cercarial stage was used as the calibration standard, and data were presented as the mean ± standard deviation.

Adults *S. japonicum* collected from infected mice (42 day infection) were in vitro cultured in the same medium and under the same conditions as described previously [11]. In some experiments, worms were cultured as pairs ($n = 10$) or as individuals ($n = 10$ for each sex) for 3- or 9-day periods. In other experiments, pairs ($n = 10$ each) were cultured for 3 days, separated for 3 days and re-paired for 6 days. qPCR was used to assess *Sj-riok-1* transcription in individual worms or pairs, employing in vitro cultured individual female and male worms from pairs as calibration standard.

4.4. Double-Stranded RNA Interference (RNAi)

Sj-riok-1 (1155 bp, including the RIOK-1 domain) and *egfp* (620 bp) cDNAs were amplified by PCR using primer pair dsRNA-riok1-F/dsRNA-riok1-R and dsRNA-egfp-F/dsRNA-egfp-R (Supplemental Table S1), respectively, employing the following conditions: 94 °C for 3 min, then 35 cycles at 94 °C for 40 s, 60 °C (for *Sj-riok-1*) or 62 °C (for *egfp*) for 40 s and 72 °C for 2 min, and final extension step at 72 °C for 10 min. These two cDNAs were each cloned into pMD-19T and their identities confirmed by sequencing. Each of the plasmids was used in the production of dsRNA which was subsequently employed in RNAi using an established soaking method [44]. The transcript levels of dsRNA-treated worms were assessed by qPCR (using untreated females as the calibrator/reference), and worms were microscopically examined for morphological alterations.

4.5. Treatment with Toyocamycin

Adult *S. japonicum* (couples) were perfused from mice and maintained in vitro at 37 °C for 96 h in a 5% CO_2 atmosphere in the culture medium (same as used in the paring experiment) which contains the inhibitor toyocamycin (APExBIO®, Houston, TX, USA) dissolved in dimethyl sulfoxide (DMSO) and was refreshed daily. In preliminary experiments, both 28- and 35-do couples were treated with varying concentrations (100, 300, 500 nM, and 1 μM) of toyocamycin in vitro for 96 h (Supplementary Figure S1). Control groups were incubated with equal volume of DMSO-only under the same conditions. IC50 concentrations and egg count as well as worm viability were determined every 24 h (Supplementary Figure S1). Worm viability was scored as recommended by WHO-TDR [64] normal motility (3), reduced motility (2), minimal motility (1), and no movement/dead (0).

4.6. Microscopic Examination of Worms and Statistical Analyses

Worms were examined by bright field microscopy (BFM) and CLSM as previously described [44]. The length, width, and the area of the ovaries were measured by BFM. Data are representative of the mean $\pm$ SD of at least three independent experiments. Statistically significant differences were analyzed by GraphPad Prism v.5.01 (San Diego, CA, USA). A two-way ANOVA with Bonferroni's multiple comparison test was used for analyzing worm viability and egg production of the different groups including the re-paired worms. One-way ANOVA with multiple comparison of the Tukey-test was used for the other experiments. Values of $p \leq 0.05$ were considered as statistically significant.

5. Conclusions

Our findings represent first evidence for the pairing-dependent influence of *Sj-riok-1* in the development and reproductive maturation of adult female *S. japonicum*. This study contributes towards the understanding of the reproductive processes in *S. japonicum* and suggests RIOK-1 as a potential target as selective anthelmintic therapeutic approach.

Supplementary Materials: The following are available online at https://www.mdpi.com/article/10.3390/pathogens10070862/s1, Table S1: DNA sequences of oligonucleotide primers used in the present study. Figure S1: Preliminary data of toyocamycin on viability and egg production of 35- and 28-do adult worms of *Schistosoma japonicum*.

Author Contributions: Conceptualization, M.H., C.G.G. and R.B.G.; methodology, M.N.M., Q.Y. and Y.L.; validation, M.N.M. and M.H.; formal analysis, M.N.M., L.Z. and W.D.; resources, X.H. and X.L.; data curation, M.N.M.; writing—original draft preparation, M.N.M. and R.B.G.; writing—review and editing, M.N.M., M.H., C.G.G. and R.B.G.; supervision, M.H.; project administration, M.H.; funding acquisition, M.H. All authors have read and agreed to the published version of the manuscript.

Funding: Funding support was obtained from grants from the National Key Basic Research Program (973 program) of China (no. 2015CB150300) and the National Natural Science Foundation of China (NSFC) (no. 31872462) (M.H.). Funding from the Australian Research Council (ARC) and the National Health and Medical Research Council of Australia (NHMRC) is gratefully acknowledged (RBG). Mudassar Niaz Mughal was a recipient of a Chinese Government Scholarship from China Scholarship Council (CSC). Funding of the LOEWE CenterDRUID, which is part of the excellence initiative of the Hessian Ministry of Science, Higher Education and Art (HMWK), is gratefully acknowledged (C.G.G.).

Institutional Review Board Statement: Animal experimentation and associated procedures were approved by the Animal Ethics Committee of Huazhong Agricultural University (ethics ID HZAUMO-2015-028, 28 September 2015) according to the Regulations for the Administration of Affairs Concerning Experimental Animals of Hubei Province, China.

Informed Consent Statement: Not applicable.

Conflicts of Interest: The authors declare no conflict of interest. The funders had no role in the design of the study; in the collection, analyses, or interpretation of data; in the writing of the manuscript; or in the decision to publish the results.

References

1. Hanks, S.K.; Quinn, A.M.; Hunter, T. The protein kinase family: Conserved features and deduced phylogeny of the catalytic domains. *Science* **1988**, *241*, 42–52. [CrossRef]
2. Rauch, J.; Volinsky, N.; Romano, D.; Kolch, W. The secret life of kinases: Functions beyond catalysis. *Cell Commun. Signal.* **2011**, *9*, 23. [CrossRef] [PubMed]
3. Manning, G.; Whyte, D.B.; Martinez, R.; Hunter, T.; Sudarsanam, S. The protein kinase complement of the human genome. *Science* **2002**, *298*, 1912–1934. [CrossRef]
4. LaRonde-LeBlanc, N.; Wlodawer, A. The RIO kinases: An atypical protein kinase family required for ribosome biogenesis and cell cycle progression. *Biochim. Biophys. Acta* **2005**, *1754*, 14–24. [CrossRef] [PubMed]
5. Breugelmans, B.; Ansell, B.R.E.; Young, N.D.; Amani, P.; Stroehlein, A.J.; Sternberg, P.W.; Jex, A.R.; Boag, P.R.; Hofmann, A.; Gasser, R.B. Flatworms have lost the right open reading frame kinase 3 gene during evolution. *Sci. Rep.* **2015**, *5*, 9417. [CrossRef]

6. Campbell, B.E.; Boag, P.R.; Hofmann, A.; Cantacessi, C.; Wang, C.K.; Taylor, P.; Hu, M.; Sindhu, Z.-U.-D.; Loukas, A.; Sternberg, P.W.; et al. Atypical (RIO) protein kinases from *Haemonchus contortus*—Promise as new targets for nematocidal drugs. *Biotechnol. Adv.* **2011**, *29*, 338–350. [CrossRef]

7. Breugelmans, B.; Jex, A.R.; Korhonen, P.K.; Mangiola, S.; Young, N.D.; Sternberg, P.W.; Boag, P.R.; Hofmann, A.; Gasser, R.B. Bioinformatic exploration of RIO protein kinases of parasitic and free-living nematodes. *Int. J. Parasitol.* **2014**, *44*, 827–836. [CrossRef] [PubMed]

8. Mendes, T.K.; Novakovic, S.; Raymant, G.; Bertram, S.E.; Esmaillie, R.; Nadarajan, S.; Breugelmans, B.; Hofmann, A.; Gasser, R.B.; Colaiácovo, M.P.; et al. Investigating the role of RIO protein kinases in *Caenorhabditis elegans*. *PLoS ONE* **2015**, *10*, e0117444. [CrossRef]

9. Ashrafi, K.; Chang, F.Y.; Watts, J.L.; Fraser, A.G.; Kamath, R.S.; Ahringer, J.; Ruvkun, G. Genome-wide RNAi analysis of *Caenorhabditis elegans* fat regulatory genes. *Nature* **2003**, *421*, 268–272. [CrossRef]

10. Sönnichsen, B.; Koski, L.B.; Walsh, A.; Marschall, P.; Neumann, B.; Brehm, M.; Alleaume, A.-M.; Artelt, J.; Bettencourt, P.; Cassin, E.; et al. Full-genome RNAi profiling of early embryogenesis in *Caenorhabditis elegans*. *Nature* **2005**, *434*, 462–469. [CrossRef] [PubMed]

11. Shan, J.; Wang, P.; Zhou, J.; Wu, D.; Shi, H.; Huo, K. RIOK3 interacts with caspase-10 and negatively regulates the NF-κB signaling pathway. *Mol. Cell. Biochem.* **2009**, *332*, 113–120. [CrossRef]

12. Baumas, K.; Soudet, J.; Caizergues-Ferrer, M.; Faubladier, M.; Henry, Y.; Mougin, A. Human RioK3 is a novel component of cytoplasmic pre-40S pre-ribosomal particles. *RNA Biol.* **2012**, *9*, 162–174. [CrossRef] [PubMed]

13. Read, R.D.; Fenton, T.R.; Gomez, G.G.; Wykosky, J.; Vandenberg, S.R.; Babic, I.; Iwanami, A.; Yang, H.; Cavenee, W.K.; Mischel, P.S.; et al. A kinome-wide RNAi screen in Drosophila glia reveals that the RIO kinases mediate cell proliferation and survival through TORC2-Akt signaling in glioblastoma. *PLoS Genet.* **2013**, *9*, e1003253. [CrossRef]

14. Rollinson, D.; Knopp, S.; Levitz, S.; Stothard, J.R.; Tchuem Tchuenté, L.-A.; Garba, A.; Mohammed, K.A.; Schur, N.; Person, B.; Colley, D.G.; et al. Schistosomiasis: Number of people treated worldwide in 2013. *Relev. Epidemiol. Hebd.* **2015**, *90*, 25–32. [CrossRef]

15. McManus, D.P.; Dunne, D.W.; Sacko, M.; Utzinger, J.; Vennervald, B.J.; Zhou, X.-N. Schistosomiasis. *Nat. Rev. Dis. Prim.* **2018**, *4*, 13. [CrossRef] [PubMed]

16. Colley, D.G.; Bustinduy, A.L.; Secor, W.E.; King, C.H. Human schistosomiasis. *Lancet* **2014**, *383*, 2253–2264. [CrossRef]

17. Cioli, D.; Pica-Mattoccia, L.; Basso, A.; Guidi, A. Schistosomiasis control: Praziquantel forever? *Mol. Biochem. Parasitol.* **2014**, *195*, 23–29. [CrossRef]

18. Xiao, S.H.; Catto, B.A.; Webster, L.T. Effects of praziquantel on different developmental stages of *Schistosoma mansoni* in vitro and in vivo. *J. Infect. Dis.* **1985**, *151*, 1130–1137. [CrossRef]

19. Pica-Mattoccia, L.; Cioli, D. Sex- and stage-related sensitivity of *Schistosoma mansoni* to in vivo and in vitro praziquantel treatment. *Int. J. Parasitol.* **2004**, *34*, 527–533. [CrossRef]

20. Grevelding, C.G.; Langner, S.; Dissous, C. Kinases: Molecular stage directors for *Schistosome* development and differentiation. *Trends Parasitol.* **2018**, *34*, 246–260. [CrossRef]

21. de Andrade, L.F.; Mourão, M.d.M.; Geraldo, J.A.; Coelho, F.S.; Silva, L.L.; Neves, R.H.; Volpini, A.; Machado-Silva, J.R.; Araujo, N.; Nacif-Pimenta, R.; et al. Regulation of *Schistosoma mansoni* development and reproduction by the mitogen-activated protein kinase signaling pathway. *PLoS Negl. Trop. Dis.* **2014**, *8*, e2949. [CrossRef] [PubMed]

22. Walker, A.J.; Ressurreição, M.; Rothermel, R. Exploring the function of protein kinases in schistosomes: Perspectives from the laboratory and from comparative genomics. *Front. Genet.* **2014**, *5*. [CrossRef]

23. Beckmann, S.; Quack, T.; Burmeister, C.; Buro, C.; Long, T.; Dissous, C.; Grevelding, C.G. *Schistosoma mansoni*: Signal transduction processes during the development of the reproductive organs. *Parasitology* **2010**, *137*, 497–520. [CrossRef]

24. Kiburu, I.N.; LaRonde-LeBlanc, N. Interaction of Rio1 kinase with Toyocamycin reveals a conformational switch that controls oligomeric state and catalytic activity. *PLoS ONE* **2012**, *7*, e37371. [CrossRef] [PubMed]

25. Yuan, W.; Lok, J.B.; Stoltzfus, J.D.; Gasser, R.B.; Fang, F.; Lei, W.-Q.; Fang, R.; Zhou, Y.-Q.; Zhao, J.-L.; Hu, M. Toward understanding the functional role of Ss-RIOK-1, a RIO protein kinase-encoding gene of *Strongyloides stercoralis*. *PLoS Negl. Trop. Dis.* **2014**, *8*, e3062. [CrossRef]

26. Yuan, W.; Zhou, H.; Lok, J.B.; Lei, W.; He, S.; Gasser, R.B.; Zhou, R.; Fang, R.; Zhou, Y.; Zhao, J.; et al. Functional genomic exploration reveals that Ss-RIOK-1 is essential for the development and survival of *Strongyloides stercoralis* larvae. *Int. J. Parasitol.* **2017**, *47*, 933–940. [CrossRef]

27. Nag, S.; Prasad, K.; Bhowmick, A.; Deshmukh, R.; Trivedi, V. PfRIO-2 kinase is a potential therapeutic target of antimalarial protein kinase inhibitors. *Curr. Drug Discov. Technol.* **2013**, *10*, 85–91. [CrossRef]

28. Vanrobays, E.; Gleizes, P.E.; Bousquet-Antonelli, C.; Noaillac-Depeyre, J.; Caizergues-Ferrer, M.; Gélugne, J.P. Processing of 20S pre-rRNA to 18S ribosomal RNA in yeast requires Rrp10p, an essential non-ribosomal cytoplasmic protein. *EMBO J.* **2001**, *20*, 4204–4213. [CrossRef] [PubMed]

29. Angermayr, M.; Roidl, A.; Bandlow, W. Yeast Rio1p is the founding member of a novel subfamily of protein serine kinases involved in the control of cell cycle progression. *Mol. Microbiol.* **2002**, *44*, 309–324. [CrossRef] [PubMed]

30. Granneman, S.; Petfalski, E.; Swiatkowska, A.; Tollervey, D. Cracking pre-40S ribosomal subunit structure by systematic analyses of RNA-protein cross-linking. *EMBO J.* **2010**, *29*, 2026–2036. [CrossRef]

31. Widmann, B.; Wandrey, F.; Badertscher, L.; Wyler, E.; Pfannstiel, J.; Zemp, I.; Kutay, U. The kinase activity of human Rio1 is required for final steps of cytoplasmic maturation of 40S subunits. *Mol. Biol. Cell* **2012**, *23*, 22–35. [CrossRef]

32. Geerlings, T.H.; Faber, A.W.; Bister, M.D.; Vos, J.C.; Raué, H.A. Rio2p, an evolutionarily conserved, low abundant protein kinase essential for processing of 20 S pre-rRna in *Saccharomyces cerevisiae*. *J. Biol. Chem.* **2003**, *278*, 22537–22545. [CrossRef]

33. Ceron, J.; Rual, J.-F.; Chandra, A.; Dupuy, D.; Vidal, M.; van den Heuvel, S. Large-scale RNAi screens identify novel genes that interact with the *C. elegans* retinoblastoma pathway as well as splicing-related components with synMuv B activity. *BMC Dev. Biol.* **2007**, *7*, 30. [CrossRef]

34. Simpson, K.J.; Selfors, L.M.; Bui, J.; Reynolds, A.; Leake, D.; Khvorova, A.; Brugge, J.S. Identification of genes that regulate epithelial cell migration using an siRNA screening approach. *Nat. Cell Biol.* **2008**, *10*, 1027–1038. [CrossRef]

35. Strunk, B.S.; Loucks, C.R.; Su, M.; Vashisth, H.; Cheng, S.; Schilling, J.; Brooks, C.L.; Karbstein, K.; Skiniotis, G. Ribosome assembly factors prevent premature translation initiation by 40S assembly intermediates. *Science* **2011**, *333*, 1449–1453. [CrossRef]

36. Esser, D.; Siebers, B. Atypical protein kinases of the RIO family in archaea. *Biochem. Soc. Trans.* **2013**, *41*, 399–404. [CrossRef]

37. Ferreira-Cerca, S.; Kiburu, I.; Thomson, E.; Laronde, N.; Hurt, E. Dominant Rio1 kinase/ATPase catalytic mutant induces trapping of late pre-40S biogenesis factors in 80S-like ribosomes. *Nucleic Acids Res.* **2014**, *42*, 8635–8647. [CrossRef]

38. Vanrobays, E.; Gelugne, J.; Gleizes, P.; Caizergues-Ferrer, M. Late cytoplasmic maturation of the small ribosomal subunit requires RIO proteins in *Saccharomyces cerevisiae*. *Mol. Cell. Biol.* **2003**, *23*, 2083–2095. [CrossRef]

39. Cohen, P.; Alessi, D.R. Kinase drug discovery—What's next in the field? *ACS Chem. Biol.* **2013**, *8*, 96–104. [CrossRef] [PubMed]

40. Keiser, J.; Utzinger, J. Food-borne trematodiases. *Clin. Microbiol. Rev.* **2009**, *22*, 466–483. [CrossRef] [PubMed]

41. Chai, J.Y. Praziquantel treatment in trematode and cestode infections: An update. *Infect. Chemother.* **2013**, *45*, 32–43. [CrossRef] [PubMed]

42. Young, N.D.; Jex, A.R.; Li, B.; Liu, S.; Yang, L.; Xiong, Z.; Li, Y.; Cantacessi, C.; Hall, R.S.; Xu, X.; et al. Whole-genome sequence of *Schistosoma haematobium*. *Nat. Genet.* **2012**, *44*, 221–225. [CrossRef] [PubMed]

43. Lu, Z.; Sessler, F.; Holroyd, N.; Hahnel, S.; Quack, T.; Berriman, M.; Grevelding, C.G. *Schistosome* sex matters: A deep view into gonad-specific and pairing-dependent transcriptomes reveals a complex gender interplay. *Sci. Rep.* **2016**, *6*, 31150. [CrossRef]

44. Zhao, L.; He, X.; Grevelding, C.G.; Ye, Q.; Li, Y.; Gasser, R.B.; Dissous, C.; Mughal, M.N.; Zhou, Y.-Q.; Zhao, J.-L.; et al. The RIO protein kinase-encoding gene *Sj-riok-2* is involved in key reproductive processes in *Schistosoma japonicum*. *Parasit. Vectors* **2017**, *10*, 604. [CrossRef]

45. Bairoch, A. The PROSITE dictionary of sites and patterns in proteins, its current status. *Nucleic Acids Res.* **1993**, *21*, 3097–3103. [CrossRef] [PubMed]

46. Finn, R.D.; Mistry, J.; Tate, J.; Coggill, P.; Heger, A.; Pollington, J.E.; Gavin, O.L.; Gunasekaran, P.; Ceric, G.; Forslund, K.; et al. The Pfam protein families database. *Nucleic Acids Res.* **2010**, *38*, D211–D222. [CrossRef]

47. Voss, H.; Benes, V.; Andrade, M.A.; Valencia, A.; Rechmann, S.; Teodoru, C.; Schwager, C.; Paces, V.; Sander, C.; Ansorge, W. DNA sequencing and analysis of 130 kb from yeast chromosome XV. *Yeast* **1997**, *13*, 655–672. [CrossRef]

48. Klein, S.L.; Strausberg, R.L.; Wagner, L.; Pontius, J.; Clifton, S.W.; Richardson, P. Genetic and genomic tools for *Xenopus* research: The NIH Xenopus initiative. *Dev. Dyn.* **2002**, *225*, 384–391. [CrossRef]

49. Vandenberg, L.N.; Blackiston, D.J.; Rea, A.C.; Dore, T.M.; Levin, M. Left-right patterning in *Xenopus* conjoined twin embryos requires serotonin signaling and gap junctions. *Int. J. Dev. Biol.* **2014**, *58*, 799–809. [CrossRef]

50. LaRonde-LeBlanc, N.; Wlodawer, A. A family portrait of the RIO kinases. *J. Biol. Chem.* **2005**, *280*, 37297–37300. [CrossRef] [PubMed]

51. LaRonde-LeBlanc, N.; Guszczynski, T.; Copeland, T.; Wlodawer, A. Structure and activity of the atypical serine kinase Rio1. *FEBS J.* **2005**, *272*, 3698–3713. [CrossRef]

52. Lin, X.; Kaul, S.; Rounsley, S.; Shea, T.P.; Benito, M.I.; Town, C.D.; Fujii, C.Y.; Mason, T.; Bowman, C.L.; Barnstead, M.; et al. Sequence and analysis of chromosome 2 of the plant *Arabidopsis thaliana*. *Nature* **1999**, *402*, 761–765. [CrossRef]

53. Elkon, R.; Milon, B.; Morrison, L.; Shah, M.; Vijayakumar, S.; Racherla, M.; Leitch, C.C.; Silipino, L.; Hadi, S.; Weiss-Gayet, M.; et al. RFX transcription factors are essential for hearing in mice. *Nat. Commun.* **2015**, *6*, 8549. [CrossRef] [PubMed]

54. Young, J.M.; Trask, B.J. V2R gene families degenerated in primates, dog and cow, but expanded in opossum. *Trends Genet.* **2007**, *23*, 212–215. [CrossRef]

55. Guderian, G.; Peter, C.; Wiesner, J.; Sickmann, A.; Schulze-Osthoff, K.; Fischer, U.; Grimmler, M. RioK1, a new interactor of protein arginine methyltransferase 5 (PRMT5), competes with pICln for binding and modulates PRMT5 complex composition and substrate specificity. *J. Biol. Chem.* **2011**, *286*, 1976–1986. [CrossRef]

56. Sasaki, T.; Matsumoto, T.; Yamamoto, K.; Sakata, K.; Baba, T.; Katayose, Y.; Wu, J.; Niimura, Y.; Cheng, Z.; Nagamura, Y.; et al. The genome sequence and structure of rice chromosome 1. *Nature* **2002**, *420*, 312–316. [CrossRef] [PubMed]

57. Matthews, B.B.; Dos Santos, G.; Crosby, M.A.; Emmert, D.B.; St Pierre, S.E.; Sian Gramates, L.; Zhou, P.; Schroeder, A.J.; Falls, K.; Strelets, V.; et al. Gene model annotations for *Drosophila melanogaster*: Impact of high-throughput data. *G3 Genes Genomes Genet.* **2015**, *5*, 1721–1736. [CrossRef]

58. Wang, J.; Czech, B.; Crunk, A.; Wallace, A.; Mitreva, M.; Hannon, G.J.; Davis, R.E. Deep small RNA sequencing from the nematode *Ascaris* reveals conservation, functional diversification, and novel developmental profiles. *Genome Res.* **2011**, *21*, 1462–1477. [CrossRef] [PubMed]

59. Hu, M.; Laronde-Leblanc, N.; Sternberg, P.W.; Gasser, R.B. *Tv*-RIO1—An atypical protein kinase from the parasitic nematode *Trichostrongylus vitrinus*. *Parasit. Vectors* **2008**, *1*, 34. [CrossRef]

60. Wang, X.; Chen, W.; Huang, Y.; Sun, J.; Men, J.; Liu, H.; Luo, F.; Guo, L.; Lv, X.; Deng, C.; et al. The draft genome of the carcinogenic human liver fluke *Clonorchis sinensis*. *Genome Biol.* **2011**, *12*. [CrossRef] [PubMed]

61. Zheng, H.; Zhang, W.; Zhang, L.; Zhang, Z.; Li, J.; Lu, G.; Zhu, Y.; Wang, Y.; Huang, Y.; Liu, J.; et al. The genome of the hydatid tapeworm *Echinococcus granulosus*. *Nat. Genet.* **2013**, *45*, 1168–1175. [CrossRef] [PubMed]

62. Protasio, A.V.; Tsai, I.J.; Babbage, A.; Nichol, S.; Hunt, M.; Aslett, M.A.; de Silva, N.; Velarde, G.S.; Anderson, T.J.C.; Clark, R.C.; et al. A systematically improved high quality genome and transcriptome of the human blood fluke *Schistosoma mansoni*. *PLoS Negl. Trop. Dis.* **2012**, *6*, e1455. [CrossRef] [PubMed]

63. Livak, K.J.; Schmittgen, T.D. Analysis of relative gene expression data using real-time quantitative PCR and the 2(-Delta Delta C(T)) Method. *Methods* **2001**, *25*, 402–408. [CrossRef] [PubMed]

64. Ramirez, B.; Bickle, Q.; Yousif, F.; Fakorede, F.; Mouries, M.-A.; Nwaka, S. Schistosomes: Challenges in compound screening. *Expert Opin. Drug Discov.* **2007**, *2*, S53–S61. [CrossRef] [PubMed]

MDPI

St. Alban-Anlage 66

4052 Basel

Switzerland

Tel. +41 61 683 77 34

Fax +41 61 302 89 18

www.mdpi.com

Pathogens Editorial Office

E-mail: pathogens@mdpi.com

www.mdpi.com/journal/pathogens